1 MONTH OF
FREE
READING

at

www.ForgottenBooks.com

By purchasing this book you are eligible for one month membership to ForgottenBooks.com, giving you unlimited access to our entire collection of over 1,000,000 titles via our web site and mobile apps.

To claim your free month visit:

www.forgottenbooks.com/free73642

ISBN 978-0-666-02253-0
PIBN 10073642

MEMOIRS OF THE GEOLOGICAL SURVEY

OF THE

UNITED KINGDOM.

THE

JURASSIC ROCKS

OF

BRITAIN.

VOL. V.

THE MIDDLE AND UPPER OOLITIC ROCKS OF ENGLAND (YORKSHIRE EXCEPTED).

BY

HORACE B. WOODWARD, F.G.S.

PUBLISHED BY ORDER OF THE LORDS COMMISSIONERS OF HER MAJESTY'S TREASURY.

LONDON:

PRINTED FOR HER MAJESTY'S STATIONERY OFFICE,
BY EYRE AND SPOTTISWOODE,
PRINTERS TO THE QUEEN'S MOST EXCELLENT MAJESTY.

And to be purchased, either directly or through any Bookseller, from
EYRE AND SPOTTISWOODE, EAST HARDING STREET, FLEET STREET, E.C.; or
JOHN MENZIES & Co., 12, HANOVER STREET, EDINBURGH, and
90, WEST NILE STREET, GLASGOW; or
HODGES, FIGGIS, & Co., LIMITED, 104, GRAFTON STREET, DUBLIN.

1895.

Price Seven Shillings and Sixpence.

339449

PREFACE.

THIS volume which, like its two predecessors, is the work of
Mr. H. B. Woodward, completes the general description of the
Jurassic Rocks of England and Wales. It includes an account
of the strata from the Oxfordian to the Purbeckian divisions,
with some remarks on the probably marine equivalents of the
Purbeck group in Lincolnshire.

As in previous volumes of this Memoir, full references are
given to the labours of others who have worked at the formations
described in the following chapters. In early days the broad
outlines of the geology of the Middle and Upper Oolites were
admirably sketched by Webster, Sedgwick, Buckland, De la Beche,
Lonsdale, Fitton, Mantell, and John Phillips. Among the later
more detailed investigations, special prominence may be given to
the elaborate work of Prof. J. F. Blake and Mr. Hudleston on the
Corallian rocks; to the separate papers by Prof. Blake on the
Kimeridge Clay and Portland Beds; and to the work of Thomas
Roberts on the Corallian clays. These observers have done much
to elucidate the minute stratigraphy and palæontology of the
formations to which they have so sedulously devoted their
attention.

Our information regarding the Oxfordian and Purbeckian
groups has not hitherto been summarised, and many details on
this subject are now for the first time published. I may specially
refer to the account of the Kellaways strata at Kellaways, and
to the lists of fossils from that and other localities in the south-
west of England.

The Geological Survey has not hitherto published any
memoir descriptive of the higher Oolitic strata in Dorset,
although the admirable Maps and Sections by H. W. Bristow,
and the detailed sections of the Purbeck Beds by that author
and Edward Forbes, have depicted some of the main features of
the geology of that region.

In the preparation of this volume full advantage has been
taken of previous Memoirs of the Survey descriptive of the strata
in the Midland area, especially those by Prof. Hull, Prof. Green,
Prof. Judd, Mr. Jukes-Browne, and Mr. Strahan.

e 82428·   750.—11/95.   Wt. 5487.   a 2

In the field Mr. Woodward has received help from Prof. Prestwich and Mr. James Parker (of Oxford), from the Rev. W. R. Andrews (formerly of Teffont Evias), from the late Robert Damon (of Weymouth), and from Mr. A. M. Wallis (of Portland). The fossils collected during the progress of his work have been named by Messrs. Sharman and Newton ; while Mr. Teall has examined and described microscopic sections of the rocks.

As in the previous volumes special attention has been given in the following pages to the subject of Economic Geology.

The illustrations, where not otherwise stated, are original. As before, we are indebted to the Council of the Geological Society and to Dr. Henry Woodward for the use of clichés, the sources of which, as well as of others taken from previous Survey Memoirs, are notified in the List of Illustrations.

The full Bibliography which forms an Appendix has been compiled with the assistance of Mr. Fox-Strangways, and will doubtless be helpful to those who seek further special information on the Jurassic Rocks of England and Wales.

It is intended that a concluding volume on the Jurassic Rocks of Scotland and Ireland shall be published after those of Scotland have been mapped in detail.

ARCH. GEIKIE,

Geological Survey Office,   Director General.
28, Jermyn Street,
London,
15th October, 1895.

TABLE OF CONTENTS.

ILLUSTRATIONS.

MAP

Note.—The concealed area of Lower Oolitic rocks has been coloured a lighter tint than the exposed area. There are two inliers of Lias between Peterborough and Oakham.

THE

MIDDLE AND UPPER OOLITIC ROCKS

OF

ENGLAND

(YORKSHIRE EXCEPTED).

CHAPTER I.

INTRODUCTION.

The Middle and Upper Oolitic Rocks.

The Middle and Upper Oolitic formations come to the surface at intervals across the country from Dorsetshire to Lincolnshire; but the outcrop, more especially of the higher members of the series, is concealed over considerable tracts by the Cretaceous strata. There is evidence that prior to the deposition of these newer formations, the Oolitic strata were tilted, and in some cases rucked up and faulted, so that the Cretaceous deposits in their extension westwards were laid down on diverse members of the Oolitic Series. This unconformable overlap (or overstep) is marked in the case of certain Lower Cretaceous or Neocomian strata, not only by the Lower Greensand resting on different formations, but by the occurrence in it of derived fossils. The overlap of the Upper Cretaceous strata was even more widespread, and this was accompanied by some destruction both of the Lower Cretaceous strata, and of the underlying Oolitic formations.

Thus the underground course of the Oolitic rocks is concealed by irregular sheets of Lower Greensand, and by a more persistent covering of Chalk and other Upper Cretaceous strata. The anticline of the Wealden area brings the Upper Oolitic rocks to the surface near Battle; and the presence of these and other members of the Oolitic series below ground in the south-east of England, has been made known through several deep borings. (*See* Figs. 144, 145, pp. 298, 299.)

The map that accompanies this Memoir shows the areas over which the various Oolitic rocks occur at the surface, and also their probable underground extent beneath the covering of Cretaceous and newer rocks.

It seems likely that during the Jurassic period there were tracts of land in portions of what are now the eastern counties, where Palæozoic rocks directly underlie the Cretaceous strata ; and possibly there was land also over portions of the Hampshire and Wiltshire area. Godwin-Austen long ago remarked that " The Oolites of Yorkshire and Lincolnshire were dependent on a land which lay to the east,"* and the results of deep borings in the South-east of England lend support to the view.

The following are the chief formations to be described :—

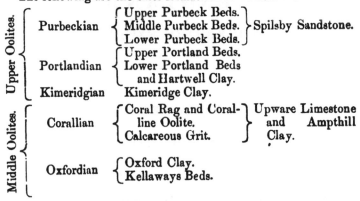

Tables showing the local divisions and leading zones were given in the general account of the Oolitic Series ; they need not therefore be repeated here, but they will be dealt with more particularly in the chapters on each formation. Particular descriptions of oolite and pisolite, and of the probable method of formation of oolitic rocks, together with a general account of the fossils of the Oolitic Series, have likewise been given in a previous volume.†

The total thickness of the Middle and Upper Oolitic Series varies from about 900 to about 2,200 feet ; that of the Middle Oolites being from 450 to 800 feet, and of the Upper Oolites, 450 to 1,400 feet.

The series includes deposits of considerable economic value ; among them the Portland Stone, the Swanage Stone and Purbeck Marble ; the Westbury iron-ore ; clays that are extensively dug for brick-making, as well as some local deposits of gypsum.

---

* Quart. Journ. Geol. Soc., vol. xii. p. 64.
† Memoir on the Lower Oolitic Rocks of England, pp. 6, 24 ; see also Fox-Strangways, Jurassic Rocks of Yorkshire, vol. i. p. 399. For a general account of the Jurassic System, see Memoir on the Lias of England and Wales (Yorkshire excepted), p. 1 ; and Fox-Strangways, op. cit., p. 7.

The Middle and Upper Oolitic strata form a series that is intimately connected, and together they constitute the Upper Jurassic division. From the Oxfordian base to the Portlandian formation there is evidence of transition; while the succeeding Purbeck Beds are superimposed with a general conformity as regards their stratification, but they exhibit marked changes in their method of formation. The Oxfordian Beds are, for the most part, distinctly marked off from the underlying Cornbrash, but they are connected palæontologically, so that there is no evidence of any break in the series, as we pass upwards from the Lower Oolites.

The mass of the Middle and Upper Oolites consists of marine deposits; of thick clays that are separated by stone-beds, which, in the case of the Corallian Beds especially, are of a variable and inconstant character. The Purbeck Beds show the incoming of terrestrial, freshwater, and estuarine conditions, that are attended by some evidences of local unconformity with the Portland Beds, but by no marked discordance that would suggest any great break in the series.

These freshwater conditions were continued in the Wealden strata, which, in the south of England, blend intimately with the earlier Purbeck deposits, so that we have a passage from Jurassic to Neocomian or Lower Cretaceous formations.

In parts of Lincolnshire, and again in Yorkshire, we have evidence of marine beds on the border-lines between the Jurassic and Cretaceous strata, and it is consequently a debatable matter as to how closely correlation can be carried between the formations of the north and south of England. In the district under consideration it seems probable that the Spilsby Sandstone of Lincolnshire, may be classed in point of age with the Purbeck Beds.

The general relationship between the Purbeck and Wealden strata has led many geologists to group them together. Nor does the local evidence justify our parting them.

It is only on wider grounds that the division between Jurassic and Cretaceous is taken at this horizon. No general system of classification will be found equally applicable to all areas. With the Trias and Lias, with the Lias and Oolites, we find in some regions such a gradual passage, that no definite plane of demarcation can be drawn. Thus, in places we have to draw an imaginary line between Lias and Oolites in the midst of a transitional series! So in separating the Purbeck and Wealden formations, we may be content to allow the claims of general convenience and of the palæontological succession established in Europe, to rule the limits of our principal chronological systems.

With the several subdivisions, the case is different; such formations must be based on their stratigraphical features, even while we admit that the actual boundaries between the rock-groups vary more or less in horizon in different areas. The evidence in general shows that we cannot adopt zonal groupings, that will correspond also with the main stratigraphical divisions over wide areas. It is, however, important to maintain the original divisions of our strata taken from the typical localities; and in the case of the Middle and Upper Oolites, we have the Kellaways Beds and Oxford Clay, the Kimeridge Clay, the Portland and Purbeck Beds, all of them named from localities in the area under consideration. To modify our grouping to suit the stratigraphy elsewhere is not justifiable, nor can any precise plan be adopted that will serve all the requirements of field-geology and palæontology over extended areas.

CHAPTER II.

MIDDLE OOLITIC.

OXFORD CLAY AND KELLAWAYS ROCK.

(OXFORDIAN.)

GENERAL ACCOUNT OF THE STRATA.

The Oxford Clay takes its name from the county of Oxford, where it enters extensively into the structure of the ground; moreover, it is exposed in the neighbourhood of the City of Oxford in several brickyards which appropriately furnish some of the largest inland sections of the formation.

The term "Clunch Clay and Shale" was used by William Smith (1815–16), because the upper part of the formation comprises hard clay rising in lumps, called "Clunch"; while deeper down it becomes blacker and laminated, under which circumstances it is called "Shale."[*] The term Oxford, Forest, or Fen Clay was used by Buckland in 1818; and this compound name was suggested from the occurrence of wooded tracts; such as the old Forest of Braydon, and from the development of the Clay beneath portions of the Fenland.[†] The general term of Oxfordian was used by D'Orbigny in 1844.

The Oxford Clay consists of bluish or greenish and lead-coloured clay, which usually becomes brown or yellowish-grey at the surface. In many places, and more especially in the upper part, the bedding in the clay is not distinct, and is marked only by bands of septaria and occasional layers of earthy limestone; but lower down the formation is usually more and more shaly, and this character is maintained even when the beds come near the surface. The shale is grey and reddish-brown or purple in colour. Both clay and shale are often calcareous, and occasionally bituminous; and a good deal of lignite is sometimes met with.

Septaria occur at irregular intervals and they may attain a large size, having a diameter of 3 feet or more. Small irregular reddish-coloured ironstone-nodules are found in places in the upper beds; these also are septarian. The beds contain much iron-pyrites and selenite; and many of the fossils are of a pyritic character.

Towards the base of the Oxford Clay there occur inconstant beds of sand and calcareous sandstone, the latter often appearing

---

[*] Parkinson described "Clunch" as a soft chalk-like stone, found towards the top of the clay (Organic Remains, vol. iii. p. 446). Clunch has also been described as a provincial term for "indurated clay, when glossy, unctuous, and tending to a slaty texture;" while when dull and smooth it is termed "Clod." W. Phillips, Selection of Facts, Geol. England and Wales, p. 61.

[†] W. Phillips, *Ibid.* (Table by Buckland.)

in the form of huge concretionary masses or "Doggers" that attain a thickness of 7 feet or more. These sandy beds alternate with clayey or loamy beds: and usually at the bottom of the formation there is a clay bed, some 10 or 12 feet thick, that rests on the Cornbrash.

Indurated sandy beds, highly fossiliferous, but of little thickness, were formerly quarried for road-metal near Kellaways, not far from Chippenham in Wiltshire as early as 1699 the fossils were noticed by Lhwyd,[*] and a century later they attracted the attention of William Smith, who introduced the name Kellaways (or Kelloway's) Stone into geological literature. The term Kellaways Beds is now usually applied to the strata which constitute an irregular and lithologically not everywhere persistent, basement-bed to the Oxford Clay. The thickness varies from a few feet to upwards of 50 feet. From a palæontological point of view the horizon is important for the fossils mark a stage that is not solely dependent on the lithological characters of the Kellaways Rock of Kellaways.

The term Oxfordian[†] may be used generally to include the Oxford Clay and the Kellaways Beds. The total thickness of the formation varies from about 300 to nearly 600 feet, but the sections as a rule do not afford many points of stratigraphical interest; the organic remains alone are calculated to attract the geologist.

The Oxfordian strata rest conformably upon the Cornbrash, and may be said to merge gradually upwards into the Corallian series. Palæontologically the beds both below and above are intimately connected. Common to the Cornbrash and Kellaways divisions we find such forms as *Ammonites macrocephalus, A. Bakeriæ, Goniomya v.-scripta, Modiola bipartita, Pecten lens, Rhynchonella varians, Waldheimia obovata,* &c. ; and belonging to the Oxford Clay and Corallian Rocks we find *Ammonites cordatus, A. perarmatus, Belemnites abbreviatus, Cerithium muricatum, Goniomya v.-scripta, Gryphæa dilatata, Myacites recurvus, Ostrea gregaria, Pecten lens, Thracia depressa, Trigonia elongata, T. irregularis,* &c.

Organic Remains.

Among the fossils of the Oxfordian strata we find remains of the Saurians, *Megalosaurus, Ichthyosaurus, Cimoliosaurus (Plesiosaurus),* and *Pliosaurus.* Mr. James Parker has a Chelonian bone from the upper beds of the Oxford Clay near Oxford. Fishes are not particularly abundant : *Asteracanthus, Hybodus,* and *Lepidotus* being the less rare genera.

Cephalopoda are abundant, and often beautifully preserved : and the Ammonites, as in the case of the argillaceous beds of the Lias, furnish evidence of a general sequence of forms that appears

---

* Lhwyd then recorded " *Nautilites modiolaris* " and other fossils from " Calloway Bridge ;" Lithophylacii Britannici Ichnographia, 8vo. Lond. 1699, p. 18.

† Phillips has employed the term Oxonian, Geol. Oxford, p. 295.

to be well marked throughout the country, as well as in other parts of Europe, although the limitations are not to be definitely fixed. Belemnites are likewise abundant, especially large specimens of *Belemnites Oweni*, the ordinary form of which, described by S. P. Pratt, was regarded by Phillips as the same as that named later as *B. puzosianus* by D'Orbigny.* Gasteropods, though not numerous in species, are occasionally abundant; of these *Alaria trifida*, *Cerithium muricatum*, and *Pleurotomaria depressa* may be mentioned. Of the Lamellibranchs, the commoner forms are *Astarte carinata*, *Avicula inæquivalvis*, *Cucullæa concinna*, *Goniomya v.-scripta*, *Gryphæa bilobata*, *G. dilatata*, *Modiola bipartita*, *Myacites recurvus*, *Nucula ornata*, *Ostrea flabelloides*, *Pinna mitis*, *Thracia depressa*, and *Trigonia* (several species). *Gryphæa dilatata* is characteristic of the upper beds of the Oxford Clay, and large specimens are almost everywhere abundant, as in the clay-cliffs north of Weymouth, and in brickyards opened in this portion of the formation. Lower down smaller forms of *Gryphæa* occur; and it is difficult to distinguish many of these from the *G. bilobata* that is characteristic of the Kellaways Rock.

Brachiopoda on the whole are rare; but *Rhynchonella varians*, *Waldheimia obovata*, and *W. impressa* are locally abundant. The Crustacea include species of *Eryma*, *Glyphea*, *Magila*, and *Mecocheirus*. Macrurous Crustacea, as remarked by Mr. J. Carter, are fairly abundant, while Brachyura are rare.†

Of Annelides the form *Serpula vertebralis* is characteristic. Echinodermata are rare, though occasionally Urchins, Crinoids, and Star-fishes are met with. Corals are extremely rare, the form *Anabacia complanata* being the only species recorded. No Sponges have been found. A number of Foraminifera are known; but no recognizable species of Plants, although foliage of *Araucarites*, abundant remains of Coniferous wood, and also of Cycads have been found in places.

On the whole the mass of the strata indicates fairly deep water, so far as terrigenous deposits are concerned. The abundance of drift-wood or lignite, in the Oxford Clay of Trowbridge, led Mantell to speak of the formation as of "fluvio-marine" character ;‡ but the occurrence of such material is no definite indication of the proximity of land.§

Zones.

In determining the chief palæontological divisions, or zones, it is most desirable to adopt, as indices, those forms which are widely

---

* British Belemnitidæ, 1865, p. 120. Prof. Pavlow on the other hand keeps the two forms distinct and adopts the spelling *B. puzosi*, Bull. Soc. Imp. Nat., Moscow, 1892.

† Quart. Journ. Geol. Soc., vol. xlii. p. 558.

‡ Phil. Trans., 1848, p. 172.

§ See Memoir on the Lower Oolitic Rocks, p. 2.

distributed, even if locally we find other species more abundant.
Acting on this principle, the following zones are adopted :—

	Zones.	Leading Lithological Characters.
Oxford Clay.	Ammonites cordatus.	Clays with occasional septaria and ironstone-nodules.
		Clays and shaly clay, with pyritic fossils. (Sub-zone of *Am. Lamberti*.)
	Am. ornatus.	Shales with pyritic fossils. (Sub-zone of *Am. Jason*.)
Kellaways Beds.	Am. calloviensis.	Alternations of clays and sands with concretionary masses of calcareous sandstone (Kellaways Rock).
		Clay (Kellaways Clay).

On the Continent the name Kellaways Rock was employed in the
modified form of "Kellovian," by D'Orbigny in 1844; and later on this
was changed to Callovian (1849).* The term, however, has come to
embrace wider time-limits than are included in our Kellaways Beds, so
that the Continental usage is by no means synonymous with our own.
Abroad the equivalents of our Cornbrash are in places so linked with the
equivalents of the Kellaways Beds, that the two are sometimes grouped
together with the Callovian, and marked as the zone of *Ammonites macrocephalus.* The "*Macrocephalus-bett*" of Oppel, although not taken to
include our Cornbrash, comprehended the zones of *A. macrocephalus* and
A. calloviensis. In this country the former species is characteristic of the
Cornbrash, although it ranges into the Lower Oxfordian strata; but we
may most conveniently group our Kellaways Beds generally in the zone
of *A. calloviensis.*

It has been pointed out by Mr. Hudleston and others, that in Yorkshire
the beds placed in the Kellaways division include higher stages than do
their supposed equivalents in Wiltshire. The fact is that the occurrence
of sandy and gritty bands is not confined to any restricted horizon;
indeed in parts of Dorsetshire such bands are hardly to be detected at all
in the Kellaways division. Even in Yorkshire it seems probable that
the development of rock-bands takes place on varying horizons, for we
find that not only *A. calloviensis, A. modiolaris,* and other true Kellaways
fossils are recorded from the local "Kellaways Rock," but likewise *A.
Jason, A. Lamberti,* and even *A. perarmatus.* On the other hand there
are recorded from the Oxford Clay of Yorkshire, *A. athleta, A. Mariœ,
A. Lamberti, A. cordatus,* and *A. perarmatus ;* fossils which show that in

* Paléontologie Française, Terrains Jurassiques, Tome 1, p. 608.

places, at any rate, the Oxford Clay of that county includes strata belonging to the zones of *A. ornatus* and *A. cordatus*.[*]

On the Continent the Callovian formation has sometimes been so extended as to take in beds with *A. Jason*, and occasionally even beds with *A. Lamberti*. This seems a quite unwarranted stretching of a formation to suit local stratigraphy, and in defiance of its original significance.

Thus Oppel includes,[†] above his "*Macrocephalus-bett*," first the "*Anceps-bett*," and then the "*Athleta-bett*." These zones of *A. anceps* and *A. athleta* were grouped together as "Schichten des *Amm. ornatus*" (*Ornaten-thone* of Quenstedt, 1843). They include the "*Ornati* group" of Ammonites, *A. Jason*, and var *Gulielmi*, *A. Elizabethœ*, *A. Duncani*, &c.; forms which characterize our Lower Oxford Clay; and we may group very properly this portion of the formation as the zone of *Ammonites ornatus*.

Linking together the lower and upper portions of the Oxford Clay are the beds with *Ammonites Lamberti*—a fossil that is sometimes given as a zonal index. Locally in this country it is convenient to speak of the zone or sub-zone of *A. Lamberti*.

Beds equivalent to the highest portion of our Oxford Clay were placed by Oppel with the Corallian formation, the lower portion of which he termed the Oxford group.[‡] This group was spoken of as the zone of *A. biarmatus* (a form considered to be near to *A. perarmatus*,[§] but not otherwise recognized in this country); and among the characteristic fossils, Oppel notes *A. Lamberti* and *A. cordatus*. The last-named species, as well as *A. perarmatus*, serve to connect the Oxford Clay with the Corallian Rocks, but we may be content to group the highest stage of our Oxfordian formation as the zone of *Ammonites cordatus*. It is a species that characterizes the uppermost beds of clay. It is a variable form, and appears to be linked with *A. Lamberti* and *A. vertebralis*; but its employment as a zonal index is very widely adopted.

The ranges of some species of *Ammonites* are subject to much local variation; thus *A. Bakeriœ* was found by T. Roberts with *A. cordatus* in Huntingdonshire; and *A. plicatilis* (a Corallian species) is recorded from the Kellaways Beds of Yorkshire, and also by Mr. E. T. Newton, from the Oxford Clay in the Swindon boring, and elsewhere. These occurrences, natural enough, do not interfere with the general succession of forms, and the prevalence of particular species at certain horizons.

The basement-clay, or Kellaways Clay, that, over great part of the country, occurs below the sandy beds of the Kellaways Rock, was recognized by William Smith (see p. 29), and was noticed by Oppel[||] as the "Clay with nodules" (of Trowbridge), beds described by R. N. Mantell, and said to contain *Ammonites gowerianus*, *A. Kœnigi*, *A. modiolaris*, and *A. macrocephalus*.

I have found this clay present, wherever sections occur, throughout the country, having generally a thickness of 10 or 12 feet, and occasionally as much as 20 feet. It is a grey or greenish clay, not usually of a shaly character, and it contains a good deal of selenite. Fossils are by no means abundant, and they are often crushed; they include *Ostrea*, *Waldheimia*, *Rhynchonella varians*, *Serpula tetragona*, and *S. vertebralis*.

---

[*] See Fox-Strangways, Jurassic Rocks of Yorkshire, vol. i. pp. 274, 277, 295.

[†] Oppel, Die Juraformation, pp. 507, 519 ; see also A. Riche, Etude Stratigr. Jurassique Inf. du Jura Mérid., Ann. Univ. Lyon, vol. vi. 1893.

[‡] Oppel, Die Juraformation, pp. 506, 517, 534, 624, &c.

[§] Fox-Strangways, Jurassic Rocks of Yorkshire, vol. i. p. 298.

[||] Die Juraformation, p. 517.

This clay is no doubt approximately on the same horizon as certain dark shales, 6 to 8 feet thick, that overlie the Cornbrash limestone in Yorkshire. Mr. Fox-Strangways remarks that "These which are the so-called 'Clays of the Cornbrash' are very similar in petrological character to the lower part of the Oxford Clay, and no doubt foreshadow the physical conditions of that deposit." With regard to the fossils he adds that "most of the species have a wide range, and are consequently of little value."* They include *Ammonites macrocephalus*, *Avicula echinata*, *Modiola cuneata*, *Waldheimia lagenalis*, &c.

FIG. 1.   FIG. 2.

Rhynchonella varians, *Schloth*, 1½.

Ammonites macrocephalus, *Schloth*, ½.

The following are the more abundant and characteristic fossils of the Oxfordian strata, the species mentioned being prevalent in, but not necessarily confined to, particular zones:—

Zone of *Ammonites calloviensis.*

(Kellaways Beds.)

Ammonites Bakeriæ (Fig. 6).
—— calloviensis (Fig. 5).
—— gowerianus.
—— Kœnigi (Fig. 4).
—— macrocephalus (Fig. 1).
—— modiolaris (Fig. 3).
Belemnites Oweni (Fig. 14).
Alaria trifida (Fig. 7).
Arca æmula.
—— subtetragona.
Avicula ovalis.
Cardium cognatum.

Corbula Macneilli.
Goniomya v.-scripta.
Gryphæa bilobata (Fig. 8).
Isocardia minima.
Lucina despecta.
—— rotundata.
Modiola cuneata.
Myacites recurvus.
Pecten fibrosus.
—— lens.
Pholadomya acuticosta.
Rhynchonella varians (Fig. 2).

* Jurassic Rocks of Yorkshire, vol. i. p. 263 ; see also H. B. Woodward, Memoir on the Lower Oolitic Rocks, pp. 431, 434.

LOWER OXFORDIAN FOSSILS.

FIG. 3.

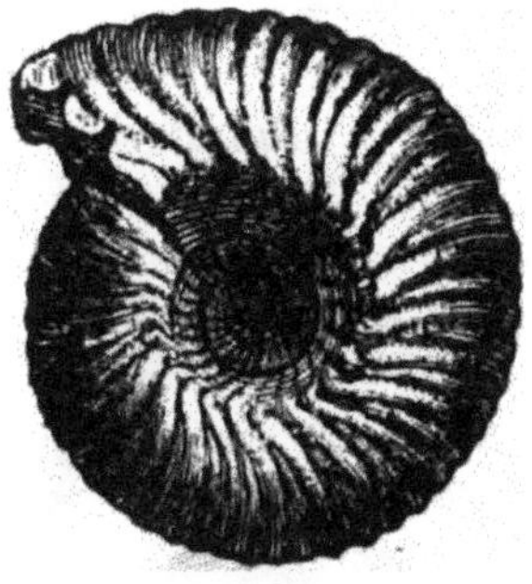

FIG. 4.

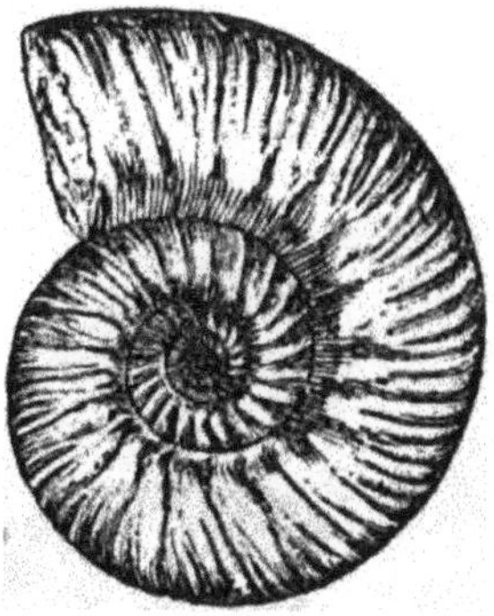

FIG. 5.

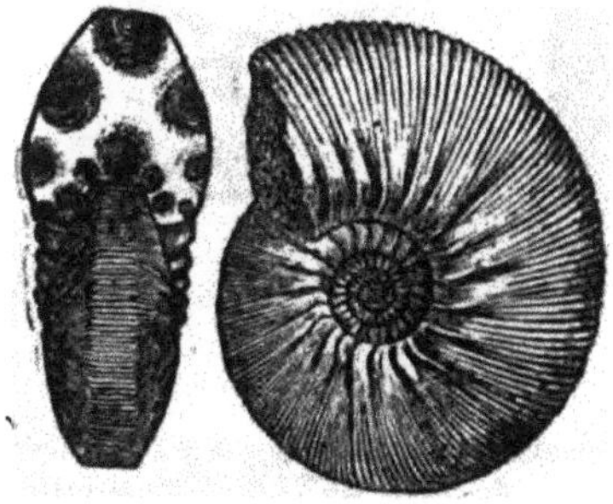

FIG. 7.

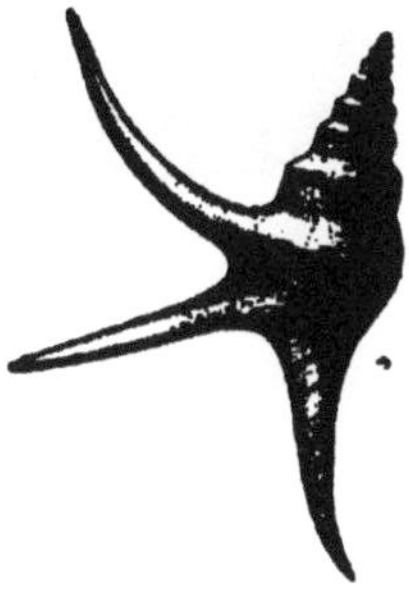

FIG. 6. ·

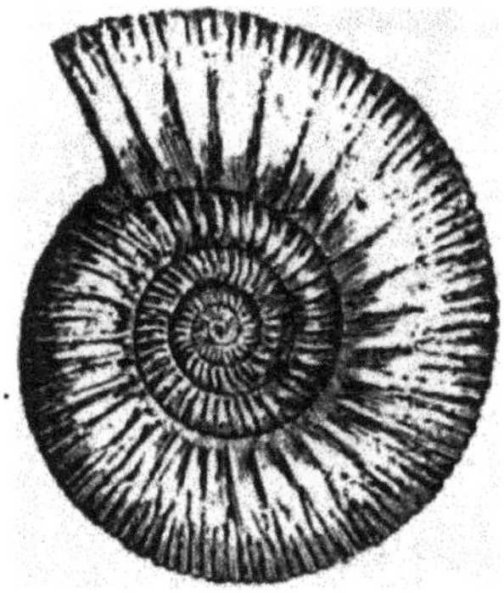

FIG. 8.

FIG. 3. Ammonites modiolaris, *Lhwyd*, ½.
  ,, 4.   ,,   Kœnigi, *Sow.* ¾.
  ,, 5.   ,,   calloviensis, *Sow.* ¾.
  ,, 6.   ,,   Bakeriæ, *d'Orb.* ½.
  ,, 7. Alaria trifida, *Phil.*
  ,, 8. Gryphæa bilobata, *Sow.* ½.

MIDDLE OXFORDIAN FOSSILS.

FIG. 9.   FIG. 10.

FIG. 11.   FIG. 12.

FIG. 9. Ammonites Duncani, *Sow.* 1½.
 ,, 10.   ,,   crenatus, *Brug.* 2⅜.
 ,, 11.   ,,   Jason, *Rein.* ½.
 ,, 12.   ,,   Lamberti, *Sow.* ⅜.

UPPER OXFORDIAN FOSSILS.

FIG. 13. FIG. 16.

FIG. 15. FIG. 17. FIG. 14.

FIG. 13. Ammonites cordatus, *Sow.* ⅔.
„ 14. Belemnites Oweni, *Pratt,* ½.
„ 15. —— hastatus, *Blainv.* ½.
„ 16. Gryphæa dilatata, *Sow.* ¼.
„ 17. Serpula vertebralis, *Sow.* Nat. size.

Zone of *Ammonites ornatus.*

Ammonites anceps.
—— athleta.
—— crenatus (Fig. 10).
—— Duncani (Fig. 9).
—— Elizabethæ.
—— hecticus.
—— Jason (Fig. 11).
—— —— var. Gulielmi.
—— Lamberti (Fig 12).
—— Mariæ.
Belemnites Oweni.

Belemnoteuthis antiquus.
Cerithium muricatum.
Arca subtetragona.
Avicula inæquivalvis.
—— ovalis.
Cucullæa concinna.
Gryphæa dilatata (small forms)
   (Fig. 16).
Leda Phillipsi.
Nucula ornata.

Zone of *Ammonites cordatus.*

Ammonites cordatus (Fig. 13).
—— —— var. excavatus.
—— Lamberti (Fig. 12).
—— perarmatus (Fig. 25.).
—— vertebralis.
Belemnites hastatus (Fig. 15).
—— Oweni (Fig. 14).

Gryphæa dilatata (large forms)
   (Fig. 16).
Modiola bipartita.
Thracia depressa.
Waldheimia impressa.
Serpula vertebralis (Fig. 17).

CHAPTER III.

OXFORD CLAY AND KELLAWAYS ROCK.

LOCAL DETAILS.

Weymouth.

The Oxford Clay is well exposed in the cliffs bordering Weymouth Bay, in the railway-cuttings and cliffs adjoining Radipole Lake, and along the shores of the East Fleet. Putting together the observations made at different points we find the following sequence :—

Zone of
*Ammonites
cordatus.*

4. Bluish-grey clay with small cement-stones, and reddish coloured irregular septarian nodules ("kidney-stones"). Large specimens of *Gryphæa dilatata* are especially abundant, also *Ammonites cordatus*, *A. vertebralis*, *A. perarmatus*, *Modiola bipartita*, *Myacites recurvus*, &c.

Zone of
A. ornatus.

3. Clays with small pyritic Ammonites: *A. athleta*, *A. hecticus*, *A. Lamberti*, *A. Mariæ*, and *Belemnites hastatus*, *Alaria trifida*, *Avicula inæquivalvis*, *Cardium Crawfordi*, *Nucula ornata*, *Thracia depressa*, *Pentacrinus*, &c.

2. Grey, greenish-grey and lilac rusty shales and paper-shales, with fragile fossils: *Ammonites Elizabethæ*, *Avicula ovalis*, *A. inæquivalvis*, small *Gryphæa*, &c.

Zone of
A. calloviensis
(Kellaways
Beds).

1. Clay with occasional thin bands of flaggy calcareous sandstone, and large septaria: *Ostrea* (small). [*A. modiolaris*, *A. macrocephalus*, *A. Kœnigi*, &c., have been recorded from Weymouth.]

The total thickness of the Oxford Clay in this area is about 500 feet, calculated in drawing the section, Fig. 52, p. 93, but there are no data for stating it with precision. In this estimate the upper clays (No. 4) would be about 250 feet thick.

The most easterly exposure of the Oxford Clay is on the coast between Black Head and Redcliff Point, to the west of Osmington. The beds rise in an anticline, and support on either side of it the Corallian Rocks, whose lowest member, the Nothe Grits, may be seen to rest immediately on the Oxford Clay. The Grits contain in some abundance *Gryphæa dilatata* and varieties of it, a form so characteristic of the upper beds of the Oxford Clay. In the central portion of this cliff, we have lower beds exposed (No. 3) with numerous small pyritic *Ammonites* and other fossils; the higher beds (No. 4) here, and elsewhere along the coast towards Weymouth, contain very many specimens of *Gryphæa dilatata*, including the finest examples I have seen. Other fossils, especially *Modiola bipartita*, occur in these beds, and occasionally impressions of Plants may be found.

At Redcliff Point a mass of Oxford Clay (with *G. dilatata*), capped by traces of the Nothe Grits, is faulted against the Corallian Rocks, the headland being protected by tumbled blocks of these newer strata.

The upper beds of the Oxford Clay again appear at the base of Redcliff, and they form the cliffs below Jordon Hill.* They consist of marly clay, which is shaly in places; it is generally of a dark slaty-blue colour at the base, becoming greenish-grey, brown or reddish-brown towards the surface. Below the coast-guard station at Jordon Hill, beds of bluish-grey clay, with small hard cement-stones, are to be seen dipping northward towards a fault, which is indicated in the clay by veins of calc-spar showing slickensides.

Reddish nodules with white calcareous veins, known as "kidney-stones," occur on the beach at Redcliff and Jordon Hill; they consist of hard cream-coloured septarian cement-stone with a shell of reddish-brown ironstone, and are derived from the Oxford Clay.

Casts of *Modiola* and *Astarte* may be found in these nodules. Damon notes the presence of a Serpula-bed, with *Vermilia sulcata*, in the clay of Jordon or Furzy Cliff. At Green Hill, to the south of Lodmoor, there are low cliffs of Oxford Clay, of a yellowish-brown colour, in which septaria occasionally occur. Damon records from these beds bones of Saurians, and scales of *Lepidotus*, also the following species :—†

Ammonites athleta.	Belemnites Oweni.
—— cordatus.	Cerithium Damonis.
—— gowerianus.	Gryphæa dilatata.
—— Kœnigi.	Nucula ornata.
—— Reginaldi.	Serpula vertebralis.
—— vertebralis.	Vermilia sulcata.
Belemnites hastatus.	Pentacrinus Fisheri.

Judging from the list, we here have portions of the zones of *A. ornatus* and *A. cordatus*; but the beds are now much concealed by débris, shingle, and by the extension of the esplanade.

The upper beds of the Oxford Clay were exposed at the Weymouth Brick, Tile, and Pottery Works, on the western side of Radipole Lake and north of the railway. Here blue and mottled clay was exposed, yielding *Ammonites cordatus*, *A. perarmatus*, and *Gryphæa dilatata*. Messrs. Blake and Hudleston note also *Ostrea gregaria* and *Serpula tricarinata* from the upper beds at Weymouth.‡

On the margin of the Lake we find clays with selenite and ferruginous cement-stones, yielding *Ammonites Lamberti*, *A. Mariæ*, *G. dilatata*, &c.

Shales were exposed further north, but they were better seen on the opposite side of the Lake, in the railway-cutting north of Weymouth Station. There we have grey and greenish rusty

---

* The site of the Roman Station, *Clavinium*.
† Geology of Weymouth, 1884, p. 30.
‡ Quart. Journ. Geol. Soc., vol. xxxiii. p. 363.

shales and paper-shales, with bands of small cement-stones; and these beds have yielded *Ammonites Elizabethæ, Avicula ovalis, Nucula ornata,* &c., but the specimens are so fragile that I failed to preserve any of those I obtained.

On the north-eastern shores of the Lake we find the lowest beds, consisting of shales with bands of small oysters, *Gryphæa,* and thin layers of hard flaggy and sandy limestone—a poor representative of the Kellaways Rock. Large septaria occur here.

At Radipole Damon obtained remains of a new Starfish, *Ophioderma weymouthiensis.*† He has given a long list of fossils found in the Oxford Clay at Radipole and at Ham Cliff, but as the particular horizons are not noted, it is unnecessary to enumerate them now, as they will be included in the general Catalogue of Fossils, in the Appendix.

It was remarked by Buckland and De la Beche that under the Barracks at Radipole, the Oxford Clay at one time "presented at low-water, a pavement of large and beautiful Septaria, known provincially by the name of Turtle Stones. The veins of these Septaria are filled with yellow semi-transparent calcareous spar, often passing into a rich deep-brown colour; their beauty, when polished, has, within these few years, caused the greater part of them to be taken up and manufactured into slabs and tables."‡

Referring to these septaria Damon has remarked that "microscopic examination has detected the laminæ of the hardened clay to be parallel to the laminæ of the shale or marl in which the stones are enclosed. In some, where the crystallisation is incomplete, and a space remains, water is found, which either originated with the formation, or has been drawn by capillary attraction into minute fissures from the exterior." He also drew attention to the finding in the Oxford Clay of Radipole of "the trunk of a fossil tree, 3 feet in length, 6 inches in diameter, from the centre of which the spar radiates to the exterior of the trunk, which is surrounded by similar spar. From this outer coating there are other radiating septa passing into the indurated clay or matrix, forming when cut, a circular section of Septaria, having the fossil trunk for its centre."§

Damon records a number of fossils that were met with in deepening Weymouth Harbour. The upper beds of the Oxford Clay have been exposed at low-water on the south shore of the Harbour, beneath the Nothe Grits (Corallian). The species include *Ammonites cordatus,* and *A. vertebralis,* but also other fossils, some of which are characteristic of the Kellaways Beds. Hence it is not improbable that among them were forms drifted down from the lower beds that outcrop on the borders of Radipole Lake.

---

* Many fossils were observed by Capt. L. L. B. Ibbetson, soon after the cuttings were made.
† Supp. to Geology of Weymouth, Ed. 3. 1888, Plate XVII.
‡ Trans. Geol. Soc., Ser. 2, vol. iv. p. 28.
§ Geol. Weymouth, 1884, pp. 26, 27.

FIG. 18.

Section along the borders of East Fleet, from near East Fleet village to Small Mouth, South of Weymouth.
Distance about 3 miles.

N.W.
East Fleet.

Tidmoor.
Point.

Near Wyke Regis.

S.E.
Small
Mouth.

4. Clays } Oxford Clay.
3. Shales }
2. Cornbrash.
1. Forest Marble.

12. Kimeridge Clay.
11. Sandsfoot Grits.
10. Sandsfoot Clay.
9. Trigonia Beds.
8. Osmington Oolite. } Corallian Beds.
7. Bencliff Grits.
6. Nothe Clay.
5. Nothe Grits.

The Oxford Clay is exposed in low cliffs on the borders of the East Fleet, but the beds are not well shown throughout the tract, there being intervals where no section occurs.

West of Wyke Regis the upper portion of the Oxford Clay is exposed below the Nothe Grits, and there *Ammonites cordatus*, *A. vertebralis*, *Modiola bipartita*, *Myacites recurvus*, and large specimens of *Gryphæa dilatata* are met with on the foreshore, together with ironstone and other nodules. To the north-west, many specimens of *G. dilatata* are met with in the promontory of Furzedown. Further on clays with ironstone-nodules occur east of Tidmoor Point; *Pentacrinus* likewise occurs, and a few large specimens of *G. dilatata.*

At the western part of Tidmoor Point, clays with selenite are exposed, and numerous fossils are met with, including many small ferruginous Ammonites, *A. athleta, A. hecticus,* and *A. Lamberti, Belemnites hastatus, Cerithium, Avicula inæquivalvis, Gryphæa, Nucula ornata,* and *Pentacrinus.* These beds and portions of those below, belong to the zone of *Ammonites ornatus.* Rounding the point we find, on the northern side, blue clay weathering brown, with small species of *Gryphæa,* and *Belemnites.* In the low cliff and on the shore in the centre of the bay, shales are exposed.

East of the point south of East Fleet, shales with septaria are shown in the cliff, and at the point itself there is a platform of clayey shale with a few fossils; *Avicula inæquivalvis, Gryphæa,* and *Ostrea.* On the western side of the point the beds undulate a little and here lilac rusty shales, and paper shales with selenite, fragile Ammonites, *Arca* and other small bivalves occur; also large septaria, one of which measured 4 feet across. Beyond, the beds are obscured by recent accumulations, and the junction with the Cornbrash is faulted; hence we have no information here with regard to the Kellaways Beds.

At the brickyard south of West Chickerel, the lower Oxford Clay (zone of *Ammonites ornatus*), consisting of brownish-grey rusty shales, was exposed. *Avicula ovalis,* and many small fossils were to be found.

On the northern side of the Weymouth anticline, between Broadway and the Swannery at Abbotsbury, we find a long vale of Oxford Clay, but no sections of interest, with the exception of one to the south of Rodden. There the Cornbrash and Kellaways Beds were shown in a pit on the east side of the road leading towards Langton Herring. The Kellaways Beds consisted of sandy and shelly shale, overlaid by bluish mottled clay, with *Waldheimia* and *Serpula.*[*]

The Ridgeway cutting of the Great Western Railway, which crosses the great fault north of Upway, brought to light a mass of Oxford Clay and also beds or "blocks" of Cornbrash, the occurrence of which was first pointed out by C. H. Weston. (See Fig. 132.) The Rev. O. Fisher had indeed

---

[*] See Memoir on the Lower Oolitic Rocks of England, p. 436.

previously seen the section, in 1846, and he has remarked on the abundance of the *Gryphæa dilatata.** Weston obtained a number of fossils from the Oxford Clay, and these which were named by Sowerby, included the following species :—†

Ammonites arduennensis.	Gryphæa dilatata.
—— Mariæ.	Modiola bipartita.
—— perarmatus (catena).	Thracia depressa.

It is curious that the Oxford Clay, judging by the above fossils, should be represented by its middle and upper portions, while the Cornbrash occurred in proximity; but the facts indicate that long previous to the great fault, the Jurassic strata were displaced by an earlier fault, which brought the Purbeck and Wealden strata against the Cornbrash and Oxford Clay ; and that the beds were denuded prior to the overlap of the Upper Greensand and Chalk. Alongside the fault no doubt the Oxford Clay and Cornbrash were disarranged. This view of the case I brought before the Geologists' Association during an excursion in 1889 made under the guidance of Mr. Hudleston, who remarks that " the old line of disturbance seems to have become a line of weakness in post-Cretaceous times, although the direction of the throw must have been reversed."‡ In the section across Abbotsbury drawn by Mr. Strahan (Fig. 52, p. 93) it will be noted that the Forest Marble on the north is brought abruptly against Kimeridge Clay, &c., on the south, so that from the Oxford Clay of the Ridgeway cutting there is a descending series beneath the Cretaceous rocks as we proceed to the south-west. Over the area to the north, the Oxford Clay directly underlies the Cretaceous rocks, perhaps beneath much of the ground south and south-west of Dorchester. We find an inlier of Oxford Clay, mapped by Bristow, at East Compton ; and again in the higher part of the Bride or Bredy valley, there is Oxford Clay, the occurrence of *Gryphæa dilatata* being noted to the south-east of Long Bredy. This latter tract is very much faulted.

A boring made at East Compton in 1855 passed through rock and sands (Upper Greensand) to a depth of 170 feet (including 40 feet sunk), when a small amount of water rose to the level of the boring (40 feet from the surface). Some sandy beds, but mainly blue clayey material, were afterwards penetrated to a depth of 80 feet (in Oxford Clay probably), and no further supply of water was obtained.

The most westerly tract of Oxford Clay in South Dorset is that exposed near Chilcombe. *Gryphæa dilatata* was noted by Bristow on the western side of Hammerdon Hill, while on the south side there was an old brickyard, which showed bluish-grey and purplish shales and clays with septaria. Beds of this character were penetrated to a depth of about 40 feet near Lower Sturthill.§

---

* Barnes, Guide to Dorchester, 1850 or 1851, pp. 30–82.
† Quart. Journ. Geol. Soc., vol. iv. p. 250 ; vol. viii. p. 116.
‡ Proc. Geol. Assoc., vol. xi. p. lii.
§ See Barrett, Geology of Swyre, &c., 1878, p. 12. The beds were stated to be Lias, with *Gryphæa incurva* and *Hippopodium ponderosum ;* these were probably the Oxfordian *Gryphæa* and *Modiola.*

Vale of Blackmore.

Oxford Clay probably occurs directly beneath the Cretaceous rocks between Cerne Abbas and Maiden Newton, although it is most likely that the underground tracts are considerably faulted. Near Rampisham and again to the north of Corscomb, small areas of Oxford Clay are shown on the Geological Survey Map; and to the east and north-east of these tracts we enter upon the main outcrop of Oxford Clay, an area which under the name of the vale of Blackmore extends in a broad belt to Wincanton, and thence onwards into Oxfordshire.

In this tract the beds are opened up here and there for the manufacture of bricks, tiles, and drain-pipes, and occasionally, as at Melbury Osmund, for the making of ornamental bricks and flower-pots.

The upper beds consist of blue and brown clay, with ferruginous nodules and occasional septaria. These were exposed in the Holnest brickyard situated to the north-east of Totnell corner, where I obtained *Ammonites vertebralis*, *Exogyra*, and *Gryphæa dilatata*; and in the brickyard north of Marnhull Street, where *G. dilatata* also occurs. This same *Gryphæa* is found at Chetnole, and in the beds worked to the west of Remedy south of Hilfield Lane, and at Middlemarsh, south of Holnest. At the brickyard north of the Buck's Head, east of Melbury Osmund, and in that of Kingstag, near Haselbury Common, I obtained no fossils. Specimens, however, have been obtained to the south of Pulham, but no section was to be seen at the time of my visit in 1885. Some records of well-borings in this neighbourhood are given in the chapter on Water-supply (p. 332).

In the lower Oxfordian beds there are but few openings. At Weston, between Sutton Bingham and Closeworth, the following section was exposed :—

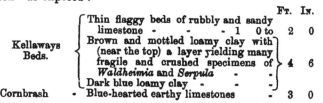

		Ft.	In.
Kellaways Beds.	Thin flaggy beds of rubbly and sandy limestone - - - - 1 0 to	2	0
	Brown and mottled loamy clay with (near the top) a layer yielding many fragile and crushed specimens of *Waldheimia* and *Serpula* - -	4	6
	Dark blue loamy clay - - -		
Cornbrash	Blue-hearted earthy limestones -	3	0

This section corresponds with that seen between Rodden, south of Portisham, and Langton Herring. A somewhat similar section of clays overlying Cornbrash limestones, and filling pipes in the rock, was exposed in a quarry south-east of the Barn, between Closeworth and Melbury Osmund.

Further on, past Sturminster Newton and Stalbridge to Wincanton, I have no records of wells or borings. *Ammonites Lamberti* has been found at Lydlinch, and *Gryphæa dilatata* at Bagber. The upper beds of the Oxford Clay have been worked in a brickyard south of Fifehead Magdalen; and they were shown at the entrance to the Kingsmead Tunnel on the London and South-Western Railway, east of Buckhorn Weston.

Wincanton to Frome.

From the neigbourhood of Wincanton the vale of Oxford Clay extends along the foot of the Cretaceous escarpment of Penzlewood (Pen Selwood) and Stourhead to Witham Friary, Longleat, and Rodden, near Frome.

Near South Brewham, a boring in search of coal was made many years ago, in spite of remonstrances from William Smith. The speculators, as we are told by John Phillips, proceeded at a ruinous expense through the Oxford or "clunch" clay, and the Kellaways Rock, with *Gryphœa bilobata*, till they entered the rocks of the Oolitic Series.[*] It is probable that the Kellaways Rock forms the gentle escarpment that rises above the Cornbrash to the south of South Brewham.

In more recent years (1867–68) another fruitless search for coal was made by J. Oxley, about a quarter of a mile south of Witham Bridge, Witham Hole, and to the north-east of Witham Friary. The details of the boring have been published by Prof. Prestwich,[†] but, from his grouping of the strata, which was very doubtfully given, and also from that of H. W. Bristow,[‡] I venture seriously to differ. The record is as follows :—

BORING AT WITHAM FRIARY.		THICKNESS.	DEPTH.
		FT. IN.	FT. IN.
Oxford Clay and Kellaways Beds.	Surface clay - - -	10 0	—
	Blue clay with *Gryphœa* - -	90 0	100 0
	Blue marls - - -	50 0	150 0
	Blue rock - - - -	1 2	151 2
	Thin layers of sandstone and limestone with *Trigonia* -	48 10	200 0
	Light grey marls - - -	20 0	220 0
Cornbrash.	Brown and blue sandstone and limestone rock - -	15 0	235 0
	Sandstone [? limestone] - -		
Forest Marble.	Light green marls - -	16 0	251 0
	Dark grey marls - - -		
	Whitish marls - - -	30 0	281 0
	Limestone - - -	6 0	287 0
	Marl slate ? - - -	3 0	290 0
	Hard limestone granular and oolitic - - - -	3 0	293 0
	Marl slate ? - - -	3 0	296 0
	Sands and sandstones, alternating in beds 18 inches thick - -	20 0	316 0

The boring was continued to a total depth of 600 feet ; but the details are not known, and there was no sign of Coal-measures. Prof. Prestwich had included the beds to a depth of 287 feet, with the Oxford Clay; and Bristow regarded the beds from depths of 251 to 287 feet as Cornbrash. The Cornbrash is, however, exposed at Witham Friary, and there

---

[*] Memoirs of William Smith, p. 66 ; Conybeare and Phillips, Geol. England and Wales, p. 195 ; and Townsend, Character of Moses, p. 128.

[†] Report Royal Coal Commission, vol. i. pp. 65, 163.

[‡] Geology of East Somerset, &c. (Geol. Survey), p. 48.

is reason to think that 220 feet is quite a sufficient thickness to assign to the Oxford Clay; tho lithological definitions of the borer must naturally be interpreted.

We have thus evidence of the Kellaways Rock in this region; and indeed its presence near Witham Friary and Marston Bigot was long ago noticed.* The upper beds of the Oxford Clay, dark grey and rusty clays, yielding *Gryphæa dilatata*, and *Myacites*, were worked in a brickyard south-east of Upper Holt, on the southern side of Witham Park.

The lower beds of the Oxford Clay were exposed in the railway-cutting south of Blatchbridge, near Frome, and again between Feltham and Fulbrook Farms, where clay, with nodules, and small specimens of *Gryphæa dilatata* were observed. At the brickyard on Rodden Hill, blue shaly clay with selenite was exposed; and shaly and sandy clays with ironstone-nodules were seen near the old mill by Rodden Down. No fossils, however, were to be found.

Trowbridge and Melksham.

The area extending from Road to Melksham and Laycock Abbey forms another portion of the great vale of Oxford Clay. In a brickyard east of Road, clays with septaria were exposed to a depth of 8 feet, the lower beds being grey and the higher mottled yellow, grey, and red, with a surface-soil of brown loamy clay. *Avicula inæquivalvis, Gryphæa* (small specimens), *Ostrea,* and a cast of *Trigonia irregularis* were here obtained. These beds probably overlie the Kellaways Rock, which was stated by Townsend to occur on Road Common.†

The neighbourhood of Trowbridge has afforded several interesting sections of the Kellaways Beds. Some of these, opened up during the construction of the Wilts and Somerset (Great Western) Railway, were described by Reginald N. Mantell, the Resident Engineer of the line.‡

Overlying the Cornbrash on the south side of Trowbridge there were seen, in ascending order: blue clay, Kellaways Rock, and then clays and bituminous shales with remains of Saurians, *Lepidotus,* &c., many Ammonites, *Belemnoteuthis,* and other fossils. Dr. Mantell, who described some of the fossils, mentioned the occurrence of a fossil tree in the blue clay overlying the Kellaways Rock, and observed that " Trunks and branches of Coniferous trees, from 10 to 20 feet in length, and from a few inches to upwards of a foot in diameter, were abundant; a few traces of the foliage of Cycadaceous plants, and of Araucariæ, were likewise met with."§

The beds, indeed, like those of Christian Malford, yielded fossils by the hundred; and the strata, when first laid bare, were seen literally to be studded " with the nacreous cones of Belemnoteuthes, and the splendid iridescent shells of Ammonites; while here and there Belemnites of large size, with their phragmocones attached,

---

* Townsend, Character of Moses, p. 143.
† Character of Moses, p. 148.
‡ Quart. Journ. Geol. Soc., vol. vi. p. 312.
§ G. A. Mantell, Phil. Trans., 1848, p. 172.

were lying in relief."* Tiny Ammonites and Belemnites occurred in clusters, and with these were associated the plant-remains before mentioned.

The following series of beds, recorded by R. N. Mantell, was exposed in the railway-cutting near Studley :—

		Ft.	In.
	Crumbly shaly marl, with *Serpula vertebralis* - - - -	3	0
	Shaly marl, with *Belemnoteuthis* (abundant) - - - -	5	0
	Strong slaty clay, with pyritic septaria; and numerous examples of *Ammonites, Belemnites,* &c. - -	9	0
Oxford Clay.	Slaty clay with many specimens of *Alaria (Rostellaria)*, and driftwood -	2	0
Zone of	Rock with *Am. Reginaldi* - -	1	0
Ammonites	Slaty clay with *Ammonites, Belemnites,*		
ornatus.	*Belemnoteuthis* (rare), and lignite -	5	0
	Crumbly stone - - - -	3	0
	Strong clay with fossils - -	2	0
	Slaty clay, with *Ammonites Jason,* *Belemnites, Belemnoteuthis,* and lignite - - - -	5	0
	Clay - - - -	10	0
Kellaways Beds.	Sandstone and sand, &c.		
	Clay - - - [10 0 to 12 0]		
	Cornbrash.		

South of Upper Studley, Trowbridge, the following section was to be seen in a disused brickyard :—

		Ft.	In.
	Clay, brown at top, and mottled red in places - - - - -		
	Unctuous pale grey clay with selenite	6	0
Kellaways Beds.	Hard grey sandy limestone or calcareous gritty sandstone, with band of septarian limestone and irregular shelly limestone; many fossils -	1	0
	Ochreous clay.		

A little further south, about half-way between this brickyard and Southwick Bridge, another clay-pit has been opened in grey, brown, and red mottled clay, with small ochreous galls, resembling the top beds in the above section and also those displayed near Road.

Probably the lowest clay was exposed in the brickyard west of Lower Studley, and south-west of Trowbridge : it contains a band of septaria. To the north-west of the railway-station a brickyard clearly shows these lower beds as follows :—

		Ft.	In.
	Hard flaggy and calcareous sandstone and sandy layers, with many fossils, (as at Upper Studley).		
Kellaways Beds.	Unctuous blue clay (weathering yellowish-brown) with much selenite and a band of septaria; seen to a depth of - - - -	10	0
	The Cornbrash would probably be reached in a foot or two.		

---

* G. A. Mantell, Phil. Trans., 1848, p. 174 ; and R. N. Mantell, Quart. Journ. Geol. Soc., vol. vi. p. 314.

FIG. 19.

Section along the Great Western Railway at Trowbridge, Wiltshire. (After R. N. Mantell.)
Distance about one mile.

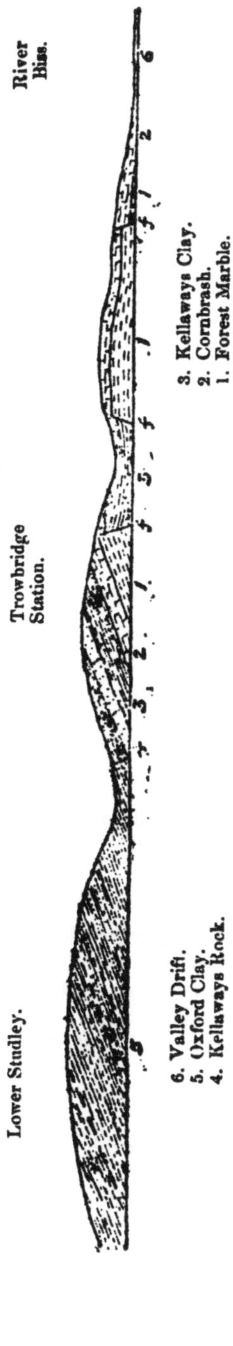

Lower Studley.

Trowbridge Station.

River Biss.

6. Valley Drift.
5. Oxford Clay.
4. Kellaways Rock.

3. Kellaways Clay.
2. Cornbrash.
1. Forest Marble.

The accompanying section (Fig. 19) shows the general arrangement of the strata. The precise horizon of the Oxfordian clay in the trough-fault is not quite certain. Mantell records from it *Ammonites Kœnigi,* so that it probably belongs to the lower beds.

The following fossils have been obtained from the Oxfordian Beds of Trowbridge :—

Kellaways Beds.

× Ammonites calloviensis.	× Gryphæa bilobata.
—— Kœnigi.	× —— dilatata (small).
—— macrocephalus.	Isocardia minima.
—— modiolaris.	Leda Phillipsi.
Ancyloceras calloviense.	Modiola bipartita.
Alaria bispinosa.	× Myacites recurvus.
Cerithium muricatum.	× Ostrea gregaria.
× Arca Quenstedti.	Pecten demissus.
—— subtetragona.	× —— fibrosus.
Astarte carinata.	× —— lens.
× Avicula braamburiensis.	Pholadomya acuticosta.
× —— ovalis.	—— Murchisoni.
Cardium cognatum.	× Trigonia.
× Corbula Macneilli.	Waldheimia ornithocephala.
× Cucullæa concinna.	× Crustacean (claw).

Shales (Zone of *Ammonites ornatus*).

Ammonites athleta.	Belemnoteuthis antiquus.
—— Chamusseti.	Belemnites abbreviatus.
—— gowerianus.	—— Oweni.
—— hecticus var. lunula.	Nautilus.
—— Jason.	Avicula expansa.
—— Kœnigi.	Leda Phillipsi.
—— modiolaris.	Modiola bipartita.
—— Reginaldi.	Serpula.

Most of these species were collected by R. N. Mantell, and named by Morris ;[*] those marked × were collected by Mr J. Rhodes and myself.

Fig. 20.

Bluish-grey clays with much selenite were exposed in the brickyards north of Islington, Trowbridge; and Kellaways Rock was noted by Wm. Smith, in the Kennet and Avon Canal near Lady-down Farm, between Trowbridge and Staverton.

Cypridea spinigera, *Sow.*

Prof. T. Rupert Jones informed me that a curious Ostracod, like *Cypridea spinigera* of the Wealden Beds, and recorded also from Oligocene strata, had been obtained from a dark-coloured clay brought by R. N. Mantell from one of the Trowbridge railway-cuttings (Fig. 20). No Microzoa, however, were to be found in specimens of the Oxford Clay (from this neighbourhood) which I forwarded to Prof. Jones for examination.[*]

The lower Oxfordian beds were exposed in a brickyard at Holt, where Mr. Rhodes obtained *Myacites recurvus* and *Arca Quenstedti*? Lonsdale notes Kellaways Rock at Whitley, near Melksham.[†]

The record of a well sunk in 1816, for the Melksham Spa Company, afforded an interesting section of the middle and lower beds of the Oxfordian formation of this district.[‡] It may be summarized as follows :—

		Ft.	In.
Oxford Clay and Kellaways Beds.	Dark-coloured marl, which " broke into rhomboids" - - -	90	0
	Marl, with *Nautilus*, and many pyritic Ammonites; and towards the base lignite and septaria - - -	132	0
	Band of septaria - - -	2	0
	Marl - - - -	50	0
	Gritty stone - - - -	2	6
	Marl - - - -	18	0
Cornbrash -	Light-coloured stone, with partings of clay - - - -	20	0
Forest Marble	Alternate strata of stone and clay -	23	6
	Pale stone - - - -	2	6
	Hard stone, separated by beds of clay and sand - - - -	11	0
		351	6

An earlier record of a well sunk at Holt was published in 1728 by the Rev. J. Lewis :[§] this I have elsewhere quoted, in connexion with the occurrence of saline waters, which were encountered, at a depth of about 10 feet, in the Kellaways Beds.[‖] It is possible that the Cornbrash was touched at the bottom of the well.

The upper beds of the Oxford Clay were shown in a brickyard to the north-east of Seend, and in another known as the Bromham brickyard, on the south side of Sandridge Hill. The beds yield *Gryphæa dilatata* in abundance.' The Bromham brickyard showed about 12 feet of grey, lead-coloured, and yellow streaky clay, with selenite, ironstone-nodules, and septaria; here Saurian bones are sometimes obtained, but fossils are not abundant.

Reference has been made by Prof. Hull to the narrow outcrop of the Oxford Clay near Laycock.[¶] The most likely explanation seems to be that of a fault—perhaps a continuation of one marked on the Geological Survey Map north of Atford.

---

[*] Jones and Sherborn, Geol. Mag., 1886, pp. 273, 274.
[†] Trans. Geol. Soc., Ser. 2. vol. iii., p. 275.
[‡] H. B. W., Quart. Journ. Geol. Soc., vol. xlii. p. 301 ; see also Britton and Brayley, Beauties of Wiltshire, vol. iii. p. 221.
[§] Phil. Trans., vol. xxxv. p. 489.
[‖] Quart. Journ. Geol. Soc., vol. xlii. p. 302 ; Memoir on the Lower Oolitic Rocks of England, p. 514.
[¶] Geol. parts of Wilts, &c. p. 19.

The Kellaways Rock has been observed near Ray (or Rey) Bridge, Laycock. Specimens of calcareous sandstone from this locality are preserved in the Museum of Practical Geology; they contain *Ammonites Kœnigi, Isocardia minima, Rhynchonella varians,* and *Waldheimia obovata.*

Chippenham, Kellaways, and Christian Malford.

The various beds of the Oxford clay, including representatives of the Kellaways Rock, have been exposed in several brickyards in the great vale that extends from Chippenham to Oxford. In this area is included the old Forest of Braydon, of which but small remnants now exist in the country between Malmesbury and Wootton Bassett. Judging from the evidence I have been able to gather, the following appear to be the chief local divisions :—

		Ft.	In.
Oxford Clay	Grey and lead-coloured clays, with occasional gritty bands, iron-stone nodules, septaria, and selenite: *Gryphœa dilatata* (abundant) -	300	0
	Dark laminated clays, shales, and marls, with occasional septaria, and pyritic fossils -	200	0
Kellaways Beds.	Bands of loamy clay, sand and calcareous and shelly sandstone, fossiliferous - - - 30 0 to	40	0
	Clay, ochreous and sandy, and blue clay, with much selenite and an occasional band of septaria 12 0 to	20	0
	About -	560	0

At Foxham, 4 miles north-east of Chippenham, a boring for water was carried to a depth of 420 feet in Oxford Clay, according to information furnished to H. W. Bristow by John Spencer, of Bowood. "Rock" was then reached, but whether this was Kellaways Rock or Cornbrash was not stated. The full thickness of the Oxford Clay is not present at that particular locality; indeed it cannot be less than 500 feet.[*]

The neighbourhood of Chippenham is a classic district for Oxfordian fossils; the localities of Kellaways and Christian Malford being famous in the annals of Geology.

The name " Kelloways Stone " was applied by William Smith to the thin fossiliferous beds of calcareous sandstone, that occur near the base of the Oxford Clay, at Kellaways, about 2½ miles north-east of Chippenham in Wiltshire.

A description of the rock, which was stated to be from 3 to 6 feet in thickness, was published by the Rev. Joseph Townsend, from observations made conjointly with Smith ; and it is recorded now the rock had been distinctly traced across country from South Brewham to beyond Malmesbury. Nevertheless, a few of the

---

[*] See also J. Buckman, Quart. Journ. Geol. Soc., vol. xiv. p. 124.

localities given by Townsend are erroneous, as the Lower Calcareous Grit was in some cases confused with the Kellaways Rock.[*]

In reference to the Kellaways Rock ("Kelloways Stone"), William Smith remarked that neither from thickness nor consistence can it be properly called a formation, and it should be regarded only as one of the divisions in the great clay-district (Oxford Clay), "there being beneath this stone another stratum of clay, which is the boundary of the great stony district called the Stonebrash Hills " (Cornbrash, &c.). He states that the course of the stone is known only by the few excavations in it, chiefly for road-materials, which in a country abounding so much with clay are very scarce. The excavations were few and shallow, and they held water. As remarked by Townsend, the Kellaways Rock " first attracted our notice, at Kelloway Bridge, in the high road from Calne, by Bremhill, to Malmesbury ;" and Smith describes it as " brown rubbly stone, with sandy exterior ; irregular lumps, bluer and harder within, composed chiefly of Organized Fossils."[†] Lonsdale, who subsequently saw some of the sections, notes that the rock is a calcareous sandstone, and that while fossils are occasionally so numerous as to constitute nearly the whole of the stratum, they are often wanting.[‡]

In 1886 when seeking for evidence of the Kellaways Rock, I learnt that the small quarries have long been abandoned and obscured, but I was fortunate in finding a section in the banks of the Avon, near a foot-bridge about half a mile below Kellaways Mill ; and subsequently in company with Mr. Rhodes we found another exposure in the banks of a deep water-course, on the western side of the river, and a little nearer to the road that leads to Peckingel Farm. The sections showed the following beds :—

		FT.	IN.
Alluvium.	Loam · · · 1 0 to 3		0
Kellaways Beds.	Buff sand and fissile calcareous sandstone, with lignite and many fossils 1 0 to 2		0
	Brown sandy clay passing down into blue clay, with small *Gryphæa* ; seen to depth of · · · · · ·		5 0

William Smith remarked that " In most parts the surface and soil of this stratum [Kellaways Rock] differs but little from that of the clay courses on each side, and that little distinction is still further partially obscured by the soil of the calcareous alluvium which is common to the clay vale district." This was notably the case at South Cerney, south of Cirencester, where the extent of the Kellaways Rock was not until lately known. The fact. however, suggests that where marked hills overlook the Cornbrash, or in other words, where there is a gentle escarpment of Oxfordian Beds above that formation, there Kellaways Rock may reasonably be expected. Smith observed that the rock " nowhere forms any

* Character of Moses, pp. 103, 127, 148.
† Strata identified by Organized Fossils, 1816, p. 23.
‡ Trans. Geol. Soc., Ser. 2, vol. iii. p. 260.

characteristic surface, or rarely a hill or other feature which is distinguishable to any but those who know where to look for the Stratum. Several small commons in North Wilts, rather sandy and springy, seem to be of the soil formed by the outcrop of this Stratum, whose course is but partially defined ;" while hollow-ways, "in roads across the course of this Stratum, but little sunk beneath the surface of the adjoining lands, seem to indicate its site." He adds, that "Selenite is very abundant in the clay above it ; bituminous wood, and a brown aluminous earth below it. There is great reason to believe that the mineral waters of the lower part of the clay vale series are from this stone, or some contiguous part of the clay above or below it."* This was the case at Holt, as previously mentioned.

The Kellaways Rock no doubt helps to form the feature that rises from near Peckingel Farm on the borders of the Avon, and extends above Cocklebury Farm and across the Great Western Railway to the north-east of Chippenham Station. Sandy layers are shown in the cutting of the Calne branch-railway near Chippenham, but the banks are mostly obscured.

According to Bristow's notes, sands with an occasional inter-rupted bed of stone (Kellaways Rock), were found, at depths of 7 to 12 feet, beneath clay and surface-soil at the Chippenham Cemetery. The beds were also exposed to the south of the town, in a cutting at the Rifle Butts, between the river and a branch of the Wilts and Berks Canal, east of Rowden Farm. There grey micaceous clay with sandy layers occurs ; *Ostrea gregaria*, and *Gryphæa bilobata* are abundant, and *Belemnites sulcatus, Avicula inæquivalvis*, and *A. ovalis* likewise occur. The locality was mentioned by William Smith.

Other indications of Kellaways Rock, at Tytherton Lucas, Christian Malford, and Dauntsey House, were mentioned by Smith ; and Brinkworth Common, and Blick's Hill, near Malmes-bury, were noted by Townsend. It is probable that at some of these places the beds were proved in the sinking of wells.

The occurrence of Kellaways Rock was proved by Mr. T. Holloway, of Chippenham, at Little Somerford (Somerford Parva) ; and two Ammonites were kindly sent by him to the Museum at Jermyn Street. These proved to be *Ammonites Bakeriæ* (9 ins. diam.) and *A. gowerianus*. They were obtained from a depth of 7 feet, in the drainage of a new burial ground, south-west of the village, and north of the turning to King's Mead Mill. The rock was a hard calcareous sandstone 2 ft. 6 ins. thick, embedded in sand, and below there was evidence of clay.

The following is a list of the fossils collected from the Kellaways Beds of Kellaways and Chippenham, by Mr. John Rhodes and myself :—

Ammonites Bakeriæ.	Ammonites Mariæ.
—— calloviensis.	—— modiolaris.
—— gowerianus.	Ancyloceras calloviense.
—— Kœnigi.	Belemnites Oweni.

* Strata identified by Organized Fossils, 1816, p. 23.

Belemnites sulcatus.
Alaria trifida.
Bourguetia (" Phasianella ").
Cerithinm Culleni.
Natica punctura.
Turbo Archiaci.
Anatina.
Arca æmula.
—— Quenstedti.
—— subtetragona.
Astarte carinata.
Avicula braamburiensis.
—— inæquivalvis.
—— Münsteri.
—— ovalis.
Cardium cognatum.
—— Crawfordi.
Corbula Macneilli.
Cucullæa concinna.
Gervillia.
Goniomya v.-scripta.
Gryphæa bilobata
—— dilatata.

Isocardia minima.
Lima.
Lithodomus.
Lucina despecta.
—— rotundata.
Modiola bipartita.
—— cuneata.
Myacites calceiformis.
—— decurtatus.
—— recurvus.
Nucula ornata.
Ostrea gregaria.
Pecten fibrosus.
—— lens.
Pholadomya acuticosta.
Pinna.
Trigonia.
Unicardium.
Rhynchonella varians.
Terebratula.
Waldheimia ornithocephala.
—— —— var. calloviensis.
Serpula tetragona.

The clays above the Kellaways division were shown in a brick-yard north of Middle Lodge, south-east of Chippenham, on the borders of the Wilts and Berks Canal ; but there are few sections to be seen in the beds, and nowhere have I seen the rich fossil-layers of the zone of *Ammonites ornatus* opened up in this part of the country. Fortunately the geologists of old have left their records.

The following section of " the quarry at Christian Malford " was noted by Lonsdale :—*

		FT.	IN.
Oxford Clay.	Pale lead-coloured clay, streaked with yellow.		
Kellaways Beds.	Rotten, rubbly stone, highly charged with oxide of iron, and enclosing few organic remains	5	0
	Sandstone, abounding with fossils	3	0
	Sand	4	0
	Clay		

The precise situation of this opening was not given, but the beds evidently come below the rich fossiliferous shales of Christian Malford, that were brought to light in 1841 during the construction of the Great Western Railway. Of the many fossils then obtained from the Oxford Clay, the majority were procured from pits and trenches dug on either side of the railway, between Wootton Bassett and Chippenham, for the purpose of obtaining material for embankments.

The following section of the beds was given by J. Chaning Pearce :—†

		FT.	IN.
Valley Drift.	Soil	2	0
	Gravel	8	0
Oxford Clay.	Four or five bands of laminated clay, alternating with sandy clay, almost entirely composed of broken shells	6	0
	Clay, containing *Gryphæa bilobata*		

* Trans. Geol. Soc., Ser. 2, vol. iii. p. 261.
† Proc. Geol. Soc., vol. iii. p. 598.

The fossils, which were obtained chiefly from the laminated clay, included Fishes, Mollusca, Crustacea, and Lignite. Hundreds of Ammonites, especially those of the " Ornati group," were obtained, and although for the most part in a flattened condition, many were otherwise perfect, and the form of the aperture, as seen in examples of *A. Jason* and its allies, was beautifully shown. Some new species were described by S. P. Pratt.[*]

Specimens from Christian Malford thus came to be distributed amongst the principal collectors of the time, including the Marquis of Northampton, Dr. Mantell, and William Cunnington, and examples are to be seen in most museums, including those of Bath, Bristol, and Devizes. Many of the fossils were obtained and developed with great skill by William Buy, of Sutton Benger; and among those were some of the most perfect specimens of Belemnites. Another form was named *Belemnoteuthis antiquus* by J. Chaning Pearce;[†] concerning this, S. P. Woodward remarked that " In the fossil calamary of Chippenham, the shell is preserved along with the muscular mantle, fins, ink-bag, funnel, eyes, and tentacles with their horny hooks."[‡]

Among the fossils from Christian Malford, the following may be enumerated:—

Aspidorhynchus euodus.	Ammonites modiolaris.
Lepidotus macrocheirus.	——Sedgwicki.
Leptolepis macrophthalmus.	Belemnites hastatus.
Ammonites Bakeriæ var fluctuosus.	—— Oweni.
—— Comptoni.	Belemnoteuthis antiquus.
—— Duncani.	Alaria bispinosa.
—— Elizabethæ.	—— trifida.
—— gowerianus.	Patella.
—— hecticus var. lunula.	Avicula ovalis.
—— Jason.	Leda Phillipsi.
—— —— var. Gulielmi.	Pholadomya deltoidea.
—— Kœnigi.	Mecocheirus Pearcei.
—— macrocephalus.	Amphiura Pratti.

It is not improbable, as the fossils were obtained at various spots along the railway, near Christian Malford, that some specimens were derived from the zone of *Ammonites calloviensis*, although the greater portion of them were found in beds that belong to the zone of *A. ornatus*.

Ammonites Duncani and *A. Achilles* have been recorded from Dauntsey. A brick-yard by Dauntsey Station was opened in reconstructed loamy and gravelly beds with derived fossils from the Corallian rocks, &c.

---

[*] Ann. Nat. Hist., vol. viii. (1841), p. 161.
[†] Charlesworth's London Geol. Journ., 1847, p. 75; Cunnington, *Ibid.*, p. 97. The species named by Pearce was described as *Belemnites Oweni*, by Owen, Phil. Trans., 1844, p. 65; see also Mantell, *Ibid.*, 1848, p. 176, and 1850, p. 393.
[‡] Manual of the Mollusca, Part 1, 1851, p. 75.

Wootton Bassett and Malmesbury to Cirencester.

About a mile west of Wootton Bassett, a brickyard by Old Park Farm, on the north side of the Great Western Railway, showed the following section :—

Oxford Clay.
{ Brown clayey soil.
Grey slightly calcareous clay.
Greyish-brown clay with bands of brown gritty calcareous stone.

These beds, which were opened to a depth of about 18 feet, belong to the upper division of the Oxford Clay. The following fossils were collected by Mr. Rhodes :—

Ammonites cordatus.	Myacites recurvus.
Gryphæa dilatata.	Nucula ornata.
Ostrea.	Thracia depressa.
Modiola bipartita.	

At a brickyard at Lower Stanton, south of Corston, where the Kellaways Beds overlie the Cornbrash, I obtained from the Basement Clay some crushed shells of *Ostrea, Waldheimia* and *Rhynchonella varians*, together with *Serpula vertebralis*; and from the Kellaways Rock, *Waldheimia ornithocephala.**

North of Malmesbury, a section at a brickyard showed eight feet of grey clay with selenite, and few fossils. Professor Hull noted the presence of the Kellaways Rock in this locality.†

A trace of the Kellaways Beds (to which I have elsewhere referred) was shown to the north-west of Kemble Junction, in a lane-cutting north of Great Barn.‡

In the neighbourhood of Cirencester, the Kellaways Rock was observed by William Smith on the banks of the Thames and Severn Canal near South Cerney (Cerney Wharf); and not far from this spot a fine section of the beds was opened up in 1883, in the cutting of the Midland and South-Western Junction Railway, west of Cerney Wharf. Attention was first directed to this section by the late Prof. Allen Harker, of Cirencester, and I had the pleasure of visiting it under his guidance. The section was as follows :—

			Ft.	In.
	Gravel (in places)			
	Brown clay - - - -		4	0
	Grey clay - - - -		0	4
	Ferruginous sandy beds with hard nodules - - -		3	0
Kellaways Beds.	Ferruginous brown sands - -		4	6
	Layer of huge "Doggers"	6 0 to	7	0
	Yellow loamy sands - -		8	0
	Clay - - - - -		3	6
			30	4

The Doggers form the most conspicuous feature in this section ; they are large spheroidal masses of calcareous sandstone, blue-hearted, but weathering yellow, and very hard. They are fissile in places, and are evidently due to a local cementation of the sand,

---

* See also Memoir on Lower Oolitic Rocks of England, p. 441.
† Geol. parts of Wilts, &c. pp. 18, 19.
‡ Memoir on the Lower Oolitic Rocks of England, p. 442.

that commenced around organic remains. Nests of Brachiopods (*Waldheimia obovata*) occur in many of the masses, together with Ammonites, shelly layers with small *Ostrea, Avicula,* and other fossils ; also lignite.   Some irregular cavities occur in the stone, and there are likewise fucoidal markings. A full account of the strata was published by Professor Harker,* who mentions that one of the large Doggers, weighing about 1¼ tons, 4 ft. 6 ins. in diameter and 1 ft. 9 ins. in thickness, has been placed in the garden attached to the Royal Agricultural College, near Cirencester.

Prof. Harker gives the following analysis of the calcareous sandstone :—

Calcium carbonate -	34·35
Ferric oxide	1·94
Manganese and Alumina -	·64
Sodium chloride · ⎱	traces.
Magnesium carbonate ⎰	
Silica (Sand)	60·74
	97·67

He remarks that after exposure, some examples of the Doggers weathered rapidly and successive coatings became separated from the central core; and that " lying about in the cutting were numbers of these regular shaped cores, called ' cannon balls ' by the workmen." It is remarkable that the Engineers in their pre-liminary trial-borings along the line of railway failed to meet in any quantity these huge masses of rock, and they were "not revealed until the excavations for the cutting had proceeded some considerable length."

The accompanying illustration is taken from a photograph published by Prof. Harker; it represents the general aspect of the cutting, looking northwards. Heavy rains had washed away the loose sand, leaving the Doggers standing out prominently. It is not until we reach the neighbourhood of Bedford that similar masses are seen in the Kellaways Beds.

FIG. 21.

Section along the Midland and South-Western Junction Railway at South Cerney, near Cirencester.   (From a Photograph.)

* Prcc. Cotteswold Club, vol. viii. p. 176.

The following fossils were obtained from the Kellaways Rock of South Cerney, by Prof. Harker, Mr. Rhodes, and myself: those marked × being recorded by Prof. Harker :—

× Ammonites calloviensis.
× —— Chamusseti.
× —— gowerianus.
× —— Kœnigi.
—— macrocephalus.
—— modiolaris.
× Belemnites hastatus.
× —— Oweni.
× Nautilus hexagonus.
Cerithium Culleni.
× Pleurotomaria depressa.
Dentalium.
Astarte robusta ?
—— ungulata.
Avicula Münsteri.
Cardium.
Cucullæa.
Goniomya v. scripta.
× Gresslya peregrina.

× Gryphæa bilobata.
× —— dilatata.
× Isocardia minima (tenera).
Lima.
Lucina despecta.
× Modiola bipartita.
× Myacites recurvus.
× Ostrea flabelloides.
Pecten fibrosus.
× —— lens.
Pinna mitis ?
× Pholadomya deltoidea.
× —— Phillipsi.
Trigonia.
× Unicardium sulcatum.
× Rhynchonella varians.
× Waldheimia obovata.
—— ornithocephala.
Pentacrinus Fisheri ?

That the sandy basement-portion of the Oxfordian formation attains a great local thickness is shown by the evidence of a boring at South Cerney, and the deep sinking at Swindon: the Kellaways Beds there are proved to be from 60 to 75 feet thick.

A boring at the Manor House, South Cerney, details of which were communicated to me in 1888 by Prof. E. Kinch, proved the following strata :—

		Ft.	In.
	Gravel	18	0
	Sandy clay	31	0
	Calcareous sandstone	2	0
	Sandy clay	3	0
	Blue clay	15	0
Oxford Clay and Kellaways Beds (75 feet ?)	Blue sandy clay	21	0
	Calcareous sandstone with *Ostrea, Terebratula,* Lignite	8	0
	Shelly marl	6	0
	Calcareous clay	7	0
	Fine sand (yielding water)	3	0
	Blue calcareous sandstone passing down to pure clay	10	1
		124	0

The Kellaways Rock forms in places a well-marked feature, which may be traced to the south of Sharncott, as well as near South Cerney. It extends northwards towards Shells Grove and Worms Farm, Siddington St. Peter, where sections (opened for brick-yards) showed ochreous sand and sandy clay, resting on stiffer clay with septaria and small specimens of *Gryphæa*. The fossils at these localities are, many of them, poorly preserved.

The brickyard to the east of Shells Grove showed blue sandy clay, from which Mr. Rhodes obtained the following specimens:—

Ammonites (fragment).	Myacites recurvus.
Arca.	Nucula.
Avicula inæquivalvis.	Ostrea (with markings of Tri-
Cardium.	gonia).
Gryphæa.	Pecten.
Modiola.	Crustacean (claw).

Higher beds of Oxford Clay were exposed in the brickyard a little north of Cerney Wharf, where brown and purple shaly clay, ferruginous in places, was exposed to a depth of about 15 feet. Small Ammonites and Belemnites occur here; and Mr. Rhodes obtained *Alaria trifida, Avicula,* and *Lucina.* Fine crystals of selenite are found in the Oxford Clay at Ashton Keynes.

The occurrence of a small faulted tract of Kellaways Beds at Lewis Lane, Cirencester, has been recorded by Prof. Harker; the beds were proved in a well-boring, as follows*:—

						FT.	IN.
	Made ground	-	-	-	-	7	0
	Gravel	-	-	-	-	12	0
Kellaways	Fine sand	-	-	-	-	2	0
Beds.	Dark brown clay	-	-	-	1	6
	Blue clay	-	-	-	-	22	10
Cornbrash.							

Specimens of *Ammonites macrocephalus, Belemnites Oweni, Avicula inæquivalvis, Modiola bipartita, Myacites recurvus, Ostrea* (or *Gryphæa*), *Terebratula intermedia?* and *Waldheimia ornithocephala,* were identified by Mr. G. Sharman from specimens collected by J. H. Taunton. Many small concretions occurred in the clay.

Shallow sections of Oxford Clay were opened up along the railway near Cricklade, and again the clay was shown in cuttings between Woodwards Bridge and Elbro Bridge, near Haydon, south-east of Little Blunsdon.

A furze-covered sandy hill near the Foss Farm, south of Driffield, was considered by John Bravender to indicate the Kellaways Rock;† and it is probably developed in the ridge south of Poulton Farm and the old Church. Many fossils, including bones of *Cimoliosaurus,* were formerly obtained from the brickyard, east of the Farm, where the basement Oxfordian clays were exposed.

Prof. Phillips described and figured under the name of *Ammonites superstes* a portion of a "Cordate" Ammonite from Minety. This must have come from the Oxford Clay, though noted by him as from Kimeridge Clay.‡

One of the best sections of the Oxford Clay in Wiltshire was that furnished in a deep well-sinking at Swindon, made in 1883–85, by the Great Western Railway Company under the direction of Captain William Dean. The Oxford Clay and Kellaways Rock were shown to have a thickness of 572 feet 9 inches, the details of the strata being noted by Messrs. W. H. Stanier, A. R.

---

* Proc. Cotteswold Club, vol. x. p. 187.
† S. P. Woodward, *Ibid.*, vol. i. p. 4.
‡ Geol. Oxford, pp. 332, 333.

Elliott, and myself. A full account being elsewhere published,[*] it is only needful here to give the following summary:—

		Ft.	In.
Corallian Beds.			
Oxford Clay	Clay with shaly and marly beds, septaria, and occasional gritty bands	466	10
	Clay with bands of septaria, laminated, bituminous, and shelly clays and shales.	43	9
Kellaways Beds.	Laminated and shelly clays with occasional septaria and greenish calcareous sands and sandstone	49	0
	Dark grey laminated clay, shelly in places	12	4
	Grey gritty rock and sand (at junction with Cornbrash)	0	10
		572	9

A large collection of fossils was made from the Oxfordian beds, these were identified by Mr. E. T. Newton, and they are tabulated in the accompanying list. More detailed references to the depth of some of the specimens would have been valuable. Broadly speaking, however, the record furnishes evidence of the Kellaways fauna at the base, and of the incoming in succession of the "*ornatus*" and "*cordatus*" types of Ammonites; and it is satisfactory to find that the ordinary succession of Ammonite-forms is maintained, even if the species are not limited to the often rigid and therefore unnatural definitions of a zone.

Thus in the upper 288 feet, *Ammonites cordatus* and *A. cordatus* var. *excavatus* are met with. This division would therefore correspond with the "*cordatus*-clays," although the two Ammonites occur a little lower down, together with other species.

In the next 220 feet, *A. crenatus*, *A. Duncani*, *A. Jason*. *A. Kœnigi*, *A. Lamberti*, *A. Mariœ*, and *A. plicatilis* occur : the last-named a Corallian form ! This division no doubt represents the "*ornatus*-clays" of the Lower Oxfordian. In the next 18 feet we find *A. Jason* var. *Gulielmi*, and *A. calloviensis*. The lowest 44 feet, comprising alternations of clays, sands, and sandstones, yielded *A. Kœnigi*, *A. Bakeriœ*, and *A. modiolaris*, found also higher up; likewise *A. gowerianus*. The last-named two species are usually regarded as Callovian forms.

The development of sandy beds at the base of the Oxford Clay is interesting. On lithological evidence, about 44 feet may be assigned to the Kellaways Rock ; but if we consider *Ammonites calloviensis* to belong exclusively to this rock, then at least 18 feet more may be grouped with the Kellaways division. The evidence, however, agrees with that furnished in other localities, that the Kellaways Rock is but an irregular and impersistent sandy basement-bed of the Oxford Clay, locally fossiliferous.

Palæontologically the beds may roughly be grouped as follows :—

		FEET.
Zone of *Ammonites cordatus*.	A. cordatus, A. cordatus var. excavatus and A. perarmatus	300
Zone of *A. ornatus*.	Am. Lamberti, A. crenatus, A. Duncani, A. hecticus, &c.	100
	A. Jason, A. Bakeriœ, A. Mariœ, A. Kœnigi, A. modiolaris, &c.	110
Zone of *A. calloviensis*.	A. calloviensis, A. gowerianus, and A. modiolaris	62

* Quart. Journ. Geol. Soc., vol. xlii. p. 247.

List of Fossils from the Oxfordian Beds of Swindon.	Depths in Feet.						Particular depths at which Species were found.
	529 to 572	512 to 528	511	322 to 510	288 to 510	1 to 288	
Ammonites Bakeriæ	x	-	-	-	x	-	361–532
—— —— var. fluctuosus	-	-	-	-	x		
—— calloviensis	-	x	x	-	-	-	511–528
—— cordatus	-	-	-	-	x	x	to 353
—— —— var. excavatus	x	-	-	-	x	x	312
—— crenatus	-	-	-	-	-	-	314–363
—— Duncani	-	-	-	-	x	-	322–411
—— gowerianus	x	x	-	-	-	-	528–544
—— hecticus	-	-	-	-	x		386
—— —— var. lunula	-	-	-	-	x		362
—— Jason	x	-	-	-	x	-	388–550
—— —— var. Gulielmi	-	x	-	-	x		370–517
—— Kœnigi	x	-	-	-	x		
—— Lamberti	-	-	-	-	x	-	304–397
—— macrocephalus	-	-	-	x	-	-	304–336
—— Mariæ	-	-	-	x	-	-	426
—— modiolaris	x	-	x	x	-	-	312–550
—— perarmatus	-	-	-	x	-	x	274–318
—— plicatilis!	-	-	-	x	-		336–353
Belemnites abbreviatus?	-	-	-	x			
—— Oweni	x	x	-	x			
—— sulcatus	-	-	-	x			
Belemnoteuthis antiquus	-	-	-	x			
Nautilus hexagonus	-	-	-	x			
Actæon retusu	-	-	-	-	x		
Alaria trifida	-	x	-	-	x	x	
Cerithium Damonis	x						
—— muricatum	-	-	-	-	x		
—— sp.	-	-	x	-	-		
Pleurotomaria reticulata	-	-	-	-	x		
Arca Quenstedti	x	-	-	-	x		
——, sp.	-	x	-	-	x	x	
Astarte carinata	x	x	-	-	x		
—— ovata	-	-	-	-	/		
Avicula ovalis	x						
—— inæquivalvis	x	x	-	-	x		
Cardium, sp.	-	-	-	-	x		
Corbula, sp.	-	-	-	-	x		
Goniomya v.-scripta	-	-	-	-	x	x	
Gryphæa dilatata	?	x	-	-	x	x	
—— bilobata	-	-	-	-	x		
Inoceramus	-	-	-	-	x		
Isocardia minima	x						
Leda Phillipsi	x						
Modiola bipartita	x						
——, sp.	-	x	-	-	x		
Myacites recurvus	x	x	-	-	-	x	
Nucula ornata	-	-	-	-	x	x	
Ostrea, sp.	-	-	-	-	x		
Pecten, sp.	x	x	-	-	x	x	
Perna mytiloides	x						
Pinna mitis	..	x	-	-	x	x	
Plicatula?	x						
Pholadomya acuticosta	x	x	-	-			
—— paucicosta	?	x	-	-			
—— sp.	-	-	-	-	x	x	

List of Fossils from the Oxfordian Beds of Swindon.	Depths in Feet.					
	529 to 572	512 to 528	511	322 to 510	288 to 510	1 to 288
Thracia depressa - - -	×	-	-	-	×	
Trigonia irregularis - - -	×	?	-	-	-	×
Unicardium, sp. - - -	-	-	-	-	×	
Rhynchonella, sp. - - -	-	-	×			
Serpula tetragona - - -	×					
—— vertebralis - - -	-	-	-	-	×	
Anabacia complanata - -	-	-	-	-	×	

CHAPTER IV.

OXFORD CLAY AND KELLAWAYS ROCK.—(LOCAL
DETAILS—*continued.*)

Cricklade, Bampton, and Woodstock.

From Cricklade to Lechlade and Bampton we follow the
broad vale of Oxford Clay, that borders the Thames onwards to
the neighbourhood of Oxford.   The full thickness of the formation
may be between 500 and 600 feet,* but it diminishes in the
direction of Oxford.   To the north-east of Bampton and near
Curbridge Common, we find rising ground that is suggestive of
the Kellaways Rock, but there are no sections to show the strata.

On the west side of Aston Sheep Common, east of Bampton,
an excavation for a brickyard displayed bluish-grey and
yellowish clay to a depth of about 8 feet.   Large specimens of
Gryphæa dilatata, and *Belemnites aripistillum* occur.   *G. dilatata*
was met with beneath the Alluvium, at Fyfield Marsh on the
south side of the Thames, as I was informed by Mr. James
Parker.

Outliers of Oxford Clay occur in Wychwood Forest, and beds
of grey shaly clay have been worked for brick-making on the
west side of Leafield.   To the east of Combe, near Woodstock,
there is also a brickyard, but the beds do not appear fossiliferous.
The higher grounds here have in places thin coverings of gravel
with quartz and quartzite pebbles.   The elevated tract of
Ramsden and Finstock Heaths in Wychwood Forest, affords an
extensive view of the country ; and there the Oxford Clay is capped
with sand and gravel.   It is possible that sandy beds belonging to
the Kellaways division may also occur at some of these localities.†

On the new branch railway to Woodstock, the Kellaways Beds
were shown at one spot, above the Cornbrash, as follows :—‡

		Ft.	In.
Kellaways Beds.	Fine yellow and grey sands - -	5	0
	Dark bluish-grey clay, and stiff mottled grey and brown clay with "race" in the lower part; and much ferruginous matter at the base - -	10	0
	Thin layer of sand overlying seam of clay - - - - -	0	5
Cornbrash.			

Witney to Oxford.

In the neighbourhood of Oxford the full thickness of the
Oxfordian Beds may be 450 feet.   I judge this to be the case

* Greater thicknesses have been assigned by Phillips, Geol. Oxford, p. 298 ; and
Hull, Explan. Hor. Sec. (Geol. Survey), Sheet 59, p. 3.
† See also Hull, Geol. Woodstock, p. 26.
‡ See Memoir on the Lower Oolitic Rocks of England, p. 303.

from drawing a diagram section from Burford through Witney and Wytham Hill to Oxford. (See Fig. 22, p. 44.) There are four borings to guide one in drawing a section, although an exact interpretation of the records of these borings is not possible.

Details of the Burford boring have been previously published—the site of it was about 350 feet above sea-level.[*]

Details of a boring at the County Police Station, Witney, made in 1883–84, by Mr. Edward Margrett (of Reading), were communicated to me by Mr. J. H. Blake. The boring, situated about 300 feet above sea-level, was carried to a total depth of 270 feet, and the strata passed through are of sufficient interest to be recorded in detail. They may be grouped (though not without doubt) as follows :—

WITNEY BORING.		THICKNESS.		DEPTH.	
		FT.	IN.	FT.	IN.
Cornbrash and Forest Marble 43 ft.	Shaft (cylinders, &c.) query strata	25	6	—	
	Rock and clay	17	6	43	0
Great Oolite and Stonesfield Beds 141 ft. 6 ins.	Shelly rock	2	6	45	6
	Blue clay	0	6	46	0
	Rock	3	0	49	0
	Blue clay	0	3	49	3
	Grey rock	1	9	51	0
	Clay and sand (with a little water)	0	3	51	3
	Grey rock	2	0	53	3
	Clay and sand	0	6	53	9
	Grey rock	0	3	54	0
	Green clay	1	6	55	6
	Hard rock with crystals	5	0	60	6
	Blue clay	0	3	60	9
	Blue rock	7	9	68	6
	Blue clay	0	3	68	9
	Rock	6	9	75	6
	Blue clay	2	0	77	6
	Rock	3	0	80	6
	Rubble	2	0	82	6
	Blue rock	5	0	87	6
	Blue clay	4	0	91	6
	Gravel	0	6	92	0
	Rubble clay	5	6	97	6
	Hard rock	4	0	101	6
	Black clay	1	0	102	6
	Green clay	5	0	107	6
	Freestone	2	6	110	0
	Blue clay	4	0	114	0
	Rubble clay	1	6	115	6
	Blue clay	3	0	118	6
	Soft rock	2	6	121	0
	Blue clay	3	6	124	6
	Hard rock in layers (? laminated)	1	6	126	0
	Green clay	2	6	128	6
	Black rock	1	9	130	3
	Blue clay	3	3	133	6
	Rock	2	0	135	6
	Blue clay	3	6	139	0
	Hard rock	0	6	139	6
	Blue clay	4	0	143	6

[*] See Memoir on the Lower Oolitic Rocks of England, p. 372.

WITNEY BORING.	THICKNESS.		DEPTH:	
	FT.	IN.	FT.	IN.
Hard rock - • • -	1	0	144	6
Blue clay • • •	3	0	147	6
Rubble rock - • • -	1	9	149	3
Blue clay • • • -	4	0	153	3
Hard rock • • • -	0	3	153	6
Blue clay • • ◡ -	2	3	155	9
Hard rock • • • -	7	9	163	6
Black clay • • • -	1	3	164	9
Soft rock • • • -	12	6	177	3
Black clay • • • -	1	6	178	9
Hard rock • • • -	2	3	181	0
Brown clay • • • -	3	6	184	6
Inferior Oolite ⎰ Hard rock • • • -	1	6	186	0
Series ⎱ Freestone with hard veins - •	26	6	212	6
30 feet. ⎰ Blue clay and sand - • •	2	0	214	6
⎰ Black rock • • • -	4	6	219	0
Clay - • • • -	1	6	220	6
Black rock • • • -	2	0	222	6
Blue clay • • • -	1	6	224	0
Rock and clay • • -	13	0	237	0
Lias ⎰ Clay - • • • -	16	0	253	0
55 ft. 6 ins. ⎱ Rock - • • • •	2	0	255	0
Light-coloured clay - • -	9	0	264	0
Clay - • • • -	3	0	267	0
Hard rock • • • -	1	0	268	0
⎱ Sand (water bearing) • -	2	0	—	
Total depth • • -	270	0	—	

The water-level was 5 ft. 6 ins. below the surface.

A boring at Wytham was made in 1829, it was situated a little west of Whitenham Lodge ; near by Wytham Hill rises to a height of about 540 feet. This boring proved the following beds :—

		FT.	IN.
Alluvial Beds,	⎰ Loamy ground (1) - • •	12	0
15 feet.	⎱ Quicksand (2) - • • •	3	0
Oxford Clay	⎰ Clunch, &c. (3-20) • • •	240	0
and	⎰ Dark blue rock (21) • • •	3	6
Kellaways	⎰ Dark parting clunch (22) - • •	0	6
Rock,	⎰ Dark blue rock (23) • • •	2	6
258 feet.	⎱ Dark clunch (24) - • • •	11	6
Cornbrash.	⎰ Strong blue rock with partings of clunch ⎱ (25-29).	19	0
Forest Marble.	Grey rock and clunch (30-32) • -	24	6
Great Oolite.	Light rock, &c. (33-49) • • -	96	0
Inferior Oolite.	Rock (50) • • • •	35	6
Upper Lias.	Mingled ground (51-55) - • •	14	6
Middle Lias.	⎰ Ironstone (56) - • -		
Lower Lias.	⎰ Clunch mixed with ironstone (57-58) -⎱	170	6
	⎱ Dark clunch (59) - • • •		
		633	0

Full details of this boring have been published by John Phillips, and the numbers above given in brackets refer to the strata recorded by him.[*] The age of the beds has also been discussed by Prof. Prestwich,[†]

* Geol. Oxford, p. 296.

† Geological Conditions affecting the Water Supply to Houses and Towns, Oxford, 1876, p. 29 ; Geol. Mag. 1876, p. 237 ; see also H. B. Woodward, Lias of England and Wales, pp. 230, 269.

who states that no specimens were preserved. The above grouping differs in some respect from that of Prof. Prestwich. I take the "dark blue rocks" (21 and 23), to indicate the presence of the Kellaways Beds, although no exposures of this sub-division have till recently been noticed in the area. I have, however, seen evidence of the beds in the railway near Woodstock, and also near Bicester; while their presence near Kirtlington was suggested by Phillips, from the occurrence there of *Ammonites Kœnigi.*[*]

The boring at St. Clement's Brewery, Oxford, was made in 1836, and is said to have passed through 265 feet of strata belonging to the Oxford Clay, and 135 feet of strata, grouped generally, with the Great Oolite.[†]

Oxford city is for the most part built on valley-gravel, which lies irregularly on the Oxford Clay. The brickyards in the neighbourhood show chiefly the upper and middle portions of the Oxford Clay. The upper beds may be seen in pits between St. Bartholomew's and St. Clement's, north of Cowley Marsh, where I obtained *Ammonites vertebralis, Belemnites excentricus, B. Oweni,* and *Gryphæa dilatata.* Other species from this locality in the Oxford Museum, and in the collection of Mr. James Parker, include *Ammonites cordatus, A. Lamberti, Belemnites hastatus, B. sulcatus, Rhynchonella varians, Pentacrinus Fisheri;* also *Asteracanthus* and *Hybodus,* Saurian and Crustacean remains (*Glyphea*).

Other sections have been opened on the Marston Road, and again at Pear· Tree Hill, on the Woodstock Road.

At Summertown, north of the city, there is an immense pit, where, beneath 8 or 10 feet of valley-gravel, the Oxford Clay, consisting of bluish-grey or lead-coloured clay with occasional scattered septaria, has been dug to a depth of about 40 feet. The clay is slightly calcareous and many of the Ammonites are pyritized. *Ammonites cordatus* occurs in the upper part, together with *Gryphæa dilatata,* and lower down we find *A. Duncani, A. Lamberti, Belemnites hastatus, B. sulcatus,* &c. Some of the septaria contain lignite. I also obtained *Ammonites calloviensis, A. macrocephalus. A. Williamsoni,* and *Nucula ornata.* It would thus seem from the evidence of the fossils, that portions of the chief zones of the Oxford Clay are here represented.

The occurrence of fossils belonging to the three main zones is confirmed by specimens in the Oxford Museum and in the collection of Mr. Parker. These include, in addition to some of those mentioned above, *Ammonites athleta, A. crenatus, A. gowerianus, A. Jason, A. Mariæ,* &c.[‡] No doubt the beds at Summertown should be grouped mainly as the zone of *Ammonites ornatus*— regarding that as a general palæontological horizon.

North-east of Wolvercot there is another very large pit showing 30 to 40 feet of Oxford Clay, with gravel on top, similar to the section at Summertown. The septaria are scattered, there being

---

[*] Quart. Journ. Geol. Soc., vol. xvi. p. 117.

[†] See notes attached to Geological Map of the Environs of Oxford, by Andrew D. Stacpoole, 1848 ; Geology of parts of Oxfordshire and Berkshire. by Hull and Whitaker, p. 5 ; and H. B. Woodward, Lower Oolitic Rocks of England, p. 513.

[‡] See also Phillips, Geol. Oxford, p. 304.

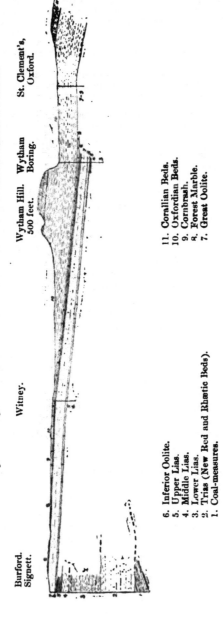

FIG. 22.

Section from Burford across Wytham Hill to Oxford.   Distance about 18 miles.

Burford.
Signett.

Witney.

Wytham Hill.   Wytham
500 feet.   Boring.

St. Clement's,
Oxford.

11. Corallian Beds.
10. Oxfordian Beds.
 9. Cornbrash.
 8. Forest Marble.
 7. Great Oolite.

6. Inferior Oolite.
5. Upper Lias.
4. Middle Lias.
3. Lower Lias.
2. Trias (New Red and Rhætic Beds).
1. Coal-measures.

no conspicuous band of them. Among the fossils *Gryphæa dila-tata* and *Belemnites sulcatus* are most abundant. *Ammonites calloviensis* also occurs.

Oxford to Buckingham.

From Oxford the Oxford Clay continues in a broad vale through Otmoor and between Bicester and Quainton to Steeple Claydon in the neighbourhood of Buckingham, where we enter a region in which the beds are largely concealed by Drifts.

The brickyard south of the Bicester railway-station showed 12 feet of dark blue and grey shaly clay with selenite ("isinglass" of workmen) and pyrites; covered by 2 feet of yellowish sandy loam and sand. These clays and the overlying sandy beds, resembling those seen on the Woodstock railway and at Akeley brickyard north of Buckingham, evidently belong to the Kella-ways Beds. The sand was formerly dug for mortar-making, and is now mostly worked out. Rock (Cornbrash) was reached a foot below the base of the pit.* Red bricks, tiles, and drain-pipes are manufactured. Formerly 17 or 18 men were engaged at the brickyard : at the time of my first visit, in 1886, only one man was employed! Pressed bricks that are machine-made, as at Peterborough and other places, take away the trade from the smaller brickyards, which will in time most probably be abandoned, to the serious loss of geologists.

To the south-east we find, at Ambrosden and Blackthorn Hill, an inlying mass of Cornbrash—one of a series of inliers that range from Islip in a north-easterly direction.

South-east of Fringford there is a brickyard in the Oxford Clay; and clay was proved to a depth of 50 feet, without reaching the bottom, about a quarter of a mile further south. About 1¼ miles south-east of Goddington Church, Prof. Green noted beds of pale calcareous and slightly concretionary clay yielding the following fossils, which were determined by Mr. Etheridge :—†

Ammonites cordatus	Astarte carinata.
—— Elizabethæ.	Avicula ovalis.
—— Jason.	Inoceramus.
	Lucina.

North-west of the railway-station, at Quainton Road, a brick-yard on the west side of the railway, showed about 10 feet of grey marly clay becoming bluish grey below, with slightly indurated marly bands, and decomposed pyritic nodules. All sorts and conditions of *Gryphæa dilatata*, broad and narrow varieties, were very abundant. Red bricks, tiles, and drain-pipes are here manufactured.

* See also Green, Geol. Banbury, p. 40 ; and Proc. Geol. Assoc., vol. xiii. p. 71.
† Geol. Banbury, pp. 40, 41.

Prof. Green has recorded, from "Quainton," the following fossils :—

Ammonites canaliculatus.	Gryphæa dilatata.
—— Duncani.	Nucula elliptica.
Belemnites hastatus.	Pecten fibrosus.
Alaria.	Pinna mitis.

He has also recorded from Ludgarshall :—[*]

Ammonites cordatus.	Belemnites abbreviatus.
—— Lamberti.	Gryphæa dilatata.
—— Mariæ.	Modiola bipartita.

He mentions that Oxford Clay was exposed in the two railway-cuttings south-east of Buckingham station; and that in a brick-yard near Padbury Mill, dark blue shaly clay, with septaria and nodules of limestone, was to be seen. From Padbury he obtained *Ammonites calloviensis, A. Jason, Belemnites,* &c. So that in the vale from Bicester to Buckingham we have evidence of the zone of *A. calloviensis,* followed by those of *A. ornatus* and *A. cordatus.*

Winslow and Fenny Stratford.

To the north-east of Quainton and Steeple Claydon, we enter a region where the upper boundary of the Oxford Clay is marked, on the Geological Survey Map, as "wholly conjectural." This applies to the country south of Winslow and near Stewkley, and is due partly to the concealment of the strata by Drift, and partly to the disappearance of the Corallian rock-beds. Nor is the lower boundary of the Oxford Clay to be traced with great certainty considering that from Goddington, by Gawcott and Padbury near Buckingham, onwards to the north of Whaddon Chase, and by Little Woolston, to the south of Newport Pagnell, the junction with the Cornbrash is "everywhere hidden by Drift."

The higher boundary with the Kimeridge Clay will be more particularly discussed in the chapter on Corallian rocks, for there is evidence that those beds are represented mainly by argillaceous strata in the country extending from Quainton to Ampthill, and the neighbourhood of St. Ives.

At Akeley brickyard, to the north of Buckingham, as mentioned previously, there is evidence of the Kellaways division, and it is interesting to find that Prof. Green noted *Ammonites Kœnigi* from this exposure.[†]

Referring to the outliers at Akeley and east of Leckhampstead, which, excepting near the brickyard just mentioned, were coloured as Cornbrash on the Geological Survey Map, Prof. Green remarks, "The ground on which these two outliers lie is high enough to leave room for a capping of Oxford Clay. No sections, however, were obtained to prove the presence of this formation,

---

[*] Geology of Banbury, pp. 42, 45.
[†] *Ibid.,* p. 41 ; see also Memoir on Lower Oolitic Rocks of England, p. 450.

and all the evidence seemed to show that the drift was quite thick enough to make up that part of the hill which lay above the outcrop of the Cornbrash. The ground has been therefore coloured as Cornbrash all over, but it must not be forgotton that patches of Oxford Clay may lie upon the hill tops beneath the drift covering."[*] It is therefore interesting to learn that recent trial-borings made in search of water by Mr. Baldwin Latham have proved that Oxford Clay does occur beneath the Drift between Maids Moreton and Akeley.

At the Old Copse brickyard, north of Wicken, in Whittlewood Forest, there was a section showing stiff grey clay, covered by Drift. The clay yielded only one *Serpula*, to my search, but in general characters it resembles Oxford Clay. The bed was noticed by Prof. Green as dark-blue shaly clay very finely laminated, and he stated that it had been proved to a depth of 36 feet without reaching the bottom. While grouping it with the Great Oolite he remarked that he had nowhere else in the district seen Great Oolite clays of the same nature and thickness.[†] If I am right in classing it with the Oxford Clay there is little doubt that it must occupy a faulted tract.

Again, rather more than a mile and a half north-west of Akeley, a boring was made midway between Tile House Farm and Tile House Wood, by Messrs. Easton, and Anderson. A shaft was sunk 21½ feet, and the rest bored. The strata passed through were as follows :—

BORING NEAR AKELEY.		THICKNESS.		DEPTH.	
		FT.	IN.	FT.	IN.
	12. Soil, &c.	7	6		
Drift.	11. Gravel about	8	6		
	10. Blue clay about	9	0	25	0
Oxfordian.	9. Sandy clay	40	0	65	0
	8. Stiff clay	25	0	90	0
Great Oolite ?	5. Hard stone	18	0	108	0
Upper Estua-	4. Clay	22	0	130	0
rine Series.	3. Black Stuff	1	0	131	0
	2. Greenish clay	10	0	141	0
Northampton Beds ?	1. Hard stone rock	9	0	150	0

There is considerable difficulty in interpreting this record, owing to the absence of beds that can clearly be classed with the Cornbrash and Forest Marble or Great Oolite Clay. The "Hard Stone" is too thick to be Cornbrash, judging from the local evidence, and the underlying clays appear too thick to be included with the Great Oolite Clay. It seems to me not unreasonable to conclude that the fault observed by Prof. Green to the north of Akeley,[‡] and which trends in a north-

* Geol. Banbury, p. 35.
† *Ibid.*, p. 22.
‡ Prof. Green has given a diagrammatic section of the fault nearer to Akeley, Geol. Banbury, p. 41.

westerly direction, extends to Tile House Wood, and that in the
boring it was passed through, as shown in the accompanying
diagram-section.

FIG. 23.

*Diagram-section near Tile House Farm, north-west
of Akeley, Buckinghamshire.*

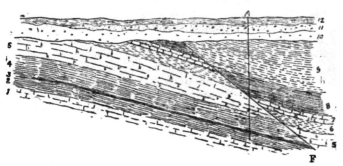

7. Cornbrash.
6. Great Oolite Clay.
5. Great Oolite.
4. ⎫
3. ⎬ Upper Estuarine Series.
2. ⎭
1. Northampton Beds.

A. Boring.
F. Fault.
12. Soil.
11. Gravel.
10. Boulder Clay.
9. ⎫
8. ⎬ Oxford Clay.

In a brickyard at Tinker's End, Winslow, there was exposed
from 12 to 15 feet of bluish-grey clay, with many small Ammonites,
including *Ammonites crenatus, A. Duncani, A. Lamberti,* and
A. Mariæ; also *Gryphæa dilatata, Ostrea,* and *Serpula tricarinata.*
I was informed that fossils are mostly found deep down in the
pit, those near the surface being fragile and decayed. 2 or 3
feet of irregular sand and fine gravel occur on top of the Oxford
Clay. Red and yellow bricks, tiles, and drain-pipes are manufactured.

A well sunk in the market-place at Winslow, through Drift
and Oxford Clay, reached rock-beds at a depth of 238 feet.
From information given me by Mr. W. H. Dalton, it is probable
that Kellaways Rock and Cornbrash were subsequently penetrated,
for the boring was carried to a depth of 279 feet.

Between Winslow and Great Horwood, there was a brickyard,
where Prof. Green noted, beneath 1 foot of Drift Gravel,
dark blue clay with *Ammonites Jason,* and *Gryphæa dilatata.*
He observed also the Oxford Clay, beneath a few feet of gravel,
in the brickpit at Swanbourn, and there *G. dilatata* and *Ostrea
gregaria* were met with.

A well boring at Bletchley Junction (1887) was carried to a depth
of 419 feet, through Boulder Clay, Oxford Clay, and Kellaways
Beds. Saline waters were met with in the Kellaways Beds. The
chief point of interest connected with the boring was the finding of

granitic rock, possibly portions of pebbles or boulders embedded in the Kellaways Series. This granitic rock was stated to occur at depths of 378 ft. 5 ins. to 400 feet and at depths of 401 to 407 feet 2 inches : in both cases being underlaid by clay of Oxfordian character.* Mr. Cameron assigns a thickness of 54 feet to the Kellaways Beds.

It may be mentioned that brackish water was found in the boring at Brick-kiln farm, S.E. of Stony Stratford, at a depth of 221 feet ; the bed reached was probably Middle Lias.† The thickness of Oxfordian Beds at this spot was about 60 feet.

A brickyard on Watling Street, two miles S.E of Stony Stratford, was noted many years ago by Mr. Whitaker, who observed blue, shaly, and bituminous clay, with blue phosphate of iron, and much selenite. *Pecten* and some other fossils were found, also a Belemnite converted into selenite. Cornbrash was seen to crop out near by, so that the clayey beds belonged to the basement portion of the Oxford Clay.

North of Little Woolston, the following section was noted by Prof. Green, in a brick-pit :—

		Ft.	In.
	Gravel - - - - -	1	0
	Blue clunchy clay - - -	3	0
Kellaways Beds.	Yellow ferruginous sand - -	0	6
	Blue sandy laminated clay, bored into for - - - - -	20	0

Leighton Buzzard to Ampthill and Bedford.

From Leighton Buzzard onwards to Biggleswade and Potton the upper portion of the Oxford Clay is extensively concealed by coverings of Lower Greensand and Gault ; and the lower portion, and its junction with the Cornbrash, from Newport Pagnell by Newton Blossomville and Carlton Hill to Bedford, is much obscured by Boulder Clay. Many brickyards in the Oxford Clay are to be found in the area, as near Woburn Station, Wavendon, Lower Shelton, Shelton Green, and Wootton.

The Lower Oxfordian Beds have been exposed in brickyards at Tickford End, and further on near Caldecot Mill, to the S.E. of Newport Pagnell, and on the road to North Crawley. J. H. Macalister obtained many fossils from near Newport Pagnell, where shales with occasional stony bands and septaria, with lignite, selenite, and much pyrites, have been exposed.‡ *Ammonites calloviensis* was obtained, also *Alaria* ; and higher up *Cerithium muricatum*, *Ammonites Jason*, and many compressed Ammonites, have been found. Prof. Green has recorded some of these fossils, and also the occurrence of Saurian bones.

Gryphæa dilatata has been found at various points in the clay exposed beneath the Woburn Sands (Lower Greensand), from

---

* Hull, Geol. Mag., 1887, p. 139 ; Jukes-Browne, Geol. Mag., 1889, p. 356 ; Cameron, *Ibid.* 1892, p. 69 ; B. Thompson, Journ. Northamptonshire Nat. Hist. Soc., vol. v. p. 20.
† See Memoir on the Lower Oolitic Rocks of England p. 391.
‡ Geologist, vol. iv. p. 214.

Rushmere Pond, north of Leighton Buzzard, to near Great and Little Brickhill.* Prof. Green has noted that at the brickyard near Rushmere Pond (Nares Gladly), the *Gryphæa* was of large size and was accompanied by *Ammonites cordatus, Belemnites hastatus,* and *Ostrea gregaria,* in some abundance, together with *Ammonites anceps, A. Bakeriæ,* and *Serpula tricarinata.* The Oxford Clay is no longer worked at this locality, loam from the Lower Greensand being brought from a short distance to the Rushmere brickyard. *Ammonites Bakeriæ* is also noted from the upper beds of Oxford Clay at St. Ives (p. 55).

A boring made about half-a-mile S.S.E. of Buttermilk farm, near Little Brickhill, was carried to a depth of 133 feet, through Lower Greensand into Oxford Clay.

At Fenny Stratford the clay has been worked in several places beneath the valley-gravel, and many fossils have been obtained. Here there is much pyrites, and the fossils are mostly replaced by it. I obtained *Ammonites Jason* var. *Gulielmi, Cerithium muricatum, Avicula, Leda Phillipsi, Nucula ornata,* and Lignite.

In the brickyards at Aspley Guise, where *Ammonites crenatus* occurs, and near Ridgmont, Mr. Whitaker observed bands of limestone in the Oxford Clay. He found many fossils near Ridgmont, including *Gryphæa dilatata;* and at Crawley Kiln, between Ridgmont and Husborne Crawley, there was much pyrites and also selenite in the clay.

Here we approach the border-line of the Oxford and Corallian (Ampthill) Clays, further reference to which will be made in the chapter dealing with the Ampthill Clay.

A boring at the Rectory, Northill, three miles north-west of Biggleswade, was made in 1877 by Messrs. Le Grand and Sutcliffe; and the following details were communicated to me by Mr. Whitaker :—

BORING AT NORTHILL.		THICKNESS.		DEPTH.	
		FT.	IN.	FT.	IN.
Boulder Clay - - - - - -		104	0		
Oxford Clay.	Green Clay - - -	12	0	116	0
	Blue clay - - -	10	0	126	0
	Blue clay and shells - -	9	0	135	0
	Dark green clay - - -	13	6	148	6
	Black stone - - -	4	6	153	0
	Greenish clay and shells - -	20	0	173	0
	Live sand - - - -	9	0	182	0
	Sandy blue clay - - -	9	0	191	0
	Sand rock - - -	7	4	198	4
	Blue clay and shells - -	2	6	200	10
	Rock and blue clay - -	1	9	202	7
	Limestone - - -	2	8	205	3
	Sandy blue clay - - -	3	0	208	3
	Blue stone - - -	3	6	211	9
	Sandy clay - - -	4	10	216	7
	Limestone - - -	4	0	220	7
	Sandy clay and stone - -	3	0	223	7

* Teall, Potton, and Wicken Phosphatic Deposits, 1875, pp. 27, 43.

I have no information about the fossils, but it is not improbable that portions of the upper beds grouped with the Oxford Clay, may represent the Ampthill Clay, and be of Corallian age.

Bedford.

The country around Bedford affords some fine sections of the Kellaways Beds, to which attention was first directed by Mr. A. C. G. Cameron, during the course of his work on the Geological Survey; and I had the advantage of his guidance in visiting the sections.

As in other tracts, where we have evidence, in Wiltshire and Oxfordshire, the mass of the sandy Kellaways Rock is separated from the Cornbrash by clays, sometimes spoken of as the 'Lower Oxford Clay,' but more properly Kellaways Clay, that attains a thickness of from 7 to 10 feet, and can in places be distinctly shown on the Geological Survey Map. This lower clay is described by Mr. Cameron as a dark clay with much selenite and " race."

Good sections of the Kellaways Beds have been exposed at the Southend Ballast Pit, beneath the valley-gravel, on the south-west side of Bedford.* Here we find masses of concretionary sandstone weathering into sand, and exhibiting curious fucoidal and tubiform markings. White and brown sands with similar concretionary masses of stone, were shown resting on clay, and Cornbrash, &c., in the brickyard ¾ mile north-west of West End, Stevington. Fine sections of Kellaways Rock were also shown at Howard's brickyard, at Lodge Hill, south of Clapham, and north-west of Bedford. There we have about 10 feet of stiff greenish-grey clay, and brown and bluish-grey loamy sand, with iron-pyrites and irregular concretions or doggers of calcareous sandstone. The stone contains *Myacites recurvus* in abundance, and occasional shelly layers with *Gryphœa bilobata* and *Belemnites.*

The widening of the Midland Railway north of Oakley Station, in 1889, opened up a fine section of the Kellaways Beds, which I noted, in company with Mr. Cameron, as follows :—

		Ft.	In.
Oxford Clay and Kellaways Beds.	Grey and brown clay passing down into dark grey clay with pyritic layers; *Ammonites* (pyritic and iridescent), *Belemnites, Gryphœa bilobata,* Saurian bones, Lignite, and occasional nodules of shelly stone -	15	0
	Large concretionary masses or doggers of grey-hearted calcareous sandstone, with ochreous coating, and fucoidal markings, and occasionally a yellow (sulphurous) efflorescence ; *Avicula inœquivalvis, Gryphœa bilobata, Myacites recurvus, Belemnites Oweni, Ammonites,* and Lignite - - - - Yellow grey and greenish sands -	10	0

---

* I was informed by Mr. Cameron that clay from the cutting at Ampthill has been tipped over the waste ground here.

Mr. Cameron has noted the occurrence in these beds of an
indurated seam of sandy marl, which exhibited the structure
known as " cone-in-cone."

The bottom sands are like those recently exposed in the cutting
of the Woodstock railway, and the doggers in the overlying strata
were in places so numerous that a navvy remarked " they might
have been used as stepping-stones at the time of the Flood."
Some of these huge stones, according to Mr. Cameron, measure
30 feet in circumference : others are united so as to form twin
stones in the form of the figure 8.*

Mr. Cameron has traced an outlying mass of Cornbrash,
Kellaways Rock, and Oxford Clay over Yardley Chase, in the
neighbourhood of Yardley Hastings. It is interesting to learn
that Wm. Smith in 1819, in company with John Phillips, noted
the occurrence of Kellaways Rock further north near Bozeat
(Boziate Hill) where they found *Ammonites modiolaris*; and
drew a section showing above the Cornbrash, an outlier of clay
(Kellaways Clay) surmounted by Kellaways Rock.†

The following species have been obtained by Mr. Cameron,
Mr. J. Rhodes, and myself, from the Kellaways Beds near
Bedford :—

Ammonites gowerianus.	Myacites recurvus.
—— macrocephalus.	—— securiformis.
Ancyloceras calloviense.	Ostrea.
Belemnites Oweni.	Pecten demissus.
Avicula inæquivalvis.	—— lens.
—— Münsteri.	Trigonia.
Gryphæa bilobata.	

Bedford to Huntingdon and St. Ives.

The Oxford Clay occupies a wide area to the north and east of
Bedford, extending over much of Huntingdonshire, into the
borders of Cambridgeshire ; forming the foundation of the greater
part of the Fenland, and appearing in islands at Ramsey, Whittlesey, and Thorney. It extends into Northamptonshire, near
Thrapston, with outlying masses between Brigstock and Oundle.
Over great part of this area it is thickly covered with Drift,
chiefly Boulder Clay, and there can be no more tenacious a soil
than that in some of the tracts formed of Oxford Clay and
Boulder Clay.

A tiny outlier of the basement or Kellaways Clay overlies the
Cornbrash in a faulted tract at Stowe-nine-churches, in North-
amptonshire: the section was noted by Mr. Beeby Thompson.‡
About 5 feet of blue clay was seen, but no fossils were obtained.

Returning to the main mass of Oxfordian strata it may be
mentioned that Kellaways Beds have been observed by Mr.
Cameron at Risely; but I have no records of any section of the

* Geol. Mag., 1892, p. 66.
† Memoirs of W. Smith, p. 93.
‡ Journ. Northamptonshire Nat. Hist. Soc., vol. vi. p. 294 ; see also Memoir on
Lower Oolitic Rocks of England, p. 398.

strata around Kimbolton, where so much Drift covers the Oxford Clay. (See p. 305.)

The Oxford Clay has been largely worked for brickmaking at St. Neot's and in the adjoining parish of Eynesbury. I visited the sections in 1893, in company with Mr. Cameron.

At the Eynesbury Brick and Tile Works, the pit showed grey and yellowish marly clay, with bands of hard earthy limestone; and a somewhat similar section was shown at Gallow Hill, on the east side of the road leading to Little Barford. At the Gallow Hill section, which adjoins the Great Northern Railway, the uppermost band of hard earthy limestone, was long ago described by Prof. Seeley,[*] as the " St. Neot's Rock," and he observed that it was exposed at the base of the railway. More recently the sections have been fully described by Thomas Roberts,[†] who rightly remarks that the St. Neot's Rock "does not differ markedly from the other calcareous bands which occur throughout the Oxford Clay," and on this account he deemed it unnecessary to retain the term St. Neot's Rock. Summarizing the evidence furnished by the brickyards near St. Neot's, he gave the following section :—

		Ft.	In.
	Soil, gravel, and boulder clay (in places).		
	Greyish-brown clay - - -	7	0
	Greyish sandy limestone (St. Neot's Rock) - - - - -	0	11
	Bluish clay - - - -	17	0
	Sandy limestone - - -	0	8
	Blue clay - - - -	4	6
Oxford Clay	Sandy limestone - - -	0	6
	Blue clay - - - -	8	0
	Sandy limestone - - -	0	10
	Blue clay - - - -	5	0
	Sandy limestone - - -	0	6
	Blue clay - - - -	18	0
	Sandy limestone.		

The beds belong to the zone of *Ammonites ornatus*, as shown by the following list of fossils :—[‡]

Pliosaurus Evansi.
Ammonites athleta.
—— Duncani.
—— Jason.
× —— —— var. Gulielmi.
Belemnites hastatus.
—— obeliscus.
—— Oweni.
—— —— var. verrucosus.
Cerithium muricatum.

Astarte robusta.
Cucullæa concinna.
Gryphæa bilobata.
× —— dilatata.
Pecten fibrosus.
Trigonia costata.
× Rhynchonella varians.
× Serpula tricarinata.
× Webbina irregularis.
Lignite.

The specimens of *Ammonites Jason* var. *Gulielmi* were very abundant, and highly charged with pyrites.

---

[*] Ann. Nat. Hist., ser. 3. vol. viii. p. 504.
[†] Jurassic Rocks of Cambridge, 1892, pp. 11–14.
[‡] Most of these are recorded by T. Roberts; those marked × were obtained by Mr. Cameron and myself.

North of the railway-station at Sandy, the upper beds of the Oxford Clay have been well exposed in a brickyard, which I visited in company with Mr. Cameron. The section showed about 15 feet of stiff grey racy clay, with ferruginous concretions (probably decomposed pyritic fossils) and selenite ; there was also a band of earthy limestone (6 to 8 inches thick) not unlike the St. Neot's Rock, that readily shivers when exposed to the weather. Septarian nodules occur near the top of the beds exposed. Near by the sands of the Lower Greensand are shown above the Oxford Clay.

Among the fossils *Ammonites cordatus,* of variable character, is abundant, and also the large Belemnite, *B. Oweni* var. *puzosianus.* The following species have been obtained :—[*]

×Ammonites cordatus.	Alaria trifida.
×—— Lamberti.	Avicula inæquivalvis.
—— Mariæ.	× Gryphæa dilatata (large speci-
—— trifidus.	mens).
Belemnites hastatus.	Nucula ornata ?
×—— Oweni.	× Rhynchonella varians.

Brickyards have been opened to the east of Godmanchester. In one of the pits near this village, T. Roberts noted, 18 feet of Boulder Clay ; and below, a series of dark blue tenacious clays, with nodules of pyrites, and near the top, a band of greyish sandy limestone. The beds of Oxford Clay, beneath the Drift, were shown to a depth of 24 feet, and from them Roberts obtained the following fossils :—[†]

Ammonites Achilles.	Ammonites oculatus
—— Bakeriæ.	Belemnites Oweni.
—— cordatus.	Gryphæa dilatata.

These beds must be high up in the Oxford Clay, so that the occurrence of *A. Bakeriæ,* which is recorded also from Leighton Buzzard and St. Ives, is noteworthy.

The most important section in this neighbourhood is that in the now deserted brickyard west of St. Ives. This section I examined in 1889, and noted the beds as follows :—

		Ft.	In.
	Brashy clay, with *Gryphæa dilatata.*		
	Grey marly clay - - -	4	0
	Band of grey shelly and calcareous gritty rock - - - 0 3 to	0	8
Oxford Clay	Marly clay, passing down into darker shaly clay with selenite, and *G. dilatata;* and below grey marly clay with small *Ammonites Mariæ, Pinna,* &c. - - - - -	18	0
	Grey gritty and calcareous rock, with calcareous and pyritic concretions 0 3 to	0	8
	Grey marly clay, *G. dilatata* (abundant); formerly worked to depth of [Calcareous bed].	16	0

---

[*] Most of these have been recorded by T. Roberts, Jurassic Rocks of Cambridge, p. 15 : those marked × were found by Mr. Cameron and myself.
[†] Jurassic Rocks of Cambridge, p. 15.

T. Roberts noted that the St. Ives Rock, to which reference will subsequently be made (p. 141), was found above the clay a little to the west of the pit; its position was 8 feet above the top band of rock, and the full thickness of Oxford Clay, observed by him, was 40 feet. He noted also the abundance of *Waldheimia impressa* in the clay just above the bottom calcareous bed, and the horizon is marked as the "zone of *Waldheimia impressa*."[*]

Many Crustacea were obtained from the Oxford Clay of St. Ives by Mr. T. George, and these and other specimens have been described by Mr. James Carter.[†] They include species of *Eryma, Eryon, Glyphea, Magila, Mecocheirus, Goniocheirus, Pseudastacus*, and *Pagurus*. Nowhere else in England has so varied an assemblage been found.

The fossils generally, and the stratigraphical position of the Oxford Clay of St. Ives, indicate the zone of *Ammonites cordatus*. At the same time Ammonites of lower zones have been found and notably *A. Bakeriæ, A. athleta*, and *A. crenatus*. The fullest and most authentic list of fossils is that of T. Roberts, derived from specimens now in the Woodwardian Museum: it includes the following species :—[‡]

Ammonites Achilles.
—— athleta.
—— babeanus.
—— Bakeriæ.
—— cordatus.
—— —— var. excavatus.
—— crenatus.
—— Eugeni.
—— Goliathus.
—— hecticus.
—— Jason.
—— lophotus.
—— Mariæ.
—— oculatus.
—— perarmatus.
—— rupellensis.
—— trifidus.
Belemnites abbreviatus.
—— hastatus.
—— Oweni.
Belemnotcuthis.
Nautilus calloviensis.
Alaria trifida.
Cerithium Damonis.
Astarte.
Avicula inæquivalvis.
Cardium Crawfordi.

Cucullæa concinna.
Exogyra nana.
Gryphæa dilatata.
Isocardia.
Leda lachryma.
Lima rigida.
Modiola bipartita.
Nucula elliptica.
—— nuda.
—— ornata.
—— turgida.
Ostrea gregaria.
Perna.
Pholadomya Phillipsi.
Pinna mitis.
Thracia depressa.
Trigonia elongata.
Rhynchonella lævirostris.
—— varians.
Terebratula oxoniensis.
Waldheimia impressa.
Serpula tricarinata.
—— vertebralis.
Vermilia sulcata.
Acrosalenia.
Pentacrinus.

During the construction of the Great Northern Railway, many fossils were obtained from the Walton cutting, near Wood

* Jurassic Rocks of Cambridge, pp. 17, 18; see also Bonney, Cambridgeshire Geology, p. 10, and Penning and Jukes-Browne, Geology of Cambridge, p. 6.
† Quart. Journ. Geol. Soc., vol. xlii. p. 54 l.
‡ Jurassic Rocks of Cambridge, p. 17; see also Penning and Jukes-Browne, Geology of Cambridge, p. 7.

Walton, north of Huntingdon; among these were *Ammonites athleta, A. cordatus* and var. *excavatus.*[*]

Mr. Cameron informs me (1893) that a shaft (in search of minerals) was lately sunk at Abbots Ripton to a depth of about 180 feet through Drift, into the Oxford Clay without reaching the base of the formation.

The higher grounds as at Alconbury Hill are mainly formed of Boulder Clay, and in places they give rise to tracts known as wolds, as Molesworth, Brington, and Barnwell Wolds, and Leighton Bromeswold. Westwards over the area at Yelden and Caldecot to near Thrapston the ground is again largely covered with Drift, and we have no records of sections.

In the area to the east of St. Ives, and north-west of Cambridge, we find the Oxford Clay consisting of bluish-grey clay with septaria, selenite and occasional bands of sandy limestone. *Gryphæa dilatata* occurs plentifully in the upper beds, which were formerly worked to the north of Willingham.[†] The area is however much covered with Drift deposits.

A few sections have been noticed near Long Stanton, in the brickyard at Fenton, north-west of Pidley, and in railway-cuttings at Over and Bluntisham; but our information concerning them is scanty, although a number of fossils were obtained when the cuttings were made.[‡]

The Oxford Clay was proved to a depth of 300 feet at Bluntisham and 200 feet at Over, but we have no record of the full thickness. Messrs. Penning and Jukes-Browne have indeed estimated it at 700 feet,[§] but that would include the Corallian Clays, and even then the estimate is I think excessive—probably 500 feet would be nearer the mark.

The Oxford Clay appears at Ramsey, and extends beneath the Alluvial beds of the Fenland, northwards to Whittlesey and Peterborough.

Thrapston and Oundle to Peterborough.

The Oxfordian Beds in Northamptonshire and north Huntingdonshire were divided as follows, by Prof. Judd :—[||]

Zone of *Ammonites cordatus.*	f. Clays with Ammonites of the group of the *Cordati.* Exposed at the Forty-foot Bridge brickyards, south of Ramsey Mere.
	e. Clays with Ammonites of the group of the *Ornati.* Dark blue clays with nodules of pyrites, and numerous pyritic Ammonites, including *A. ornatus, A. Duncani, A. Bakeriæ,* and *A. athleta,* and also *Waldheimia impressa.* Dug in brickyards about Whittlesey, at Thorney, and Eye Green.

---

* Catalogue of Fossils in Museum of Practical Geology, 1865, pp. 239, &c.
† Geol. parts of Cambridgeshire and Suffolk, p. 6.
‡ J. K. Watts, Rep. Brit Assoc. for 1852, Sections, p. 63.
§ Geol. Cambridge, pp. 5, 163, 167.
|| Geol. Rutland, p. 232.

	d. Clays with *Belemnites hastatus*. Blue clays with many fossils found in Division c., but characterized by the abundance of *B. hastatus*. Dug at Werrington, Ramsey, and Eyebury.
Zone of *Ammonites ornatus*.	c. Clays with *Belemnites Oweni*. Dark blue clays and shales, with *B. Oweni*, often of gigantic size. *Gryphæa dilatata* occurs, but is more plentiful in the beds above. Saurians and Fishes occur; and masses of lignite, sometimes converted into jet are found. Exposed in brickyards at Standground, Fletton and Woodstone, near Peterborough; and at Conington, Luddington, and Great Gidding.
	b. Clays with *Nucula*. Laminated blue shales, with compressed Ammonites, and *Nucula nuda*. Dug at Haddon, Holme, south of Peterborough, and at Eyebury to the north-east.
Zone of *Ammonites calloviensis*.	a. Kellaways Sands, Sandstones, and Clays. Alternations of light-coloured sandy and sometimes pyritous clays, with irregular beds of whitish sand—the latter fossiliferous and frequently cemented by calcareous matter into a friable rock: with dark blue clay at the base. *Belemnites Oweni*, *Avicula inæquivalvis*, *Gryphæa bilobata*, &c. Beds dug for brickmaking at Warmington, Oundle, Southwick, Benefield, and again at Dogsthorpe, Uffington, and Kate's Bridge, near Thurlby.

Commencing in the western portion of the area we find a large and straggling outlier of the Lower Oxfordian Beds, and this stretches from near Sudborough, north-west of Thrapston, to the west of Oundle, and northwards to the neighbourhood of Southwick. The sections have been described by Prof. Judd. Thus at the brickyard at Sudborough we find clayey beds like those near Peterborough, and they yield in abundance large specimens of *Belemnites Oweni;* also Ammonites and Saurian remains.

At Oundle brickyard the Kellaways Beds are shown, and they comprise pale sandy clays, with bands of hard sandy and ferruginous rock, that often thin out within short distances.

At Benefield and again near Southwick there are brickyards that show the Kellaways Beds, yielding large *Belemnites Oweni*, also *Avicula expansa, Gryphæa bilobata, Serpula,* &c.

A small outlier of the beds was proved at Brigstock Park, in a deep well at one of the farm-houses; but, as Prof. Judd remarks, owing to the thick covering of Boulder Clay, the boundaries of the outlier, as shown on the Geological Survey Map, are purely hypothetical.*

The lower beds of the Oxford Clay were well exposed in the Wigsthorpe cutting of the Northampton and Peterborough

---

* Geol. Rutland, pp. 233, 234, 239.

Railway. This cutting was examined by Prof. Morris and Capt. Ibbetson, who described the section in the following terms:— "The Oxford Clay is well seen in the Wigsthorpe cutting, near Thorpe Aychurch, and is marked by zones of Septaria frequently containing fossils, *Am. Kœnigi*, &c., the lower part of the section being thin slaty clays full of *Ammonites Jason* or *Elizabethæ* much compressed, *Belemnites, Avicula*, and numerous bivalves."[*]

Prof. Judd states that at the brickyard of Ashton, lying on the opposite side of the Nene Valley to that of Oundle, Oxfordian strata, probably a little higher in the series than those of Oundle and Benefield, are worked. These consist of dark-blue clays containing *Nucula nuda*, many fragments of *Belemnites*, crushed *Ammonites* and large quantities of wood converted into jet. He mentions that near Haddon Church a well sunk at the new parsonage penetrated the Oxford Clay to a depth of more than 30 feet, but no water was obtained. The clay brought up was dark coloured and highly laminated. It contained many fossils, including *Belemnites Oweni, Ammonites ornatus*, and *Nucula nuda*. The fossils were all crushed and very imperfectly preserved. At Haddon brickyard similar clays with *Nucula nuda* and *Belemnites*, were dug.[†]

A cutting on the Great Northern Railway, between Farcett and Yaxley (described by Prof. Judd), exposed a considerable thickness of light-blue Boulder Clay full of chalk detritus, with some irregular gravelly beds intercalated in it. At the bottom of this Drift occur the dark-blue Oxfordian Clays containing very large septaria and yielding many fossils, among which were the following :—

Belemnites Oweni.	Ammonites Duncani.
—— hastatus.	—— cordatus.
Ammonites excavatus.	Gryphæa dilatata.
—— athleta.	Serpula vertebralis.

At Conington brickyard Prof. Judd observed blue Oxfordian clays, which yield abundant specimens of *Belemnites Oweni*, and, somewhat rarely, examples of *Gryphæa dilatata*. Similar clays have been dug at the Holme brickyard.

Extensive brickyards have been opened up near Peterborough, at Fletton, Standground, and Woodstone, comparing well in size with the brickyards near Oxford. In one of the openings at Fletton, I noted 20 feet of clay with gravelly pockets on the top. The clay is lighter near the surface, and darker grey below, and rusty at the joints. Large cement-stones occur occasionally, and there is a good deal of lignite, which is sometimes converted into jet. Prof. Judd remarks that the lignite occurs in great masses, which were evidently portions of drift timber.

About 30 years ago the fossils of the Oxford Clay were zealously collected by Dr. Henry Porter, of Peterborough ; since that date Mr. Alfred N. Leeds and Mr. Charles Leeds, have

---

[*] Rep. Brit. Assoc. for 1847, 1848, Sections, p. 127.
[†] Judd, Geol. Rutland, pp. 233–235, 237, 288.

added very considerably to the knowledge of the vertebrate fauna, many specimens obtained by them having been described by Mr. Hulke, Prof. Seeley, and Mr. Smith Woodward.

The clay is of a very tenacious character, to which Dr. Porter bore personal testimony, for on riding on one occasion into a brickyard at Standground, his horse took fright and plunged into a great heap of clay recently dug, "and on trying to extricate himself only got deeper into the mire, until both horse and rider were completely stuck fast in the middle "; and they required the services of half-a-dozen men with ropes, and of many buckets of water, before they were unmired.*

Among the fossils the most abundant form is *Belemnites Oweni*, of which large examples occur. Some, according to Porter, 10 or 12 inches in length, and 2 or 3 inches in circumference. (See p. 318.) Many iridescent and other crushed Ammonites occur at the base of the workings.

The following is a list of Fossils from the Oxfordian Beds of Peterborough :—

Reptilia.

Camptosaurus Leedsi.
Omosaurus durobrivensis.
Pelorosaurus (Ornithopsis) Leedsi.
Sarcolestes Leedsi.
Suchodus durobrivensis.
Ophthalmosaurus icenicus.
Cimoliosaurus durobrivensis.
—— eurymerus.
—— Richardsoni.
Peloneustes philarchus.

Fishes.

Hypsocormus Leedsi.
—— tenuirostris

Leedsia problematica.
Lepidotus latifrons.
Asteracanthus ornatissimus var. flettonensis.

Mollusca.

Ammonites arduennensis.
—— Bakeriæ.
—— hecticus.
—— macrocephalus.
—— ornatus.
Belemnites Oweni.
Gryphæa bilobata.

Brachiopoda.

Rhynchonella varians.

Prof. Seeley has given a brief note of a well sunk at the gas-works at Peterborough, where the strata proved were as follows :—†

	Ft.	In.
Blue clay - - - - - - -	24	0
Fine grey sand, with at base bones of *Ornithopsis* (= *Pelorosaurus*) - - - - -	12	0
Clay.		

This record must refer to beds near the base of the Oxfordian Series.‡

Ramsey, Whittlesey, and the Fenland.

At Ramsey and Whittlesey the Oxford Clay rises in low "islands" in the Alluvium of the Fenland, and the same

---

* Geology of Peterborough, 1861, p. 61 ; see also Judd, Geol. Rutland, p. 235.
† Quart. Journ. Geol. Soc., vol. xlv. p. 391.
‡ The record of a boring at New England, Peterborough, is given in the Memoir on the Lower Oolitic Rocks of England, p. 202.

formation no doubt forms the foundation over a considerable area, to the west of Wisbech and Holbeach. It is possible that the insulated tracts of Chatteris and March may be outliers of Kimeridge Clay based on Corallian and Oxford Clays; but the older clays are concealed beneath the Alluvium, and in some cases by Boulder Clay. The evidence obtained here and there is not always of a conclusive nature, because Jurassic fossils are so plentiful in the Boulder Clay, and some of the Oxford Clay fossils recorded from March, may have been so derived.[*]

In a few places we have direct evidence of the Oxford Clay in brickyards, as pointed out by Prof. Judd. Thus south of Bury, near Ramsey, dark blue clays, with much pyrites, yielded the following fossils, and show that there the upper portions of the Oxford Clay occur :—

Saurian bones.
Ammonites cordatus.
—— Lamberti.

Belemnites hastatus.
—— Oweni.
Gryphæa dilatata.

A somewhat similar assemblage of fossils was met with in the two clay-pits at Forty-foot Bridge, where many varieties of *Ammonites Lamberti* occur, and *Belemnites hastatus* is very abundant. Prof. Judd notes that a bed of hard rock, 8 or 10 inches in thickness, was found at a depth of 15 feet in one of the pits.[†] At Ramsey there are several brickyards where the clay is locally termed "Galt," as is the case generally with clay in Cambridgeshire. A boring made in the Cricket Field at Ramsey (1885), to a depth of 303 feet failed to find water; nor could it reasonably have been expected at such a depth, as the sinking was commenced in the Upper Oxfordian Beds, and considering the argillaceous nature of the Great Oolite divisions, no reliance could be placed on them as water-bearing strata. The details of the sinking, communicated to me by Mr. A. C. G. Cameron, were as follows :—

		FEET.
	Alluvial soil	} 128
	Clay	
	Rock	} 33
	Clay	
Oxford Clay	Hard rock (pyrites)	20
and	Clay	22
Great Oolite	Chalk and rock	8
Series?	Clay (moist)	46
	Rock	11
	Clay	15
	Rock	20
		303

Prof. Judd states that to the north-west of Whittlesey the clays are dug in a very extensive pit; they are of a deep-blue colour and contain much pyrites and lignite. The fossils indeed

* See List in Skertchly's Geology cf the Fenland, p. 317; and Whitaker, Geol. S.W. Norfolk and N. Cambridgeshire, p. 9.
† Geol. Rutland, p. 237.

are often so thickly encrusted with pyrites that it is impossible to determine their species. Among them are the following :—

Ammonites athleta.	Gryphæa dilatata (very abundant).
Belemnites hastatus.	Serpula.
—— Oweni.	

The large clay-pits at Whittlesey have yielded great numbers of beautiful specimens of *Ammonites,* especially those belonging to the *Ornatus* group, including the following species :—

Ammonites Bakeriæ.	Ammonites Elizabethæ.
—— Comptoni.	—— Jason.
—— Constanti.	—— ornatus.
—— cordatus (variety).	—— tatricus.
—— Duncani.	

Belemnites, bones of Saurians, and specimens of *Gryphæa dilatata* also abound in these pits.

At Eastrea brickyard, clays, somewhat higher in the Oxfordian series, are exposed ; these yield :—

Saurian remains.	Belemnites Oweni.
Ammonites Lamberti.	Rhynchonella varians.
Belemnites hastatus.	

At the brickyard at Eyebury a number of interesting vertebrate remains have been collected by Mr. Leeds. The clays at this place have yielded a considerable number of specimens of *Waldheimia impressa.*

Near Eye several pits opened in the Oxfordian strata, expose beds of blue clay with *Gryphæa dilatata,* and numerous *Belemnites* and *Ammonites.* At Eye Green a thin ferruginous stony seam occurs in the midst of the Oxford Clay. At this place there are found in the clays great numbers of *Ammonites* in all stages of growth, including many varieties of *Ammonites Jason,* and *A. ornatus : Belemnites Oweni* is very rare here, while *B. hastatus* (*gracilis*) is abundant. *Waldheimia impressa* also occurs at this locality.

At Thorney there is an excellent section of the Oxford Clay, it being here dug to a considerable depth. *Ammonites* of the *Ornatus* group are abundant, but the specimens are usually encrusted with pyrites. *Gryphæa dilatata* occurs in prodigious numbers, but *Belemnites Oweni* is very rare.*

The occurrence of Oxford Clay at some depth beneath the surface at Lynn, in Norfolk, was suggested by Fitton, who in 1827 noted the finding of *Gryphæa dilatata* together with other fossils in a well sunk at Allen's Brewery.† Further reference will be made to this record and to a boring at March, in the account of the Corallian Beds. (See p. 147.) *Ammonites calloviensis, Astarte ovata,* and *Gryphæa dilatata* have been obtained from the Fens near March.‡

---

* The above notes are by Prof. Judd, Geol. Rutland, pp. 238, 239.
† Trans. Geol. Soc., ser. 2. vol. iv. p. 316.
‡ Specimens in Wisbech Museum.

Peterborough to Bourn.

North of Peterborough there are brickyards at Dodsthorpe or Dogsthorpe, where the strata, as described by Prof. Judd, comprise light and dark-blue clays, often mottled, becoming in some places very sandy and passing in others into light brown sands which are somewhat indurated. The sandy rock here does not appear to form regular beds in the clay, but to constitute nests and irregular lenticular masses. The beds belong to the Kellaways division, and include at their base, 7 feet of hard blue "dicey" clay. In the sandy stone, great numbers of specimens of *Gryphæa bilobata*, and *Belemnites Oweni*, including individuals of all ages occurred. In the basement-clay there was a band crowded with fossils including *Ammonites macrocephalus*, *Nucula nuda*, *Corbula*, and *Rhynchonella*. Beneath this clay the Cornbrash was proved.

At the westernmost of the two mills at Werrington a brickyard exhibits beds of Oxford Clay, overlaid by thick masses of Boulder Clay with patches of gravel at its base. The Oxford Clay here yielded bones of *Ichthyosaurus*, also *Ammonites Duncani* and *Belemnites Oweni*. In the clay-pit at the east end of the village of Werrington the clays yield rather numerous fossils, including *Ammonites*, *Belemnites hastatus* (very abundant), *B. Oweni*, *Gryphæa dilatata*, *Nucula nuda*, and *Serpula vertebralis* (abundant).

The above sections were described by Prof. Judd.[*] In the Casewick cutting of the Great Northern Railway, described by Prof. Morris in 1853, some interesting sections of the lowest beds of the Oxfordian series were exposed. Resting upon the Cornbrash there was 10 feet of dark laminated unctuous clay, with grey-brown sandy ferruginous clay; the dark clay contained *Ammonites macrocephalus*, abundantly, as well as *Modiola bipartita*, *Trigonia*, *Thracia depressa*, *Nucula nuda*, and Saurian bones. The brown sandy clay, which passed into ferruginous rock, contained many well preserved fossils, the most abundant being as follows :—

Ammonites calloviensis.	Gryphæa bilobata.
Belemnites Oweni.	Lima rigidula.
Nautilus.	Pecten demissus.
Avicula expansa.	Pholadomya acuticosta.
Gresslya peregrina.	

These fossils indicated the Kellaways rock which had not previously been noticed in the district.[†]

Prof. Judd mentions that at Kate's Bridge four miles south of Bourn there are two pits in the Kellaways strata.[‡] In one of these, the thickness of clay overlying the Cornbrash is only 6 feet, the beds consisting of light-blue sandy clay containing *Belemnites*, *Gryphæa*, *Avicula*, and other shells.

---

[*] Judd, Geol. Rutland, pp. 235, 236.
[†] Quart. Journ. Geol. Soc., vol. ix. p. 333.
[‡] Judd, Geol. Rutland, p. 237.

At Bourn, in a brickyard, the following section was noted by Mr. S. B. J. Skertchly :—

		Ft.	In.
	Soil - - - - -	1	6
	Clay becoming sandy and yellow below	1	6
Kellaways	Light-blue and yellow mottled sand -	2	0
Beds.	Light-coloured, laminated clay -	2	6
	Sandy rock (very irregular) - -	4	0
	Light-blue clay - - -	8 to 9	0
Cornbrash.	- - - - -		

The bed of sandy rock contained *Belemnites Oweni*, *Avicula expansa*, *Gryphœa bilobata*, and Lignite.

In the cuttings of the Bourn and Saxby Railway which I examined in 1892, some interesting sections were exposed.* (See Fig. 24, p. 64.) Where this new railway leaves the Essendine and Bourn Branch, south-west of Bourn, there were shallow cuttings that showed grey and brown clay, with loamy beds above, and in places a thin gravelly soil with flints, &c. A temporary brick-yard, opened on the north side of the railway, about half-a-mile from Bourn station, showed 10 feet of bluish-shaly clay with small decomposed septaria and pyrites, and on top 2 or 3 feet of yellow loam. Here the bedding was nearly flat, further west there was a slight westerly dip, and higher beds were exposed by the footbridge. The section was as follows :—

Kellaways	Loam.
Beds.	Thin calcareous sandstone with plant-like
11 feet seen.	markings.
	Loam.
	Blue clay.

Still further west the cutting is deeper, and the calcareous sandstone, a flaggy blue-hearted rock, is two feet or more in thickness, and contains in abundance *Belemnites* and *Gryphœa bilobata*. Above this rock-bed there is a considerable thickness of dense grey shaly clay with selenite, lignite, and numerous flattened Ammonites. Here I obtained a Saurian bone, *Ammonites Jason* var. *Gulielmi*, *Belemnites Oweni*, some Gasteropods, *Arca*, and *Gryphœa bilobata*. These beds, therefore, belong to the zone of *Ammonites ornatus*. About 10 feet above the rock-bed, before-mentioned, there is a band of huge septaria, oval and spheroidal masses, in some instances 3 × 2 × 1 feet in size, and mostly unfossiliferous. They occur practically on one horizon, though the band was slightly irregular.

The railway-tunnel is excavated in Oxford Clay, which is over-laid by about 12 feet of Chalky Boulder Clay. On the western side of the tunnel, the cutting showed this Drift resting on about 12 feet of Oxford Clay, which here has an easterly dip. As the beds rise, so we find the band of septaria further west in the cutting, and at a lower level there are the loamy beds belonging to the Kellaways division, not very clearly exposed, with the

---

* Other sections of the Lower Oolites have been described in the Memoir on the Lower Oolitic Rocks of England, pp. 208, 421, 455 ; and those of the Lias, in the Memoir on the Lias of England and Wales, pp. 170, 238, 281.

Cornbrash at the base. The outcrop of the Kellaways Beds forms a gentle escarpment above the Cornbrash to the north of Lound.

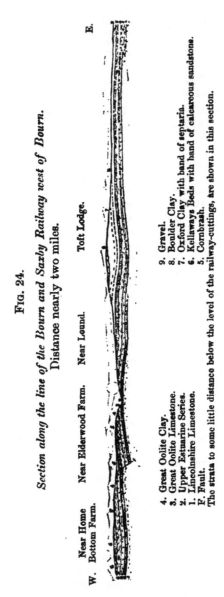

FIG. 24.

Section along the line of the Bourn and Saxby Railway west of Bourn.
Distance nearly two miles.

W.   Near Home Bottom Farm.   Near Elderwood Farm.   Near Lound.   Toft Lodge.   E.

9. Gravel.
8. Boulder Clay.
7. Oxford Clay with band of septaria.
6. Kellaways Beds with band of calcareous sandstone.
5. Cornbrash.
4. Great Oolite Clay.
3. Great Oolite Limestone.
2. Upper Estuarine Series.
1. Lincolnshire Limestone.
F. Fault.
The strata to some little distance below the level of the railway-cuttings, are shown in this section.

The synclinal structure shown in the cuttings confirms that depicted on the Geological Survey Map, as outcrops of Cornbrash are marked in places both north and south of Bourn, on the eastern side of the tract of Oxford Clay. Clays and sands belonging to the Kellaways division were again shown above the Cornbrash to the north of Elderwood Farm and to the south of Home Bottom Farm.

In the accompanying section I have drawn the strata below the level at which they are exposed in the railway-cuttings, so as to exhibit more clearly the general structure of the country.

Bourn to Sleaford, Lincoln, and Brigg.

The Kellaways Beds may be persistently traced in the country north of Bourn, by Sleaford towards Lincoln; indeed, according to the observations of Messrs. W. H. Dalton and W. H. Holloway, they occupy a considerable area between Bourn and Folkingham, and occur in the outliers to the west of the main mass, near Edenham and Ingoldsby.

Sections were exposed in brickyards near Morton and Kirkby Underwood, and the beds were proved in a well at Aswarby Hall (18 feet), and in the railway-cutting north of Sleaford. Clays and shales indicating the zone of *Ammonites Jason* and higher beds were seen near Burton Pedwardine and Mareham, south-east of Sleaford.[*] Purplish shaly beds were shown in the brickyard (now abandoned) north-east of Sleaford. These occur above the Kellaways division, and they reminded me of beds that occupy the same position near Weymouth. South-east of Digby finely laminated clays yielding *Ammonites, Avicula, Nucula,* &c., were observed by Mr. Dalton, and in the adjoining railway-cutting he obtained *Cerithium muricatum* in abundance.

We have but little information respecting the higher zones of the Oxford Clay in this area, for they do not often appear at the surface; indeed over much of the ground they are concealed by Drift or buried beneath the Alluvium of the Fenland.

In the area north of Kirkby Green, and thence along the outcrop east of Lincoln, the Kellaways Beds have been separated from the Oxford Clay on the Geological Survey Map. They include at their base here as elsewhere, a band of clay, proved at Sudbrook Holme to be 7 feet thick, and increasing in thickness further north. These "Basement Clays" as they are termed by Mr. W. A. E. Ussher are rightly grouped by him as equivalent to the "Avicula Shales" or "Cornbrash Clay" of Yorkshire.[†] (See p. 10.) The full thickness of the Kellaways Beds at Sudbrook is about 25 feet.

Sections of the Kellaways Rock have been observed east of Nocton Hall, where sands and soft sandstone occur, with *Belemnites* and *Gryphæa bilobata.* At other localities along the outcrop we find an occasional pit or ditch where sand, buff, brown, and pale-greenish in colour, has been exposed. Occasional beds of calcareous and shelly sandstone are met with.

---

[*] Jukes-Browne, Geol. S. W. Lincolnshire, pp. 70–72; *see* also H. B. Woodward, Memoir on Lower Oolitic Rocks, p. 426.

[†] Geol. Lincoln, p. 73.

At Timberland there is a large brickyard, where, beneath Fen Deposits and Boulder Clay, the Oxford Clay of a dark slate colour was exposed to a depth of 6 feet. The following species were obtained by Mr. Jukes-Browne:—[*]

Ammonites Lamberti.	Belemnites.
—— Mariæ.	Gryphæa dilatata.
—— oculatus.	

The highest beds consisting of stiff blue clay were shown to the depth of 40 feet in a pit near Bardney, and here the following fossils have been found :—

x Ichthyosaurus (vertebra).	x Cerithium muricatum.
Ammonites arduennensis ?	x Gryphæa dilatata.
x —— cordatus.	x Pinna.
x —— Eugenii.	x Serpula sulcata.
x —— excavatus.	x —— tricarinata.
—— perarmatus.	

The species marked x were noted by Mr. T. Roberts. The identifications of some of the species recorded in the Geological Survey Memoir have been considered doubtful by Mr. Roberts.

At Langworth brickyard, north-east of Lincoln, a somewhat similar assemblage has been met with, including *Am. Lamberti*, suggestive of the presence of that zone, together with the lower part of that of *A. cordatus*. The species noted by Mr. Roberts are as follows :—

Ammonites cordatus.	Alaria bispinosa.
—— Eugenii.	Avicula Münsteri.
—— hecticus.	Gryphæa dilatata.

Am. Lamberti was obtained here by Mr. Dalton, while *Am. perarmatus* has been obtained at Rand.

South-west of Bishop's Bridge, between Glentham and West Rasen, a well was sunk through 6 feet of mottled clay into grey shaly clay, yielding *Am. Duncani, A. canaliculatus, A. Lamberti,* and other fossils, specimens of which were obtained by Mr. Ussher.[†] In a brickyard north-west of Kingerby, *Belemnites hastatus* was found ; while a somewhat lower horizon was indicated in the brickyard east of Bishop's Norton, where Mr. Roberts obtained *Belemnites Oweni* in abundance, together with *Cerithium muricatum,* and *Gryphæa dilatata.*[‡]

Northwards through Brigg to the shores of the Humber the Oxfordian beds are largely concealed by the Alluvium of the Ancholme valley.

A boring at Brigg, described by Mr. Ussher, showed the following beds :—

						Ft.	In.
	Drift &c.	.	.	.	.	40	0
Oxford Clay	Blue Shale	.	.	.	.	42	0
and	Sandstone Rock	.	.	.	.	2	0
Kellaways Rock.	Blue Shale	.	.	.	.	18	0
Cornbrash	Limestone Rock	.	.	.	.	3	0

[*] Geol. Lincoln, p. 77.
[†] *Ibid.*, pp. 77, 78.
[‡] Roberts, Quart. Journ. Geol. Soc., vol. xlv. pp. 548, 549.

At Gander Hill, south-east of Hibaldstow, there appeared to be little more than 2 feet of clay between the Cornbrash and Kellaways Rock, but wherever evidence could be obtained, the clay was persistent.

At Winterton Holme, the beds are masked by Drift. While, however, the Kellaways Rock appears to be so thin beneath Brigg, yet, according to Mr. Ussher, it develops further south into a series of sands irregularly consolidated into rock, and may attain a thickness of 20 feet, while the overlying Oxford Clay is probably about 300 feet thick.

The indurated beds in the Kellaways group generally yield *Gryphæa bilobata* and *Belemnites.*

The exposures in the Oxford Clay in this part of Lincolnshire, are not very important. Mr. Ussher records *Posidonomya* and *Leda Phillipsi* from bluish-grey shaly clay at Black Dyke, east of Waddingham. The most northerly exposure south of the Humber is at the east end of Winterton Holme Hill, west of Scabcroft, where the lower part of the Oxford Clay, blue clay with limestone concretions, is shown ; and the following fossils were obtained :—*

Avicula inæquivalvis.	Serpula tetragona.
Gryphæa dilatata.	

Sussex and Kent.

The occurrence of Oxfordian strata in the south-east of England was first made known by the results of the Sub-Wealden Boring. There near Battle the Oxford Clay was proved beneath Corallian and newer beds. Strata indicating a like succession, including also Lower Oolites, have been penetrated in the Dover Boring.† At Chatham again, Oxford Clay has been definitely proved, and in this case directly beneath the Lower Cretaceous Strata. Hence we may assume that a line from Battle to Dover, where the Upper Oolites occur, corresponds roughly with the general strike of the underground Jurassic rocks; and that the beds have a south-easterly dip, so that in passing from Battle to Chatham lower strata come successively beneath the Cretaceous coverings. Finally at Streatham and Richmond the Lower Oolites are present, and as we proceed towards Faringdon in Berkshire, we again come upon the higher portions of the Oolitic series, as represented in the Diagrams. (Figs. 144 and 145, pp. 288, 289.) Thus there is evidence beneath this great tract of a denuded anticline of Jurassic rocks covered by Lower Cretaceous and newer formations.

With regard to the Oxford Clay in the Sub-Wealden Boring but little can be said. The lowest cores, as remarked by Mr. Topley, are only one-inch in diameter, and therefore the opportunity of obtaining recognizable species of fossils was limited. At

* Geol. North Lincolnshire, &c., pp. 91–98.
† See Memoir on the Lower Oolitic Rocks of England, p. 362, and J. F. Blake, Annals of British Geology for 1892, pp. xxv. and 112.

the base of the lowest bed of oolitic limestone "Dark shale" was reached, but no fossils were recorded from it. Above this limestone, there was a thick bed of shale and this yielded the following fossils which were identified by Mr. Etheridge:—

	Depth from surface. Feet.
Ammonites Chamusseti - - - -	1866
—— (near to vertebralis) - - - -	1869

The specimen identified as *A. Chamusseti*, is a fragment only : the species according to D'Orbigny is very near to *A. cordatus*, but being found in the Callovian strata, he assigned the distinct name to it.[*]

Mr. Topley at first thought it "possible that the lowest 60 feet of the Boring may be *Oxford Clay*; but this point is doubtful."[†] Later on he doubtfully assigned 117 feet to this formation. According to the detailed statement of the Aqueous Works and Diamond Rock-Boring Co., the boring was carried to a depth of 1,906 feet and the lowest 120 feet were entirely in shale. (See p. 347.) It is unfortunate that these differences exist. Having regard to the position of the above fossils and the general character of the strata, I am disposed to agree with the last interpretation and to group 120 feet with the Oxford Clay, being the beds proved from 1,786 feet to the bottom of the boring.

With regard to the deep boring at Chatham Dockyard (1880–84), the discovery of Oxford Clay was announced in 1884 by Mr. Whitaker, from an examination of the fossils by Messrs. G. Sharman and E. T. Newton. In this boring, after passing through Alluvium, Gravel, Thanet Sand, Chalk, Gault, and Lower Greensand, stiff grey clay was reached at a depth of 943 feet, and penetrated for a depth of about 20 feet. The clay contained pyritic and phosphatic (?) nodules, and yielded the following fossils :—[‡]

Ammonites crenatus.	Crustacean claws and limbs.
—— hecticus ?	Bairdia.
—— Lamberti.	Serpula vertebralis.
—— plicatilis !	Cidaris.
Belemnites.	Pentacrinus Fisheri.
Alaria trifida.	Cristellaria crepidula.
Astarte.	—— rotulata.
Pecten.	Lignite.

The evidence concerning the Jurassic rocks in the Dover Boring has at length been published by Prof. W. Boyd Dawkins. From his statement we learn that beds of Oxfordian character were probably touched at a depth of 769 feet below high-water mark, and penetrated to a depth of 188 feet. The occurrence of *Gryphæa dilatata* and of *Rhynchonella varians* var. *socialis*, together with *Belemnites*, as well as the stratigraphical position of the strata, support the view that this series of clays, shales, and

[*] Pal. Française Terrain Jurassiques, Tome i. p. 437.
[†] Dixon, Geol. Sussex, Ed. 2, p. 155.
[‡] Whitaker, Guide to Geol. London, Ed. 4, pp. 19, 21; J. W. Judd and C. Homersham, Quart. Journ. Geol. Soc., vol. xli. p. 526; Whitaker, *Ibid.*, vol. xlii. pp. 30, 32; vol. xliii. p. 197; and Geology of London, vol. i. p. 42.

marls, with bands of limestone and sandstone, belongs to the Oxfordian group. (See p. 344.)

In connexion with the irregular overlap of the Lower Cretaceous rocks over the denuded plane of Jurassic rocks in the south-east of England, it is interesting to note that the Lower Oolitic fossils *Terebratula fimbria* and *Rhynchonella oolitica,* and the Oxfordian fossils *Ammonites crenatus* and *A. Lamberti,* were found by Godwin-Austen ; while remains of Liassic and Rhœtic Fishes have been obtained by Mr. C. J. A. Meyer, from the Lower Greensand near Godalming. These fossils, derived from the Jurassic rocks, occur in a pebble-bed that lies at the base of the Bargate Stone-beds grouped by Mr. Meyer with the Folkestone Beds and by Mr. Topley with the Hythe Beds.[*]

---

[*] Godwin-Austen, Quart. Journ. Geol. Soc., vol. xii. p. 71 ; Meyer, on the Lower Greensand of Godalming (Geol. Assoc., 1868, printed separately) ; Geologist, vol. vi. p. 52 ; Geol. Mag., vol. i. p. 249 ; Topley, Geology of the Weald, p. 122 ; and F. Chapman, Quart. Journ. Geol. Soc., vol. l. p. 677.

CHAPTER V.

CORALLIAN.

GENERAL ACCOUNT OF THE STRATA.

Between the Oxford Clay and the Kimeridge Clay there is a variable series of rock-beds that are well shown in the cliffs near Weymouth, and extend inland to the neighbourhood of Oxford. They are again seen at Upware, near Cambridge; but over chief part of the area from the north-east of Oxford to north Lincolnshire this intermediate formation is represented mainly by clays. Further on in Yorkshire the rock-beds are prominently developed; and they are again shown in the far north, at Brora, in Sutherlandshire. In mass this division consists of calcareous sandstone and sand, termed "Calcareous Grit"; and of oolite, pisolite, and rubbly coral-rock known as "Coral Rag"; but in the early days of geology the clayey equivalents were not recognized as part of the series.

Townsend, who derived his information from William Smith, divided the beds, in 1813, into Calcareous Grit, with overlying Superior Oolite. The latter term, however, was too general a one, being applicable not only to the Portland Stone, but to the Great Oolite; hence Smith, later on (1815–16) spoke of the beds as Coral Rag and Pisolite, and for many years, even to the present day, the name "Coral Rag," as adopted by Conybeare, has been used for the formation as a whole.*

Buckland, in 1818, divided the beds into Calcareous Grit, Coral Rag, and Upper or Oxford Oolite; but this upward sequence, derived from sections near Oxford, was of very local application.†

A more general classification was introduced in 1829,‡ by John Phillips, who divided the beds in Yorkshire, as follows:—

> Upper Calcareous Grit.
> Coralline Oolite.
> Lower Calcareous Grit.

These subdivisions were adopted by the Geological Survey, for the strata from Dorsetshire to Berkshire, although in the process of mapping, the three portions, or distinct representatives of them, could not always be identified.

D'Orbigny had, in 1849, proposed the name Corallian for the formation,§ and this is now generally adopted; because, as a group-name, it is better than that of Coral Rag, for the variable strata that occur between the Oxford and Kimeridge Clays.

---

* Conybeare and Phillips, Geol. England and Wales, 1822, p. 185.
† Table appended to the Geology of England and Wales, by William Phillips, 1818.
‡ Geology of Yorkshire, Part I.
§ The name Corallian was taken from the *calcaire corallien* of Thurmann, D'Orbigny, Pal. Française, Terrains Jurassiques, Tome i p. 609.

Our present knowledge of the Corallian Beds is most largely
due to the researches of Prof. J. F. Blake and Mr. W. H.
Hudleston, who, in 1877, gave a very full description of the strata
and their fossils.* They showed the need of adopting local
stratigraphical divisions, so that the nature of the beds and their
changes might be better understood. In Yorkshire, especially,
the threefold division of Phillips, was found to be inadequate, and
in Dorsetshire other very local divisions were found necessary,
and to these reference will subsequently be made.

In our area from Dorsetshire to Lincolnshire, two broad
divisions can, as a rule, be made in the Corallian Beds.

The lower portion includes occasional beds of clay, but consists
mainly of calcareous sandstone, and of sands with large doggers
or concretionary masses of sandstone. Small pebbles of lydite
occur in certain localities.

The upper portion, which on the whole is more variable in
character, comprises a series of limestones, oolitic, pisolitic, and
shelly, with beds of coral-rag and occasional sandy and clayey
beds.

The Corallian Beds may thus be grouped as follows :—

| Upper Corallian. | Upper Calcareous Grit, Upper Coral Rag and Ironstone. Coral Rag and Coralline Oolite. | Zone of *Ammonites plicatilis*. |
| Lower Corallian. | Lower Calcareous Grit - - | Zone of *Ammonites perarmatus*. |

Where fully developed, as in Dorsetshire, the total thickness of
the series is about 200 feet. In parts of Wiltshire, it is about
100 feet, and the thickness diminishes as we proceed into
Oxfordshire.

In Dorsetshire there is perfect conformity between the
Corallian Rocks, and both Oxford and Kimeridge Clays. As the
beds are traced through Wiltshire into Berkshire and Oxfordshire,
the junction of Corallian Beds and Oxford Clay remains conformable, but there is a comparatively abrupt change from the Corallian
Beds to the Kimeridge Clay, and near Oxford there are indications of some pause between the deposition of the strata.
Further north too, where the rock-beds are but thinly represented, there may have been some local erosion.

Formation of the Strata—Coral Banks.

Broadly speaking, the formation affords evidence of sandy
sediments, followed by calcareous shelly beds, current-bedded
oolites, and coral-growths.

Mr. Hudleston has remarked on the probability of oolitic
beds being formed in one area, and gritty beds in another, so
that oolitic grains were sometimes drifted.† Thus we find, as
near Highworth, sandstones with oolitic grains. There is also

* Quart. Journ. Geol. Soc., vol. xxxiii. p. 383; Geol. Mag. 1878, p. 91.
† Proc. Geol. Assoc., vol. iv. pp. 384, 385.

abundant evidence of reconstruction among the upper beds in Berkshire, where pebbles of oolite and limestone bored by Lithodomi, as well as lydite pebbles, occur on two or more horizons in the strata.

The beds altogether exhibit evidence of shallow water. Ripple-marks and worm-burrows are found on some of the sandy layers. Lignite, too, is not uncommonly preserved in the strata. We have, in fact, more pronounced indications of the proximity of land than we find elsewhere in the older Oolitic strata of the south-west of England, if we except the Inferior Oolite of the Mendip Hills. The land-area probably lay to the south or south-east.

It has been mentioned that in the Jurassic rocks there are occasional Coral-banks of limited extent, but little or no evidence of particular reefs.* The term "reef" has however been applied by some geologists to the lenticular bands of Coral-rock that occur here and there in the Lower Oolites, and in the Corallian Beds. The "Coral Rag," according to Dr. Wright, "has long been considered to be the great metropolis of Jurassic Corals; and although in certain regions reef-structures are found, the specimens are for the most part collected from beds that crop out at the surface of arable lands."† These remarks apply to the celebrated locality of Steeple Ashton, in Wiltshire. In other parts the "reefs" described by Messrs. Blake and Hudleston, consist of bands of coral-limestone, rubbly coral-rock, and clay. Layers of large masses of *Thamnastræa*, *Thecosmilia*, and *Isas-træa*, bored by *Lithodomus*, are met with here and there; and the coral-rock itself is often changed into crystalline limestone, in which the organic structure is no longer visible. In many cases the coral-growth was *in situ*, in other cases some of the specimens appear to have been drifted.‡ Duncan spoke of the Corallian Beds as affording evidence of the last "reefs" of the British Area.§

The association of oolite and Coral-reefs in modern days, has been already pointed out; and it is interesting to find that, as is the case with the Inferior Oolite and the Great Oolite, there are bands of coral-growth associated with the more oolitic beds in the Corallian formation. The same is the case with the Lower Lias at Applecross, in Ross-shire. Reference has also been made to the formation of Pisolite, and to the minute tubular structures discovered by Mr. E. Wethered, and which he has referred to *Girvanella*.||

The general character of the Corallian Beds is shown in the diagram, Fig. 56, p. 107.

---

* Memoir on the Lias of England and Wales, p. 16.
† Proc. Cotteswold Club, vol. iv. p. 154.
‡ Quart. Journ. Geol. Soc., vol. xxxiii. pp. 288, 291, 293, &c.; see also Fox-Strangways, Jurassic Rocks of Yorkshire, pp. 398, &c.
§ Rep. Brit. Assoc. for 1869, p. 165.
|| H. B. Woodward, Memoir on the Lower Oolitic Rocks of England, pp. 14, &c. Wethered, Geol. Mag., 1889, p. 196, and Quart. Journ. Geol. Soc., vol. li. p. 196; G. F. Harris, Proc. Geol. Assoc., vol. xiv. p. 59.

Microscopic Structure of Corallian Rocks.

Microscopic sections of the rocks have been examined by Mr. J. J. H. Teall, who reports as follows:—*

In the Osmington oolite the grains show concentric and radiate structure with *Girvanella*-structure (?) in the nuclei of one or two grains. Quartz grains act as nuclei in several instances, and there are small black grains scattered through some of the oolite grains. Fragments of shells also occur, and the matrix is of clear crystalline calcite and calcareous mud.

The Todbere freestone showed oolite grains with ill-defined structure, also pellets, quartz-grains, &c., in a matrix of fine-grained crystalline calcite.

The oolite of Westbury, in Wiltshire, showed pellets, oolite grains, Foraminifera, &c., in a fine-grained matrix of crystalline calcite.

The oolitic and pisolitic limestones of Stower and Keevil, showed, in addition to the oolite and pisolite grains, organic fragments, &c., in a matrix of crystalline calcite. Foraminifera occur at Keevil, and some of the oolite grains are compound.

The margin of some of the pisolitic grains from Stower, shows a very minute form of *Girvanella*-structure, and a similar structure, somewhat more obscure, occurs in the centre of some grains.†

Very fine *Girvanella*-structure occurs in the pisolite of Sturminster Newton. Mechanical picking-up of foreign matter must also have played a part in the growth of the pisolite, because the grains contain quartz-fragments.

In the limestone of Wheatley, fragments of organic remains, and pellets (of calcareous mud ?) occur in a matrix of crystalline calcite.

In the Bencliff Grits of Osmington, and in the Lower Calcareous Grit of Seend and Calne, the rock consists of quartz-grains, in a matrix of crystalline and granular calcite. Organic fragments were seen in the rocks from Seend and Calne. In the Bencliff rock, the quartz-grains are angular.

In the Upper Corallian sandstone of Highworth, there occur in addition to quartz-grains, oolite and pisolite grains, and organic fragments, cemented by calcite. (See p. 119.)

The top bed of Coral Rag at Ringstead Bay, an iron-shot earthy limestone, showed brown grains of ferruginous oolite and angular fragments of quartz in a fine-grained calcareous matrix.

The Iron-ore of Westbury, in Wiltshire, showed reddish-brown oolitic grains with concentric structure. The nucleus in one grain is a fragment of another grain. The matrix is yellowish-green, and contains numerous small detached grains of colourless calcite.‡

Zones.

The Corallian Beds may be divided into two broad palæontological groups—the zones of *Ammonites perarmatus* and *A. plicatilis*.§

As before mentioned (p. 9), Oppel‖ included the Corallian Rocks in his "Oxford Group," the lower zone being that of *Ammonites biarmatus*¶ (equivalent generally to that here grouped as the Upper Oxford Clay, or zone of *A. cordatus*); and the upper zone being that of *Cidaris florigemma*, that embraced the main mass of our Corallian Rocks, or the zones of *A. perarmatus* and *A. plicatilis*.

---

* See also Sorby, Quart. Journ. Geol. Soc., vol. xxxv. (Proc.) p. 80.
† See Plate I. Fig. 1, in Memoir on the Lower Oolitic Rocks of England, p. 27.
‡ See Plate II. Fig. 12, in Memoir on the Lower Oolitic Rocks of England, p. 29.
§ See also Fox-Strangways, Jurassic Rocks of Yorkshire, vol. i. p. 301.
‖ Die Juraformation, p. 624.
¶ *A. biarmatus*, Ziet., was regarded by D'Orbigny as the same as *A. perarmatus*, Sow.

It is well known that *Cidaris florigemma* prevails in the higher portion of the zone of *A. plicatilis* in this country, and especially in the upper Coral Rag ; but it is not confined to that position.   On the Continent, moreover, two "zones" of this Echinoid have locally been constituted, so that its vertical range is too wide to render it of particular value as a restricted zonal index.   Corals and Echinoids are preserved locally according to the conditions.   Thus it is that we find an "Upper Coral Rag" in some places, or an Upper "zone" of *Cidaris florigemma.*

Ammonites perarmatus is conveniently taken as marking a zone equivalent to the Lower Corallian Beds, although it does occur also in the Oxford Clay together with *A. cordatus.*   Both species moreover have been recorded from Upper Corallian Beds.

In defining the boundary between Corallian Rocks and Kimeridge Clay we must be guided mainly by stratigraphical considerations, admitting that in conformable strata the actual boundary may be taken at somewhat varying chronological horizons.

Thus some difficulty has been felt in the Weymouth district in grouping the Upper Coral Rag of Osmington and the Abbotsbury Ironstone with the Corallian Beds, because the fossils show affinities with beds that on the Continent have been placed in the Kimeridge Group.   Damon went so far as to term the Upper Coral Rag the "Kimeridge Grit."[*]

We find however that Continental geologists are not all of one opinion on questions of nomenclature,[†] and it is obviously wrong to alter classifications based on the strata developed in this country, to suit the varied stratigraphy of different districts on the Continent.   We may object therefore to place any portion of our Corallian Rocks with the Oxford or Kimeridge formations, as we object to include our Portland strata with the Kimeridge Beds. Correlations may be made by means of zones ; and zones themselves may be grouped together irrespective of formations.   The zones, however, that are included in the Oxfordian group should be restricted to those belonging to the strata that gave name to the group.

The zones marked by Oppel, above that of *Cidaris florigemma* were (in ascending order) those of *Diceras arietina, Astarte supracorallina,* and *Pterocera oceani.*   The zone of *Diceras* is regarded as part of the Corallian, and would be included in our zone of *Ammonites plicatilis.*   The zone of *Astarte* being based on a limestone-fauna, cannot be compared with our formations, suffice it to say that it contains an admixture of Corallian ard Kimeridge Clay fossils ; and if the beds had been developed in this country, they would doubtless have been linked with the Corallian.   The zone of *Pterocera oceani* contains essentially a Kimeridge Clay fauna, and it is noteworthy that although the genus has been recorded from the Abbotsbury Ironstone, its identification from beds in this country is now considered to have been erroneous. (See p. 152.)

---

* Geology of Weymouth, 1884, p. 65.

† See for instance, S. Nikitin, Die Cephalopoden fauna der Jurabildungen, 1884, pp. 72, 73.   He therein includes the Kimeridgian zone of *Ammonites alternans* in the "Oxford" strata.

Organic Remains.

Among the fossils of the Lower Calcareous Grit perhaps the most remarkable is the *Ammonites perarmatus*, which occurs more especially in Wiltshire, Berkshire, and Oxfordshire, with the chambers only preserved in the form of casts. Specimens of this kind were named "*Ammonites catena*."* The interior of the Ammonites was filled with ferruginous and calcareous mud or sand, and the shell and septa have subsequently been dissolved away, leaving only the casts of chambers in a loose form or linked together so as to resemble a rough tapering chain. *Ammonites cordatus* also occurs in the form of casts.

Under the name *Rhaxella perforata*, Dr. G. J. Hinde has described certain small siliceous sponges, whose skeleton is entirely built up of aggregated masses of minute globate spicules. Attention had previously been directed by Sorby to the occurrence of these organisms in the Corallian Rocks of Yorkshire, and subsequently Prof. J. F. Blake noticed the same forms (under the name "*Renulina sorbyana*") in equivalent rocks of Dorsetshire and Wiltshire, at Sturminster Newton and Hillmarton.† The nature of the detached spicules was not, however, understood until Dr. Hinde had studied them.‡ He observes that a few Calcisponges have been found in the Coral Rag of Lyneham, Wiltshire.

The more noteworthy fossils in the Upper Corallian Rocks, include, in addition to the Ammonites and Belemnites, many Gasteropods, such as *Nerinæa*, *Chemnitzia*, and *Bourguetia*. The large *Bourguetia* or "*Phasianella*" *striata* appears to be identical with a form found in the Inferior Oolite of Cleeve Hill, near Cheltenham. Beds, rich in *Trigonia* are found, and of these the pavements on the foreshore near Weymouth are striking examples. Brachiopoda are, as a rule, comparatively rare; and it is remarkable that few Polyzoa have been obtained.

The most abundant of the Echinoderms is *Echinobrissus scutatus;* more interesting perhaps are the clusters of *Hemicidaris intermedia* that have been found at Calne; while among other forms *Pygaster umbrella*, and species of *Pygurus* are locally abundant, and spines of *Cidaris florigemma* are very frequently met with.

Among Corals, the "Honeycomb Coral" *Isastræa explanata*, so abundant at Headington, also *Thecosmilia annularis*, and *Thamnastræa arachnoides*, are those more commonly to be obtained.

Saurian remains are rarely found; and remains of Fishes, such as *Asteracanthus*, and *Lepidotus*, are only occasionally met with in the Corallian Rocks.

The occurrence in the lower beds of *Ammonites cordatus, A. perarmatus, Belemnites hastatus, Gryphæa dilatata,* &c. links them

* See Sowerby, Mineral Conchology, vol. v, p. 21.
† Quart. Journ. Geol. Soc., vol. xlvi. p. 54; Micros. Journ., vol. xv. p. 362.
‡ British Jurassic Sponges, pp. 192, 212.

with the Oxford Clay ; while the occurrence in the upper beds of
Ammonites decipiens, *A. mutabilis*, *Ostrea deltoidea*, and *Rhynchonella inconstans*, links them with the Kimeridge Clay.

The common and characteristic fossils of the Lower and Upper
Corallian Beds include the following species, but they are not all
of them confined to the horizons indicated :—

CORALLIAN FOSSILS.	LOWER CORALLIAN. Zone of Ammonites perarmatus.	UPPER CORALLIAN. Zone of Ammonites plicatilis.
Ammonites cordatus	1	—
—— decipiens	—	2
—— perarmatus (Fig. 25)	1	—
—— plicatilis (Fig. 26)	—	2
Nautilus hexagonus	1	—
Belemnites abbreviatus (Fig. 27)	1	·
Bourguetia (Phasianella) striata (Fig. 28)	—	2
Cerithium muricatum	1	2
Chemnitzia heddingtonensis (Fig. 29)	—	2
Cylindrites Luidi	1	—
Littorina muricata	—	2
Natica corallina	—	2
Nerinæa Goodhalli	—	2
Pleurotomaria reticulata (Fig. 30)	—	2
Astarte ovata	—	2
Cucullæa corallina	1	2
Exogyra nana (Fig. 34)	1	2
Gervillia aviculoides (Fig. 35)	1	2
Gryphæa dilatata	1	
Hinnites tumidus (Fig. 37)	—	2
Lima elliptica	1	2
—— pectiniformis	—	2
—— rigida	—	2
Lithodomus inclusus	—	2
Myacites decurtatus	1	2

CORALLIAN FOSSILS.	LOWER CORALLIAN. Zone of *Ammonites perarmatus.*	UPPER CORALLIAN. Zone of *Ammonites plicatilis.*
Ostrea gregaria (Fig. 38) - - -	1	2
—— solitaria - - - - -	1	2
Pecten articulatus (P. vimineus) (Fig. 36) - -	—	2
—— fibrosus - - - - -	1	
—— lens - - . - - -	1	2
Perna mytiloides - - - - -	—	2
Pinna pesolina - - - - -	—	
Trichites - - - - -	—	
Trigonia clavellata (Fig. 32) - - -	—	x
—— Meriani (Fig. 31) - - -	—	
—— monilifera (Fig. 33) - - -	—	2
—— perlata - - - -	1	2
Serpula sulcata - - - -	—	x
—— tricarinata - - - -	1	2
Cidaris florigemma (Figs. 39, 40) - - -	—	2
—— Smithi (Fig. 41) - - -	—	x
Echinobrissus scutatus (Fig. 42) - - -	—	
Hemicidaris intermedia (Fig. 43) - - -	—	ʋ
Pygaster umbrella - - - -	—	2
Comoseris irradians (Fig 45) - - -	—	2
Isastræa explanata (Fig. 46) - - -	—	∩
Montlivaltia dispar (Fig. 44) - - -	—	2
Thamnastræa arachnoides (Fig. 47) - -	—	2
—— concinna - - - -	—	2
Thecosmilia annularis (Fig. 48) - - -	—	2

CORALLIAN CEPHALOPODS AND GASTEROPODS.

FIG. 25.　　　　　　　　　FIG. 28.

FIG. 26.

FIG. 27.　　　FIG. 29.

FIG. 30.

Fig. 25. Ammonites perarmatus, *Sow.*, ⅓.
  ,, 26. ——plicatilis, *Sow.*, ⅓.
  ,, 27. Belemnites abbreviatus, *Mill.*, ⅓.
  ,, 28. Bourguetia (" Phasianella ") striata, *Sow.*, ⅔.
  　　　　　　ndomelania heddin   nensis, *Sow.*, ⅓.

CORALLIAN LAMELLIBRANCHS.

FIG. 31.

FIG. 32.

FIG. 33.

FIG. 34.

FIG. 35

FIG. 36.

FIG. 37.

FIG. 38.

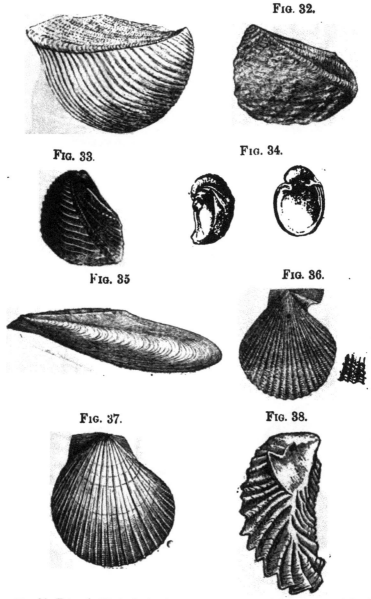

FIG. 31. Trigonia Meriani, *Ag.*, ¼.
 „ 32. —— clavellata, *Sow.*, ½.
 „ 33. —— monilifera, *Ag.*, ⅓.
 „ 34. Exogyra nana, *Sow.*, Nat.
        size.
 „ 35. Gervillia aviculoides, *Sow.*, ⅓.

FIG. 36. Pecten articulatus, *Schloth.*,
        Nat. size.
 „ 37. Hinnites tumidus, *Ziet.*, ⅔.
 „ 38. Ostrea gregaria, *Sow.*, Nat.
        size.

CORALLIAN ECHINOIDS.

FIG. 39.   FIG. 41.

FIG. 40.

FIG. 42.   FIG. 43.

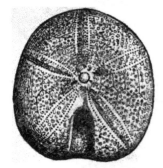 

FIG. 39. Cidaris florigemma, *Phil.*, ⅓.
  ,, 40. —— —— spine, ⅔.
  ,, 41. —— Smithi, *Wright*, ⅓.
  ,, 42. Echinobrissus scutatus, *Lam.*
  ,, 43. Hemicidaris intermedia, *Flem.*

CORALLIAN CORALS.

FIG. 44.   FIG. 45.

FIG. 46.   FIG. 48.

FIG. 47.

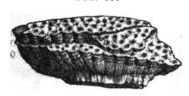

FIG. 44. Montlivaltia dispar, *Phil.*
,, 45. Comoseris irradians, *M. Edw.*
,, 46. Isastræa explanata, *Goldf.*, ⅓.
,, 47. Thamnastræa arachnoides, *Park.*, ½.
,, 48. Thecosmilia annularis, *Flem.*, ⅔.

CHAPTER VI.

CORALLIAN.

LOCAL DETAILS.

Weymouth.

Nowhere in our district are the Corallian Beds so well exposed as in the neighbourhood of Weymouth. They are shown in the cliffs that extend from the Nothe Fort at Weymouth to Sandsfoot Castle, and also for a short distance along the shores of the East Fleet. They are again displayed in the cliffs that extend along the northern side of Weymouth Bay, past Osmington to Ringstead Bay. Throughout they furnish rich fossil-beds, and they have naturally attached much attention from geologists.

Sedgwick, in 1820, examined the strata, and subsequently gave a short account of the principal subdivisions; he also noted that certain fossils passed from the Oxford Clay to the "Coral Rag," and from the "Coral Rag" to the Kimeridge Clay.*

Fitton, in 1827, briefly referred to the beds as "Weymouth Strata, and Oxford Oolite," the term Weymouth clay and sand being applied to the upper portion of the Corallian Series. He also gave a short list of fossils from the neighbourhood of Weymouth.†

A more particular account of the formation was given in 1830 by Buckland and De la Beche, who described the beds as "Coralline, or Oxford Oolite, and Calcareous Grit."‡

Robert Damon, in 1860, gave a general account of the rocks and fossils, and this he considerably amplified in 1884.§

In the meanwhile, in 1877, Messrs. Blake and Hudleston published their full and excellent account of the Corallian Rocks; and therein described and named the chief subdivisions, near Weymouth, with particular lists of their fossils.||

Adopting the names given by Messrs. Blake and Hudleston for the local divisions of the Corallian Rocks, we have the following sequence:—

		Ft. In.	Ft. In.
Kimeridge Clay.			
Upper Corallian	Upper Coral-rag and Abbotsbury Iron-ore - -		
	Sandsfoot Grits (named from Sandsfoot Castle, south of Weymouth) - -	15 0 to 25 0	
	Sandsfoot Clay - -	12 0 to 40 0	
	Trigonia Beds - - -	12 0 to 16 0	
	Osmington Oolite - -	45 0 to 60 0	
Lower Corallian	Bencliff Grits (named from Bencliff, south of Weymouth) - -	10 0 to 35 0	
	Nothe Clay (named from the Nothe Fort, Weymouth) -	30 0 to 40 0	
	Nothe Grits - - -	20 0 to 35 0	
Oxford Clay.			

---

* Ann. Phil., vol. xxvii. (ser. 2, vol. xi.) p. 346.
† Trans. Geol. Soc., ser. 2, vol. iv. pp. 105, 232.
‡ *Ibid.*, p. 24.
§ Geology of Weymouth, 1860, ed. 2, 1884.
|| Quart. Journ. Geol. Soc., vol. xxxiii. p. 262.

FIG. 49.

Section of the Cliffs south of Weymouth.

Distance two miles.

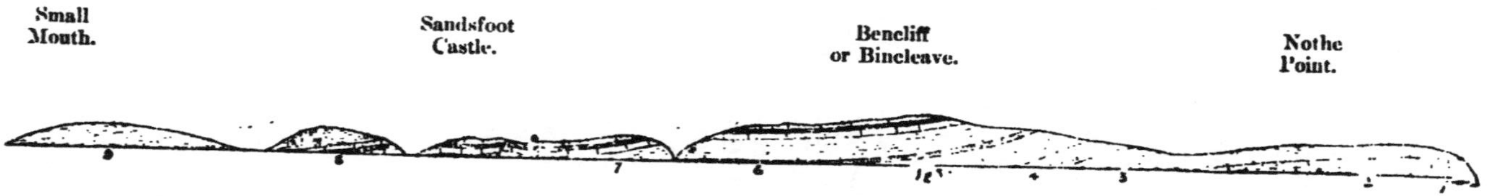

Small
Mouth.

Sandsfoot
Castle.

Bencliff
or Bincleave.

Nothe
Point.

9. Kimeridge Clay.
8. Sandsfoot Grits (with band of clay).
7. Sandsfoot Clay.

6. Trigonia Beds.
5. Osmington Oolite.
4. Bencliff Grits.

3. Nothe Clay.
2. Nothe Grits.
1. Oxford Clay.

Beds 2 to 8 are Corallian.

Although the subdivisions vary much in thickness, the total thickness of the Corallian Series at Weymouth and Osmington appears, according to my measurements, to be much the same— about 200 feet—the actual thickness recorded at Weymouth being 196 feet, and at Osmington 203 feet. Some of the subdivisions shade one into the other so that there is no marked plane of division between them, and varying thicknesses, as in the case of the Bencliff Grits and Nothe Clays, may be assigned locally to each of them.

Commencing on the western side of Ringstead Bay and proceeding towards Osmington Mill, we find low cliffs of Kimeridge Clay (blue clays that contain layers of *Ostrea deltoidea*, and a band of shelly limestone with *Exogyra nana*, &c.) capped in places by gravel. Underlying the Kimeridge Clay, a thin band of grey earthy and iron-shot limestone crops out to the east of the plantation and to the west of the boat-house; it is partially slipped and also slightly faulted and repeated in places. This is the " Kimmeridge Grit " of Robert Damon; the " Upper Coral Rag " of Blake and Hudleston. It is a richly fossiliferous bed from which there is no difficulty in obtaining many fossils.* Teeth, scales, and coprolites of Fishes, and remains of *Asteracanthus ornatissimus*, are recorded by Damon; and Prof. Blake has found *Eryma Babeaui*, and other fossils mostly obtained from the " Kimmeridge Passage-beds " near Sandsfoot Castle.

This interesting bed is sometimes concealed by slips or by recent marine accumulations, and the details of the strata down to the Trigonia Beds, were not clearly to be noted during my visit in 1884. The bed may however be traced in places along the upper part of the cliff to the flag-staff belonging to the Coast-guard station, a little east of Osmington. A little below this point I saw the bed in 1893, when in company with Mr. Strahan. Underlying this fossiliferous bed, there are very ferruginous beds, clays with bands of nodular ironstone (equivalent probably to the Abbotsbury iron-ore), and these rest on one or two bands of red calcareous sandstone with *Belemnites* (Sandsfoot Grits), below which are the Sandsfoot Clays, not clearly exposed.

From the base of these clays the sequence is well shown down to the Nothe Grits, in the cliffs and in the ledges on the foreshore extending westwards to the cascade near Osmington Picnic Inn.

The beds rise a little in a gentle anticlinal to the west of this point, where they are faulted against the Kimeridge Clay, &c. The sequence is generally similar to that of the Weymouth Cliffs, shown in Fig. 49, p. 83.

The sequence from Ringstead Bay to near the Cascade at Osmington is as follows :—

---

* The specimens obtained by myself are included in the list on p. 95. Lists have been published by R. Damon, Geology of Weymouth, 1884, p. 65 ; J. F. Blake, Quart. Journ. Geol. Soc., vol. xxxi. p. 214; Blake and Hudleston, *ibid.*, vol. xxxiii. p. 272; and W. Waagen, Versuch einer allgemeinen Classification der Schichten des oberen Jura (Munich), 1865.

		Fr.	In.
Kimeridge Clay	Clay with layers of *Ostrea deltoidea, Rhynchonella inconstans, Serpula intestinalis,* &c. Band of shelly limestone with *Exogyra nana.* Clay	2	0
Upper Coral Rag and Sandsfoot Grits.	Grey earthy limestone. Fossil-bed	0	6
	Ferruginous beds: clay with bands of nodular ironstone, and gritty calcareous bands. *Ostrea deltoidea*	14	0
	Red grits (one or two beds) and grey, gritty, and sandy beds - 2 0 to	4	0
Sandsfoot Clay	Clays with *O. deltoidea* - 10 0 to	12	0
Trigonia Beds	Grey, and more or less iron-stained and oolitic limestones, with abundant fucoidal markings; divided in places by 4 ft. of yellowish-grey oolitic marl, and thin clay bands; with *Ammonites, Ostrea solitaria, Pecten fibrosus, Perna mytiloides, Pholadomya æqualis, Trigonia Meriani, T. clavellata,* &c. -	15	9
Osmington Oolite	Rubbly and nodular limestones, alternating with bands of oolite (more or less shelly), oolitic marl, and occasional pisolite, the most conspicuous band of pisolite being about 6 feet from the base. *Pecten, Chemnitzia, Nerinæa, Ostrea, Trigonia, Echinobrissus dimidiatus,* and plant-stem. The basement-bed is a calcareous and oolitic grit, with tubiform and fucoidal markings - - - about	60	0
Bencliff Grits	Brown false-bedded sands, with large concretionary masses of calcareous sandstone, or " Doggers," and flaggy beds, showing ripple-marks, with laminated clay and sand, and seams of carbonaceous clay : seen to a thickness of - - - - -	11	6
Nothe Clay	Clays [the full thickness of these clays could not here be estimated, owing to the junction with the overlying Bencliff Grits being hidden by slips; and the beds are at one point faulted].		
Nothe Grits	Greyish calcareous grits, shales, marly clays with bands of nodular calcareous sandstone: *Gryphæa dilatata* abundant, *Ostrea, Serpula,* &c. Seen to depth of - - - -	12	0

West of Osmington, in the cliffs of Black Head, we again have a grand exhibition of the Corallian Beds, surmounted by the dark Kimeridge Clay. The beds occupy a synclinal curve, faulted on the eastern side against the Kimeridge Clay, where the lowest beds seen (west of the fault) are the Bencliff Grits. Westwards the entire sequence may be seen down to the Oxford Clay, while on the foreshore the beds are well exhibited in the ledges formed by the Trigonia Beds, the Osmington Oolite, &c., which run westwards curving slightly inwards towards the cliffs.

Immediately west of the steps leading to the beach from the Picnic Inn, there are indications of the upper Corallian Beds, although at times they are almost entirely concealed by slipped masses of the Kimeridge Clay. The beds are more or less veined with calc-spar (the result of disturbance), and the Belemnites and Lamellibranchs are occasionally bent. The lower Corallian Beds down to the Osmington Oolite occur successively on the fore-shore.

FIG. 50.

Section of the Cliff west of Black Head, near Osmington, Dorset.

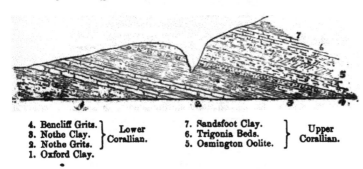

4. Bencliff Grits.	Lower Corallian.	7. Sandsfoot Clay.	Upper Corallian.
3. Nothe Clay.		6. Trigonia Beds.	
2. Nothe Grits.		5. Osmington Oolite.	
1. Oxford Clay.			

The Section of the Corallian Beds in the cliffs of Black Head, west of Osmington, is as follows :—

		FT.	IN.
Kimeridge Clay	Blue clays. Band of gritty limestone with *Exogyra nana, Serpula,* &c. Clays with *Ostrea deltoidea, Rhynchonella inconstans,* &c. - - -	3	6
Upper Coral Rag	Grey earthy and iron-shot limestone, irregular and nodular. Fossil-bed -	0	9
Sandsfoot Grits	Clays with ironstone nodules, gritty calcareous bands, and cement stones: *Ostrea deltoidea, Rhynchonella.* Thickness variable from about - 9 0 to	12	0
	Red, brown, and greenish-grey oolitic and calcareous and ferruginous sandstones, with quartz-grains: *Ostrea deltoidea* (in layers), *Belemnites, Serpula,* &c. - - - -	7	6
Sandsfoot Clay	Bluish-grey clay with layers of *Ostrea deltoidea* - - - -	11	0
	Sandy and ferruginous clays, with sandy iron-shot nodules - -	4	0
	Bluish-grey marly clay with pale earthy cement stones: *Ostrea deltoidea* -	10	0
Trigonia Beds	Red, grey, and purplish-coloured calcareous gritty and ferruginous rocks, in irregular beds, shelly and sparry, and more or less oolitic, with bands of oolitic marl and clay: *Ammonites plicatilis, Nerinœa, Gervillia, Trigonia clavellata,* and Corals - -	14	0

		Ft.	In.
Osmington Oolite	Irregular bluish nodular oolitic lime-stones and marls, with *Chemnitzia*-bed in upper part, that forms a prominent whitish ridge along the foreshore. *Ammonites, Pleurotomaria*	10	0
	Shelly oolite and oolitic marly clays	5	6
	Clays and nodular earthy limestones (thicker in places)	6	0
	Oolite, shelly, ferruginous, and pisolitic (at base): *Ostrea,* plant-remains, and tubiform markings	6	6
	Bluish-grey oolitic marl, with impersistent band of pale earthy limestone	2	0
	False-bedded oolite with shelly layers, *Pecten,* and vertical tubiform markings at base	9	0
	Oolitic marly bed	0	10
	False-bedded shelly oolite, with tubiform markings	1	0
	Rubbly earthy limestones and marls, oolitic in places	10	0
	Grey iron-shot oolite and pisolite	2	0
	Marly oolitic clay, with irregular band of sandy and shelly oolite	6	0
Bencliff Grits	Laminated sands and clays, and indurated yellow sands, with doggers of hard blue calcareous sand-tone (some 6 feet in diameter) Flaggy grits	10	0
Nothe Clay	Bluish sandy clays: *Ostrea, Pleurotomaria,* and *Serpula,* near base	40	0
Nothe Grits	Alternations of irregular and rubbly-looking calcareous grits (much iron-stained), and bluish marly clay	12	0
	Bluish-grey marly clay	8	0
	Bluish and yellowish-grey, slightly calcareous sandstones, with impure cement-stone nodules; fucoidal markings on the surfaces of the beds, the top layer a massive shelly bed. *Gryphæa dilatata, Pleurotomaria,* &c.	15	0
Oxford Clay	Bluish-grey clay, with calcareous nodules near top.		

The total thickness of the Corallian Beds is here seen to be about 203 feet.

The junction of the Bencliff Grits with the beds above and below was well shown in a ravine on the western side of Black Head. (See Fig. 50.)

Passing the anticlinal of Oxford Clay, we come again upon the Nothe Grits and Clays on the eastern side of Redcliff Point, and rounding the headland of faulted Oxford Clay, we come to Redcliff, the upper part of which, formed of the Osmington Oolite resting on the Bencliff Grits, has a brown or yellowish-brown appearance. The beds appear to be slightly faulted (or slipped) on the west, and the cliffs exhibit the strata down to the Oxford Clay. They are, however, not very accessible, and the strata in the central portion of the cliff are much obscured by slips.

On the shore here, as also to the east of Osmington cascade, we find many tumbled Doggers from the Bencliff Grits, huge masses of calcareous sandstone from 3 to 6 feet or more in diameter, and of various forms, some having an irregular bulging shape, and others of a flattened spheroidal form with protuberances and hollows. Sometimes they exhibit false-bedding with oblique layers capped by horizontal layers. Small cavities here and there in the rock are due to the dislodgment of iron-pyrites. The grits which contain shelly layers with Lamellibranchs, are here much thicker than further east, and present a section somewhat similar to that seen at Bencliff, south of Weymouth. These beds appear to merge downwards into the Nothe Clays which, as at Weymouth, contain indurated bands of sandy limestone.

At Redcliff the Nothe Grits comprise about 20 feet of massive beds of brown calcareous sandstone, more or less concretionary, with some nodules or doggers of calc-grit, also fucoidal markings, and a clayey division about 4 feet thick. The beds here yield *Ammonites, Chemnitzia, Gryphæa dilatata, Homomya, Myacites securiformis, Ostrea gregaria, Pecten fibrosus*, &c. The junction with the Oxford Clay is well shown as follows :—

		FT. IN.
	Nothe Grits.	
Oxford Clay	Sandy clays with hard limestone nodules : passing down into indurated calcareous clay	4 0
	Bluish-grey sandy clay with thin band of earthy limestone.	
	Blue clay.	

Passing to the south side of Weymouth Harbour we find the Nothe Grits exposed above the Oxford Clay in the brickyard to the west of the Union Workhouse. The sandy and slightly calcareous clay resting on the Nothe Grits, has been dug for brickmaking, but the upper part of the clay contains too many fossils to be a good brickearth. These Lower Corallian Beds extend along the slopes to the Nothe Point. They consist of sands and calcareous grits with quartz grains and fucoidal markings. They are overlaid by the Nothe Clay, here about 30 feet thick, which extends along the Nothe promontory, and the Grits are exposed beneath them on the foreshore to the south of the Nothe Fort.

A well sunk through the clays near Rodwell House obtained from the Nothe Grits sufficient water to supply a great part of old Weymouth.

The Nothe Grits here yield *Belemnites, Chemnitzia, Gryphæa dilatata, Pholadomya, Myacites decurtatus, Ostrea, Pecten, Trigonia, Lignite*, &c.

The Nothe Clay, consisting of bluish-grey marly and shaly clays with indurated bands of calcareous sandstone and selenite, is exposed in the little bay south-west of the Nothe : here the thickness is estimated at 40 feet by Messrs. Blake and Hudleston. The beds yield *Belemnites, Ostrea, Pecten fibrosus, Thracia*, &c.

Towards Bencliff (or Bincleave) these clays are overlaid by the Bencliff Grits and Osmington Oolite, in the following sequence:—

		Ft.	In.
Osmington Oolite	Hard fucoidal bed with scattered oolitic grains - - - - -	2	0
Bencliff Grits	Fucoidal gritty beds, with clayey nodules - - - - -	4	0
	Pale grey earthy nodular beds and bluish clays - - - -	12	0
	Hard band of calcareous sandstone -	1	6
	Brown sand and sandstone, with streaks of laminated clay, becoming more clayey towards the base, where there occur hard doggers of calcareous sandstone - - - -	18	0
Notho Clay	Blue clay.		

Proceeding southwards, the Osmington Oolite with *Chemnitzia heddingtonensis*, follows on in upward succession, surmounted by the Trigonia Beds, and these form the foreshore of the Western Ledges, and we walk for some distance over a pavement of Trigonias, like that exposed in places beneath the cliffs of Black Head, Osmington. *Trigonia clavellata* is the more abundant form, and slabs containing 30 or 40 specimens have been obtained from near Osmington.* Much time and labour have been expended in cleaning them, and it is by no means easy to obtain a perfect example from the exposed portions of the strata.

The Trigonia Beds are overlaid by the Sandsfoot Clay, well seen in Castle Cove north of Sandsfoot Castle, and exposed also

Fig. 51.

Section south of Sandsfoot Castle, Weymouth.

Upper Corallian.	4. Red and brown ferruginous sand and sandstones.	Sandsfoot Grits.
	3. Blue sandy clay, 6 feet.	
	2. Red and brown ferruginous sandstones, 11 feet.	
	1. Blue sandy clay - - - -	Sandsfoot Clay.

* See J. Buckman, Trans. Dorset Nat. Hist. Club, vol. ii. p. 19; also Frontispiece to Damon's Geology of Weymouth, 1884; and J. C. Mansel-Pleydell, Proc. Dorset Nat. Hist. Club, vol. iii. p. 111.

in the cliffs to the north and south. The Sandsfoot Clay yields the Kimeridge oyster (*Ostrea deltoidea*) in abundance, and was indeed included with the Kimeridge Clay by Sedgwick. It is, however, more consistent with the stratigraphy to include the overlying Red Grits with the Corallian, on the same principle as we include the Nothe Grits in that series; the one having Kimeridgian and the other Oxfordian affinities.

The Sandsfoot Grits form the somewhat striking though low cliffs at Sandsfoot Castle. (See Fig. 51, page 89.) They are divided by a band of blue sandy clay about 6 feet thick, that yields *O. deltoidea*, and might in places be confused with the thicker band of Sandsfoot Clay below. As remarked by Messrs. Blake and Hudleston, these beds "are so coloured by iron in various combinations and states of oxidation, as to seem much richer in that metal than they really are."

Curious fucoidal markings or branching stem-like forms traverse both bands of the Sandsfoot Grits. Referring to these, Messrs. Blake and Hudleston speak of "the strange interlacing fucoid or sponge-growths," which "weathering out in a purplish tint upon a greenish ground, give a very curious aspect to the surface;" and they conclude that here we have "the actual spot on which colonies of fucoids and sponges luxuriated."[*] So far as I am aware, no organic structure is to be detected in these fucoidal and branching structures. They are found here and there in many formations; in the Cornbrash, in the Purbeck Beds (see p. 233), and in the Red and White Chalk of Hunstanton. The origin of the so-called *Spongia paradoxica* of Hunstanton, has been discussed by Prof. Hughes, who considered that the structure was due to concretionary action.[†] It presents appearances similar to those of the fucoidal markings in the Sandsfoot Grits, but these must still remain enigmas, for the absence of definite organic structure in such markings is not considered to be proof of inorganic origin.

The Sandsfoot Grits overlie the Sandsfoot Clay on the southern side of Bencliff, and they are slightly faulted at the gap in Castle Cove and again to the south of the Castle. As we proceed to Small Mouth the beds are overlaid by the Kimeridge Clay, which contains fossiliferous ironstone-nodules near the base; but the state of the cliffs at the time of my visit obscured all evidence of the Fossil-bed ("Kimeridge Grit") that at Osmington marks the top of the Corallian Beds.

Mr. E. T. Newton obtained remains of a Crocodilian jaw from the Sandsfoot Grits, west of Sandsfoot Castle. It was found in a block together with *Goniomya literata* and *Pinna lanceolata*.[‡]

The general section of these beds from near Small Mouth northwards, may be stated as follows:—

---

[*] Quart. Journ. Geol. Soc., vol. xxxiii. p. 271.
[†] Ibid., vol. xl. p. 273.
[‡] Ibid., vol. xxxiv., p. 400.

		Ft.	In.
Kimeridge Clay.			
	Red and brown ferruginous sands and sandstones, shelly in places, with layers of *Ostrea deltoidea*, and fucoidal markings - - - 8 0 to	7	0
	Blue sandy clay, with hard band near base; layers of *O. deltoidea* - -	6	0
Sandsfoot Grits (See Fig. 51.)	Red and brown ferruginous and fucoidal sandstones (slightly calcareous), and greenish clayey sand; with much concretionary iron-ore, and occasional decomposed oolitic grains; beds tinged a purple colour towards the base : they thicken towards the north : *Belemnites, Pecten, Ostrea deltoidea, Serpula intestinalis,*[*] and Plant-remains - - -	11	0
Sandsfoot Clay	Blue, sandy, and calcareous clay, large specimens of *O. deltoidea* abundant : *Astarte, Serp. intestinalis.* Near the base there are one or two layers of fissile calcareous sandstone with *Belemnites, Exogyra nana,* and small cement-stone nodules - - -	38	0
Trigonia Beds	*Trigonia*-bed : bluish-grey sandy limestone with *Trigonia clavellata, Gervillia aviculoides, Nerinœa, Exogyra,* &c. - - - - -	1	6
	Irregular bands of shelly and sandy limestone, with partings of shaly and sandy marl : *Ammonites, Natica corallina, Chemnitzia, Exogyra nana, Myacites decurtatus, Mytilus pectinatus, Pecten fibrosus, Pholadomya œqualis,* Lignite, &c. - - - -	10	6
Osmington Oolite.	Limestones and shales with *Nerinœa*-bed at base - - - - -	7	0

The Osmington Oolite consists here of bands of more or less shelly oolite, alternating with softer rubbly beds and oolitic marls and clays. The total thickness appears to be 45 feet, a diminution from the thickness assigned to the beds at Osmington, due in part to the grouping of more beds with the Bencliff Grits. This, however, is merely a matter of convenience Messrs. Blake and Hudleston include only 21 feet as Bencliff Grits at Bencliff. Moreover they include with the Trigonia Beds the *Nerinœa*-bed and associated layers, which seem to me more appropriately linked with the Osmington Oolite. A good section of the Trigonia Beds, Osmington Oolite, and Bencliff Grits, was noted by them in the railway-cutting at Rodwell.[*]

Passing from the Royal Victoria Inn along the margin of the Fleet, the succession of the beds from the Kimeridge Clay down to the Oxford Clay may be traced, but the strata are not sufficiently exposed to admit of detailed measurements, for there is much slipped ground in places.

The Sandsfoot Grits with *Lima pectiniformis*, were exposed in the low cliffs south-west of Rymead Cottages, and beneath them the blue Sandsfoot Clay with masses of *Ostrea deltoidea*, was

* Quart. Journ. Geol. Soc., vol. xxxiii. p. 267.

exposed. The Trigonia Beds were shown south-west of Wyke, by the bay, and in the cliffs between two lanes leading from Wyke to the Fleet. Westwards we have a good section of the Osmington Oolite, rubbly and shelly oolites, and earthy limestone, occasionally pisolitic, and with marls and, here and there, sandy shales. *Chemnitzia, Nerinæa, Natica corallina, Littorina muricata, Lucina, Modiola bipartita, Pecten fibrosus, Cidaris,* and *Echinobrissus scutatus* were here obtained.

The Bencliff Grits, with deggers and ironstone-concretions, were seen east and west of the main road from Rodwell to Wyke and the Fleet. They were underlaid by clay with nodules, the Nothe Clay ; and below by shelly nodular and fucoidal calcareous grits, the Nothe Grits. These yielded *Belemnites, Gryphæa dilatata,* &c.

Traced westwards from Osmington, on the northern side of the Weymouth anticline, we find here and there evidence of the divisions shown so well at Osmington, but some of them are much reduced in thickness. In the cutting of the Abbotsbury railway near Broadway, the Sandsfoot Grits and Clay were seen overlying the Trigonia Beds. The latter exposed to a depth of 9 feet consisted of sandy, shelly, and partially oolitic limestone, with pockets of brown clay washed into fissures from the overlying beds. They yielded *Ammonites cordatus, Cerithium, Bourguetia, Pleurotomaria, Astarte, Cucullæa,* and *Ostrea.* Damon records *Ceromya excentrica* from this locality. Messrs. Blake and Hudleston note that the Nothe Grits at Broadway appear to be represented by a band of ferruginous sandstone about 2 feet thick, and this is still further reduced south of Abbotsbury.[*]

Higher beds of Corallian Rocks are well shown at Linton Hill, south-east of Abbotsbury. The Osmington Oolites are quarried on the south-west side of the hill, where they are seen to comprise 18 feet of brown oolite with shelly layers. Following the scarp to the north we find still higher beds of shelly and coarse-grained oolite, with rubbly and marly beds. They yield *Ammonites, Pleurotomaria, Ostrea solitaria, Pecten lens, Pinna ampla* (abundant), and *Echinobrissus scutatus* (abundant). Above come the shelly iron-stained and partially oolitic limestones of the Trigonia Beds, yielding *Natica, Cylindrites, Isocardia, Myacites decurtatus,* and large *Pinna.* Messrs. Blake and Hudleston note also *Ammonites plicatilis, Ceromya orbicularis (inflata), Anatina, Goniomya v.-scripta, Mytilus jurensis, Pygaster umbrella,* and *Acrosalenia decorata.*[†]

Still further on we cross a hollow suggestive of the Sandsfoot Clay, and come to a scarp of red rocks consisting of very ferruginous sands and sandstone, with iron-shot grains in the upper beds. A thickness of about 15 feet is seen, and the beds evidently represent the Sandsfoot Grits.[‡]

We have no clear sections of the Lower Corallian beds in the hills on the south and south-west of Abbotsbury. Oolites outcrop

---

[*] Quart. Journ. Geol. Soc., vol. xxxiii. p. 264.
[†] *Ibid.*, p. 269.
[‡] See also Blake and Hudleston, Quart. Journ. Geol. Soc., vol. xxxiii. p. 268.

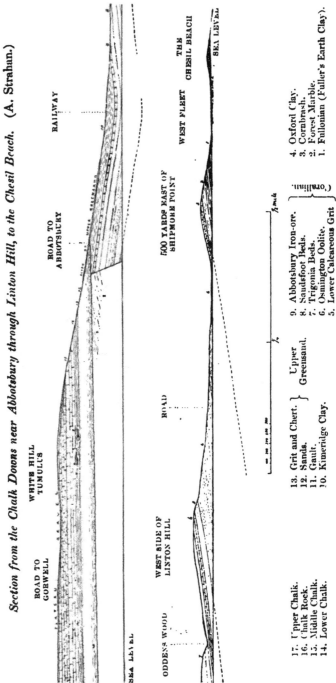

FIG. 52

Section from the Chalk Downs near Abbotsbury through Linton Hill, to the Chesil Beach. (A. Strahan.)

17. Upper Chalk.
16. Chalk Rock.
15. Middle Chalk.
14. Lower Chalk.

13. Grit and Chert. ⎫
12. Sands.   ⎬ Upper Greensand.
11. Gault.   ⎭
10. Kimeridge Clay.

9. Abbotsbury Iron-ore. ⎫
8. Sandsfoot Beds.   ⎬ Corallian.
7. Trigonia Beds.   ⎭
6. Osmington Oolite.
5. Lower Calcareous Grit

4. Oxford Clay.
3. Cornbrash.
2. Forest Marble.
1. Fullonian (Fuller's Earth Clay).

on the south of St. Catherine's Chapel. The lowest bed seen in the hill west of Abbotsbury was a ferruginous grit, and above there could be traced a series of oolitic beds, grey and occasionally iron-stained, together with bands of flaggy calcareous sandstone and clay. The higher portions of this hill are formed of the iron-stone, and the same beds appear to the north of St. Catherine's Chapel, dipping towards the village of Abbotsbury.

The occurrence of iron-ore at Abbotsbury attracted the notice of Sedgwick,[*] and it has been referred to by Buckland and De la Beche[†] and subsequent geologists. (See p. 74.) It occurs in the upper part of the Corallian Beds bordering the Kimeridge Clay. West of West Elworth Farm the ploughed fields present a very red appearance, and the scarps of Corallian rocks on the west side of Linton Hill are very much iron-stained. (See p. 306.)

The oolitic iron-ore is exposed in most of the lane-cuttings north and west of Abbotsbury village, on the high grounds above the Earl of Ilchester's mansion, and on the northern slope of the hill on which St. Catherine's Chapel stands. The western part of the village is built on the iron-ore. The best sections to be seen are in the red lanes north of the village. Here the rocks dip towards the village, which evidently lies in a syncline of the beds, as they rise again to the south. (See Fig. 52.)

The red lanes are excavated in loose red and brown oolitic iron-ore, with some sand and sandstone and irregular vertical and other bands of ironstone, and flaggy oolitic ironstone on top. A thickness of at least 20 feet is shown.[‡]

The more prevalent fossils include *Ammonites, Cardium, Myacites, Trigonia, Chemnitzia, Waldheimia, Rhynchonella,* and *Serpula. Ostrea deltoidea* and lignite also occur.

North of Abbotsbury the Oolites are much faulted, and the outcrops of many of the divisions are concealed by the overlapping of the Cretaceous Rocks. Corallian Beds reappear from beneath this covering of newer strata, in the Vale of Bride (or Bredy), at Long Bredy, and south of Litton Cheney. Near the last-named village, oolite was shown in the road-cutting, but the stone is not worked in the neighbourhood.

So faulted are the Oolites generally in this neighbourhood that it is impossible to indicate with any certainty their course beneath the Dorsetshire Downs from Long Bredy to Cerne Abbas. The outcrop at Abbotsbury has been shifted northwards to Long Bredy, and it is modified underground not only by east and west faults, but probably also by north and south faults, that date prior to the Cretaceous overlap. The main underground course of the Corallian Beds may be from below Dorchester to Cerne Abbas.

The following list of fossils from the Corallian Beds of Weymouth is based mainly on the work of Messrs. Blake and Hudleston, the species collected by myself being added. The vertebrate remains are not included, as their particular horizons are not as a rule noted, and it will be sufficient to refer to the general list in the Appendix for records of Fishes and Saurians.

---

* Ann. Phil., vol. xxvii. p. 350.
† Trans. Geol. Soc., ser. 2, vol. iv. p. 27.
‡ A further account of the iron-ore is given in the Chapter on Economic Products, p. 323.

LIST OF FOSSILS FROM THE CORALLIAN ROCKS, NEAR WEYMOUTH.

	Oxford Clay.	Nothe Grit.	Nothe Clay.	Bencliff Grit.	Osmington Oolite.	Trigonia Grit.	Sandsfoot Clay.	Sandsfoot Grit.	Upper Coral Rag.	Abbotsbury Iron Ore.	Kimeridge Clay.
	×	1	2	3	4	5	6	7	8	9	×
Ammonites achilles	×							7			
—— Berryeri									8		×
—— cordatus	×	1	2			5					
—— cymodoce, var.									8		
—— decipiens								7	8	9	×
—— Hector									8	9	
—— mutabilis									8	9	
—— perarmatus	×	1			4						
—— plicatilis						5	6		8		
—— pseudomutabilis									8		×
Nautilus hexagonus	×		2						8		×
Belemnites abbreviatus	×	1							8		
—— hastatus	×	1									
—— nitidus								7	8	9	×
—— sp.						5					
Alaria ? Deshayesea						5					
—— seminuda						5					
—— trifida	×		2								
—— sp.										9	
Bourguetia (Phasianella) Buvignieri					4						
—— (——) striata									8		
Cerithium limæforme					4						
—— muricatum	×	1				5					
—— septemplicatum							6				
Chemnitzia ferruginea										9	
—— heddingtonensis					4	5					
——. See also Pseudomelania.											
Cylindrites elongatus						5					
Littorina muricata		1									
—— var. near to pulcherrima						5	6	7	8		
Natica Clio						5					
——— Clytia					4	5					
—— corallina					4	5				9	
—— Eudora										9	

	Oxford Clay.	Nothe Grit.	Nothe Clay.	Bencliff Grit.	Osmington Oolite.	Trigonia Grit.	Sandsfoot Clay.	Sandsfoot Grit.	Upper Coral Rag.	Abbotsbury Iron Ore.	Kimeridge Clay
	×	1	2	3	4	5	6	7	8	9	×
Nerinæa Desvoidyi	–	–	–		–	5	–	–	–	–	–
—— Goodhalli	–	–	–		–	5	–	–	–	–	–
—— sp.	–	–	–		–	–	–	–	–	9	–
Pleurotomaria Münsteri	–	·1	–		←	5	–	7	–	–	–
—— reticulata	×	–	–	–	–	5	6	–	8	9	×
Pseudomelania Delia	–	–	–	–	–	–	–	–	–	9	–
—— pseudolimbata	–	–	–	–	–	–	–	–	–	9	–
——. See also Chemnitzia.											
Turbo exiguus	–	–	–	–	–	–	6	–	–	–	–
Turritella jurassica	–	–	–	–	–	5	–	–	–	–	–
Dentalium cinctum	–	–	–	–	–	5	–	–	–	–	–
Anatina	–	–	–	–	–	5	–	–	–	–	–
Anomia radiata	–	1	2	–	–	–	–	–	–	–	–
Arca æmula	×	–	2	–	–	–	–	–	–	–	–
—— sublata	–	–	–	–	–	–	–	–	–	9	–
—— sp.	–	–	–	–	–	–	6	–	–	–	–
Arcomya	–	–	–	–	–	5	–	–	–	–	–
Astarte extensa	–	–	–	–	–	5	–	–	–	–	–
—— ovata	×	–	–	–	–	5	–	7	8	–	–
—— polymorpha	–	–	–	–	–	5	–	7	–	–	–
—— supracorallina	–	–	–	–	–	–	6	7	–	–	×
Avicula ædiligensis	–	–	–	–	–	–	–	7	–	–	–
—— ovalis	×	1	–	–	–	–	–	–	–	–	–
—— pteropernoides	–	–	–	–	–	5	6	–	–	–	–
—— Struckmanni	–	–	–	–	–	5	–	–	–	–	–
Cardium cyreniforme	–	–	–	–	–	5	–	–	–	–	–
—— delibatum	–	–	–	–	–	–	6	7	–	9	–
Ceromya excentrica	×	–	–	–	–	5	–	–	–	–	–
—— orbicularis	–	–	–	–	–	5	–	–	–	–	–
Corbula Deshayesea	–	–	–	–	–	–	6	–	–	–	–
Cucullæa contracta	–	–	2	–	4	–	–	–	–	–	–
—— corallina	–	–	–	3	–	5	6	–	–	–	–
—— superba	–	–	–	–	–	–	6	–	–	–	–
Cypricardia glabra	–	–	–	–	–	5	–	–	–	–	–
Cyprina tancrediformis	–	1	–	–	–	–	–	–	–	–	–
Exogyra nana	×	–	2	3	–	5	–	7	8	–	–
—— spiralis	–	1	–	–	–	–	–	–	–	–	×

	Oxford Clay.	Nothe Grit.	Nothe Clay.	Bencliff Grit.	Osmington Oolite.	Trigonia Grit.	Sandsfoot Clay.	Sandsfoot Grit.	Upper Coral Rag.	Abbotsbury Iron Ore.	Kimeridge Clay.
	×	1	2	3	4	5	6	7	8	9	×
Exogyra virgula	—	—	—	—	—	—	—	—	—	9	×
Gervillia aviculoides	—	\	—	—	3	—	5	—	—	—	—
Goniomya literata	×	—	—	—	—	—	—	7	—	—	—
— marginata	—	—	—	—	—	—	6	—	—	—	—
— v. scripta	×	—	—	—	—	5	—	7	—	—	—
Gresslya peregrina	×	—	—	—	—	—	—	7	—	—	—
Gryphaea dilatata	×	1	2	—	—	—	—	—	—	—	—
Hinnites abjectus	×	1	—	—	—	—	—	—	—	—	—
— tumidus	×	—	—	—	—	5	6	7	—	—	—
Inoceramus	—	—	—	—	—	—	—	—	—	9	—
Lima elliptica	—	—	2	—	—	5	—	—	—	—	—
— pectiniformis	×	—	—	—	—	—	—	/	8	—	—
— rigida	×	—	—	—	—	5	—	7	8	—	—
— subantiquata	—	—	—	—	4	—	—	—	—	—	—
Lucina aliena	—	—	—	—	—	5	—	—	—	—	—
— moreana	—	—	—	—	4	—	—	—	—	—	—
— substriata	—	—	—	—	—	—	—	7	—	—	—
— sp.	—	1	—	3	—	—	—	—	—	—	—
Modiola bipartita	×	—	2	—	—	—	—	—	8	—	—
— imbricata	×	—	—	—	—	—	—	—	8	—	—
— subaequiplicata	—	—	—	—	—	5	—	—	—	9	—
Myacites decurtatus	×	1	—	—	—	5	—	—	—	—	—
— jurassi	×	1	—	—	—	5	—	—	—	—	—
— recurvus	×	—	—	—	—	—	—	—	8	—	—
— sp.	—	—	—	—	4	—	—	7	—	9	—
Myoconcha	—	—	—	—	—	—	—	—	8	—	—
Mytilus jurensis	—	—	—	—	—	5	—	—	—	—	×
— pectinatus	×	—	—	—	—	5	—	—	—	—	×
Nucula Menkei	×	—	—	—	—	—	6	—	—	—	×
— nuda	×	—	—	—	—	5	—	—	—	—	—
Opis corallina	—	—	—	—	—	5	—	—	8	—	×
— Phillipsi	—	1	—	—	4	5	—	—	—	—	—
Ostrea deltoidea	—	—	—	—	—	5	6	7	—	—	×
— duriuscula	—	—	—	—	—	—	—	7	—	—	×
— gregaria	—	1	2	—	—	—	—	—	8	—	v
— solitaria	—	—	—	—	—	—	—	—	—	—	—
Pecten articulatus	\	—	—	—	—	—	6	—	8	—	—

	Oxford Clay.	Nothe Grit.	Nothe Clay.	Bencliff Gri .t	Osmington Ooli e.	Trigonia Grit.	Sandsfoot Clay.	Sandsfoot Grit.	Up er Coral Rag.	Abbotsbury Iron Ore.	Kimeridge Clay.
	×	1	2	3	4	5	6	7	8	9	×
Pecten distriatus	-	-	-	-	-	-	-	7	-	-	-
—— fibrosus	×	1	2	-	4	5	-	-	-	-	-
—— intertextus	×	-	-	-	-	5	-	-	-	-	-
—— lens	×	-	-	-	-	-	6	-	-	-	×
—— midas	-	-	-	-	-	-	6	7	-	9	-
—— qualicosta	-	-	-	3	4	5	-	-	-	-	-
—— virdunensis	-	-	-	-	-	5	-	-	-	-	-
Perna mytiloides	×	-	-	-	-	-	6	-	-	-	×
—— quadrata	×	1	2	-	-	5	-	-	-	-	-
Pholadomya aequalis	-	1	-	-	-	-	6	-	-	-	×
—— concinna	-	1	-	-	-	-	-	-	-	-	-
—— decemcostata	-	-	-	-	-	5	-	-	-	-	-
—— hemicardia	-	-	-	-	-	-	6	7	-	-	-
—— paucicosta	-	-	-	-	-	5	-	-	8	-	-
Pinna ampla	×	-	-	-	4	-	-	-	-	-	-
—— granulata	-	-	-	-	-	5	-	-	-	-	-
—— pesolina	-	-	-	-	-	-	6	7	-	9	-
Pleuromya donacina	-	-	-	-	-	-	-	-	-	9	×
—— tellina	-	-	-	-	-	-	-	7	-	9	×
—— Voltzi	-	-	-	-	-	-	-	-	-	9	-
Plicatula fistulosa	×	-	-	-	-	5	-	-	-	-	-
—— semiarmata	-	-	-	-	-	5	-	-	-	-	-
Protocardium isocardioides	-	-	-	-	-	5	-	-	-	-	-
Sowerbya Deshayesea	-	-	-	-	-	5	-	-	-	-	-
—— triangularis	×	-	-	-	4	5	-	-	-	-	-
Taucredia curtansata	×	-	-	-	4	-	-	-	-	-	-
—— disputata	-	-	-	-	4	-	-	-	-	-	-
—— planata	-	-	-	-	4	-	-	-	-	-	-
Thracia depressa	×	-	2	-	-	-	6	7	-	-	×
Trigonia clavellata	-	-	-	-	4	5	6	-	-	-	-
—— corallina	-	-	-	3	-	-	-	-	-	-	-
—— Hudlestoni	-	-	-	-	4	-	-	-	-	-	-
—— Meriani	-	-	-	-	-	5	-	-	-	-	×
—— monilifera	-	-	-	-	-	5	-	-	-	9	×
—— muricata	-	-	-	-	-	-	-	7	-	-	×
—— perlata	-	1	2	-	-	-	-	-	-	-	-
Unicardium sulcatum	-	-	-	-	-	5	-	-	-	-	×

	Oxford Clay.	Nothe Grit.	Nothe Clay.	Bencliff Grit.	Osmington Oolite.	Trigonia Grit.	Sandsfoot Clay.	Sandsfoot Grit.	Upper Coral Rag.	Abbotsbury Iron Ore.	Kimeridge Clay.
	×	1	2	3	4	5	6	7	8	9	×
Unicardium sp.			–	–	–	–	–	–	–	8	
Discina humphriesiana		–	–	–	–			7	–		×
Lingula ovalis		–	–	–	–			7	–		×
Rhynchonella corallina			–	–	–	–	–	–	–	9	–
—— inconstans				–	–	–			8	9	×
Terebratula insignis	×		–	–	–	–	–		–	9	
—— subsella			–	–	–				–	9	
Waldheimia dorsetensis			–	–	–				–	9	–
—— lampas				–	–	–		–	–	9	–
Glyphea ferruginea			–	–	–					9	–
Serpula gordialis			–	–	–	–			–		9	–
—— intestinalis			–	–	–	–			8		
—— Royeri			–	–	–	–			–	9	
—— sulcata		1	–	–	–			7	–		
—— tetragona	×	–	–	–	–			7	–		
—— tricarinata	×	1	2	–	–		–		8	–	
—— variabilis		–	–	–	–		–	–	8	–	×
Millericrinus echinatus		1	–	–	–	–	–		–		
Acrosalenia decorata				–		5	–	–			
Cidaris florigemma				–		5		7	8	–	
—— Smithi			–	–							
Echinobrissus scutatus			–	–	4	5		7		9	
Hemicidaris intermedia			–	–		5	–	–	–		
Pseudodiadema versipora			–	–		5	–	–	–		–
Pygaster umbrella				–	–	–	5	–			–
Comoseris irradians			–	–		5	–	–	–	–	
Protoseris Waltoni			–			5	–	–	–		–
Thamnastræa arachnoides			–	–		5	–	–	–		
—— concinna				–		5	–	–	8		
Thecosmilia annularis				–		5	–	–	8	–	

Vale of Blackmore.

On crossing the Dorsetshire Downs to the north of Cerne
Abbas we enter the Vale of Blackmore, and there the Corallian
Beds re-appear from beneath the thick mantle of Cretaceous
Rocks. They are seen north-east of Hilfield and at Glanvilles
Wooton, over a tract of grassy country with small enclosures and
well-timbered hedgerows. Proceeding towards Sturminster Newton, the summit of the low escarpment forms an undulating tract
of cultivated land, and near Haselbury Bryant we find quarries,
where the stone is burnt for lime and employed for building and
road-mending. There are quarries north-east of the church, near
Shorts Barn, and south-east of Zoar, and we find blue shelly and
oolitic limestones and marls, with occasional sandy layers. The
beds are pisolitic in places, and they belong to the upper portion
of the limestone-series, so well shown in the railway-cutting at
Sturminster Newton (p. 101). Thus all the strata observed, may
be grouped with the Trigonia Beds and the Osmington Oolite. I
obtained the following fossils at Haselbury Bryant :—

Belemnites abbreviatus.	Ostrea gregaria.
Natica corallina.	Pecten fibrosus.
Bourguetia (Phasianella).	Trigonia clavellata.
Pleurotomaria.	Echinobrissus scutatus.
Astarte.	Glyphea.
Myacites.	

Gasteropods are fairly abundant and fragments of lignite are not
uncommon. Other species from this locality and Mappowder,
are recorded by Messrs. Blake and Hudleston: from the latter
place they note *Rhynchonella Thurmanni*, a fossil unknown
elsewhere at so high an horizon.* (See Fig. 53.)

FIG. 53.

Rhynchonella varians, *var.* Thurmanni, *Volts*, ⅓.

The best section of the Corallian Beds in the neighbourhood is
that shown in the railway-cutting at Sturminster Newton. The
section was described by Messrs. Blake and Hudleston, and the
strata which I noted were as follows :—

		Ft.	In.
Sandsfoot Beds	6. Brown ochreous sandy loams and clays - - - -	[14	0]
	5. Bluish-grey clay; *Ostrea* - -	[8	6]

* Quart. Journ. Geol. Soc., vol. xxxiii. p. 281.

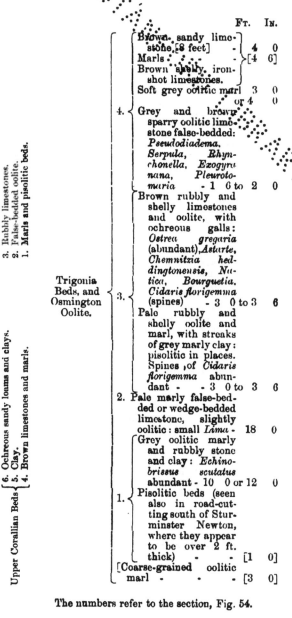

FIG. 54.—*Section in the Railway-cutting, Sturminster Newton, Dorset.*

3. Rubbly limestones.
2. False-bedded oolite.
1. Marls and pisolitic beds.

Upper Corallian Beds
{
6. Ochreous sandy loams and clays.
5. Clay.
4. Brown limestones and marls.
}

	Ft.	In.
Brown sandy lime-stone, [8 feet] -	4	0
Marls - - - } [4	6]
Brown shelly iron-shot limestones.		
Soft grey oolitic marl	3	0
or 4	0

4. { Grey and brown sparry oolitic lime-stone false-bedded: *Pseudodiadema, Serpula, Rhynchonella, Exogyra nana, Pleurotomaria* - 1 6 to 2 0

Brown rubbly and shelly limestones and oolite, with ochreous galls: *Ostrea gregaria* (abundant), *Astarte, Chemnitzia heddingtonensis, Natica, Bourguetia, Cidaris florigemma* (spines) - 3 0 to 3 6

Trigonia Beds, and Osmington Oolite.

3. { Pale rubbly and shelly oolite and marl, with streaks of grey marly clay: pisolitic in places. Spines of *Cidaris florigemma* abundant - - 3 0 to 3 6

2. Pale marly false-bedded or wedge-bedded limestone, slightly oolitic: small *Lima* - 18 0

1. { Grey oolitic marly and rubbly stone and clay: *Echinobrissus scutatus* abundant - 10 0 or 12 0

Pisolitic beds (seen also in road-cutting south of Sturminster Newton, where they appear to be over 2 ft. thick) - - [1 0]

[Coarse-grained oolitic marl - - - [3 0]

The numbers refer to the section, Fig. 54.

The thicknesses in square brackets are given on the authority of Messrs. Blake and Hudleston.*

* Quart. Journ. Geol. Soc., vol. xxxiii. p. 276.

The limestone-beds of Sturminster Newton have yielded the following fossils :—

× Ammonites plicatilis.
× Actæonina.
Alaria ? Deshayesea.
Bourguetia Bavignieri.
Ceritella.
Cerithium inornatum.
× —— limæforme.
—— muricatum.
× Chemnitzia (Pseudomelania) heddingtonensis.
Delphinula funiculata.
× Littorina muricata.
× Natica corallina.
—— dejanira.
Nerinæa.
× Pleurotomaria Agassizi.
—— reticulata.
× Arca.
Astarte aliena.
—— polymorpha.
Cypricardia glabra.
× Exogyra nana.
× Lima elliptica.
—— subantiquata.

Lucina.
Modiola.
Myacites securiformis.
× Opis corallina.
Ostrea gregaria.
× —— solitaria.
Pecten strictus.
× —— subtextularis.
Perna mytiloides.
× Pholadomya æqualis.
Quenstedtia lævigata.
Sowerbya triangularis.
Trigonia clavellata.
× Unicardium.
× Rhynchonella.
× Serpula.
Acrosalenia decorata.
× —— spinosa.
× Cidaris florigemma.
× Echinobrissus scutatus.
Hemicidaris intermedia.
× Pseudodiadema radiatum.
—— versipora.

The species marked × were collected by Mr. J. Rhodes and myself; but many of them had been previously recorded by Messrs. Blake and Hudleston, on whose authority the remaining species are given.

The uppermost beds of the Corallian Series were shown in a lane-cutting near Hole leading to Banbury Common, near Sturminster Newton. There the following sequence could be traced :—

			Ft. In.
Kimeridge Clay	{ Clay.		
	Sandy layer.		
	Clay with selenite: *Ostrea deltoidea*	-	3 0
	Nodular bed.		
Upper Corallian.	Sandsfoot Beds, &c.	{ Ferruginous beds like the oolitic iron-ore of Abbotsbury and Westbury, with band of blue-hearted sandy limestone: *Ostrea deltoidea, Pecten fibrosus, Belemnites* - }	10 or 15 feet.
		Ochreous clays : *Belemnites abbreviatus*, &c.	
	Oolitic Beds, &c.		

Messrs. Blake and Hudleston have referred to these "ferruginous fossiliferous sands and concretions," and record from them the following fossils :—*

Ammonites plicatilis ?
Belemnites nitidus.
Avicula ædiligensis.
Pecten midas.

Pinna pesolina.
Pleuromya tellina.
Serpula runcinata.

It seemed to me that the ferruginous beds would be worth opening up more clearly, so that their economic value might be tested. The general section in this area compares well with that

---

at Abbotsbury, and it is of special interest to find the ironstone represented, for we again meet with it at Westbury in Wiltshire.

The Pisolitic Beds were well shown in a bank by the road-side near the Castle south of Sturminster Newton. There they consist of small bean-shaped concretions in a marly matrix. (See p. 73.) The occurrence of this pisolitic band was marked by H. W. Bristow on the Geological Survey Map to the west of Haselbury Bryant and again near Todbere. Further north, as noted by Messrs. Blake and Hudleston, it "spreads out on the surface at Stower Provost, the village square being entirely upon it,"* and I have seen the bed between East and West Stower.

We have no sections as yet showing the junction with the Oxford Clay, this main pisolitic bed being the lowest bed visible.

Of the Nothe Grits or their equivalents we see nothing, and it is possible that this division is here represented chiefly by clays that have been included with the upper part of the Oxford Clay.

A small inlier of Corallian Rocks was at one time thought to occur to the west of Fontmell Magna, south of Shaftesbury; but the beds have been 'recognized by Mr. Jukes-Browne as Lower Greensand.

In the neighbourhood of Marnhull and Todbere the Corallian limestones develop into a good freestone which is largely quarried for building-purposes and lime-burning.

East of Marnhull Church there are quarries showing about 15 feet of buff marly oolite, that is well-bedded in layers from 1 ft. to 18 ins. thick, with marly partings. The stone is shelly in places, and when quarried it hardens and turns white on exposure. The rubbly oolite on top is used for road-metal.

A well sunk to a depth of 30 feet at the Brewery, north of Walton Elm, did not yield sufficient water for the wants. At East Stower (centre of village) water is obtained at a depth of 35 feet.

At Todbere we have fine sections, which show a remarkably false-bedded set of beds below the more regular strata of oolite just mentioned, similar to features seen in the railway-cutting near Sturminster Newton, and again in a quarry near Steeple Ashton.

There is a large quarry east of the high road at Todbere, and this showed the following section:—

	Ft.	In.
3. Rubbly oolite, shelly in places	6	0
Marly layer, a few inches, passing into oolite.		
2. Buff flaggy oolitic limestone, shelly in places: good building-stone	5	0
Hard blue shelly oolitic limestone: weathering brown	2	6
Rotten oolite and marl	0	6
1. False-bedded oolites, yielding good building-stone - 7 0 to	9	0

Coralline Oolite {

A good section of the beds has been given by Messrs. Blake and Hudleston.† This shows the even layers resting on the obliquely bedded

* Quart. Journ. Geol. Soc., vol. xxxiii. p. 279.
† Ibid., vol. xxxiii. p. 280.

freestones, and when the former have been quarried away, the lower beds might be taken to have a dip of 8° to 10° in an easterly direction. The full thickness of this lower freestone appears to be 14 feet.

FIG. 55.

Section of Corallian Beds at Todbere, Dorset.

W.

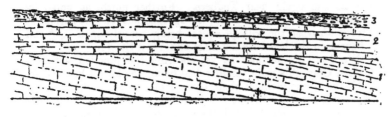

		FT.
3	Soil and brown rubbly limestone and marl - - -	2 to 6
2.	Oolitic limestones, shelly in places - -	8
1.	False-bedded oolite (the layers inclined 8° or 10° E.) -	9

The remarkable false-bedded oolite, overlaid by even layers, suggests, as Messrs. Blake and Hudleston remark, "that some time elapsed between the deposition of the two." Similar features are not uncommon in the Great Oolite and Forest Marble ;* and they may be attributed to changes in currents, accompanied in some cases by contemporaneous erosion.

From the oolites of this locality Messrs. Blake and Hudleston record the following species, which are to be found only in occasional bands :—

Ammonites plicatilis, 2.
Chemnitzia heddingtonensis, 1, 2.
Natica, 2.
Nerinæa fasciata, 1.
Exogyra nana, 1.
Ostrea solitaria, 2.

Pleuromya tellina, 1.
Trigonia clavellata, 1.
—— Meriani, 2.
Cidaris florigemma, 2.
Echinobrissus scutatus, 1, 2.

The numbers refer to 1, false-bedded freestone, and 2, the overlying oolites.

The stone beds are again quarried to the west of East Stower.

Lower Calcareous Grit was not traced south of Sturminster Newton by H. W. Bristow, and there is but feeble evidence of it even in this part of Dorsetshire. Proceeding northwards the first good indications of the beds are in the railway-cuttings on either side of the Kingsmead tunnel, between Gillingham and Templecombe, on the London and South-Western Railway. The cuttings, at the time of my visit, were too obscure to afford any detailed measurements, but on the eastern side of the tunnel, beneath rubbly beds of oolite and some thickness of clay, oolitic marl, and blue flaggy limestone, grey calcareous sandstone and sands were shown. On the western side of the tunnel, fissile calcareous grits with *Gryphœa dilatata* and *Pecten* were to be seen

---

* See Memoir on the Lower Oolitic Rocks of England, pp. 306, 370.

resting on the Oxford Clay. Messrs. Blake and Hudleston have given a detailed account of the beds, and they notice the presence of immense spheroidal doggers of calcareous grit. Probably if we assign a thickness of 22 feet to these beds of Lower Calcareous Grit, we shall not be far wrong, for I believe there is some repetition of the beds in the cutting.

The overlying Pisolitic beds, comprising about 12 feet of coarse grained oolite, pisolite, and marl, were noticed at Cucklington by Messrs. Blake and Hudleston.* The beds yielded *Ammonites plicatilis, Turbo, Lima rigida, Pecten fibrosus, Myacites decurtatus, Echinobrissus scutatus,* Corals and a Sponge.

Higher beds are seen in quarries by Langham Farm, and to the south-east of Langham, west of Gillingham. Here we find beds of shelly and oolitic limestone (blue-hearted)— irregular rubbly limestones, with bands of sandy and oolitic marl and clay, exposed to a depth of about 11 feet. The fossils include *Ammonites, Chemnitzia heddingtonensis, Natica corallina, Pleurotomaria, Exogyra, Myacites, Ostrea, Pholadomya, Trigonia,* and *Echinobrissus scutatus.*

Messrs. Blake and Hudleston note the following fossils from gritty limestone, marl, and grit, that were exposed at Langham :— *Am. cordatus, Belemnites abbreviatus, Natica clio, Bourguetia (Phasianella) striata, Gervillia aviculoides, Modiola,* and *Perna quadrata.* They include the beds with the Upper Corallian, remarking, however, that the Cephalopoda elsewhere belong generally to the lower beds.

Near Preston, north-east of Gillingham, they notice beds of sandy and flaggy oolite, about 12 feet thick, and with remains of Algæ. These beds yield fossils found also at Langham, and they are overlaid by sandy and oolitic shell-beds, which yield species of *Stylina* and *Thamnastræa.* From these uppermost beds Messrs. Blake and Hudleston record also the following species :—†

Ammonites plicatilis.	Cucullæa corallina.
Actæonina miliola.	Gastrochæna carinata.
Cerithium muricatum.	Pecten lens.
—— Pellati.	—— qualicosta.
Natica clio.	Tancredia planata.
Neridomus (Nerita) minuta.	Trigonia (clavellate).
Astarte polymorpha.	

The Lower Corallian Beds were exposed in a pit north of Higher Hatherly, south-west of Wincanton, where the section was as follows :—

FT.

Lower Corallian { Sands with concretionary masses of stone · · · · } 15 to 20
{ Sandy clay: *Ostrea flabelloides, Myacites recurvus, Gryphæa dilatata, Modiola bipartita, Ammonites cordatus* · }

Oxford Clay - Bluish clay with *Gryphæa dilatata.*

---

* Quart. Journ. Geol. Soc., vol. xxxiii. pp. 278, 279.
† *Ibid.,* vol. xxxiii. pp. 282, 283.

H 2

Comparing the Corallian Beds of North and South Dorset, we may roughly correlate them as follows (see Fig. 56):—

		FT.
South Dorset.	*North Dorset.*	
Abbotsbury Iron-ore.	Ferruginous Beds near Sturminster Newton -	15
Sandsfoot Beds.	Clays, &c. - - -	22
Trigonia Beds.	Sandy, shelly, and oolitic limestones - -	
Osmington Oolite.	Marnhull and Todbere freestone, and pisolitic beds -	58
Bencliff Grit.	{ Clays and sands with doggers (observed locally) -	
Nothe Beds.		25

In South Dorset the full thickness was found to be about 200 feet; in North Dorset it does not exceed 120 feet; of which 25 feet may be assigned to the Lower Corallian, and 95 feet to the Upper Corallian.

Longleat.

North of the great fault marked on the Geological Survey Map between Wincanton and Mere, we journey some distance before meeting with the Corallian Rocks. Here the Kimeridge Clay is represented to rest directly on the Oxford Clay, but the evidence is by no means satisfactory. Over portions of this area Mr. W. T. Aveline stated that it was not altogether clear whether the newer clay may not be Gault;[*] and I found the slopes bordering the escarpment of the Upper Greensand near Stourton, so obscured by talus that no evidence was to be obtained. Mr. Jukes-Browne, however, informed me (1889) that he found Upper Corallian Beds, in the form of shelly and partially oolitic limestones, in Longleat Park, and that the clay above it is undoubtedly Gault. The Lower Corallian Beds are not clearly exposed, while if any representative of the Westbury iron-ore be present, it is overlapped by the Gault.

Westbury.

Most important sections of the Corallian Rocks are to be seen at the Ironworks near the railway-station at Westbury in Wiltshire. There the Upper Corallian Beds yield a valuable iron-ore to which attention was directed in 1856 by Mr. G. C. Greenwell.[†] The workings extend from a little north-west of Penleigh to Westbury Field, a distance of rather more than one mile, and the ironstone is exposed in the tract north of Westbury Station, further eastwards than is shown on the Geological Survey Map. How far to the south the beds may occur in workable quantities can only be proved by trial-borings. Northwards the outcrop of the beds is shifted by a fault, but Mr. Greenwell states that the

---

[*] See Geol. E. Somerset, &c., p. 137.
[†] Proc. S. Wales Inst. Civ. Eng., 1859, vol. i. pp. 311, &c.

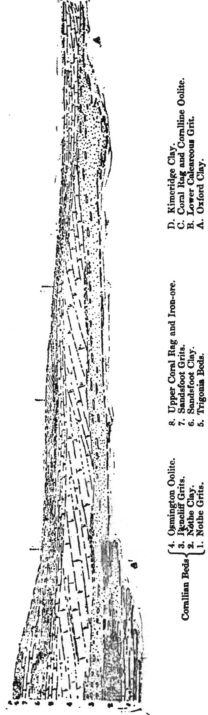

Fig. 56.

Diagram-section to show the general structure of the Corallian Beds from Weymouth to Oxford.

Distance about 100 miles.

Weymouth. Sturminster Newton. Westbury. Calne. Highworth. Faringdon. Oxford.

Corallian Beds
4. Osmington Oolite.
3. Kyncliff Grits.
2. Nothe Clay.
1. Nothe Grits.

8. Upper Coral Rag and Iron-ore.
7. Sandsfoot Grits.
6. Sandsfoot Clay.
5. Trigonia Beds.

D. Kimeridge Clay.
C. Coral Rag and Cornline Oolite.
B. Lower Calcareous Grit.
A. Oxford Clay.

ironstone was " found to extend to Steeple Ashton and the
neighbourhood. It is apparently of similar quality to that found
at Westbury, but is, so far as I have seen, much thinner and
somewhat uncertain in position."

The Ironstone dips 4° to 10° in a south-easterly direction
beneath the Kimeridge Clay, but as observed by Mr. Greenwell
" A short distance west of the outcrop of the upper limestone beds,
the strata assume a rapid westward dip, by which the ironstone
is again thrown in ; but about 20 or 30 yards further west, a pit
was sunk 20 feet in Oxford Clay, showing the existence of a large
upcast fault to the west, in fact the same fault shown on the
Ordnance Geological Map. The throw of this fault cannot be
less than 100 feet."

Messrs. Blake and Hudleston state that " A well sunk about
halfway between the town and the railway station, after passing
through about 50 feet of the Kimeridge Clay, showed that the
ore had diminished to a thickness of 2 feet."[*]

The following series may be clearly traced out in the several
workings :—

		FT.	IN.
Soil and Super-ficial Deposits (local).	Brown loamy clay and grey clay, with bits of chalk and flint, and fragments of *Ostrea deltoidea* in places, resting irregularly on bed beneath -	2	6
	Brown sand and fine gravel with flint and limestone pebbles - 3 0 to	5	0
	Brown and yellow clay with " race " oolitic ironstone and ochreous débris about	2	0
Kimeridge Clay	Dark blue and bluish-green clay with *Ostrea deltoidea* - - -	4	0
Upper Corallian	Hard greenish-coloured oolitic iron-stone with *O. deltoidea*, *Myacites*, *Pecten* and *Ammonites* - -	3	0
	Irregular masses of oolitic iron-ore with hard slightly calcareous bands, in-termingled with dark greenish and slate-coloured clay and black earthy powder : iron-pyrites, lignite, and occasional veins of calc-spar : *Ostrea deltoidea*, *Belemnites*, &c. - 7 0 to	9	0
	Light greenish-grey buff and brown sands with seams of laminated clay, casts of *Myacites*, &c. ; *Ostrea gregaria* at base - - - 4 0 to	10	0
	Rubbly beds of blue and buff marly oolite and pisolite, passing into marly stone with layers of oolitic marl and clay. The bottom beds false-bedded in places : *Ammonites*, *Chemnitzia*, *Cidaris florigemma* (spines), *Natica*, corals, and other fossils - 10 0 to	12	0
	Fine false-bedded oolite, buff at top and blue towards base, with tubiform markings at various horizons : *Echino-brissus scutatus* - - 8 0 to	10	0

---

[*] Quart. Journ. Geol. Soc., vol. xxxiii. p. 284.

FIG. 57.

Section at the Iron-works, Westbury, Wiltshire.

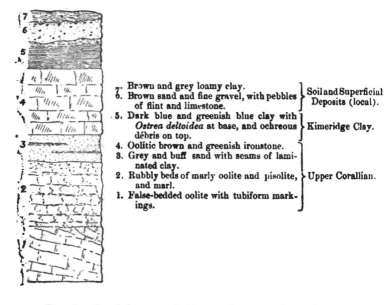

7. Brown and grey loamy clay.
6. Brown sand and fine gravel, with pebbles of flint and limestone. } Soil and Superficial Deposits (local).

5. Dark blue and greenish blue clay with *Ostrea deltoidea* at base, and ochreous débris on top. } Kimeridge Clay.

4. Oolitic brown and greenish ironstone.
3. Grey and buff sand with seams of laminated clay.
2. Rubbly beds of marly oolite and pisolite, and marl.
1. False-bedded oolite with tubiform markings. } Upper Corallian.

The details of the superficial covering vary from place to place. In one spot the blue clay overlying the Ironstone contains ferruginous layers. Where this capping of clay is absent and the Ironstone comes to the surface or is merely covered by sand and gravel, the ore is reddish-brown and rusty and much disintegrated. On the whole it is well-bedded, presenting either a reddish-brown aspect resembling the Abbotsbury iron-ore, or a dark grey or greenish appearance with dark clayey galls, and with rusty joints. Its maximum thickness is noted as 14 feet by Messrs. Blake and Hudleston. The layers of shelly stone with *Ostrea deltoidea* are thrown aside as useless.

The calc-spar that fills oblique veins in the Ironstone probably marks slight disturbances, and the material may have been derived from the calcareous matter in the overlying sands and limestone-gravel.

Of the fossils we miss the Brachiopoda, so abundant at Abbotsbury, but large Ammonites are not uncommon. In the Devizes Museum there are specimens over 1 foot in diameter. The following is a list of the fossils obtained from the Ironstone :—

× Ammonites Berryeri.
× —— decipiens.
—— plicatilis.
× —— pseudocordatus.
Belemnites.
Bourguetia (Phasianella) striata.

× Cardium delibatum.
Lima læviuscula.
—— pectiniformis.
Myacites recurvus.
—— securiformis ?
× Ostrea deltoidea.

× Pecten distriatus.
—— fibrosus.
× —— lens.
× —— midas.
× Perna quadrata.

Pholadomya æqualis.
× —— hemicardia.
Unicardium.
× Serpula.
Lignite.

The species marked × were recorded by Messrs. Blake and Hudleston[*] ; the other species were collected by myself.

The Limestone-beds yielded the following fossils :—

Ammonites plicatilis.
Bourguetia (Phasianella).
Cerithium muricatum.
Chemnitzia (Pseudomelania) heddingtonensis.
—— (——) gigantea.
Natica corallina.
Nerinæa.
Astarte ovata.
Exogyra nana.
Isocardia.

Lucina.
Ostrea gregaria.
—— solitaria.
Pecten fibrosus.
—— qualicosta.
—— vagans.
Cidaris florigemma.
Echinobrissus scutatus.
Pseudodiadema.
Corals.

The pisolitic matter was in many instances accumulated around univalves and bivalves.

The limestone is quarried for use as a flux in the furnaces; formerly the material was obtained from the Great Oolite of Bradford-on-Avon, so that when Messrs. Blake and Hudleston examined the area, there was but a poor exposure of the upper portion of the limestones.

FIG. 58.

Section north of the Railway-station, Westbury, Wiltshire.

S.E.   N.W.

6. Kimeridge Clay.
5. Ironstone, with *Ostrea deltoidea*, &c.
4. Ferruginous loam and sand.
3. Oolite with streaks of clay and pisolite. The upperm st portion, 1 ft., mainly oolitic débris almost an iron-ore, with spines of *Cidaris, Exogyra, Ostrea solitaria,* &c.
2. False-bedded oolite (not exposed).
1. Oxford Clay.

The above section was seen north of the lane to Storridge Farm, in a deep trench that was exposed in 1885, and it confirms the account given by Mr. Greenwell, as previously mentioned.

The Limestone-beds here rise in an anticline shown north-west of Westbury Station. Westwards the beds are faulted against the Oxford Clay, so that we do not see any exhibition of the Lower Calcareous Grit, but these beds were proved in a well (as noted by Mr. Greenwell), which was sunk at the Iron-works :—

---

[*] Quart. Journ. Geol. Soc., vol. xxxiii. p. 285; see also Proc. Geol. Assoc., vol. xiii. p. 138.

		Ft.	In.
8. Soil - - - - -		2	6
Upper Corallian { 7. Ironstone - - - -		12	0
6. Greenish blue and brown sand -		6	0
5. Blue pisolitic limestone, with beds of marl and sand - - -		52	8
Lower Corallian (Lower Calcareous Grit). { 4. Alternations of blue rock and sand -		7	0
3. Loamy blue sand with large hard concretion - - -		14	6
2. Blue marl - - - -		8	10
1. Rock and sand (20 gallons of water per minute) - - -		2	0

It is not clear that the base of the Corallian beds was reached, but the Oxford Clay would not be far below. Indeed we have the record of a well-sinking, noted by Messrs. Blake and Hudleston, who state that beneath 40 feet of oolitic limestone and stiff marl, "Loose sands, containing four or five beds of rocks about 1 ft. thick each" were proved to a depth of 50 feet; and " The well left off in dark hard rock, from which the water flowed abundantly."[*]

Steeple Ashton and Seend.

The beds in Wiltshire were first described in detail by Lonsdale, whose observations were mainly directed to the sections seen in the country extending from Steeple Ashton to the neighbourhood of Calne. Over this area he made an original geological survey of the Oolitic rocks on the one-inch Ordnance Maps.[†]

In the neighbourhood of Steeple Ashton we come to one of the most famous of the localities for the "Coral Rag."

The Lower Calcareous Grit was well shown in a pit west of the Seend Iron-works, where sand is obtained for moulding, for mortar, and for brick-works. The general aspect of the beds reminded me of sections in the Norwich Crag Series. The section showed about 12 feet of white sand, stained yellow and brown by iron-oxide, with seams of brown clay, impersistent and broken up, showing slight disturbances in the strata. Some portions of the sand were shelly, with fragments of *Pecten*, *Ostrea*, and *Exogyra nana*. Here and there the sand was cemented into hard stones, like greywethers. These lower beds have been described by Lonsdale, who notes a quarry at the foot of the hill, and on the south-east side of the Trowbridge road, showing 25 feet of the strata.[‡]

In the Devizes Museum there are specimens of *Ammonites cordatus*, *A. perarmatus*, *A. Sutherlandiæ*, and *A. vertebralis*, from the Lower Calcareous Grit of this locality. Other species, noted by Messrs. Blake and Hudleston, are as follows :—[§]

Nautilus hexagonus.	Arca æmula.
Cerithium muricatum.	Astarte depressa.
Littorina (Turbo) Meriani.	Avicula ovalis.
Pleurotomaria Münsteri.	—— inæquivalvis.

* Quart. Journ. Geol. Soc., vol. xxxiii. p. 285.
† Trans. Geol. Soc., ser. 2, vol. iii. pp. 261–265. The MS. Maps are preserved in the Library of the Geological Society of London.
‡ Ibid., ser. 2, vol. iii. p. 264.
§ Quart. Journ. Geol. Soc., vol. xxxiii. pp. 287, 288.

Lima elliptica.	Pecten lens.
Lucina circumcisa.	—— subtextorius.
Modiola subæquiplicata.	Sowerbya triangularis.
Opis Phillipsi.	Trigonia.
Pecten fibrosus.	

The limestone-beds are quarried at the lime-kilns south of Steeple Ashton, where the stone burns to a strong lime, which however is used for manure, as well as mortar. The quarry showed about six feet of marly oolite and pisolite, resting on a similar thickness of obliquely bedded oolite, which is used for building-purposes, large blocks of it being raised. The section presents the same general features as were shown at Todbere. (See Fig. 55, p. 104.) *Chemnitzia* was the only fossil I obtained here. The upper and pisolitic beds were shown by the road north-west of Keevil Church, and again by the lane to the south-east of the lime-kilns just mentioned near the brook, where the beds appear to be faulted against the Kimeridge Clay on the south-east of Steeple Ashton. A shallow pit here yielded *Acrosalenia, Pseudo-diadema,* and *Ostrea.*

It is above this spot and between the lane just mentioned and one further north (leading from Steeple Ashton in an easterly direction towards Hurst Mill and the Devil's Dairy) that Corals have been obtained in such abundance in the ploughed fields, as to render the locality so noted.

Formerly the strata were more extensively worked in quarries than they now are ;* and possibly the Coral-beds were opened up. As it is, specimens can only be gathered in fair abundance from the ploughed field near the junction of the two lanes, where, as Messrs. Blake and Hudleston remark, "every stone is still a coral, though the best have long since been removed." I obtained a number of examples, including *Comoseris irradians, Isastræa explanata, Stylina tubulifera, Thamnastræa arachnoides, T. concinna,* and *Thecosmilia annularis.* Many of these are much iron-stained.

There can be no doubt that the Coral Rag occurs above the mass of the limestones at Steeple Ashton. Of beds between the Coral Rag and Kimeridge Clay I saw no exposures near Steeple Ashton, nor any evidence of the iron-ore noted by Mr. Greenwell (p. 108). Messrs. Blake and Hudleston, however, mention that "the high ground round the village church is highly charged with red oxide of iron; and pits are said to have been dug here for ore." † I should rather have expected the ironstone (if present) to occur considerably to the east of Steeple Ashton, and east of Keevil. It is however possible that the faulted junction of Corallian Beds and Kimeridge Clay may be continued further to the north-west than shown on the Geological Survey Map. Messrs. Blake and Hudleston notice evidence of the Coral-bed to the south-east of East Ashton. Lonsdale had referred certain

---

* Sedgwick, Ann. Phil., vol. xxvii. (ser. 2, vol. xi.) p. 349.
† Quart. Journ. Geol. Soc., vol. xxxiii. p. 286.

beds in the district to the " Upper Calcareous Grit," but from his account of their exposure between West Ashton and Dunge, &c., they evidently belong to the Lower Corallian Beds.

On the northern side of Seend the outcrop of Corallian Beds is reduced to so narrow a limit that it seems most probable there is a fault following the local strike. This line of disturbance probably extends from the ascertained fault of Trowbridge and Semington, to the south of the Seend Iron-works and thence in an E.N.E. direction to the north of Rowde, probably between Rowde Farm and Rowdeford. On the south side of this presumed fault, between Seend and Rowde, there is no evidence of Corallian rocks, the ground by Seend Bridge being clay and probably Kimeridge Clay. Just above the pit, in Lower Calcareous Grit, near the Seend Iron-works, I noticed traces of pisolitic marls, &c., probably on the south side of the fault.

Between Rowde and Bromham the Corallian Beds are largely concealed by the Lower Greensand, but we find evidence of Lower Calcareous Grit in a lane by Rowde Wick, with *Ammonites perarmatus* and *Ostrea gregaria.* Again south of Bromham there is a narrow scarp of the Lower Calcareous Grit, consisting of sand and calcareous grit, passing down by alternations of sand and clay into the Oxford Clay.

On the south side of Sandridge Hill we find buff and brown sands with bands of calcareous sandstone dug to a depth of 8 feet. Higher beds were shown at the Westbrook limeworks (now abandoned), and there Messrs. Blake and Hudleston noticed " layer upon layer of large masses of *Thamnastræa concinna* and *Isastræa explanata,* bored by the characteristic *Lithodomus inclusus.*" (See p. 72.) This band " spreads over the surface, resting immediately upon a bed of sand, which is itself not far removed in elevation from the Oxford Clay. The spaces between the coral growths are filled with a rubbly brash, made up of comminuted materials, and sometimes with clay charged with fragments of shells. These intercoralline accumulations obtain the mastery here and there ; corals disappear, and we have great rubbly beds of shelly clay and limestone brash forming the whole reef." From these beds they record the following fossils :—*

Littorina muricata.	Ostrea gregaria.†
Exogyra nana.	Pecten lens.
Gastrochæna recondita.	—— subtextorius.
Lima densepunctata.	Cidaris Smithi.
—— pectiniformis.	Pseudodiadema versipora.
Opis Phillipsi.	

The section is of considerable interest in showing the local development of a coral bank in the lower part of the limestone series. It shows the variable nature of the beds, and confirms the view that the strata above the Lower Calcareous Grit may for stratigraphical purposes be conveniently regarded as one group.

---

* Quart. Journ. Geol. Soc., vol. xxxiii. pp. 288, 289.
† A cluster of these shells (*O. gregaria*) from Steeple Ashton is preserved in the Museum of Practical Geology.

To the north of this area the Corallian Beds are partially obscured by coverings of ferruginous sand and ironstone belonging to the Lower Greensand.

Calne

The neighbourhood of Calne affords many fine sections of the Corallian rocks.

The Lower Calcareous Grit, about 25 feet thick, has been quarried for road-metal on Derry Hill,* and sections are now to be seen near Conygre Farm, to the north-east of Calne. The beds comprise sands and loamy clays with irregular bands of ferruginous and calcareous sandstone, the latter somewhat cherty in places. Fossils are most abundant in the loamy beds, the more common being *Ammonites perarmatus, Chemnitzia, Pleurotomaria, Avicula ovalis, Exogyra nana, Goniomya v.-scripta, Gryphæa dilatata, Myacites,* and *Ostrea gregaria.*

Higher beds are exposed near the Union workhouse, where the Calne freestone is quarried beneath a capping of marl and rubble. Here we find false-bedded coarse oolite and marl, shelly in places and with occasional clay-seams, that have been opened up to a depth of nearly 20 feet.

Not many fossils were to be found at the time of my visit, but subsequently Mr. J. Rhodes saw a deeper section in the quarry and obtained many fine specimens of *Hemicidaris intermedia* for which this locality is famous. It occurs at or near the base of the freestone, where slabs containing a number of specimens with spines can sometimes be procured. Other fossils include *Echinobrissus scutatus, Pecten fibrosus,* and *Trigonia perlata.*

Messrs. Blake and Hudleston have published a list, including *Cidaris florigemma,* and they point out that the freestones merge upwards into beds of Coral Rag, with *Thamnastræa* and *Cidaris florigemma,* that may be traced near Quemerford beneath the Kimeridge Clay.†

The freestones are separated from the Lower Calcareous Grit by about 30 feet of strata. According to Messrs. Blake and Hudleston a well sunk near the Union workhouse, passed through 20 feet of blue oolitic rock and 10 feet of light hard marl.

Mr. W. A. Baily, of Cirencester, informed me that many years ago he obtained numerous specimens of the *Hemicidaris intermedia* from openings east of Calne Church. In the quarries it seems that examples are found in abundance at intervals, apparently in groups, as is the case with *Acrosalenia pustulata* in the Great Oolite of Cirencester.‡

At Hillmarton bands of Coral-rock and clay have been exposed, and Messrs. Blake and Hudleston have obtained from them many fossils characteristic of the " *Cidaris florigemma* Rag." At Goat

---

- * See also record of section at Spirt hill (Spirit hill), north-east of Bremhill, by Lonsdale, Trans. Geol. Soc., ser. 2, vol. iii. p. 264.
  † Quart. Journ. Geol. Soc., vol. xxxiii. p. 290.
  ‡ See Memoir on the Lower Oolitic Rocks of England, p. 283.

Acre to the north, false-bedded limestones quarried for building-purposes are again developed. These are described by the same authorities: the strata include layers of clay, and yield *Bourguetia* (*Phasianella*) *striata, Chemitzia heddingtonensis, Cidaris florigemma*, &c. The beds are also quarried at Preston to the north-east, where fine specimens of *Echinobrissus scutatus* and other fossils are met with.*

Rubbly Coral-rock is exposed by Hillocks Mill, north of Lyneham. In this neighbourhood the beds form a fine escarpment. To the eastwards beds of ferruginous sand have been mapped as "Upper Calcareous Grit," and this is separated from the Coral Rag by a parting of clay.†

To the south of Tockenham Wick we have an opportunity of seeing the relations of the Coral-beds to the underlying sands of the Lower Calcareous Grit, which appear to be much attenuated near Clack, Wootton Bassett and Purton. There were two quarries, the southern of which showed 9 feet of rubbly coral limestones, in irregular beds with partings of marly clay. Towards the base the beds become pisolitic and oolitic. I obtained the following fossils :—

Cerithium muricatum.	Pecten lens.
Littorina muricata.	Serpula tricarinata.
Astarte.	Cidaris florigemma (spines).
Cardium.	Hemicidaris intermedia.
Exogyra nana.	Echinobrissus scutatus.
Gastrochæna ?	Pseudodiadema versipora.
Lima duplicata.	Isastræa explanata.
—— rigida.	Stylina.
Lithodomus inclusus.	Thamnastræa Lyelli.
Ostrea gregaria.	Thecosmilia annularis.
Pecten articulatus.	Sponge.

The pit near Tockenham Wick, a little further north, showed the following beds :—

		FT.	IN.
Upper Corallian.	Irregular masses of Coral-limestone and grey slightly oolitic limestone, with *Cidaris florigemma* (spines); passing down into oolitic and pisolitic marl (1 ft. or more). The fossils are coated with calcareous matter, and include *Natica, Astarte, Ostrea*, and *Echinobrissus* - - -	3	0
	Oolitic marl with concretionary lumps of oolite: *Natica.*		
	Brown and grey sandy, oolitic, and pisolitic marl, with race; and at base, a band of ironstone-nodules with ochreous kernels - - 1 6 to	2	0
Lower Calcareous Grit.	Ferruginous sand with lignite, and buff false-bedded sand with comminuted shells, ferruginous layers, ironstone-nodules, and occasional clay-seams -	10	0

* Quart. Journ. Geol. Soc., vol. xxxiii. p. 293.
† Aveline, in Geol parts of Wilts and Gloucestershire, p. 22 ; Lonsdale, Trans. Geol. Soc., ser. 2, vol. iii. p. 262.

Somewhat similar sections at Green's Cleeve and Catcombe, to the south of Clack, are recorded by Messrs. Blake and Hudleston.* Lower Calcareous Grit is shown to a thickness of 6 feet. I was informed that near Clack, rocks that are turned up by the plough are known as Hitching stones, as the plough "hitches" against them; such rocks are very suggestive of the irregular concretionary masses in the Calcareous Grit. At Grittenham Hill the thickness of the Lower Calcareous Grit is stated by Professor Hull to be about 50 feet.†

Comparing the Corallian Beds of North Dorset with those of Westbury and Calne in Wiltshire we may roughly correlate them as follows (see Fig. 56, p. 107) :—

	North Dorset.	Westbury and Calne, Wilts.
Upper Corallian.	5. Ferruginous Beds.	5. Ironstone.
	4. Clays, &c.	4. Sands and Clays.
	3. Sandy, shelly, and oolitic limestones.	3. Rubbly oolite, coral-beds, and pisolite.
	2. Marnhull and Todbero freestone, and pisolitic beds.	2. Calne freestone, &c.
Lower Corallian.	1. Clays and sands with doggers.	1. Lower Calcareous Grit.

The general sequence compares well with that of the Weymouth district. Bed 1. may represent the Nothe Beds and Bencliff Grits; bed 2. the Osmington Oolite; bed 3. the Trigonia Beds; bed 4. the Sandsfoot Beds; and bed 5. the Abbotsbury iron-ore.

Wootton Bassett and Swindon.

In the neighbourhood of Wootton Bassett the Corallian Rocks are faulted on the south against the Kimeridge Clay, and we have no sections to show the complete sequence of the strata.

The Upper Division consists of a variable series of coral-limestones, with bands of clay, together with oolitic and pisolitic marls and limestones; it is about 30 feet thick. The Lower Division is not well exposed, it comprises sands with clay-seams, ironstone-nodules and occasional bands of calciferous shelly sandstone. It appears at most to be 10 feet thick, but in some places it may be less, for it has not been recognizable on the ground, and the Geological Survey Map shows the Upper Division resting on the Oxford Clay.

A well-boring at the Beaufort Brewery, north of the railway-station at Wootton Bassett, proved a thickness of 30 feet of Corallian Rocks (base not reached), beneath 84 feet of the Kimeridge Clay. Water no doubt was obtained from the Lower Calcareous Grit: the beds passed through being described as coral-rock, and thin bands of rock and clay (mostly rock).

---

* Quart. Journ. Geol. Soc., vol. xxxiii. p. 294.
† Geol. parts of Wilts and Gloucestershire, p. 21.

At Banner's Ash, north of Wootton Bassett, we find the rubbly grey coral-limestones and clay beds ; and these are again well exposed in quarries north-west of Purton Church.

The details vary a good deal, but the general section at Purton is as follows :—

		Ft.	In.
Upper Corallian.	Rubbly-grey earthy coral-limestones and clay beds, with *Lima pectiniformis, Lithodomus inclusus, Ostrea, Pecten, Trigonia, Cidaris florigemma,* &c.	9	0
	Marly pisolitic beds, indurated in places		
	Thick irregular bands of shelly and pisolitic limestone and coral-rock alternating with softer pisolitic beds	10	0
	Shelly oolitic limestones and earthy and shelly limestones (used for building-stone) : *Cidaris florigemma*		

The beds at Banner's Ash were regarded by Messrs. Blake and Hudleston as "in the form of a true reef, with massive Thamnastræan corals in full development, and with an abundant fauna, including *Nerita Guerrei.*" I obtained *Cidaris florigemma, Thecosmilia annularis,* and *Lima rigida.*   From the lower beds at Purton, Messrs. Blake and Hudleston record *Littorina muricata, Avicula ovalis, Lima rigida, L. rudis, Ostrea, Pecten articulatus (vimineus), Perna mytiloides,* and *Cidaris florigemma.*[*]

A cutting on the Midland and South-Western Junction Railway east of Sparsholt Farm (where the Great Western Railway crosses) showed a good section of the Upper Corallian rocks.   South of the bridge (see Fig. 59) the beds consist of grey earthy limestone and clay, with layers of compact coral-rock, with *Thecosmilia annularis,* Sponge-remains, *Lithodomus inclusus,* and *Gastrochæna recondita*?

The beds are no doubt faulted on the south against the Kimeridge Clay.

North of the bridge we find at the base, beds of grey clay with bands of rubbly irregular grey marly and septarian limestone, containing *Lima pectiniformis, Exogyra nana,* and *Cidaris florigemma.*   Above there is a seam of black carbonaceous clay, and, on top, clay with two layers of stone that coalesce towards the north.   Here again *Cidaris* is met with.   The thickness near the bridge is about 12 feet, but altogether about 20 feet of Corallian beds are shown.   The occurrence of black carbonaceous bands near Faringdon was noted by Messrs. Blake and Hudleston as indicative probably of the proximity of land.[†]

---

* Quart. Journ. Geol. Soc., vol. xxxiii. p. 295 ; see also Hull, Geol. parts of Wilts and Gloucestershire, p. 22 ; and Proc. Cotteswold Club, vol. vi. p. 293.
† Quart. Journ. Geol. Soc., vol. xxxi. p. 303.

FIG. 59.

*Section on the Midland and South-Western Junction Railway,
west of Rodbourn Cheney, near Swindon.*

S.   Great Western
    Railway Bridge.   N.

Upper Corallian Beds.

Further north, near Elbro bridge, the Lower Corallian Rocks
are shown (between two bridges) resting on grey and dark blue
clay ferruginous in places, with *Gryphæa dilatata* and *Pecten
subtextorius.* The beds here have a southerly dip, and, as shown
on the Geological Survey Map, they are again faulted on the south
against the Kimeridge Clay.

A number of fossils were collected from Blunsdon by Prof.
Buckman.* The upper beds comprise rubbly limestones with
many Corals and Echinoderms. Here, again, the Lower
Calcareous Grit was not well shown, and is apparently thin.†

In the deep well at Swindon, the beds from 72 to 112 feet,
having a thickness of 40 feet, have been assigned to Corallian
Rocks by Mr. E. T. Newton and myself. They were as
follows :— ‡

		FT.	IN.
Upper Corallian.	Compact grey marly and shelly limestones, marl and clay - - -	14	7
	Alternate beds of clay and gritty limestone, shelly in places - -	7	3
	Clay, &c. - - - - -	4	0
	Compact grey limestone, shelly and oolitic in places, and calcareous grit with clay seams - - -	5	10
Lower Corallian.	Sand, clay, and oolitic sandy marl -	8	5
		40	1

Towards the base of what we included in the Kimeridge Clay
at Swindon, there was a bed of iron-shot earthy limestone, six
inches in thickness, which has yielded *Ammonites cordatus*, var.
excavatus, a well-known Lower Kimeridge fossil, though found here
in the Oxford Clay. *A. rotundus* likewise occurs, and this is a
Kimeridge form. *Ostrea deltoidea* occurs, and this species is
abundant in the Upper Corallian beds in many localities. This
fossiliferous bed reminded me of the gritty limestone of Ringstead Bay, near Weymouth, termed the "Kimeridge Grit" by
Damon, and known also as the "Upper Coral Rag."

---

* Quart. Journ. Geol. Soc., vol. xiv. p. 126.
† See Hull, Explan. Hor. Sec., Sheet 59, p. 2.
‡ Quart. Journ. Geol. Soc., vol. xlii. pp. 288, 293, 300.

The fossils found in the Swindon bed are not sufficient to enable any palæontological comparison to be made; but when we compare the Swindon section with that at Highworth, about six miles to the north-east, we might be disposed to put more than we have included with the Corallian Rocks at Swindon. Nevertheless at "Rodbourn Lane," the lane leading from Swindon towards Rodbourn Cheney, the record of a well-sinking has been given as follows:—

					Ft.
Kimeridge Clay	·	·	·	·	68
Corallian Beds	·	·	·	·	27
Oxford Clay -	·	·	·	·	36
					131

The Corallian Beds north of Swindon comprise on the whole a very argillaceous set of beds, and therefore the actual limits of the formation can only be given approximately.

At the same time it seems likely that the Lower Calcareous Grit is of variable thickness in this district, becoming perhaps more attenuated from north to south, as is the case further east near Abingdon. Phillips mentions that quicksands were reached in a well at Even-Swindon, after the Kimeridge Clay and Coralline Oolite had been penetrated; for "Then burst up a great stream of water, followed by sand in such abundance as to fill the well to a considerable height." *

Highworth.

In the neighbourhood of Highworth there are a number of interesting sections, and these afford evidence of the presence of what has locally been termed an "Upper Calcareous Grit" division. It may correspond roughly with the Sandsfoot Beds of Weymouth.

This sequence was very clearly pointed out by Lonsdale,† and subsequently by Messrs. Blake and Hudleston.‡

Summarizing the evidence obtained by sections near Highworth railway-station, the brickyard north-west of the church, and the quarry and brickyard south of the town, we find the following sequence:—

	Ft.	In.
Rubbly coral-rock, with *Thecosmilia annularis*, &c. [full thickness estimated at 6 feet]§ - - -	2	0
Brown and white current-bedded sand, with derived oolitic grains, and with layers of oolitic sandstone ("Upper Calcareous Grit") - · -	7	6

* Geol. Oxford, &c., p. 298.
† Trans. Geol. Soc., ser. 2, vol. iii. p. 262.
‡ Quart. Journ. Geol. Soc., vol. xxxiii. pp. 297, &c.
§ R. F. Tomes, *Ibid.*, vol. xxxix. p. 556.

		Ft.	In.
Upper Corallian.	Brown laminated loamy clay, passing down into blue loamy clay, with calcareous concretions in places -	5	0
	Oolitic marly clay - - -	0	6
	Oolitic and pisolitic limestone, and softer marly and sandy bands : with layer of very shelly limestone at base in places - - - -	5	6
	Rubbly irregular Coral-bed with *Thecosmilia annularis, Montlivaltia dispar*, &c. - - - -	2	0
	Shelly gritty and calcareous rock with ferruginous parting, and Corals - }	1 to	6
	Shelly oolitic layer (impersistent) - }	8	0
Lower Calcareous Grit.	Hard calcareous grit with tubiform markings : irregular band - -	1	6
	Brown and white sands, with bands of calcareous sandstone, and tubiform markings : seen to depth of - -	15	0

The species of Corals are here recorded on the authority of Mr. R. F. Tomes. Other fossils obtained by myself from the Coralline Oolite were *Chemnitzia heddingtonensis* (abundant), *Exogyra nana, Lima rigida, Myacites, Ostrea solitaria, Pecten articulatus, P, fibrosus, Unicardium, Echinobrissus scutatus*, spines ot *Cidaris*, and *Serpula*. Messrs. Blake and Hudleston record many other species, including *Ammonites cordatus, A. goliathus, A. plicatilis*, and *A. perarmatus*; a remarkable and interesting assemblage from the Upper Corallian Beds.

We have here a thickness of from 45 to 50 feet for the Corallian Beds, which agrees generally with the thickness assigned to the formation in the Swindon Well. It must be remembered, however, that we have not the complete section; Prof. Hull assigned a thickness of 100 feet to the beds, and Messrs. Blake and Hudleston have (doubtfully) assigned a thickness of 84 feet to the Lower Calcareous Grit, including passage-beds down into the Oxford Clay.*

Between Sevenhampton and Watchfield towards Shrivenham, the thickness of the beds is estimated at about 30 feet by Messrs. Blake and Hudleston.†

At Shrivenham the upper beds comprise "a local deposit of ferruginous " sand which was traced over a considerable area by Prof. Hull. This "is separated from the Coral Rag by a parting of clay," as at Lyneham.‡

Doubtless it may be difficult in some places, in the absence of the upper coral-band, to discriminate between this "Upper Calcareous Grit," and the Lower Greensand which, here and there, stretches on to the Corallian Rocks.

The record of a well-section at Shrivenham House, to the south-east of Shrivenham Church, furnishes us with some knowledge of the strata ; but the record is difficult to interpret. The

---

* Quart. Journ. Geol. Soc., vol. xxxiii. pp. 299, 300, and Table, p. 404.
† *Ibid.*, p. 296.
‡ Geol. parts of Wilts and Gloucestershire, p. 22.

details, communicated by Mr. J. H. Blake, were obtained from
Mr. Edward Margrett of Reading, who made the boring in 1887.
The section was as follows:—

		FT.	IN.
Upper Corallian (the top 30 feet included with doubt).	Loamy clay and sand · - -	9	0
	Quicksand - - - -	2	3
	Sandy clay - - - -	0	9
	Blue clay with sand - - -	18	0
	Brown clay with sand - - -	2	3
	Stiff clay with shells - - -	10	0
	Light Corallian stone - - -	0	6
	Blue rock with yellow crystals - -	10	6
Lower Corallian.	Light coloured clay - -	1	6
	Light blue clay with beds of sand; with water, which rose to about 13 feet below the surface.		

	54	9

The evidence here as at Faringdon, shows that the mass of
Corallian rock-beds is some 10 or 12 feet thick. Between
Sevenhampton and Watchfield the whole of the Corallian Beds,
as remarked by Messrs. Blake and Hudleston, may be no more
than 30 feet thick. They refer to " a considerable mass of reddish
sands, which at Shrivenham village are of importance; but of
their precise nature and thickness we have no opportunity of
judging."[*] The above record gives the nature of these beds, but
affords no actual evidence of the Corallian age of the higher
strata.

Faringdon to Abingdon.

The Corallian Beds at Faringdon have been very fully described
by Messrs. Blake and Hudleston. In the brickyard south-east of
the town, about two feet of ferruginous earth with lumps of
calcareous clay, ironstone, and fragments of *Ostrea deltoidea* and
Serpula tricarinata, was seen to intervene between the Kimeridge
Clay and the Coral Rag. The bed like that at Shrivenham
called to mind the " great development of this class of rock at
Sandsfoot Castle."

A quarry near the workhouse showed the following succession:—

		FT.	IN.
Upper Corallian.	Coral Rag (quarried for lime-burning)	6	0
	Red clay parting, and broken Rag -	0	4
	Hard limestone, pisolitic in places, with *Thecosmilia* in upper part (road-stone) - - - - -	2	0
	Nodular and shelly oolite with *Pygaster umbrella, Echinobrissus scutatus,* &c. (road-stone) - - - -	2	0
	Gritty limestone, moderately oolitic, with *Astarte ovata, Hinnites tumidus, Pecten fibrosus. P. lens, Trigonia Meriani, T. perlata* (road-stone) -	1	1

[*] Quart. Journ. Geol. Soc., vol. xxxiii. p. 296.
[†] *Ibid.*, vol. xxxiii. pp. 301, 302.

	FT. IN.

Limestone, partially oolitic, with flattened lumps of marl (road-stone) · 0 8
Iron-stained sandy and blue-hearted oolites, with lignite, *Gervillia aviculoides*, *Pecten lens*, *Perna mytiloides* (building-stones) · · · 3 0

Lower Corallian. } Sands and sandy clays - - [about 70 0]

A bore-hole at the Eagle Brewery, Faringdon, after passing through 22 feet of soil and rock-beds (limestones, &c.), was carried to a further depth of 70 feet in sands and sandy clays, with occasional indurated bands; and then into grey sandy clay (Oxford Clay ?) to a depth of 22 feet. The thickness of the Lower Calcareous Grit on the Lechlade Road near Faringdon, was estimated at about 60 feet by Prof. Hull.[*]

In quarries to the east of Faringdon, Messrs. Blake and Hudleston noticed the sandy nature of the Upper Corallian beds, with *Cidaris florigemma*, &c. They observe, moreover, that in all the quarries there is a kind of passage bed to the Lower Calcareous Grit.[†]

In a quarry described by Prof. Hull, a quarter of a mile east of Kingston Bagpuize, the lower beds of Grit were very obliquely bedded. (See Fig. 60.) He notes the pebbly nature of the beds, and estimates the total thickness of the Lower Calcareous Grit as from 20 to 80 feet, while the Upper Corallian is about 10 feet.[‡]

FIG. 60.

Quarry near Kingston Bagpuize, West of Abingdon.
(Prof. E. Hull.)

A. Upper Corallian (Coral Rag). Rubbly and shelly limestone, with casts of shells ; few Corals.
B. Lower Corallian (Calcareous Grit). Irregular beds of calcareo-siliceous rock, conglomeratic, fossils fragmentary, alternating with coarse brown siliceous sand, with pebbles of quartz, lydian-stone, &c. Beds obliquely laminated.

The sections near Highworth and Faringdon, show that there as elsewhere the Corallian Beds are exceedingly variable.

At the Lamb Inn near Fyfield the following section was to be seen :—

* Geol. parts of Wilts and Gloucestershire, p. 21.
† Quart. Journ. Geol. Soc., vol. xxxiii. p. 303.
‡ Geol. parts of Oxfordshire and Berkshire, p. 6 ; see also Explan. Hor. Sec., sheet 72, p. 7.

			Ft.	In.
Soil		12. Reddish-brown earth about	3	0
		11. Marly bed - - - -	0	6
		10. Flaggy and somewhat gritty oolitic stone - - -	1	0
		9. Oolitic sandy débris, with dark clayey streaks, and occasional lydite pebbles: *Echinobrissus*	1	0
		8. Laminated clay and sand with irregular calcareous (racy) seams, forming white bands in places	7	0
		7. Oolitic limestone: few fossils	1	6
		6. Marly oolitic layer: *Myacites*	0	6
Upper Corallian.		5. Hard pale grey and very shelly limestone, *Trigonia*-bed: *Ammonites cordatus, Trichites,* and casts of shells at base -	1	8
		4. Brown earthy rock with casts of shells, *Myacites, &c.*: passing down into blue rubbly pisolitic rock, and this merges into bed below - - - -	2	0
		3. Grey shelly limestone, oolitic and pisolitic, with rolled fragments of hard limestone, bored by Lithodomi and encrusted with Serpulæ; small pebbles of lydite and quartz	1	6
Lower Calcareous Grit.		2. Hard grey calcareous grit, impersistent - - - -	3	0
		1. Buff sands.		

Among the fossils from the beds 3 to 7, but chiefly from bed 5, are *Ammonites plicatilis, A. cordatus, Belemnites abbreviatus, Trigonia perlata, T. Meriani, Trichites, Gervillia aviculoides, Ostrea gregaria, Perna mytiloides, Hinnites tumidus, Pecten fibrosus, P. lens, Lima pectiniformis, Echinobrissus scutatus,* &c. Many of these and other species are recorded by Messrs. Blake and Hudleston.* They consider that the upper sandy and clayey beds rest unconformably on the limestones below. I could see no evidence of this in the section at the time of my visit; the only indication of a local break being at the base of bed 3.

Prof. Hull† grouped the top strata of "variegated sand and clay," five feet thick, as "Upper Calcareous Grit." There is no doubt that beds 8 to 11 may be grouped with the upper beds at Highworth, where, as pointed out by Messrs. Blake and Hudleston, they underlie a layer of Coral Rag.

Coming now to the celebrated district of Marcham we find two quarries between that village and Shippon, south of the high road between Fyfield and Abingdon. The principal quarry is to the south of Oakley House. This shows the sequence from the Coral Rag down to the Lower Calcareous Grit; while the other quarry, parted by two fields to the north-west, shows only the lower beds and basement portion of the upper division.

---

* Quart. Journ. Geol. Soc., vol. xxxiii. p. 304.
† Geol. parts. of Oxfordshire and Berkshire, p. 7.

The general section at Marcham is as follows (see Fig. 61):—

FT. IN.

Upper Corallian.

7. Red and brown earthy soil.
6. Impure and slightly oolitic lime-stone, passing down into rubbly coral-rock and clay -   -   0   0
5. Oolitic and sandy marl, passing down into earthy oolitic limestone and rubbly oolite, with *Trigonia*-bed 4, at base. This fossil-bed tapers away towards the western part of the Oakley House quarry -   -   2   6
3. Reddish brown and white shelly sands, with indurated fissile beds and laminated clay; with pebbly layer at base, when resting directly on Lower Calcareous Grit 2 6 to   4   0
2. Oolitic and somewhat flaggy and gritty limestone, 1 foot, tapering away eastwards in the Oakley House quarry, and replaced by rubbly oolite, and by a conglo-meratic band with pebbles of lydian stone, oolite and bored limestone. In the quarry to the north-west this fossil-bed is again seen, where it is 2 feet thick, and conglomeratic. It contains *Serpula, Trigonia* and many Lamellibranchs; also *Natica marchamensis.*

Lower Corallian (Lower Calcareous Grit).

1. Hard calcareous sandstone (1 ft.), *Natica*-bed (of Blake and Hudle-ston), with *N. marchamensis:* in some places united to the fossil-bed above, in other places separated by a thin bed of shelly sand (6 in.) with *Exogyra nana, Gervillia, Pecten fibrosus* and *Ostrea solitaria* - False-bedded buff and ochreous sands with ironstone-concretions, streaks of clay, pebbles of quartz and lydian stone: *Ostrea gregaria*, lignite, &c. Here and there the mass of the strata up to the base of the upper division consists of layers of fissile calcareous sand-stone occasionally ripple-marked; in other places there are only im-persistent and irregular layers and doggers of sandstone -   -   8   0

In the Upper Corallian Beds the higher *Trigonia*-bed yields *Ammonites plicatilis, Belemnites abbreviatus, Lima pectiniformis Trigonia, Ostrea gregaria, Echinobrissus scutatus, Pygaster umbrella*, Lignite, &c. The lower fossil-bed is also a *Trigonia*-bed. The presence of sandy beds among the fossiliferous Upper Corallian rocks, and the occurrence of the two fossil-beds and their impersistent character form interesting features.

The sands of the Lower Calcareous Grit at this locality have been famous, for fine examples also occur of *Ammonites perar-*

FIG. 61.—*Section at Marcham, near Abingdon.*

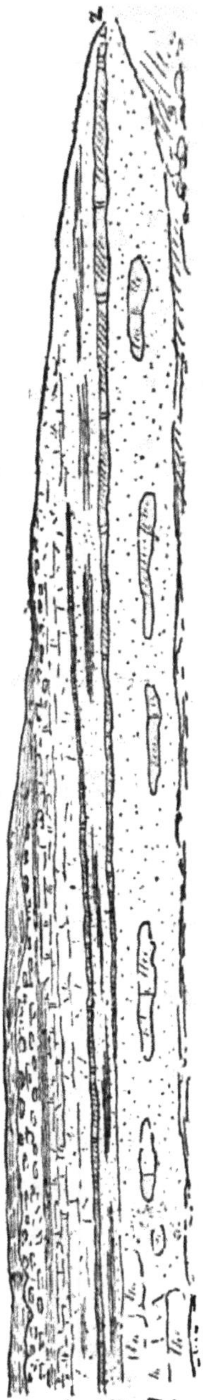

1. Sand and sandstone with concretionary masses of sandstone (Lower Calcareous Grit).
} Lower Corallian.

} Upper Corallian.

7. Red earthy soil.
6. Rubbly coral-rock and clay.
5. Marl and rubbly oolite.
4. *Trigonia*-bed.
3. Laminated clay and shelly sand.
2. Oolitic shelly limestone (fossil-bed).

matus, the casts of which were formerly described under the name of *A. catena.* Casts of *A. cordatus* are also met with. Here too *Hemipedina marchamensis* occurs.

The sections at Marcham, including a quarry at Noah's Ark further south, have been described by Messrs. Blake and Hudleston, who record many fossils.* They drew attention to the evidences of local erosion, but the lower fossil-bed was evidently not exposed at the time of their visit.

The irregular cementation of the sands of the Lower Calcareous Grit into doggers and into bands of sandstone was well shown in the quarries near Marcham.

Passing on to Abingdon our information respecting the Corallian Beds is derived chiefly from well-sections.

At Abingdon water was obtained from Corallian rocks, at a depth of about 60 feet, through Kimeridge Clay; but we have no record of the precise thickness of the strata. The water was slightly chalybeate, and impregnated with sulphuretted hydrogen.† At the water-works by the cross-road S.S.E. of Wootton the beds passed through were as follows :—‡

		Ft.	In.
Coral Rag.	Ragstone	31	6
Lower Calcareous Grit.	Solid blue stone	2	6
	Clean sand	6	0
	Solid blue stone	2	6
Oxford Clay.	Soft soapy clay	34	6

* Quart. Journ. Geol. Soc., vol. xxxiii. pp. 305, &c.; see also H. B. W., Proc. Geol. Assoc., vol. xii. p. 331.
† Rev. J. C. Clutterbuck, Journ. R. Agric. Soc., ser. 2, vol. i. p. 281.
‡ Details communicated to Mr. De Rance by Mr. A. T. Atchison, 1877.

Hence we may conclude that the full thickness of the Corallian rocks at this locality is not more than 50 feet. Further north and locally to the south, the beds appear to be much thicker.

A well was made for the Wantage Brewery Company at Wantage in 1889-90, by Messrs. C. Isler and Company. Samples of the strata were sent to the Museum at Jermyn Street and examined by Mr. Strahan and myself. He recognized the basement-bed of the Gault from specimens obtained from a depth of 279 feet.

The boring was commenced at a depth of 46 feet below an old dug well. The quality of the water from the higher strata down to 413 feet was not good, and this was shut out by 3-inch tubes. The water from the lower strata proved satisfactory after pumping, and then stood at 83 feet from the surface. There was an abundant supply.

The details of the boring are as follows :—*

	BORING AT WANTAGE.	THICKNESS.		DEPTH.	
		FT. IN.		FT. IN.	
Upper Greensand.	{ CHALKY AND SANDY BEDS [Dug well] - - - -	46	0	—	
Gault -	{ Grey slightly micaceous and calcareous HARD CLAY, with band of sandstone at depth of 68 feet - - -	229	0	275	0
	CLAY WITH SAND SEAMS: coarse quartz grains and pebbles of hard siliceous rock -	4	0	279	0
Kimeridge Clay ? 94 feet.	{ HARD GREEN SAND: grey clay with decayed shells at 280 -	8	0	287	0
	HARD DARK CLAY: with decayed shells at 320, and fibrous carbonate of lime ("beef") at 354 -	71	0	358	0
	HARD CLAY WITH SHELLS: greenish-grey clay with decayed shells at 360 - - - -	5	0	363	0
	HARD CLAY - - - -	9	0	372	0
	HARD STONY CLAY - - -	1	0	373	0
Corallian Beds ? 77 feet.	{ ROCK: white marly rock at 373 -	3	0	376	0
	CLAY: stiff grey marly clay at 377, and white marly rock at 378 - - - -	20	0	396	0
	ROCK: dark shelly clay with *Pecten* or *Lima* at 397 - -	7	0	403	0
	LIGHT GREY SAND: crumbly calcareous rock at 404 - -	1	6	404	6
	ROCK: white calcareous rock at 405 - - - -	3	3	407	9
	DARK GREEN LOAMY SAND: calcareous gritty rock at 409 -	5	3	413	0
	HARD BLUE CLAY: grey shelly clay at 416 - - - -	3	0	416	0
	ROCK: white chalky rock at 416, and bits of *Ostrea*, some pyritic, at 434 - - -	18	6	434	6
	LIGHT GREY SAND, greenish calcareous sandy bed at 435 -	3	6	438	0
	DARK GREY LOAMY SAND: greenish calcareous sandy bed at 450 -	12	0	450	0

* The information given by Messrs. Isler is put in capitals, to distinguish it from the notes which I made from the specimens.

This record unfortunately gives no precise information with regard to the Jurassic strata. The beds, classed with some doubt as Kimeridge Clay, might on lithological grounds be grouped as Purbeck, and the lower strata as Portland Beds. This view, however, I feel compelled to abandon on account of the water-bearing character of the strata; for the Portland Beds do not come to the surface over any region near by that could furnish a gathering ground for a supply of water. The strata may therefore be classed as most probably Corallian Beds and Kimeridge Clay. The "hard green sand" might perhaps form a portion of the passage-beds from Kimeridge Clay to Portland Beds, but that rock was not noticed in a record sent to me by Messrs. Isler in 1894, although given in a preliminary statement in 1889.

Further east at Shillingford, to the north of Wallingford in Berkshire, a boring was made in 1885 by Messrs. Isler and Company. The section has been described by Mr. Jukes-Browne, who records the strata as follows:—*

BORING AT SHILLINGFORD.		THICKNESS.	DEPTH.
		FT. IN.	FT. IN.
	Soil, &c. - - - -	4 0	—
	Sand and gravel - - -	15 0	19 0
Gault -	- - - - -	144 0	163 0
Lower Greensand	- -	25 0	188 0
Kimeridge Clay. 111 ft. 6 in.	Stiff grey clay and stones (? septaria) - - -	20 0	208 0
	Black clay, with shells and stones, and phosphatic nodules at 296 feet - - -	91 6	299 6
	Rock - - - -	12 9	312 3
Upper Corallian. 44 ft. 6 in.	Stony clay and shells (yellow gritty loam) - -	5 8	317 11
	Rock, with some layers of clay -	9 11	327 10
	Stony clay, with layers of rock -	8 10	336 8
	Rock, with shell-fragments (Ostrea) - - - -	7 7	344 3
Lower Corallian. 35 ft. 3 in.	Sand and stone - - -	7 4	351 7
	Alternations of hard grey rock and clay - - - -	10 4	361 11
	Dark gritty clay and shells -	10 1	372 0
	Hard grey limestone -	2 6	374 6
	Sand and stone, with water -	5 0	379 6
Oxford Clay -	Blue clay, soft and slightly mottled	4 6	384 0

The water obtained at the base of the Corallian rose to within 60 feet of the surface, and although palatable it did not yield a sufficient supply. Curiously enough that obtained from the Lower Greensand above was saline, containing 98 grains per gallon, 54 of which were chloride of sodium.

Comparing the Corallian Beds of Westbury and Calne with those of Highworth in Wiltshire and Faringdon in Berkshire, we may roughly correlate them as follows (see Fig. 56, p. 107):—

* Midland Naturalist, vol. xiv. p. 201.

	Westbury and Calne.	Highworth and Faringdon.
Upper Corallian.	Ironstone. Sands and clays. Rubbly oolite, Coral-beds, and pisolite. Calne freestone, &c.	Rubbly Coral-rag. Sands and clays. Coral limestone, oolitic, pisolitic and shelly beds.
Lower Corallian.	Lower Calcareous Grit.	Lower Calcareous Grit.

The absence of any bed that corresponds with the Corallian Oolite of Calne may be due to the local erosion of which we have evidence at the base of the Upper Corallian rocks, near Marcham. Further on towards Oxford we have but two divisions in the Corallian formation, the rubbly coral-limestones and the calcareous grit beneath. In some places these two divisions are so intimately connected, we can fix no definite plane of separation; in other places they are clearly marked off one from the other.

Cumner and Oxford to Headington and Wheatley.

To the south-east of Cumner, and near Bradley Farm, there are several quarries in the Corallian rocks. One of these, near the south-east end of Cumner village, showed bedded Coral-rock, rubbly and irregular in appearance, and with little or no clay, yielding *Thecosmilia annularis, Lima rigida*, &c. Quarries by the cross-roads and west of Bradley Farm showed the sequence as follows :—*

		Ft.	In.
Upper Corallian.	Coral-rock with interbedded tufaceous stone, like that at Wheatley - 4 0 to 5	0	
	Hard beds of Coral-rock, as at Purton, with *Lithodomus inclusus* and spines of *Cidaris florigemma*; passing down into sandy shelly and tufaceous rock of variable character (as at Wheatley) with *Pecten, Ostrea, Exogyra*, &c. -	8	0
Lower Corallian.	Brown and grey sands, shelly in places	5	0
	Calcareous sandstone, two or three thick beds - - -	3	6
	Buff sand and laminated clay - -	4	0

The beds on the whole point to very irregular accumulation. From the Lower Calcareous Grit at Bradley Farm, Messrs. Blake and Hudleston record *Ammonites perarmatus, A. vertebralis*, and a number of Gasteropods, including *Cylindrites Luidi*. This is one of the most fossiliferous beds in the Corallian series. With regard to the Coral-rag, they observe that " The reef-corals here are in a more perfect state of preservation than in any locality we know of." I obtained a fine specimen of *Isastræa explanata*.

---

* Hudleston, Proc. Geol. Assoc., vol. vi. p. 343; Blake and Hudleston, Quart. Journ. Geol. Soc., vol. xxxiii. p. 307.

West of North Hinksey there was a quarry in which I noted the following section :—

		FT.	IN.
Upper Corallian.	⎡Rubbly Coral-rock with much clay: spines of *Cidaris, Lithodomus*, &c. -	8	0
	Hard bed of sandy oolitic limestone with small pebbles : merging upwards into Coral-rock; *Ostrea solitaria, Pecten articulatus*, &c. - 1 0 to 2		0
Lower Corallian.	⎧Buff sands, with beds of hard calcare-⎩ous sandstone - - - - 10		0

The hard bed below the Coral-rock presents an aspect similar to beds of Wheatley stone.

On Wytham Hill the Lower Calcareous Grit is surmounted by the Coral Rag.* (See Fig. 22, p. 44.)

Phillips states that near Oxford the Lower Calcareous Grit is about 60 or 70 feet thick, and that " the sand is sometimes so loose as to be ' quick,' and choke the wells, which in many places are sunk to it through the superincumbent rock and clay."† The beds were exposed in the railway-cutting at Kennington, where they contained large masses of calcareous sandstone, as at Littlemoor.

West of Littlemoor Station, a cutting on the Oxford and Thame railway showed the following section :—

		FT.	IN.
Upper Corallian.	⎡Dark clay with paler marly beds - ⎤		
	Clays and soft white marly beds, with } [15		7]
	but few corals = " Coral beds " - ⎦		
	Shelly sand, with *Natica corallina*.		
	Hard calcareous sandy bed - 0 10 to 1		6
	Sand - - - - - 2		6
	Shelly calcareous grit (employed for building railway bridges) - 2 0 [to 2		6]
Lower Corallian (Lower Calcareous Grit).	⎡Buff sands with spherical and elongated concretionary masses of hard calcareous sandstone (blue hearted) 4 × 5 × 2 feet and less in size, like the " doggers " in the Bencliff Grits: occasional streaks of clay - - 15		0

Here, as at Bullingdon, we have a difficulty in making any separation between Upper and Lower Corallian Beds ; there seems to be no evidence of any local break, and the sandy strata below the Coral-beds are doubtfully referred to the Upper Corallian, though they appear to represent the fossiliferous beds of Fyfield and Marcham.

The beds undulate very much in the railway-cutting west of the station, but the general inclination is westwards. The clayey Coral-beds make heavy land on the slopes to the west, and they much resemble those seen in the railway-cutting north of Swindon. (Fig. 59, p. 118.)

---

* Hull, Geol. Woodstock, p. 26.
† Geol. Oxford, &c., p. 293.

The above section was described by Mr. E. S. Cobbold, from whose paper I have inserted the thickness of the upper clayey and marly beds.* He states that, in laying the sewage-pumping main along the road up Sandford Hill, there was a thickness of about 12 ft. 6 ins. of very hard fine-grained limestone above the marls ; it occurred in three or four layers with hardly any fossils. Above this, again, there was a stratum of very sandy limestone, 2 ft. 6 ins. thick, suggesting a trace of the Upper Calcareous Grit.

Mr. Cobbold also gives the record of a trial-hole at Sandford-on-Thames, which appears to me to have passed through 5 ft. of soil and Kimeridge Clay and about 10 feet of Upper Corallian Beds.

To the east and south-east of Oxford the Corallian Beds are divisible into two portions : an upper one of limestones largely made up of corals and comminuted shells and corals, and a lower one of sands with indurated bands and concretionary masses of calcareous sandstone (Lower Calcareous Grit).

The upper division, which in many places is a Coral-rag, is very variable in nature, in some places it consists of impersistent layers of hard grey shelly limestone (which has been employed for building-purposes) alternating with shell and coral-sand ; in other places it comprises irregular bands of coral-limestone and clay. The quarries show from 10 to 25 feet ; but the beds thicken towards Wheatley, and the total thickness is given by Phillips as 40 or 50 feet.

Conybeare thought that the upper surface of the freestone-beds, beneath the Kimeridge Clay, bore evidence of water-action.† Sedgwick observed, " May not, therefore, the coral rag and superincumbent freestone of Headington Hill together represent the central group of the Weymouth and Steeple Ashton sections ? The conjecture seems to be confirmed by the appearance of the beds in Headington quarries. In that place the top freestone supports the Kimmeridge clay ; and the separation between the two is as well defined as a geometric line. Now the instantaneous passage from one formation to another frequently indicates the absence of certain beds or deposits. May not then the upper part of the Weymouth section be wanting near Oxford."‡ Buckland and De la Beche, and later on John Phillips and others, have also remarked that the Kimeridge Clay rests on a waterworn surface of the Corallian Beds. I have not observed any distinct evidence of this, but the change in conditions is abrupt. and as remarked by Mr. Hudleston, " there appears to be a very marked break."§

There are numerous quarries at Headington, but the junction with the Kimeridge Clay was best shown in the pit to the left of the road leading from Oxford to Shotover Hill. The section was as follows :—

* See Quart. Journ. Geol. Soc., vol. xxxvi. pp. 315-317, 319 ; see also Blake and Hudleston, Ibid., vol. xxxiii. p. 310.
† Geol. Eng. and Wales, 1822, p. 189.
‡ Ann. Phil., vol. xxvii. (ser. 2, vol. xi.) p. 350. See also Buckland and De la Beche, Trans. Geol. Soc., ser. 2, vol. iv. p. 26 ; Phillips, Geol. Oxford, &c., p. 299.
§ Report of Sub-Committee on Classification, Internat. Geol. Congress, 1888.

			FT.	IN.
Kimeridge Clay.	Dark bluish-grey clay - Gray racy clay -	Saurian bones, *Ammonites, Ostrea deltoidea*, large crystals of selenite.		
Upper Corallian.	Marly layer - - - - False-bedded sandy rock (decomposed) and passing down into bed beneath -		1	0
	Hard grey shelly limestone - -		1	6
	Coral-rock, with bands of hard grey shelly limestone (building-stone); few fossils - - - - -		12	0

Here the total thickness of the Upper Corallian Beds, down to the Lower Calcareous Grit, is about 30 feet. In other quarries the beds are more fully exposed, and their variable character is manifest. We find Coral-rock, shelly sand, and lenticular layers of hard grey oolitic and shelly limestone; the whole irregular and false-bedded. In places the bottom-bed, as at Littlemoor, is a very shelly calcareous gritty rock (2 ft. to 2 ft. 6 ins. thick); and this rests on brown sands with concretionary beds of calcareous sandstone (Lower Calcareous Grit). In other places Messrs. Blake and Hudleston have noticed on top of the Grit a pebbly bed (about 8 inches thick), with rolled oolite, and nodules with *Ammonites cordatus*, and containing also *Ostrea solitaria*, *Avicula expansa*, *Lima læviuscula*, and *Natica clytia*.[*]

A quarry east of the road at Bullingdon (or Bullington) showed the following section:—

		FT.	IN.
Upper and Lower Corallian.	Rubbly coral-beds, with much clay, and with shelly sand at base; *Pleurotomaria reticulata, Exogyra nana*, &c.	3	0
	Tufaceous sandy rock - - -	0	8
	Shelly sand, with *Ammonites perarmatus, Chemnitzia, Ostrea, Pecten, Serpula*, &c. - - - -	0	8
	Shelly calcareous grit - - -	1	2
	Sandy bed - - - -	0	6
	Shelly calcareous grit, slightly oolitic -	1	6
	Sand - - - - -	1	0

Here I have hesitated to mark a division between Upper and Lower Corallian, though comparing the section with that of Littlemoor, it might be taken just above the basement sand. The presence of *Ammonites perarmatus* is interesting.

The Coral-beds yield many of the Echini (*Pygaster umbrella*), which are known to the quarrymen as "Mushrooms." Mr. James Parker has obtained some Fish-remains, *Asteracanthus* and *Lepidotus*, from this locality.

The commoner fossils of the Upper Corallian Beds of Bullingdon and Headington are noted in the accompanying list. It is curious that the *Chemnitzia heddingtonensis* is stated by Sowerby to have been received "from Heddington, near Calne,

---

[*] Quart. Journ. Geol. Soc., vol. xxxiii. pp. 308, 309.

in Wiltshire ; though he also mentions that he had found specimens about Shotover Hill (= Headington) in Oxfordshire.*

LIST OF FOSSILS FROM THE UPPER CORALLIAN BEDS OF HEADINGTON AND BULLINGDON.

Ammonites plicatilis.
× Belemnites abbreviatus.
Bourguetia (Phasianella) striata.
Chemnitzia (Pseudomelania) heddingtonensis.
× Natica clytia.
Pleurotomaria reticulata.
Astarte ovata.
Avicula expansa.
Corbicella lævis.
× Gervillia aviculoides.
× Exogyra nana.
× Goniomya v.-scripta.
× Isocardia.
Lima elliptica.
× —— læviuscula.
× —— rigida.
× Lithodomus inclusus.
× Myacites.

Myoconcha Sæmanni.
× Ostrea solitaria.
× Pecten articulatus.
× —— fibrosus.
—— lens.
Perna mytiloides.
× Pholadomya æqualis.
× Trigonia.
× Serpula tricarinata.
× Cidaris florigemma.
× Echinobrissus scutatus.
× Hemicidaris intermedia
× Pygaster umbrella.
Pygurus.
× Isastræa explanata.
× Thamnastræa arachnoides.
× Thecosmilia annularis.
Lignite.

× Species so marked were obtained by myself; many of these and others have been recorded by Phillips, Geol. Oxford, &c. p. 300; by Blake and Hudleston, Quart. Journ. Geol. Soc., vol. xxxiii. p. 309 ; Hudleston, Proc. Geol. Assoc., vol. vi. p. 342; and E. S. Cobbold, Quart. Journ. Geol. Soc., vol. xxxvi. p. 314.

North-east of Headington, at Holton and near Stanton St. John, rag-beds have been quarried, and there Messrs. Blake and Hudleston notice the abundance of Echinoderms, *Cidaris flori-gemma*, *Echinobrissus scutatus*, and *Pseudodiadema versipora* ; while Corals seem to be absent or very scarce.†

The section given by Prof A. H Green is as follows :—

		FT.	IN.
Upper Corallian	Rubbly limestone, full of broken fossils	3	0
	Yellow and brown sandy clay, with irregular beds of concretionary limestone, full of broken shells, *Ostrea sandalina* (= *Exogyra nana*), &c.	10	0
	Hard, dark-blue limestone	6	0

The sands of the Lower Calcareous Grit have been largely worked between Elsfield and Beckley.‡

Large quarries to the north-west of the village of Wheatley have been opened in comminuted-shell sand, cemented into rock-beds in places, and passing into hard shelly limestone, not unlike some varieties of Cornbrash. Occasional clayey, sandy, and oolitic beds occur, and the thickness may be from 30 to 50 feet. The strata are much false-bedded and they yield but few fossils. I obtained spines of *Cidaris florigemma*, *Chemnitzia*, and *Exogyra*

---

* Mineral Conchology, vol. i. p. 86.
† Quart. Journ. Geol. Soc., vol. xxxiii. p. 310.
‡ Green, Geol. Banbury, &c., pp. 43, 44.

nana; and Messrs. Blake and Hudleston note also *Ammonites plicatilis* and *Belemnites abbreviatus.* They remark, " We are here, then, presented with the deposits which were formed on the extreme edge, not only of the coral reef, whose thickness would not account for so much false dip (which amounts in the most easterly quarry to 12°, and is even marked at 18° at another spot on the [Geological Survey] map), but probably of the Lower Calcareous Grit sandbank also—a conclusion which is supported by the fact of its apparent sudden termination eastwards." They add that "the Corallian formation has died out, not gradually, but suddenly, and the normal pelolithic formation reigns supreme."*

This termination has been attributed to the unconformable overlap of the Kimeridge Clay,[†] but the evidence favours the view that the rock-beds may be largely represented in point of time by sediments of an argillaceous character. We find the same kind of abrupt ending in connexion with the Upware Limestone, and although some amount of local and contemporaneous erosion may have taken place, there was no great interval attended by upheaval and subaërial denudation. In Yorkshire, near Birdsall, as pointed out by Mr. Fox-Strangways, the Kimeridge Clay rests with marked unconformity on the Upper Corallian rocks, and further south it reposes directly on the Lower Calcareous Grit.[‡]

---

* Quart. Journ. Geol. Soc., vol. xxxiii. p. 311.

† Sedgwick, Ann. Phil., 1826, p. 350; Fitton, Trans. Geol. Soc., ser. 2, vol. iv. p. 274; Hull, Geol. parts of Oxfordshire and Berkshire, p. 6.

‡ Jurassic Rocks of Yorkshire, vol. i. pp. 369, 405; and Geol. country N.E. of York, &c., p. 23.

CHAPTER VII.

CORALLIAN.

LOCAL DETAILS—*continued.*

AMPTHILL CLAY.

Wheatley to Quainton.

North-east of Wheatley the calcareous rocks of Corallian age abruptly terminate; nor, excepting at Studley and Arngrove, near Boarstall, have any representatives of the Lower Calcareous Grit been observed in this neighbourhood beyond Wheatley and Stanton St. John.

It is true that "Lower Calcareous Grit" was so marked on the Geological Survey Map from Holton by Worminghall to Oakley, and from Dorton by Westcot to near Quainton. This area is, however, a tract of clay-land, and it was the opinion of Mr. T. R. Polwhele that along the belt so marked there were argillaceous equivalents of the Lower Corallian Beds; while Prof. Green suggested that these clayey beds might be the equivalent of the Tetworth (Ampthill) Clay described by Prof. Seeley.*

At Studley "a sort of argillaceous chert" was described by Prof. Phillips, containing *Ammonites, Pinna lanceolata,* &c. According to Prof. Green "This bed runs with a good escarpment by Arngrove Farm to the north of Gravel Pit Farm, beyond which point we lose sight of it altogether. It dips gently to the east, but whether it runs under the clay beds to the east of it, or passes into a clay, it is not easy to say. Immediately below this stone we find Oxford Clay with *Gryphœa dilatata,* and it is therefore without doubt the bottom bed of the Calcareous Grit." He records from it *Ammonites cordatus, A. vertebralis, Pecten fibrosus,* &c. He further remarks that, "To the east of this is a band of very marked light-blue clay, somewhat sandy, and in many places crowded with *Ostrea sandalina* [*Exogyra nana*]. This bed ranges through Worminghall, Oakley, and Boarstall, and was again found with its characteristic fossil at Westcot, and one mile west of Waddesdon Field. On the strength of this evidence a belt of this clay has been drawn between the Kimeridge and Oxford Clays up to the last-named spot, from whence it has been supposed to thin away towards the north-east, and it has been looked upon as the representative of the Coralline Oolite, (1) from its position, lying as it does between the Oxford and Kimeridge Clays, and differing in mineral character from both; (2) from the abundance of *Ostrea sandalina* [*Exogyra nana*] found in it, that shell being plentiful in the lower part of the Coral Rag."

Lists of fossils from Ickford, Worminghall, Oakley, and Westcot are given by Prof. Green,† but the particular localities

---

and horizons are not sufficiently distinguished for present purposes. Further research in the district is desirable.

Quainton to Ampthill and Gamlingay.

On the Geological Survey Map, from Quainton eastwards, the boundary between Oxford and Kimeridge Clay has been carried, in a "wholly conjectural" manner, through Stewkley Village to near Leighton Buzzard, where the beds are mainly concealed by Drift, and where but few sections are to be seen.

It is now, however, generally considered that over this tract, and indeed from the north-eastern part of Oxfordshire through Buckinghamshire, Bedfordshire, Cambridgeshire, and Lincolnshire, the Corallian Beds are represented, in point of time, for the most part by clays. Traces of Lower Calcareous Grit were marked on the Geological Survey Maps at Papworth St. Everard and at Houghton Hill, west of St. Ives, but with the exception of the mass of limestone at Upware there is no prominent development of Corallian rock-beds in this region.

Our information concerning the area near Grandborough and onwards to Stewkley and Leighton Buzzard is meagre. About 1¼ miles west of North Marston Church, Prof. Green found *Gryphæa dilatata* plentifully in the ditches and drain-cuttings, and this shows that the clay at Grandborough is Oxford Clay. The same fossil was found N.N.W. of Mains Hill Farm. On the other hand, Kimeridge Clay was worked in the brickyard south of Stewkley. Leighton Buzzard itself is probably situated over Corallian clays; and from this point the strike of the Jurassic rocks is to the north-east, the beds being, however, concealed by the Lower Greensand of Woburn. The Corallian clays re-appear in the vale near Ampthill.

In 1861 Prof. H. G. Seeley drew attention to a number of bands of rock that occur at different places in this great clay-area. He suggested that the name Fen Clay[*] should be used as a general term for all the strata from the Oxford to the Kimeridge Clays; while for the intervening clay, which, in his opinion, replaced the Corallian formation, he proposed the name Bluntisham Clay, afterwards altered to Tetworth Clay,[†] and finally changed to Ampthill Clay.[‡]

The name AMPTHILL CLAY is now generally adopted for the mass of Corallian Clays, which have been so well shown in the cuttings of the Midland Railway at Ampthill, in Bedfordshire. Our further knowledge of the beds is due most largely to the labours of Thomas Roberts, of Cambridge,[§] whose early death has been such a serious loss to science.

---

[*] Sowerby originally used the term Forest or Fen Clay for the Oxford Clay. Mineral Conchology, vol. ii. p. 129.

[†] Geologist, vol. iv. p. 460; Ann. Nat. Hist., ser. 3, vol. viii. p. 503; vol. x. p. 108; and Rep. Brit. Assoc. for 1861 (1862), Sections, p. 132.

[‡] Index to Fossil Remains of Aves, &c., 1869, p. 109.

[§] Jurassic Rocks of the neighbourhood of Cambridge, 1892.

. Prof. Seeley pointed out that in this division of Ampthill Clay the fossils show an admixture of forms belonging to both Oxford and Kimeridge Clays. Of these the more prominent are the following :—

Ammonites achilles.	Belemnites abbreviatus.
—— cordatus.	—— nitidus.
—— —— var. excavatus.	Gryphæa dilatata.
—— plicatilis.	Ostrea deltoidea.
—— vertebralis.	—— discoidea (Fig. 62).

FIG. 62.

Ostrea discoidea, *Seeley*, ¼.
Ampthill Clay.

This species has not before been figured, but the example now taken for illustration, has been seen and approved by Prof. Seeley.

The cuttings north of Ampthill railway-station have exposed the following strata, which have a general inclination to the south (see Fig. 63) :—

		Ft.	In.
Glacial Drift	Boulder clay (seen in places) - -	3	0
Kimeridge Clay	{ Clay and dark blue shale, with *Ammonites biplex, Ostrea deltoidea*, &c. -	10	0
Ampthill Clay	{ Septarian band, with *Ostrea discoidea* in and beneath it. Grey marly shale and stiff clay, with selenite; line of calcareous nodules, with *Ischyodus*, and rusty band, with *Ammonites cordatus, Alaria, Arca subtetragona, Ostrea discoidea, Thracia* -	50	0
	Band of pale earthy limestone - -	0	8
	Grey and yellow marly clay - -	6	0
	Rubbly rock-bed (like the basement-bed at Gamlingay), with *Alaria, Exogyra nana, Ostrea gregaria, Pinna lanceolata, Trigonia, Webbina irregularis* -	4	6
Oxford Clay	{ Clays with "race," selenite; *Ammonites perarmatus, Belemnites hastatus,* and *Gryphæa dilatata.*		

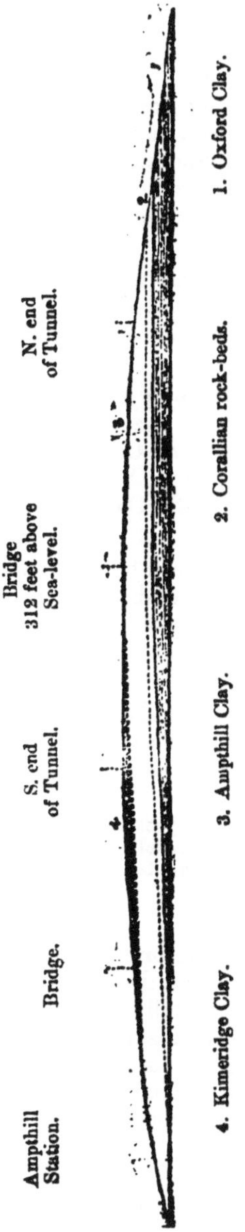

FIG. 63.—*Section along the Midland Railway, north of Ampthill Station.*

Ampthill
Station.

Bridge.

S. end
of Tunnel.

Bridge
312 feet above
Sea-level.

N. end
of Tunnel.

4. Kimeridge Clay.

3. Ampthill Clay.

2. Corallian rock-beds.

1. Oxford Clay.

The following invertebrate fossils have been recorded from the Ampthill Clay of Bedfordshire and Cambridgeshire :—*

Ammonites achilles.
—— cawtonensis.
× —— cordatus.
—— —— var. excavatus.
× —— plicatilis.
—— vertebralis.
× Belemnites abbreviatus.
—— nitidus.
Nautilus hexagonus.
Alaria bispinosa.
Pleurotomaria.
Arca longipunctata.
—— rhomboidalis.
× —— subtetragona.
Astarte supracorallina.
Cardium.
Corbula Deshayesea.
Cucullæa contracta.
× Exogyra nana.
× Gryphæa dilatata.
Leda.
Lima pectiniformis.
Lucina aliena.
Myacites decurtatus.
Nucula Menkei.
Ostrea deltoidea.
× —— discoidea.
× —— gregaria.
× Pecten articulatus.
—— fibrosus.
—— lens.
—— Thurmanni.
× Pinna lanceolata.
Plicatula.
Thracia depressa.
Trigonia clavellata.
—— paucicosta.
Discina humphriesiana.
Serpula intestinalis.
—— tetragona.
—— tricarinata.
Cidaris florigemma.
—— Smithi.
× Pentacrinus.
Webbina irregularis.

× Specimens obtained by Mr. A. C. G. Cameron, Mr. Rhodes, and myself, from Ampthill and Gamlingay.

There is more evidence of a mingling of Oxford and

* This list is based on that of T. Roberts, Jurassic Rocks of Cambridge, p. 49. See also Quart. Journ. Geol. Soc, vol. xlv. p. 556; Blake and Hudleston, *Ibid.*, vol. xxxiii. p. 313; J. Saunders, Geol. Mag., 1890, pp. 117, 127.

K 2

Kimeridge Clay forms than we find to be the case in the area where the Corallian stone-beds are developed. This is natural enough, for the conditions of sedimentary deposition were practically maintained throughout this great argillaceous series from Oxfordshire to Lincolnshire.

Traced beyond Ampthill the boundary of the Oxford and Kimeridge Clays is largely concealed for some distance by the Cretaceous Rocks of Biggleswade, Sandy, and Potton ; and it is interesting to note that no Corallian fossil, such as characterize the stone-beds, are recorded from the coprolite-beds of Brickhill and Potton, where so many other derived Jurassic fossils occur.[*]

In the great mass of clays there occur at various horizons calcareous and gritty beds : some in undoubted Oxford Clay, as at St. Neots, others on or about the junction of the two Clays, and some in the Kimeridge Clay.

Particular attention was drawn to these by Prof. Seeley, who from observations in brickyards and from the evidence of wells, noted the occurrence of rock-beds in the Fen Clays at Tetworth, Gamlingay, Papworth St. Everard (High Papworth), Bourn (south-east of Caxton), Elsworth, Boxworth, Conington, St. Ives, Holywell (?) and Bluntisham.[†]

In the absence of fossil evidence, little can be said about some of these rocks, but from clay overlying rock-beds at Gamlingay (Gamlingay Clay), Elsworth, and Bluntisham, *Gryphœa dilatata*, together with *Ostrea deltoidea*, have been recorded, and at Gamlingay *Ammonites biplex* was also found.

Tetworth, near Everton, is about 6 miles S.S.E. of St. Neots, in Huntingdonshire, and from a brickyard on the top of the hill, Prof. Seeley recorded the following fossils :—

Ammonites achilles.	Lima pectiniformis
Belemnites abbreviatus (excentricus).	Ostrea deltoidea.
Gryphœa dilatata.	Serpula tetragona.

To the south-east lies Gamlingay, and in brickyards worked on the northern side of the London and North-Western (branch) Railway, near Gamlingay Bogs, the Ampthill Clay has been well exposed beneath coverings of Lower Greensand. The section at the Belle Vue Steam Brick Works has been recorded by T. Roberts ; to this I have added a few particulars, some of which were communicated by Mr. Cameron, as follows :—[‡]

---

* See Teall, Potton and Wicken Phosphatic Deposits, 1875 ; and W. Keeping, Fossils of Upware and Brickhill, 1883.

† Ann. Nat. Hist., ser. 8, vol. viii. p. 504, and vol. x. p. 100 ; Hughes, Proc. Geol. Assoc., vol. viii. p. 401. See also Penning and Jukes-Browne, Geol. Cambridge, pp. 5, 167 ; T. Roberts, Quart. Journ. Geol. Soc., vol. xlv. p. 547.

‡ Jurassic Rocks of Cambridge, p. 87.

		Fr.
Lower Greensand.	} False-bedded ochreous sand - - -	20
Ampthill Clay -	⎧ Greyish black clays, shaly at the top, and	
	nipped up in places - - - -	11
	Grey argillaceous limestone, shivered in places	
	8 in. to	1
	Black clays - - - - -	9
	Grey argillaceous and nodular limestone (floor	
	of pit) - - - - -	1
	Clay, with two bands of stone (no water),	
	⎩ proved to depth of - - - -	33

Roberts records the following fossils from the clays above the base of the pit :—

Ammonites biplex ?	Gryphæa dilatata.
Belemnites abbreviatus.	Ostrea discoidea.
Alaria bispinosa.	Pecten fibrosus.
Cucullæa contracta.	—— Thurmanni.

The lower bed of grey argillaceous limestone is very shelly in places, and it contains pyritic nodules; it now forms the floor of the pit, and contains *Serpulæ* and *Ostrea* as noted by Roberts. From this bed Mr. Cameron and myself obtained *Ammonites*, *Exogyra nana*, *Pecten articulatus*, *Serpula tricarinata*, *Cidaris florigemma* (spine), and *Webbina irregularis*. (See p. 136.)

From the clays red and white bricks are made ; while the beds of limestone are burnt for lime.

Elsworth and St. Ives.

Proceeding some miles to the north-east, past Caxton, we come to Elsworth, and there a rock-bed occurs, to which attention was long ago directed by Lucas Barrett, who regarded it as " Upper Calcareous Grit."

The Elsworth Rock was described by Prof. Seeley in 1862, when he made excavations to prove the strata.[*] The uppermost beds were found to comprise three layers of hard whitish-grey rock, 6 to 8 in. thick, yielding *Gryphæa dilatata*. At a lower horizon the following beds were noticed :—

		Fr.	In.
[Ampthill Clay] -	⎧ Dark blue laminated clay (reddish-		
	brown at base), with minute crystals		
	of selenite, &c. *Ammonites vertebralis*,		
	⎩ *Gryphæa dilatata* - - -	6	6
Elsworth Rock, 14 feet.	⎧ Rock-bed, like that below - -	1	6
	Brown-black clay, with *Ostrea flabel-*		
	loides (*Marshi*); this bed passes into		
	sandstone - - - -	5	0
	Dark blue homogeneous limestone, with		
	ironshot oolitic grains, and iron		
	⎩ pyrites - - -3 0 to	7	0

The Elsworth Rock was then considered to be the uppermost zone of the Oxford Clay ; it is now regarded as equivalent to the St. Ives Rock; and both are grouped with the Lower Calcareous Grit, a view suggested by Messrs. Blake and Hudleston, and confirmed by Thomas Roberts.

---

* Ann. Nat. Hist., ser. 3, vol. viii. p. 504; and vol. x. pp. 98, 99, 107, 109; Sedgwick, Lecture on the Strata near Cambridge, 1861, p. 23.

With regard to the three layers of grey rock above noted, they may belong to a bed of calcareous grit mentioned by Penning and Jukes-Browne, who say, "The grit is probably continuous with a small exposure of calcareous rock, a foot or more thick, with *Gryphæa dilatata*, just west of Elsworth, a bed not to be confounded with the 'Elsworth Rock.'"[*]

Many fossils have been recorded from the Elsworth and St. Ives Rocks, by Prof. Seeley, Messrs. Blake and Hudleston, and by Messrs. Penning and Jukes-Browne. The accompanying list is that revised by T. Roberts.[†]

Corallian Fossils.	Elsworth Rock.	St. Ives Rock.	Corallian Fossils.	Elsworth Rock.	St. Ives Rock.
Hybodus grossiconus	-	2	Lithodomus	1	2
Ammonites Achilles	1	-	Lucina Beani (globosa)	1	-
— canaliculatus	1	-	Modiola bipartita	1	2
— convolutus	1	-	— cancellata	-	2
— cordatus	1	2	Myacites jurassi	1	2
— Goliathus	1	-	— recurvus	1	2
— Henrici	1	2	Myoconcha	1	2
— perarmatus	1	2	Nucula	1	2
— plicatilis	1	-	Opis angulosa	1	-
— vertebralis	1	2	Ostrea discoidea	-	2
Belemnites hastatus	1	2	— flabelloides	1	2
Nautilus perinflatus	1	-	— gregaria	1	2
Littorina Meriani	1	2	Pecten articulatus	1	2
Natica calypso var. tenuis	1	2	— fibrosus	-	2
— clymenia	1	-	— lens	1	2
Nerinæa	1	2	— vagans	1	2
Patella	1	-	Perna mytiloides	-	2
Pleurotomaria granulata	1	2	Pholadomya æqualis	1	2
— Münsteri	1	2	— concentrica	1	-
Anatina siliqua	-	2	— paucicosta	1	-
Anomia	-	2	Pinna lanceolata	1	2
Arca æmula	1	-	Placunopsis	1	-
— terebrans	1	-	Plicatula fistulosa	1	-
Astarte ovata	1	-	Thracia depressa	1	2
— robusta	1	2	Trigonia elongata	1	2
Avicula braamburiensis	1	2	— Hudlestoni	1	-
— expansa	1	-	— perlata	1	-
— inæquivalvis	1	-	Unicardium depressum	1	-
— ovalis	1	-	Terebratula insignis	1	-
— pteropernoides	1	-	Waldheimia bucculenta	-	2
Cardium Crawfordi	1	2	— Hudlestoni	-	2
Cucullæa clathrata	1	-	Serpula	1	2
— elongata	1	-	Glyphea	1	-
— oblonga	1	-	Goniocheirus cristatus	-	2
Exogyra nana	1	2	Apiocrinus	1	-
Goniomya literata	1	2	Millericrinus echinatus	1	-
Gryphæa dilatata	1	2	Pentacrinus	1	-
Hinnites abjectus	1	-	Cidaris florigemma	-	2
— tumidus (velatus)	-	2	— Smithi	1	-
Isocardia globosa	1	2	Collyrites bicordata	-	2
Lima duplicata	1	-	Holectypus depressus	-	2
— læviuscula	1	2	Pseudodiadema versipora	1	2
— pectiniformis	1	-	Thecosmilia	1	-
— rigida	1	2			

The Boxworth Rock was described by Prof. Seeley as occurring in two layers, the upper, hard and dark blue, with a few small shells; and the lower part pale brown, and composed almost entirely of shells, chiefly univalves. The rock, locally called "flint," was 18 inches thick. The fossils included the following :—

---

[*] Geology of Cambridge, p. 4.
[†] Jurassic Rocks of Cambridge, p. 25. See also Seeley, Ann. Nat. Hist., ser. 3, vol. x. p. 109; Blake and Hudleston, Quart. Journ. Geol. Soc., vol. xxxiii. p. 313; Roberts, *Ibid.*, vol. xlii. p. 543; Penning and Jukes-Browne, Geology of Cambridge, p. 7; Bonney, Cambridgeshire Geology, p. 11.

Ammonites biplex.
—— alternans?
Alaria bispinosa.

Cerithium muricatum.
Pecten lens.

The clay beneath yielded *Ostrea deltoidea*, *Gryphæa dilatata*, and a species like *O. læviuscula* named *O. discoidea*.* (See Fig. 62, p. 136.) Saurian remains also occur.

The Boxworth Rock was regarded by Prof. Seeley as on a higher horizon than the Elsworth Rock. Messrs. Penning and Jukes-Browne observe that the brickyard is in bluish-grey clay, " with two thin layers of whitish sandy limestone, separated by a foot of clay, and very fossiliferous. There is said to be a layer of septaria, several feet below these limestone bands."† At the time of my visit a bed of pale septarian limestone called "Clunch," said to be about 2 feet thick, was shown at the base of the shallow pit, and I was informed that clay had been worked beneath it.

Lately, Mr. Cameron has found a rock-bed at Hilton, south-east of Huntingdon, and this has yielded *Ammonites* and *Pholadomya*.

Another rock-bed, known as the " Red Rock of St. Ives," has been opened up in the brickyard to the west of St. Ives. There Prof. Seeley found abundant remains of a rock, which an old brickmaker told him " once extended continuously all over the pit to a thickness of 3 feet, quite at the surface, and sometimes parted into two beds by an intervening layer of clay. Where visible, the rock here is much weathered, but is the same kind of reddish-brown deposit, full of oolitic grains, which occurs at Elsworth."

Beds of rock, dipping eastward, have also been noticed in the brickyards to the north and north-east of St. Ives, but, as remarked by Prof. Seeley, " whether this rock is an extension of that of Elsworth, or another bed inferior or superior to it, is a very complex question, difficult to answer."‡

Roberts has noted the St. Ives section as follows :—§

		Ft.	In.
	Yellowish brown calcareous clay .	0	7
St. Ives Rock	Brown ferruginous limestones, with decomposed nodules of pyrites in upper part - - - .	1	6
	Brown, sandy, fossiliferous limestone, with oolitic grains of oxide of iron; nodular towards base - .	0	6
Oxford Clay -	Grey calcareous sandy clay - .	0	5

A list of fossils from the St. Ives Rock has been given previously (p. 140; see also p. 55).

Upware.

On the eastern side of the River Cam, about 6 miles south of Ely, there is an "Ancient Bank" or ridge formed of Corallian

* Ann. Nat. Hist., ser. 3, vol. x. p. 104. See also p. 107.
† Geology of Cambridge, p. 6.
‡ Ann. Nat. Hist., ser. 3, vol. x. p. 101, and vol. viii. p. 504.
§ Jurassic Rocks of Cambridge, p. 21.

limestones. It extends from the Inn known as " Five miles from Anywhere," at Upware, northwards for nearly three miles to Padney.

The occurrence of this rock was, of course, well known to agriculturists, and it was mentioned by Vancouver in 1794, who, in describing the parish of Wicken (" Wickin "), says, of this portion, " The arable land consists of a deep brown mould, upon a dry bed of ragstone."*

The attention of Sedgwick was early directed to this exposure of rock, and collections of fossils were made by him. He does not appear, however, to have been acquainted with it before 1826, but he drew the attention of Fitton to the strata, and that geologist published the earliest account of them, grouping them as portions of the " Oxford Oolite," and noting some of the fossils.†

It is remarkable that between Wiltshire and East Yorkshire this is the only known occurrence of beds that correspond in character with the Coralline Oolite ; while the nearest beds of Coral-rock are those of the neighbourhood of Oxford.

The Upware Limestone has been opened up in three quarries, and only the southern of these is occasionally worked for road-metal. The quarry east of High Fen Farm, has long been obscured. That to the south-west of the Farm is now abandoned, and partially filled with water. When I last visited the section, in 1889, it showed about 9 feet of crumbly oolitic and pisolitic limestone, not unlike the Osmington Oolite of Weymouth. About 12 feet of false-bedded rock has been opened up, and it has yielded in abundance *Echinobrissus scutatus* and *Holectypus depressus*.

Further south the oolite was shown in several places by the side of the lane leading towards the southern quarry, and by that leading from Wicken Lamas Ground towards Fen Side. In the field to the south of this lane, I picked up pieces of oolite and also some corals, suggesting that thereabouts the rock shown in the southern quarry (Coral Rag) overlies that of the northern (Coralline Oolite).

In the southern quarry, which is about half a mile north of Upware Inn, the beds are shown to a depth of about 20 feet, and consist of pale . cream-coloured oolite, bluish-grey where unweathered, and false-bedded.

In the upper, 8 or 10 feet, there are well-marked coral-beds with *Thamnastræa, Isastræa,* and *Stylina,* also *Lithodomus inclusus, Pecten articulatus,* &c. The lower beds consist largely of shell fragments, while, at the base, there were oolitic shelly and marly limestones, resembling beds exposed in the northern quarry. This sequence was inferred by Messrs. Blake and Hudleston, who have given an excellent account of the beds and their fossils.

The following list of fossils from the Upware Limestone is taken from that of T. Roberts; the species marked N. being obtained

---

* General View of the Agriculture of Cambridge, p. 134.
† Trans. Geol. Soc., ser. 2, vol. v. pp. 307, 317, pl. xa.

from the northern, and those marked S. from the southern quarry* :—

Ammonites Achilles, N. S.
—— mutabilis, S.
—— perarmatus, N.
—— plicatilis, N. S.
—— trifidus, N.
—— vertebralis *var.* cawtonensis, S.
Belemnites abbreviatus, N.
Alaria, S.
Amberleya princeps, S.
Bourguetia (Phasianella) striata, S.
Cerithium muricatum, S.
Chemnitzia heddingtonensis, S.
Emarginula Goldfussi, S.
Fissurella corallensis, S.
Littorina Meriani, N.
—— muricata, N. S.
Natica clymenia, S.
—— clytia, S.
Neritopsis decussata, S.
—— Guerrei, S.
Pleurotomaria reticulata, S.
—— sp., N.
Trochotoma tornata, S.
Trochus, S.
Anomia suprajurensis, S.
Arca æmula, S.
—— anomala, S.
—— contracta, S.
—— pectinata, S.
—— quadrisulcata, S.
Astarte aytonensis, S.
—— ovata, S.
Cardita ovalis, S.
Cardium (c.f.) delibatum, S.
Cucullæa elongata, S.
Cypricardia glabra, S.
Exogyra nana, S.
Gastrochæna moreana, S.
Gervillia angustata, S.
—— aviculoides, N. S.
Goniomya v.-scripta, S.
Hinnites tumidus, S.
Homomya tremula, S.
Isoarca grandis (multistriata), S.
—— texata, N.S.
Lima elliptica, S.
—— gibbosa, S.
—— læviuscula, S.
—— rigida, S.
—— rudis, S.
Lithodomus inclusus, S.
Lucina Beani, S.
—— moreana, S.

Modiola bipartita, N. S.
—— (c.f.) rauraciencis, S.
—— subæquiplicata, S.
Myacites decurtatus, S.
—— recurvus, S.
Myoconcha Sæmanni, S.
—— texta, S.
Mytilus jurensis, N.
—— pectinatus, S.
—— ungulatus, N. S.
Opis arduennensis, S.
—— corallina, S.
—— Phillipsi, N. S.
—— virdunensis, S.
Ostrea gregaria, S.
—— solitaria, S.
Pecten articulatus, S.
—— fibrosus, N.
—— inæquicostatus, S.
Perna subplana, S.
Pholadomya decemcostata, S.
Pinna lanceolata, S.
Pleuromya Voltzi, S.
Plicatula fistulosa, S.
Quenstedtia lævigata, S.
Trigonia Meriani, S.
Rhynchonella, S.
Terebratula insignis *var.* maltonensis, S.
Gastrosacus Wetzleri, S.
Glyphea, S.
Prosopon marginatum, S.
Serpula tetragona, S.
Vermicularia, S.
Apiocrinus polycyphus, S.
Cidaris florigemma, S.
—— Smithi, S.
Collyrites bicordata, N. S.
Echinobrissus scutatus, N. S.
Hemicidaris intermedia, S.
Holectypus depressus, N. S.
Hyboclypus gibberulus, N.
—— sp. S.
Millericrinus, S.
Pentacrinus, S.
Pseudodiadema versipora, N.
Pygaster umbrella, N. S.
Stomechinus gyratus, S.
Isastræa explanata, S.
Montlivaltia dispar, S.
Rhabdophyllia Phillipsi, S.
Stylina tubulifera, S.
Thamnastræa arachnoides, S.
—— concinna, S.
" Scyphia," S.

---

*Roberts, Jurassic Rocks of Cambridge, p. 53. The above list is revised from that given by Roberts in the Geology of parts of Cambridgeshire and Suffolk (Geol. Survey), p. 12. See also Blake and Hudleston, Quart. Journ. Geol. Soc., vol. xxxiii. p. 314, and Geol. Mag., 1878, p. 92.

The apparently isolated nature of the Corallian Beds at this locality has been remarked upon by many geologists. Prof. Seeley employing the name "Upware Limestone" regarded it as a reef; Messrs. Blake and Hudleston, and Prof. Bonney have also attributed its local occurrence rather to its "reef-like character" than to denudation.* Sedgwick, too, long ago remarked that the rock "does not form a continuous band concealed under the fens, as several wells have been sunk (between Cambridge and Lynn) through the *Kimmeridge clay* into the *Oxford clay* without passing through any beds which bore a resemblance to the Coral Rag."†

On the western side of the Upware ridge, between the two open quarries, there is a fringe of Lower Greensand and Gault, and beneath them the Kimeridge Clay was proved to rest on the Corallian limestone. The Lower Greensand rested irregularly on the eroded Kimeridge Clay, overlapping its margin and resting itself directly against the bank of Corallian Beds, so that no Kimeridge Clay was exposed at the surface. Details of this section were published by Messrs. H. and W. Keeping.

On the eastern side of the Corallian ridge no Kimeridge Clay has been proved, for in pits near Spinney Abbey Farm, the Corallian Beds were found by W. Keeping and E. B. Tawney, directly beneath the Lower Greensand at a depth of 8 feet 9 ins. from the surface. The Corallian rock was described as a "hard, gritty, bedded limestone; grey coloured, with scattered large oolitic grains."‡

The accompanying diagram is intended to show the general structure of the region at Upware; it is based on the evidence furnished by borings in the vicinity. It is probable that the Upware Limestone is to some extent, at any rate, replaced by the Ampthill Clay, as roughly represented in the diagram. When we compare the fossils enumerated from the Ampthill Clay (p. 137) with those from the Upware Limestone, their close connexion is evident.

A well-boring made (1885) at Chettering Farm, half a mile south-east of Stretham Ferry, and two and a half miles north-west of Upware, was described by Thomas Roberts, who has given the following details:—§

		Ft.	In.
Alluvial Deposits {	Peat	5	0
	Sand	3	6
Lower Greensand		13	6
Kimeridge Clay {	Black clays, with greyish limestone bands; the latter were about 1 ft. thick, and intervening clays 3 ft. 6 in. The clay contained some phosphatic nodules	36	0
	Laminated clays, in which *Ammonites alternans* was found	12	0
	Clays with black phosphatic nodules	3	0

* A fuller account of various views concerning the Corallian Beds of Upware is given in the Memoir on the Geology of parts of Cambridgeshire and Suffolk, pp. 8–11.
† Rep. Brit. Assoc. for 1845, Sections, p. 43.
‡ Neocomian Deposits of Upware and Brickhill, 1883, pp. 3, 7.
§ Jurassic Rocks of Cambridge, pp. 23, 47, 65.

		FT.	IN.
Ampthill Clay -	Black clay with iron-pyrites; with 2 feet of dark sandy limestone near base - - - - -	26	0
Lower Calcareous Grit (= Elsworth Rock).	Hard greyish limestone, crowded with oolitic grains of oxide of iron; *Pholadomya æqualis* - - -	8	0
	Light brown sandstone - - -	11	0
Oxford Clay, 81 feet.	Alternating beds of clay and limestone, the latter 8 to 9 in. thick - -	12	0
	Clays with some fragments of fossils -	57	0
	Argillaceous limestone - -	2	0
	Clay, with small pyritized Ammonite and *Ostrea* - - - -	10	0
		199	0

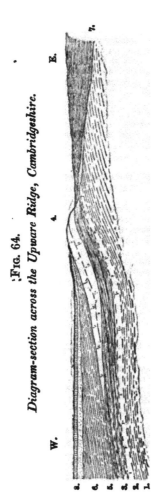

FIG. 64.

Diagram-section across the Upware Ridge, Cambridgeshire.

E.

W.

8. Alluvium.
7. Gault.
6. Lower Greensand.
5. Kimeridge Clay.

4. Upware Limestone.
3. Ampthill Clay.
2. Lower Corallian Rock-beds.
1. Oxford Clay.

A boring at Dimmock's Cote, Stretham Fen, on the western side of the Cam, and but a little distance from the Upware ridge proved the following strata :—

			FEET.
Alluvial Deposits.	{ Black earth - - - - -		3
	{ Peat - - - - -		18
Gault ? -	- Blue clay - - - -		110
Lower Greensand ?	} Rock and sand - - - -		10
			141

The Blue clay was considered by Messrs. Penning and Jukes-Browne to be Kimeridge Clay;* but it seems by no means improbable it may be Gault, with underlying Lower Greensand. On consulting with Mr. Jukes-Browne (1894), he states that 110 feet is more than the known thickness of Gault; but that the blue clay may in part be Gault.

In another boring in Stretham Fen (Mr. Feust's Farm) "Rock and black sand" were reached at a depth of 30 feet, there being only 18 feet of blue clay above, with a capping of 12 feet of sand and gravel. Roberts thought the "rock and sand" might be the same as that met with in the Chettering boring, and which there clearly appears to represent the Lower Calcareous Grit; † but it seems to me the evidence on the whole favours the view that Lower Greensand and Gault have alone been proved in the Stretham Fen, and the Geological Survey Map tends to support this view.

A boring at Wicken was carried through soil 3 feet, and then blue clay, with bands of black rock (1 foot thick), to a depth of 200 feet. Wicken village is on the Gault, but, as noted by Mr. Jukes-Browne, its thickness near by, at Soham, has been proved to be but 90 feet. The Wicken boring affords no indication of Lower Greensand, but that formation may occur in outlying portions beneath the Gault, for indeed the Gault overlaps it at Upware. Hence the Wicken boring may have been carried into Oxford Clay.

At Soham 12 feet of "red sand and rock" were found beneath the Gault. At another well at Soham Fen, sunk 120 feet to rock, the water overflowed, but was brackish.

At Mildenhall it seems not unlikely that Jurassic clay was reached at a depth of 261 ft. 6 ins. beneath Chalk, Gault, and Lower Greensand. The record of Sir H. Bunbury, quoted by Mr. Whitaker, gives "Blue clay with fragments of large shells," 9 or 10 feet thick, and penetrated at the depth of 261½ to 271 feet from the surface. This may, as suggested by Mr. Whitaker, be Kimeridge Clay; it may be Oxford Clay.

In a boring at Culford, between Mildenhall and Bury St. Edmunds, Lower Greensand was found to rest on Palæozoic rocks at a depth of 637½ feet.‡

At the Woolpack Inn, Cambridge, a boring, noted by Mr. Whitaker, was made to a depth of 433 feet. The lowest 15 feet, beneath the Lower Greensand, comprised rock, clay, white marl, and dark sand—possibly Corallian Beds.§

In 1836 a boring was made at Saffron Walden in Essex, and this was continued to a depth of 1,004 feet. The records that have been published give no satisfactory particulars of the lower strata, all we know is that 10 feet of gravel, and 265 feet of Chalk were penetrated, after which the bore-hole was continued in "chalk-marl, containing numerous shells with pyrites." It was considered likely by Mr. Whitaker that this "chalk-marl," included not only Gault, but probably Kimeridge and Oxford Clays.‖

Mr. T. Roberts has shown that the basement-bed of the Kimeridge Clay contains numerous phosphatic nodules, which

---

* Geology of Cambridge, p. 165.
† Jurassic Rocks of Cambridge, p. 59.
‡ Whitaker and Jukes-Browne, Quart. Journ. Geol. Soc., vol. l. p. 488.
§ Geology of parts of Cambridgeshire and Suffolk, pp. 111, 115.
‖ Geol. N.W. part of Essex, p. 79; Geol. Mag., 1890, p. 515. See also Proc. Norwich Geol. Soc., vol. i. p. 28.

give a well-defined upper boundary to the Ampthill Clay [*] We have therefore the following general sequence:—[†]

Kimeridge Clay, with phosphatic nodule-bed at base.
Corallian   { Upware Limestone and Ampthill Clay.
Beds.   { Elsworth Rock.
Oxford Clay.

Willingham, Bluntisham, and the Fenland.

Turning to the main line of outcrop which occurs in the neighbourhood of Willingham, we find evidence of the base of the Kimeridge Clay to the east of that village. My attention was directed in 1882 by Prof. Hughes to some sections at the west end of Balsar's Hill.[‡] There I obtained the following fossils:—

Ammonites alternans.	Ostrea deltoidea.
Avicula costata.	Serpula variabilis.
Exogyra nana.	Vermilia sulcata.
Gastrochæna ?	

Small black "coprolites" were very numerous, and these I took to belong to the Kimeridge Clay, a view confirmed by the later observations of T. Roberts.[§] The old brickyard north of Willingham appears to have been worked in Oxford Clay.

Further to the north-west we come to Bluntisham, where Prof. Seeley described the "Bluntisham Clay." At this locality in 1874 Mr. Skertchly noticed a section showing Boulder Clay, underlaid by Kimeridge Clay (?), and with Oxford Clay beneath, yielding *Gryphea dilatata* and numerous fragile Ammonites.[||]

Prof. Seeley has noted that in the bottom of the railway-cutting, west of Bluntisham, an iron-shot oolitic and shelly rock, resembling that of Elsworth, was proved. Roberts remarks that the clay now seen in the cutting is certainly the Ampthill Clay, so that probably the Bluntisham rock is Lower Calcareous Grit.[¶]

Through the Fenland area from the neighbourhood of Bluntisham northwards, the Corallian clays are probably present to the west of Chatteris and March.

At March both Oxford and Kimeridge Clay fossils have been recorded from the brickyards,[**] and further north in a deep boring (made about 1812) at Allen's Brewery, Lynn, clays were penetrated to a depth of 630 feet beneath about 50 feet of Alluvium.[††] The lower portion of these clays yielded the following species, which were recorded in 1835 by C. B. Rose:—

Ammonites decipiens.	Ostrea (Gryphæa) bullata.
—— excavatus.	Thracia (Mya) depressa.
Belemnites abbreviatus.	Serpula tricarinata.

---

[*] Quart. Journ. Geol. Soc., vol. xlv. p. 547.
[†] This agrees with that given in 1871, by Prof. Morris, Proc. Geol. Assoc., vol. ii. pp. 220, 223 (Excursion to Cambridge).
[‡] Geology of parts of Cambridgeshire and Suffolk, pp. 14, 18.
[§] Jurassic Rocks of Cambridgeshire, p. 62.
[||] Geology of parts of Cambridgeshire and Suffolk, p. 113.
[¶] Jurassic Rocks of Cambridgeshire, p. 21.
[**] Skertchly, Geol. Fenland, pp. 192, 268, 317.
[††] Phil. Mag., ser. 3. vol. vii. pp. 173, &c. See also Fitton, Trans. Geol. Soc., ser. 2, vol. iv. p. 316; and Whitaker, Geol. S.W. Norfolk, &c. p. 158.

To these Fitton added *Gryphœa dilatata* and *Ostrea deltoidea*.

Although identified so long ago, it is noteworthy that all these species are among those enumerated by T. Roberts from the Corallian clays of Lincolnshire,[*] and we are justified therefore in concluding that we have here the representatives of these beds or of the Ampthill Clay of other parts.

It is possible that Corallian clay occurs also at Denver Sluice, south-west of Downham Market, for Rose has recorded *Ammonites decipiens* from dark blue clay beneath the Alluvium at this locality.

Lincolnshire.

Passing into Lincolnshire we traverse a considerable area where the Kimeridge Clay and the passage-beds into the Oxford Clay (of Corallian Age) are concealed beneath the Alluvium of the Fenland. At Billinghay we find Oxford Clay and at Tattershall Kimeridge Clay, the intermediate ground being concealed by the Alluvium of the Witham.

Further north from Bardney onwards to near Barnetby-le-Wold an approximate boundary-line between Kimeridge and Oxford Clays has been drawn on the Geological Survey Map, but the tract is very largely covered by Drift, so that definite fossil evidence is obtained only here and there.

Referring to the southern portion of this area Mr. Jukes-Browne has stated (1888) that "the Corallian group [*i.e.*, as a rock-formation] is entirely absent, and the Oxford Clay passes up into the Lower Kimeridge without the development of any bed which can be taken as forming a line of division." He remarks that "Near Bardney the position of this zone of passage can be fixed with tolerable certainty," and he further observes that "The Lower Kimeridge Clay contains many species which are usually considered as Oxford Clay forms, such as *Ammonites plicatilis* and *Gryphœa dilatata*."[†] The northern portion of this area was examined by Mr. Ussher, and he obtained near Wrawby evidence of the palæontological boundary between the clays.[‡]

In the meantime (1889) Thomas Roberts published the results of his researches on this subject. He examined the country with the view of ascertaining whether any clays corresponding to the Ampthill Clay occur in Lincolnshire between the Oxford and Kimeridge Clays; and he came to the conclusion that there was on this horizon a band of black selenitiferous clays with occasional septaria, and with *Ostrea deltoidea* and *Gryphœa dilatata*, which correspond both lithologically and palæontologically with the clay at Ampthill, and may therefore be considered of Corallian age. The maximum thickness observed was 17 feet, and the beds were seen in brickyards west of Hawstead Hall, near Stixwould; on

---

[*] Quart. Journ. Geol. Soc., vol. xlv. p. 556.
[†] Geol. Lincoln, pp. 73, 78, 79.
[‡] Geol. North Lincolnshire, p. 99 (1890).

Bardney Common; north of South Kelsey; and west of North Kelsey railway-station. Evidence was also obtained of these beds in the railway-cuttings east of Brigg and near Wrawby. Roberts regarded the beds in the western half of the Wrawby-Bridge cutting as Corallian; there, and in another cutting a quarter of a mile north-west, the black clays with selenite were exposed. At Wrawby-Bridge these clays are succeeded by dark clays with brownish iron-stone concretions, yielding other fossils. The list of species from Wrawby published by Mr. Ussher includes also forms obtained from the base of the Kimeridge Clay; but he states that *Ammonites rotundus* has been found in association with *Gryphæa dilatata*, both at Wrawby and at South Kelsey.*

In the brickyard south-east of Bardney thin gritty bands occur in the dark shaly clay, but at the time of my visit (1889) the clay had not been dug for the past three years. A list of fossils has been published by Mr. Jukes-Browne,† and this includes some species not recorded (or differently named) by Roberts. These clays overlie the Oxford Clay with *Ammonites cordatus, A. perarmatus*, &c., and are surmounted by Kimeridge Clay, with *Ostrea deltoidea* in abundance. All the species observed by Roberts also occur in the Corallian Beds of other localities, nearly all of them occur in the Ampthill Clay, and not one of them is peculiar to either Oxford or Kimeridge Clay. The association of *Gryphæa dilatata* and *Ostrea deltoidea* is characteristic of the passage-beds between these clays. In Lincolnshire he states that the former species is most common in the lower part of the selenitiferous clays, whilst the latter is more frequently met with in the upper part. The careful observations of Roberts therefore enable us to group these intermediate clays in Lincolnshire with the Ampthill Clay. Their thickness is not less than 20 feet, but owing to the absence of any clear continuous sections their full thickness has not been ascertained.

The following is a list of fossils from the Ampthill Clay of Lincolnshire, as recorded by Roberts;‡ those marked × occur also in Bedfordshire :—

Ichthyosaurus.
× Ammonites achilles.
× —— cawtonensis.
× —— cordatus.
—— —— var. excavatus.
—— decipiens.
× —— plicatilis.
× —— vertebralis.
× Belemnites abbreviatus.
× —— nitidus.
× Alaria bispinosa.
× Arca longipunctata.
× Astarte supracorallina.

Avicula pteropernoides.
Cardium striatulum.
× Cucullæa contracta.
× Gryphæa dilatata.
Ostrea bullata.
× —— deltoidea.
× Pecten fibrosus.
Pholadomya concentrica.
× Pinna lanceolata.
× Thracia depressa.
Rhynchonella.
Serpula tricarinata.

---

* Geol. N. Lincolnshire, pp. 99–101, 104. See also J. F. Blake, Quart. Journ. Geol. Soc., vol. xxxi. p. 209.

† Geol. Lincoln, p. 79.

‡ Quart. Journ. Geol. Soc., vol. xlv. p. 556; see also Ussher and others, Geol. Lincoln, p. 79, and Geol. N. Lincolnsh, p. 101.

Sussex and Kent.

In the Sub-Wealden Boring representatives of the Corallian Beds have been recognized in certain strata passed through, but it is not possible to clearly mark off their limits. They come in the series between the depths of 1,325 and 1,806. In the detailed account of the boring (p. 347) I have limited them to the beds proved from 1,565 to 1,806 feet—a thickness of 241 feet.

Mr. Topley has remarked that "The 'true Coralline Oolite,' 17 feet thick, was met with at 1,769 feet from the surface." Mr. Hudleston was inclined to regard the beds at from 1,330 feet downwards as Corallian. *Gervillia aviculoides* was met with at depths of 1,380 and 1,381 feet; *Ammonites mutabilis* is recorded, by Mr. Etheridge, down to a depth of 1,562 feet.* These, however may belong to the Kimeridgian Beds.

Pinna lanceolata, Trichites, and *Rhynchonella pinguis* are recorded from a depth of 1,656 feet. Distinctly oolitic limestones occur in the lower beds at depths of 1,769 to 1,786 feet, and again between 1,865 and 1,880 feet. The last-named band is grouped with the Oxfordian Beds. I am indebted to Mr. E. Crane and to Mr. B. Lomax for showing me these and other specimens in the Brighton Museum. A sample from the depth of 1,775 feet was shelly oolite with *Trigonia.* Specimens from a depth of 1,587 feet showed somewhat concretionary calcareous sandstone.

In the Dover Boring 159 feet of strata (between the depths of 610 and 769 feet) are assigned to the Corallian Beds, by Prof. Boyd Dawkins. (See p. 343.) Higher up, (between 546 and 558 feet) there was found a bed of ironstone 12 feet thick, comparable with that of Westbury or Abbotsbury. Prof Dawkins remarks "It is composed of small dark brown shining grains of hydrated oxide of iron, like millet seed, imbedded in a base partly composed of calcium carbonate and partly of iron carbonate. These grains are oolitic in structure." This iron-ore occurs at an horizon more than 60 feet above the mass of oolite with *Cidaris florigemma* and Corals, and is included in the Kimeridge Clay by Prof. Dawkins. There is no evidence to prove its particular palæontological zone, though there can be little doubt that it belongs to beds on a somewhat higher horizon than the iron-ores of Westbury and Abbotsbury.

---

* Dixon's Geology of Sussex, ed. 2, pp. 155, 159.

CHAPTER VIII.

UPPER OOLITIC.

KIMERIDGE CLAY.

(KIMERIDGIAN.)

GENERAL ACCOUNT OF THE STRATA.

This formation derives its name from the village of Kimeridge, in the so-called Isle of Purbeck, the dark shales and cement-stones of which it consists, being well shown in the cliffs of Kimeridge Bay, and indeed all along the coast between Gad Cliff and St. Albans Head.

The term "Kimeridge Strata" was employed in 1812 by Thomas Webster,[*] and the term "Kimeridge Clay" was used by Buckland in 1818. The Kimeridge Clay consists of dark grey and black shaly clay, with much selenite in places. The clay weathers brown at the surface. The beds are as a rule darker, and in mass more shaly, than the Oxford Clay, and the layers of septaria are more persistent and more frequent. Here and there the shale is very bituminous, and we find occasional bands of pale and somewhat shaly limestone : but there are no marked strati-graphical divisions in the Kimeridge Clay. It varies in thick-ness from over 1,200 feet in the south of England to about 100 feet in Oxfordshire. It is one of the most persistent formations in this country, presenting similar lithological characters through-out its extent from Dorsetshire through the country to Lincoln-shire, and to the Yorkshire coast bordering the Vale of Pickering

It is a marine mud, accumulated in the English area too far from land to be affected by coarse detrital material; for we see no evidence of marginal deposits. In Sutherlandshire very angular debris is commingled with the shales of equivalent age.

The Kimeridge Clay owes its dark tint partly to carbonaceous matter, partly to bisulphide of iron.[†] The decomposition of pyrites by atmospheric agents leads to the formation of selenite, lime being present in the clay itself as well as in the fossils. In some cases the Oyster shells (*O. deltoidea*) have been converted into sulphate of lime, as noted by Dr. Kidd, at Shotover.[‡] The decomposition of the pyrites leads to the formation of limonite, which, in the case of the Headington quarries near Shotover, is washed from the Kimeridge Clay on to the Corallian stone beneath.[§]

---

[*] Englefield's Isle of Wight, p. 187, and Map (Plate 50).
[†] G. Maw, Quart. Journ. Geol. Soc., vol. xxiv. p. 357 ; see also analysis by J. D. Kendall, Trans. N. of Eng., Inst. of Mining Engineers, vol. xxxv. p. 156.
[‡] Outlines of Mineralogy, 1809, vol. i. p. 68.
[§] See Report of Excursion of Geologists' Association to Oxford, 1869, p. 6 (separately printed).

Organic Remains and Zones.

Our particular knowledge of the fauna of the Kimeridge Clay in this country is due largely to the researches of Prof. J. F. Blake.[*] The local ranges of many of the fossils render it difficult to sub-divide the formation into zones that can compare with those of the Lias. As remarked by Mr. Hudleston "There is no formation in the whole Jurassic system as developed in England which is more difficult to tabulate or understand than the *Kimeridge Clay*."[†]

Oppel included in his Kimeridge Group three zones, in ascending order, those of *Astarte supracorallina, Pterocera oceani,* and *Trigonia gibbosa*; thus including beds that in this country would be grouped on the one hand as Corallian and on the other as Portlandian. His zone of *Pterocera oceani* included the main mass of our Kimeridge Clay.[‡] Excluding the Portlandian portion, the following palæontological divisions have been also suggested by Oppel and Waagen:—[§]

Kimeridge Clay.	*Exogyra virgula* (Virgulian)	{ *Ammonites mutabilis* and *E. virgula.*
	Pterocera oceani (Pterocerian)	{ *A. alternans* and *Rhynchonella inconstans.*
	Astarte supracorallina (Astartian).	

Stratigraphically the beds with the Astartian fauna belong in this country mainly to the Corallian Beds, although *Astarte supracorallina* is stated by T. Roberts to be abundant in the Lower Kimeridge Beds at Ely.[||] The occurrence in the Upper Corallian (so-called Kimeridge Grit) of Osmington, of *Ammonites mutabilis* and *Rhynchonella inconstans* attests the close connexion of the formations; while on the continent *Exogyra virgula* is recorded from the Astartian beds as well as from the beds above. Hence these species do not well lend themselves to zonal groupings. In the above grouping no account is taken of our higher beds of Kimeridge Clay, with *Ammonites biplex, Discina latissima,* and *Lingula ovalis.*

There is no need in this country to divide the Kimeridge Clay into more than two zones for general stratigraphical purposes, and these are intimately blended. The zones and sub-zones are as follows:—

ZONES.	SUB-ZONES.	
Ammonites biplex.	Discina latissima.	
	Exogyra virgula.	
Ammonites alternans.	Ostrea deltoidea.	

---

[*] Quart. Journ. Geol. Soc., vol. **xxxi.** p. 196.
[†] Report of Sub-Committee on Classification, Internat. Geol. Congress, 1888, (reprint) p. 101.
[‡] Die Juraformation, pp. 724, 808.
[§] Versuch einer allgemeinen Classification der Schichten des oberen Jura (Munich), 1865.
[||] The Astartian Beds were grouped as Corallian by D'Orbigny; See also T. Roberts, Jurassic Rocks of Cambridge, p. 69.

KIMERIDGE CLAY SAURIANS.

FIG. 65.

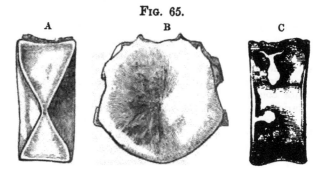

A B C

FIG. 66.

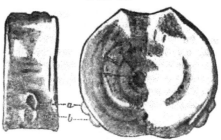

FIG. 67.

FIG. 65. Ichthyosaurus entheciodon, *Hulke*; Centrum of dorsal vertebra: A. Section, B. Anterior aspect, C. Left lateral aspect.

FIG. 66. Ichthyosaurus trigonus, *Owen*; Centrum of dorsal vertebra: *a.* Upper and *b.* Lower costal tubercle.

FIG. 67. Tooth of Geosaurus (Dacosaurus) maximus, *Plien.*

The Lower Beds, when fully developed, are about 400 feet thick, and consist of clays and dark shales with layers of cement-stone and septaria.   They yield near the base *Ammonites muta-bilis, Ostrea deltoidea,* and *Rhynchonella inconstans,* and higher up *Ammonites eudoxus, A. longispinus, A. alternans, Exogyra nana, E. virgula, Aptychus (Trigonellites) latus,* &c.

The Upper Beds, which attain a thickness of 600 or 650 feet in places, comprise dark bituminous shales and paper shales, often with white compressed shells, with many layers of cement-stone and septaria in the lower portion, and higher up occasional bands of pale calcareous shaly rock; they merge gradually upwards into the Lower Portland Beds.   They yield *Ammonites biplex, A. longispinus, Cardium striatulum, Exogyra virgula, Lucina minuscula, L. lineata, Thracia depressa, Discina latissima, Lingula ovalis,* &c.

It should be mentioned that forms identified as "*Ammonites biplex*" range from the Lower Kimeridge Clay to the Portland Beds, and that the species includes several mutations.

Fossils may be found almost everywhere in the Kimeridge Clay, where the formation is exposed, not only in the shales, but sometimes in the cement-stones.   Occasionally the Mollusca are partially replaced or coated by iron-pyrites, and most of them are delicate and brittle, and much flattened, as is usual in shale that has undergone much compression from strata now or formerly overlying it.

Remains of a Cetacean, named *Palæocetus Sedgwicki* by Prof. Seeley, were obtained from the Boulder Clay at Ely, and it has been considered that this fossil may have been derived from the Kimeridge Clay.[*]   Many Saurian remains are found in the formation, including the Dinosaurs *Omosaurus* and *Pelorosaurus,* Crocodiles such as *Dacosaurus* and *Steneosaurus,* Ornithosaurians (Pterodactyl), also *Ichthyosaurus, Pliosaurus,* and *Cimoliosaurus,* and likewise Chelonians.

Among Fishes we find *Asteracanthus, Gyrodus, Hybodus* and *Lepidotus,* but we have usually to be content with the discovery of fin-spines, palatal teeth, and scales.

The commoner Mollusca and Brachiopoda are noted in the accompanying List.   Gasteropods are not numerous, but they include *Alaria, Cerithium, Natica,* and *Pleurotomaria.*   The Crustacea include *Pollicipes.*   Insects, Echinoderms, and Corals are rare.   Foraminifera may be obtained by those who care to wash the clay and extract these tiny organisms from the residue. Few Plant-remains are to be found, but Mr. G. Murray has described a fossil Alga, belonging to the genus *Caulerpa,* which was obtained by Damon from the Kimeridge Clay of Weymouth.[†] Another specimen (also figured by Mr. Murray) was found at Weymouth by Mr. H. W. Fuller, and presented by him to the Museum of Practical Geology.

---

[*] Woods, Catalogue Type Fossils in Woodwardian Museum, 1891, p. 180.
[†] Phycological Memoirs, 1892, p. 11, and Plate v. fig. 1 ; See also Damon, Supplement to Geology of Weymouth, Ed. 3, 1888, Plate xix. figs. 12 and 12a.

KIMERIDGE CLAY FOSSILS.

FIG. 68.

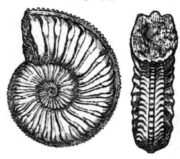

FIG. 69.

FIG. 71.

FIG. 70.

FIG. 68. Ammonites alternans, *von Buch.* ⅔.
  ,, 69. Ostrea deltoidea, *Sow.* ⅔.
  ,, 70. Thracia depressa, *Sow.* ⅔.
  ,, 71. Rhynchonella inconstans, *Sow.* 1¼.

KIMERIDGE CLAY FOSSILS.

FIG. 72. FIG. 73.

FIG. 75.

FIG. 74.

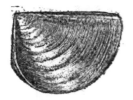 

FIG. 77.

FIG. 76.

FIG. 78.

FIG. 79.

FIG. 72. Ammonites biplex, *Sow.* ⅔.
  ,, 73. —— eudoxus, *d'Orb.* ⅔.
  ,, 74. Aptychus (Trigonellites) latus, *Park.* ⅔.
  ,, 75. Cardium striatulum, *Sow.* Nat. size.
  ,, 76. Lucina minuscula, *Blake.* Nat. size.
  ,, 77. Astarte hartwellensis, *Sow.* ½.
  ,, 78. Exogyra virgula, *Defr.* Nat. size.
  ,, 79. Lingula ovalis, *Sow.* Nat. size.

In the following list the leading fossils of the Upper and Lower Kimeridge Clay are enumerated, and the records show how impossible it is to define any limits. *Ammonites alternans* has been regarded as equivalent to *A. serratus*, and that species has been considered by some to be the same as *A. cordatus* var. *excavatus* ; it is well to bear in mind that at any rate these are closely allied forms.

List of Common and Characteristic Fossils of the Kimeridge Clay.

	Lower Beds.	Upper Beds.
Ammonites alternans (Fig. 68) - -	1	—
—— biplex (Fig. 72) -	–	2
—— Callisto -	1	–
—— cymodoce -	1	
—— decipiens - . -	1	—
—— eudoxus (Fig. 73) -	1	2
—— longispinus .	1	2
—— mutabilis -	1	—
—— orthocera -	1	
—— rotundus - - - -	1	—
Aptychus (Trigonellites) latus (Fig. 74) -	1	2
Belemnites Souichi - - -	–	2
Arca rhomboidalis -	1	2
Astarte hartwellensis (Fig. 77) -	—	2
—— lineata - -	1	2
—— supracorallina -	1	—
Cardium striatulum (Fig. 75) -	1	2
Exogyra nana Fig. 34) -	1	—
—— virgula (Fig. 78) -	1	2
Lucina minuscula (Fig. 76) -	—	2
Nucula Menkei - -	—	2
Ostrea deltoidea (Fig. 69)	1	—
Thracia depressa (Fig. 70)	1	2
Discina humphriesiana -	--	2
—— latissima - -	—	2
Lingula ovalis (Fig. 79) - -	1	2
Rhynchonella inconstans (Fig. 71)	1	—
Serpula intestinalis - -	1	—
—— tetragona -	1	2

CHAPTER IX.

KIMERIDGE CLAY.

LOCAL DETAILS.

Dorset Coast.

The Kimeridge Clay is well exhibited in the dark cliffs that extend from the base of Gad Cliff, eastwards by Kimeridge Bay to the foot of Emmit Hill and St. Alban's Head. At both ends of these cliffs we find the higher beds of Kimeridge Clay merging upwards through dark sandy clays into the Lower Portland Beds. The general structure presented to view is that of a gentle anticline, the denuded summit of which, in Kimeridge Bay, does not reveal the lowest beds of the Kimeridge Clay. In these cliffs therefore we can study only a portion of the formation. It comprises beds of black iron-stained shales with layers of cement-stone or septaria, and bituminous shales, one layer being the well-known "Kimeridge Coal." The beds stretch along the foreshore in a series of broad and slippery pavements. The cement-stones where they occur near the sea-level form the platform of Broad Bench and the Kimeridge Ledges, which serve partially to protect the coast-line, and indeed account for the irregularities which form the bay. Tumbled blocks and rolled pebbles of the septaria are scattered along the beach, while the crumbling away of small portions of the shale above occasionally reminds the pedestrian in an uncomfortable manner of the waste of the cliffs by meteoric agencies. Springs issue at numerous points along the cliff, and an occasional stream descends in the form of a cascade. The strata are displaced by numerous faults, doubtless produced during the upheaval of the strata; hence it is difficult at first to form any estimate of their thickness, although in many places the amount of throw may be calculated from the displacement of particular bands of cement-stone.* Prof. J. F. Blake has estimated that the upper beds of the Kimeridge Clay exposed near Kimeridge attain a thickness of 650 feet;† and this does not differ materially from the estimate made recently by Mr. A. Strahan.

The Kimeridge Clay at the base of St. Alban's Head and Emmit Hill is almost entirely concealed by tumbled masses of Portland Rocks. We find the first good exposures of the beds at Chapman's Pool, a bay at the foot of Renscombe. Here the black and rudely laminated shales dip gently to the south-east, and they are seen in the cliff and in the ravine through which a stream flows to sea over ledges of the shales. This is a noted locality for

---

* Some of the faults have lately been figured by Mr. Mansel-Pleydell, Proc. Dorset Nat. Hist. Club, vol. xv. p. 172.

† Quart. Journ. Geol. Soc., vol. xxxi. p. 198.

fossils, and we may find many Ammonites, some of them pyritic. The following species were collected by Mr. J. Rhodes :—

Fish remains.	Cardium striatulum.
Ammonites biplex.	Ostrea duriuscula.
Coccoteuthis.*	Pecten lens var. Morini.
Arca.	Discina iatissima.
Astarte lineata.	Rhynchonella.
Avicula.	Cidaris (spine).

FIG. 80.

*Cliff of Kimeridge Clay and " Coal workings" at Kimeridge.
(From a photograph by A. STRAHAN).*

To the above list may be added *Belemnites Souichi*, recorded by Prof. Blake. He gives a detailed section showing 651 feet of strata extending from Chapman's Pool to Hen Cliff.†

Westwards by the Encombe Cascade, and under Swyre Head the cliffs exhibit a thick mass of shales, without hard bands, excepting only a layer of banded greyish-brown stone, weathering white, which appears near the top of the cliff east of the Kimeridge "coal-works," and descends to the shore near the gap at Encombe. Bands of this character, sometimes 3 feet thick, may be seen also at Gad Cliff and Ringstead Bay, and they were noted by Fitton‡ who mentions that they are locally called "White Lias," which to some extent they resemble, although they cannot be considered as limestones. A bed of cement-stones occurs at the base of these shales, below which the more bituminous beds outcrop, passing beneath the sea-level to the east of

---

* First found by W. R. Brodie, Quart. Journ. Geol. Soc., vol. xi. p. 124.
† *Ibid.*, vol. xxxi. p. 198.
‡ Trans. Geol. Soc., ser. 2, vol. iv. p. 213.

the coal-works.   These bituminous shales are seen in the cliffs of Little Kimeridge, where they have been worked for various economic purposes.   (See p. 329.)

The best bed or "Black-stone," is a dense shale about 2 feet thick that occurs near the middle of the workings.   It contains grey pyritic nodules ; and is overlaid by 4 feet 6 ins. of clay, and again by about 16 feet of more or less bituminous shale and soil.   From the shale above it, Mr. Rhodes obtained Fish-remains, *Littorina, Astarte mysis, Cardium striatulum, Lingula ovalis* and *Serpula*; and from the Black-stone, *Ammonites biplex, Ostrea læviuscula,* and *Discina latissima.*   (See Fig. 80, p. 159.)

Westwards at Hen Cliff, we find alternations of black and iron-stained shales, having a zig-zag appearance due to the rusty joint-faces that are exposed.   Here and there alum crystals produce a white efflorescence, and there are four conspicuous bands of cement-stone, more or less continuous, and from 8 inches to 1 foot in thickness.   Small cement-stones also occur here and there in the shales.   The dip here is about 4° E.N.E.   From the shales at the base of Hen Cliff, Mr. Rhodes obtained Fish-remains, *Ammonites biplex, Astarte, Cardium striatulum, Exogyra virgula, Lucina minuscula, Ostrea duriuscula, O. læviuscula,* and *Discina latissima.*   Prof. Blake records, in addition to some of these species, *Ammonites Thurmanni, Trigonellites, Astarte lineata,* and *Arca.*[*]

On the eastern side of Kimeridge Bay the lower beds exposed on this portion of the coast are to be seen ; the beds consist of dark bluish-grey shale and paper-shale, more or less iron-stained, dipping at a low angle to the south-east.   Here Mr. Rhodes obtained Fish-remains, Coprolites, *Ammonites eudoxus, Cardium striatulum, Corbula, Ostrea duriuscula,* and *Lingula ovalis.*

On the north-western side of the Bay, the beds contain much pyrites, and the Ammonites (*A. biplex*) are pyritic.   They appear to be on a higher horizon than the beds seen south-east and south-west of the bay.

On the west side of the Bay we find shales with cement-stones and septaria dipping westwards, and the fossils correspond generally with those found on the eastern side of the Bay.   The strata, however, are faulted, and as Prof. Blake remarks "the continuation of the beds is not distinct."

From the shales at Broad Bench, Mr. Rhodes obtained the following fossils :—

Fish-remains.	Avicula inæquivalvis.
Ammonites biplex.	Cardium striatulum.
—— eudoxus.	Corbula.
—— Kapffi.	Cyprina ?
—— longispinus.	Exogyra virgula.
Coccoteuthis latipennis.	Hinnites.
Aptychus (Trigonellites) latus.	Lucina.
Alaria.	Ostrea duriuscula.
Arca longipunctata.	Pecten lens var. Morini ?
Astarte lineata.	Lingula ovalis.

Many fine Saurian remains have been obtained by Mr. J. C. Mansel-Pleydell from the Kimeridge Clay of Kimeridge.   In the

---

[*] Quart. Journ. Geol. Soc., vol, xxxi. p. 199.

Dorchester Museum there is a gigantic humerus and a complete paddle of *Pliosaurus grandis,* the latter measuring more than 6 feet in length. Two species of Pterodactyl have also been obtained, *Pterodactylus Manseli,* and *P. Pleydelli.*[*]

Proceeding towards Gad Cliff we find the higher beds of Kimeridge Clay much obscured by slipped material. Towards the upper part of the clay there is a hard band of calcareous sandy shale (3 or 4 feet thick) weathering white, and with impersistent partings of black shale; and 15 or 20 feet lower down there are two bands of sandy shale, each about 3 inches in thickness, and these also weather white, and they are separated by 3 inches of dark sandy shale with iron-pyrites.

The uppermost portion of the Kimeridge Clay is a pale-grey clay that passes downwards into dark blue clay and shale, and paper-shale, with small cement-stones. Some Saurian remains occur here, together with *Ammonites biplex,* &c.

A band that represents the Black-stone was observed in Brandy Bay by Mr. Strahan, it is a shaly coal with nodules, about 2 feet thick; and it occurs beneath the thick white bed.

Coming to Ringstead Bay we find a fine exhibition of the Portland Beds and Kimeridge Clay in Holworth Cliff, to the west of White Nore. The section is, however, not sufficiently clear to admit of detailed measurements. Towards the upper part of the Clay, as at Gad Cliff, there is a layer of white, fissile, shaly, calcareous rock 2 feet thick, with intercalated band of shale. This was noticed by the Rev. P. B. Brodie who obtained Fish-remains (*Hybodus*) from it, and found the elytron of a Beetle in the beds below.[†] Thin bands of flaggy and sandy shale may be traced here and there both above and below this white rock, and Fish-remains occur in them.

Shales and cement-stones, the latter in isolated nodules or in persistent bands, form the mass of the Kimeridge Clay in this cliff. Saurian and Fish-remains occur, and *Lucina minuscula* and other bivalves may be found in the shales. *Ammonites biplex* is not uncommon. Prof. Blake has recorded from lower beds *Ammonites cymodoce, Exogyra nana, E. virgula, Thracia depressa, Lingula ovalis,* and many other fossils.[‡]

The lowest beds of the Kimeridge Clay are well shown further west in the low cliffs near the boat-house. By the plantation on the west side of that building the following succession was observed:—

		FT. IN.	FT. IN.
Lower Kimeridge Clay.	Blue clays, &c., with *Ichthyosaurus, Astarte, Cardium, Exogyra virgula.*		
	Oyster-bed, gritty limestone with *Exogyra nana, Serpula,* &c.		
	Clays with *Ostrea deltoidea, Rhynchonella inconstans,* &c. - - - -	3　0 or 4　0	
Corallian Beds.	Irregular iron-shot sandy limestone, with *Cidaris florigemma,* &c. (Kimeridge Grit of Damon, see p. 84)	0　6 to 0　9	

---

* Proc. Dorset Nat. Hist. Club, vol. i. p. 2.
† Quart. Journ. Geol. Soc., vol. ix. p. 52.
‡ *Ibid.,* vol. xxxi. p. 212.

Prof. Blake mentions a band with *Trigonia Meriani* on top of the " Kimeridge Grit," which he includes in the " Kimeridge Passage-beds."[*]

The Kimeridge Clay is well shown in the cliffs of Black Head west of Osmington, and thence the beds extend inland by Upway to Portisham and Abbotsbury.

A shaft sunk by W. Manfield a little north-east of Portisham railway-station, was carried to a depth of over 300 feet in the Kimeridge Clay. Several beds of bituminous shale were met with, including a layer of "coal" 18 inches thick (perhaps the equivalent of the Black-stone of Kimeridge), at a depth of 135 feet 6 inches.

South of Weymouth the Kimeridge Clay is exposed in low cliffs by the railway, dipping gently towards Portland. It consists of dark slaty shales, rusty in places, with selenite and numerous bivalves and Ammonites; and layers of white shelly matter occur here, as in Kimeridge Bay and other places. There are two or more bands of septaria, isolated oval masses, with shells. Towards the base of the clay, thin reddish-brown ironstone layers occur, and lower still a band (18 inches) of ironstone with nodules of pale buff compact cement-stone. *Ostrea deltoidea, Exogyra nana,* and *Serpula intestinalis,* are abundant. *Corbula* and *Serpula variabilis* also occur. Here and there remnants of burnt shale may be found, perhaps due to some "spontaneous" ignition of the shales. (See p. 331.)

The clay forms the anchorage-ground at Portland, and lies at the base of portions of the Chesil Beach. Patches of blue clay have at times been observed when the beach has been scoured away during violent storms.

Vale of Shaftesbury.

North of the Weymouth area the Kimeridge Clay is concealed for some distance by the overlap of Cretaceous Rocks. It reappears on the northern side of the Dorsetshire Downs to the north-east of Cerne Abbas. It is, however, rarely exposed to any depth in the vale that extends from Buckland Newton by Gillingham to Tisbury and Mere. The clay here forms an undulating country with here and there gravelly patches, formed of detritus from the Chalk and Greensand. Brickyards have been opened at Gold Hill, near Child Okeford, and to the west of Shaftesbury.[†] Those at Okeford Fitzpaine and near Iwerne are in the Gault, as ascertained by Mr. Jukes-Browne. The junction with the Corallian Rocks is to be seen in a lane-cutting near Hole west of Piddle's Wood, south of Sturminster Newton, and dark shales are exposed near Anger's Farm, on the high road between Sturminster and Shilling Okeford.

The Kimeridge Clay is exposed in several of the railway-cuttings east and west of Semley between Tisbury and Gillingham.

---

* Quart. Journ. Geol. Soc., vol. xxxi. p. 213.
† See Fitton, Trans. Geol. Soc., ser. 2, vol. iv. p. 256.

West of Semley a cutting showed dark shales, grey on top, with selenite and fragments of lignite. *Thracia depressa* was obtained by Fitton, west of Shaftesbury, but few fossils have been recorded from the district. To the west of Gillingham, Station a brick-yard exposed about 10 feet of clay, bluish at the base and weathering brown towards the surface. Septaria and shelly cement-stones occur. Ferruginous and rotten specimens of *Ostrea deltoidea* are not uncommon ; there also occur *Belemnites, Modiola, Pholadomya, Rhynchonella inconstans, Serpula,* and Saurian bones.* Remains of *Ophthalmosaurus Pleydelli,* described by Mr. Lydekker, were obtained here. Red bricks, tiles, and drain-pipes are manufactured.

Mere, Westbury, and Seend.

Between Mere and Westbury, where the Cretaceous Rocks again overlap, there is no evidence of any exposure of Kimeridge Clay—the tracts so shown on the Geological Survey Map near Stourton and to the west of Penzlewood may be partly Oxford Clay and partly Gault,† and that in Longleat Park is now known from the observations of Mr. Jukes-Browne to be Gault.

At Westbury the bottom beds of the Kimeridge Clay, with *Ostrea deltoidea,* rest on the Ironstone of the Upper Corallian Series. (See p. 108.) Some Foraminifera were obtained from specimens of clay collected by the Rev. H. H. Winwood.‡

A boring for the Frome waterworks at Upper Whitborne, south-east of Corsley, reached "dark sandy clay" at a depth of 200 feet 6 inches, and was carried in it to a further depth of 13 feet, but no actual evidence of the age of this clay was obtained. In lithological character this clay resembles the upper beds of Kimeridge Clay of the neighbourhood of Potterne.

Immediately underlying the Lower Greensand, near the Iron-works at Seend, there is a very tenacious grey clay, seen to a depth of 3 feet.§ Mr. William Cunnington, in describing the Kimeridge Clay at Seend, observed septaria bored by lithodomus shells of the Lower Greensand, which rests unconformably on the Kimeridge Clay. He noted the occurrence of *Ostrea deltoidea,* and various Ammonites, as well as Saurian remains in this clay.‖ The deltoid oyster was originally found in the canal-cutting near Seend by William Smith. Here we have only the lower portion of the Kimeridge Clay. Dark blue clay with septaria was exposed in a brickyard north of Coulston. The only fossils found were Foraminifera, obtained by Prof. Rupert Jones and Mr. Sherborn from specimens of the clay forwarded to them.¶ The clay belongs to the Upper Kimeridge Clay, and this portion of the formation was shown in brickyards south-east of Worton and between

---

* Some specimens from this locality are preserved in the Museum of the Sherborne Grammar School.

† See also Fitton, Trans. Geol. Soc. ser. 2, vol. iv. p. 257.

‡ Jones and Sherborn, Geol. Mag. 1886, p. 272.

§ No microzoa were found in samples of this clay which I sent to Prof. Jones, see Jones and Sherborn, Geol. Mag., 1886, p. 272.

‖ Quart. Journ. Geol. Soc., vol. vi. p. 453.

¶ Op. cit., p. 272.

Potterne and Dues Water. At the Worton brickyard about 15 feet of slaty-grey loam was overlaid by bluish-grey loamy clay, with ferruginous and sandy layers near the top. The brickyard south of Potterne showed about 23 feet of blue sandy and slightly calcareous clay with fragile shells, *Exogyra, Ostrea,* &c. In this neighbourhood we come upon the passage-beds which lead upwards into the Portland formation.

Calne, Wootton Bassett, and Swindon.

At Rowde, north-west of Devizes, the Kimeridge Clay is again overlapped by Cretaceous deposits, but it reappears near Calne and forms a belt at the foot of the North Wiltshire Downs. Prof. Blake has recorded a number of fossils from the deep railway-cutting between Wootton Bassett and Swindon. Among these fossils were *Ammonites decipiens, Thracia depressa, Rhynchonella inconstans,* &c.* Sections unfortunately are scarce until we come to Swindon, where the beds have been extensively worked for brickmaking.

Prof. Hull remarks that in this area " towards the top, nodules of argillaceous limestone occur in layers, as well as sandy marl-stone with *Ammonites biplex,* in which the nacreous lustre is well preserved."†

A well-boring at the Beaufort Brewery, Wootton Bassett, proved a thickness of 84 feet of Kimeridge Clay before the Corallian rocks were reached. Some fossils obtained from the boring, forwarded for examination by Mr. Howard Horsell, were identified by Mr. G. Sharman, as follows : *Ammonites mutabilis, Inoceramus, Ostrea deltoidea, O. læviuscula,* and *Pholadomya læviuscula.*

The total thickness of the Kimeridge Clay near Swindon was estimated at 300 feet by Prof. Hull,‡ and I think this estimate may be considered as approximately correct. There and at Bourton to the north-east we have the Portland Beds on top in conformable succession, and thus the Kimeridge Clay is less than one-third of the thickness it attains in Dorsetshire.

Along the Midland and South-Western Junction Railway, to the south-west of North Lanes, a cutting showed brown and grey clay with septaria, yielding many specimens of *Ostrea deltoidea,* also *Exogyra virgula, E. nana,* and *Serpula tricarinata.* Above the Kimeridge Clay there was in places a thin gravelly soil, with chalk-flints and oolitic detritus. Further north there was a shallow cutting in clay, with a clayey soil, containing flints.

The upper and, perhaps, portions of the lower beds of the Kimeridge Clay are shown at the Swindon brick and tile-yards, north of Old Swindon. There we see a thickness of about 25 feet of blue shaly clay, racy near the top, with septaria and selenite. Many large iridescent Ammonites occur, as at Brill and Market Rasen. Remains of *Ichthyosaurus, Cetiosaurus,* and

---

* Quart. Journ. Geol. Soc., vol. xxxi. pp. 212, 215 ; see also p. 200.
† Geol., parts of Wilts, &c., p. 23.
‡ Explan. Hor. Sec., Sheet 59, p. 2.

Omosaurus have been obtained, and Mr. Rhodes and myself collected many of the following fossils, some of which have also been recorded by Prof. Blake :—*

Ammonites biplex.
Aptychus latus.
Dentalium Quenstedti.
Astarte ovata.
—— supracorallina.
Cardium striatulum.

Exogyra nana.
—— virgula.
Ostrea deltoidea.
Thracia depressa.
Trigonia.

The remains of *Omosaurus* were obtained in 1874 through Mr. J. K. Shopland, and were described by Sir R. Owen.[†] The humerus of this Dinosaur measured 2 feet 9 inches in length.

In the deep well at Swindon, made by the Great Western Railway Co. at the western end of their works, the following beds of Kimeridge Clay were proved :—[‡]

		Ft.	In.
	Made ground - - - -	8	0
	Bluish-grey clay, with Ammonites, Belemnites, &c. - - - - -	42	0
	Pale grey shelly limestone, with *Ostrea deltoidea* - - - -	0	1
Kimeridge Clay.	Grey and brown earthy limestone, with patches of iron-shot grains - -	0	6
	Hard brown calcareous muddy clay -	2	0
	Bluish-grey shelly clay - - -	19	6

Mr. E. T. Newton states that from 8 feet to 72 feet, Kimeridge Clay was passed through, but only a few fossils were collected; thus between 10 feet and 16 feet *Rhynchonella inconstans, Ammonites cordatus,* and *Ostrea deltoidea,* were noticed; at 45 feet *Astarte* and *Thracia depressa,* and from 45 feet to 52 feet *Belemnites* were met with.

The earthy limestone-bed with iron-shot grains, between 50 feet and 50½ feet, yielded the following fossils (see p. 118):—

Ammonites cordatus, var. excavatus (= A. serratus).
—— rotundus.
—— varicostatus.
Astarte.

Myacites recurvus.
Ostrea deltoidea.
Pecten lens?
Perna.
Pholadomya protei?

The occurrence of *Am. cordatus* is noteworthy; it is, however, recorded from the Lower Kimeridge Clay of Yorkshire.[§]

Further reference to well-borings at Swindon is made in the chapter on Water-supply (p. 336).

Faringdon to Abingdon and Culham.

In Oxfordshire and Berkshire the following succession may be traced in the Kimeridge Clay :—

Bluish-grey clay, with earthy limestone nodules and septaria, merging upwards into sandy clay of the Portlandian Beds (Hartwell Clay, &c.) about 40 feet.	Ammonites biplex. Astarte hartwellensis. Cardium striatulum. Perna. Thracia depressa. Discina humphriesiana.
Clays and shales, with septaria, about 60 feet.	Aptychus latus. Astarte lineata. Exogyra virgula. Lingula ovalis. Ostrea deltoidea. Rhynchonella inconstans.

---

* Quart. Journ. Geol. Soc., vol. xxxi. p. 211 ; see also J. Buckman, *Ibid.,* vol. xiv. p. 127.

† Mesozoic Reptilia (Palæontogr. Soc.), Part 2, p. 45.

‡ Quart. Journ. Geol. Soc., vol. xlii. pp. 288, 291.

§ Blake, *Ibid.,* vol. xxxi. p. 218 ; Fox-Strangways, Jurassic Rocks of Yorkshire, vol. ii. p. 238.

Near Woolstone, south of Faringdon, Prof. Hull reckoned that the Kimeridge Clay had a thickness of nearly 300 feet;[*] but in the neighbourhood of Oxford the thickness does not exceed 100 feet.

The base of the Kimeridge Clay has been exposed in a brickyard east of Faringdon, but we have few records of the strata in the area extending by Stanford-in-the-Vale, Denchworth, and Drayton to Abingdon.

At Cumner, where the Lower Greensand rests unconformably on the Kimeridge Clay, the thickness of the Clay was estimated at 70 or 80 feet by Prof. Prestwich. At the Chawley brickyard north of Cumner Hurst, the following section was to be seen, the, details of the lower beds, which were not exposed at the time of my visit, and the lists of fossils, being given on the authority of Prof. Prestwich :—

		Ft.	In.
Lower Greensand.	Coarse brown false-bedded quartzose sand, with concretionary ironstone -	12	0
Kimeridge Clay.	Pale grey sandy clay - - -	2	0
	Dark bluish-grey slightly calcareous clay, with greenish sandy seams, nodules of earthy limestone, and large septaria with *Ammonites biplex* and lignite - - -	18	0
	Clay, with band of fossils, yielding "*Plesiosaurus*," *Am. biplex*, *Pleurotomaria reticulata*, *Astarte*, *Cardium striatulum*, *Lima*, *Modiola bipartita*, *Myacites recurvus*, *Pecten nitescens*, *Perna mytiloides*, *Pinna lanceolata*, *Thracia depressa*, *Trigonia*, and *Serpula tetragona* - - -	14	0
	Seam of clay laminated with white sand (Iguanodon-bed), with *Camptosaurus (Iguanodon) Prestwichi*, *Aptychus latus*, *Pleurotomaria*, *Astarte*, *Exogyra virgula* (abundant) *Trigonia*, *Lingula ovalis* - - -	0	3
	Clay with large septaria, with *Dacosaurus*, *Ichthyosaurus*, *Pliosaurus*, *Astarte (Lucina) lineata* - -	4	0

The discovery of Iguanodon was announced in 1879 by Prof. Prestwich, and it was thought that the entire skeleton was probably present. The species was named by J. W. Hulke, and it was considered to be the earliest record of the genus : it is now regarded as *Camptosaurus*.

Prof. Prestwich has observed in reference to the Kimeridge Clay near Oxford, " The presence of drifted wood and of the *Iguanodon* in the Kimeridge Clay of this district, and of large Dinosaurs at Swindon, together with the great thinning of this formation as it trends to the south-west, render it probable that land in that direction was not far distant, and that that land may have been the same as that of the proximity, of which we have more distant evidence in the many quartz, slate, and metamorphic rock-pebbles present in the Lower Greensand of Faringdon, a

---

[*] Explan. of Hor. Sec., Sheet 72, p. 7.

deposit evidently formed near an old shore. This land, since submerged and covered by upper Cretaceous strata, was in all probability the prolongation of the old axis of the Mendip and Ardennes, the elevation of which took place in Permian or Triassic times."* It seems to me that the present state of the evidence indicates land to the south, south-east, and east, rather than to the south-west.

In the brickyard east of Culham, and south-east of Abingdon we have a section which shows some of the higher beds of the Kimeridge Clay. The section was as follows :—

		Ft.	In.
Valley gravel.			
Gault - -	{ Grey marly and shaly clay with ferruginous stains, passing down into sandy calcareous clay, with pebbles of quartz, &c., and small phosphatic nodules - - -	24	0
Kimeridge Clay.	{ Irregular band of greenish grey earthy limestone - - - 0 3 to 0		6
	Grey or greenish-grey sandy and slightly calcareous loam, passing down into darker sandy clay; with *Cardium striatulum* - - - -	9	0

The sandy basement-bed of the Gault was originally grouped as Lower Greensand.

Prof. Phillips mentions that the Kimeridge Clay had been exposed to a depth of about 23 feet, and that it contained a zone of Ammonites, and brown nodules with crystals of bisulphide of zinc in the cracks.†

The fossils recorded from Culham include besides Saurian and Fish-remains the following species :—

Ammonites biplex.
—— triplicatus.
Astarte hartwellensis.
Cardium striatulum.
Cucullæa.

Thracia depressa.
Perna.
Pinna granulata.
Discina humphriesiana.

Phillips remarked that " the shells of Aylesbury being abundant, we may infer that it is the upper part of the Kimeridge Clay which is here seen "; while in reference to the band of sandy calcareous rock he regarded it as " really a sandy cap of the Kimeridge Clay—perhaps the stage of a change towards the Portland series, but still to be classed with the clay." (See also p. 217.) From this band he recorded the following fossils :—

Ammonites (*cf.*) polyplocus.
Cardium striatulum.
Corbula.

Pecten arcuatus.
Thracia depressa.

At Sandford according to Mr. E. S. Cobbold " a peculiar, bright red, earthy layer from 4 to 6 inches thick " occurred at the junction of the Kimeridge Clay with the Corallian Beds. In the clay he obtained vertebræ of *Pliosaurus*, good specimens of *Rhynchonella inconstans*, and numerous crystals of selenite.‡

---

* Quart. Journ. Geol. Soc., vol. xxxvi. pp. 431, 432 ; Proc. Geol. Assoc., vol. vi. p. 343 ; see also E. S. Cobbold, Quart. Journ. Geol. Soc., vol. xxxvi. p. 319 ; and H.B.W., *Ibid.*, vol. xlii. p. 307.
† Quart. Journ. Geol. Soc., vol. xvi. p. 310, and Geol. Oxford, &c., p. 427 ; see also Hull, Geol., parts of Oxfordshire and Berks, p. 9.
‡ Quart. Journ. Geol. Soc., vol. xxxvi. p. 318.

Shotover Hill.

The brickyards on the western side of Shotover Hill, near Headington, show the Kimeridge Clay resting on the Corallian rocks and overlaid by Portland Beds. The thickness of the clay was estimated at 100 feet by Prof. Phillips, but only about 30 feet of the clay is exposed, and it may not be much more than 70 feet thick.

It consists of bluish-grey and dark blue shaly clay, weathering brown at the top, with lignite and with a few layers of septaria and much selenite. (See p. 151.) Prof. Phillips notes that about 15 feet from the base there is a septarian band yielding *Rhynchonella inconstans*, &c. ; lower down there are layers of *Ostrea deltoidea*, and beneath *Thracia depressa* and *Exogyra virgula* are found fairly abundant. At the base coprolites were first noticed by the Rev. H. Jelly.* These may help in tracing the base of the Kimeridge Clay northwards, where it rests on Oxford Clay without the intervention of any Corallian stone-beds. As noted by T. Roberts in the Cambridge area, the base of the Kimeridge Clay yields coprolites which assist in determining the boundaries of the strata. (See p. 170.)

Sowerby figures *Ostrea deltoidea* from this locality, remarking that the fossils were known as "Heddington (Headington) Oysters."

In addition to various Saurian remains, the following fossils have been recorded from the Kimeridge Clay of Shotover Hill :—

Asteracanthus ornatissimus.	Pleurotomaria reticulata.
Hybodus acutus.	Astarte ovata.
Ischyodus Egertoni.	Exogyra virgula.
Lepidotus maximus.	Ostrea deltoidea.
Ammonites biplex.	Thracia depressa.
Belemnites abbreviatus var.	Discina latissima.
excentricus.	Rhynchonella inconstans.
Aptychus (Trigonellites).	Serpula.

Prof. Phillips records *Belemnites "explanatus"* from the upper part of the Kimeridge Clay at Waterstock, near Thame, and Wheatley near Oxford, as well as from the Hartwell Clay.† The species is now regarded as *B. nitidus.*

Brill and Aylesbury.

The Kimeridge Clay has been exposed in a brickyard, by the high road, between Waterstock and Great Milton, and it was again shown in a brickyard south of Long Crendon.

At the brickyard north-west of Brill, clays are dug beneath the Portland Beds and numerous iridescent fossils, like those of Market Rasen, are obtained. The beds may be compared with

---

* Buckland, Trans. Geol. Soc., ser. 2. vol. iii. p. 232 ; see also Fitton, *Ibid.*, vol. iv. p. 278 ; Phillips, Geol. Oxford, &c., pp. 329, 413, and Quart. Journ. Geol. Soc., vol. xiv. p. 238 ; and Hull, Geol. parts of Oxfordshire and Berkshire, p. 10.
† British Belemnites, 1865, p. 128.

those seen at Culham. Prof. Green has recorded the following fossils :—*

Ammonites biplex.	Pinna granulata.
Astarte hartwellensis.	Thracia.
Cardium striatulum.	Lingula ovalis.

Near Aylesbury the Kimeridge Clay appears to merge gradually upwards into a clay containing Portlandian fossils, and known as the Hartwell Clay. On the Geological Survey Maps this clay is naturally included with the Kimeridge Clay; but as the fossils seem to indicate an horizon higher than is represented in the Kimeridge Clay of Kimeridge, the Hartwell Clay is considered in the Chapter that describes the Portland Beds. It is not possible, however, to be consistent in all our dealings with these formations, without detailed and long-continued hunting for fossils.

Thus in the brickyard at Whitchurch, we find sandy grey and bluish-grey clay: a "mild earth" well adapted for brickmaking, and fine red bricks and drain-pipes are manufactured. The beds worked appear to be the same as at Brill, but they may include representatives of the Hartwell Clay. Iridescent Ammonites, *Thracia*, and other somewhat poorly preserved Lamellibranchs were found, together with large bones of Saurians.

Prof. Green has noted the presence of Kimeridge Clay beneath the Gault at Littleworth brickyard, north of Wing. The sections have since been examined by Mr. Jukes-Browne who describes the older clay as a stiff and dark blue deposit with large septaria and lignite, and also Saurian remains, *Ammonites biplex* and *Belemnites*.

From the brickyard south of Stewkley the following fossils were obtained from the Kimeridge Clay by Mr. Whitaker and they were identified by Mr. Etheridge :—

Ammonites.	Exogyra virgula.
Aptychus.	Modiola.
Cardium striatulum.	Ostrea.

Stewkley to Cambridge.

From the neighbourhood of Stewkley north-eastwards we lose sight of the Kimeridge Clay owing to the overlap of the Lower Cretaceous strata.

Clays of Corallian age appear at Ampthill and at Tetworth to the north-east of Sandy. From Great Gransden northwards there have been found traces of the Kimeridge Clay at the base of the Lower Greensand, and thence it outcrops persistently by Knapwell and Oakington, and appears in straggling "islands" in the Fenland at Sutton, Ely, Barraway, Chatteris, March, Littleport, and Southrey.

At Knapwell there was formerly a section at a brickyard which showed the Ampthill Clay with a capping of Kimeridge Clay. The beds opened up consisted of about 15 feet of dark

---

* Geol. Banbury, p. 46.

blue shaly clay, with fine crystals of selenite, and on top about
3 feet of clay with a layer of black phosphatic nodules at the
base of it.   The fossils recorded by Messrs. Penning and Jukes-
Browne include the following :—*

Ammonites biplex.	Ostrea deltoidea (abundant).
Belemnites abbreviatus.	—— gregaria.
Avicula echinata.	Trigonia.
Exogyra nana.	Serpula.
Gryphæa dilatata.	

T. Roberts, remarking on this section, considers the nodule-
bed to mark the base of the Kimeridge Clay ; and thus many of
the fossils, including the *Gryphæa dilatata,* come from the
Ampthill Clay below.†

Reference has already been made to the Kimeridge Clay at
Willingham (p. 147), and to the relations of the beds at Upware
(p. 144).

Isle of Ely.

Further north we come to the " Isle of Ely," an irregular mass
of Kimeridge Clay which rises through the alluvial flats, and is
capped by two outliers of Lower Greensand, marked by the
towers of Ely Cathedral and the steeples of Wilburton and Haddenham ; and this " Isle " could in Norman times only be reached
by an artificial causeway across the marshes.‡

The lower beds of the Kimeridge Clay were observed by
T. Roberts in a brick-pit half a mile west of Haddenham railway-
station.   This section was as follows :—

	Ft.	In.
Soil - - - - - - -	1	6
Ferruginous clays with some calcareous nodules -	6	0
Thin grey limestones, weathering white - -	1	3
Black tenacious clays, with few fossils - -	2	0
Grey nodular limestone, with *Ammonites mutabilis,* and some gasteropods - - - -	0	9
Black clays, with *Ostrea deltoidea,* and, at base, black phosphatic nodules - - - - about	9	0

A number of fossils are recorded by Roberts, including *Ammonites biplex.*   South of Haddenham station there are two
brickyards, showing black clays with selenite and septaria.   From
these beds, which belong to "a somewhat higher horizon," Roberts
obtained *Ammonites mutabilis, Astarte supracorallina, Exogyra
virgula,* &c.§

The Kimeridge Clay has been well exposed north-east of Ely
in the large pit of Roslyn Hole, where about 30 feet of dark
shales and clays, in places bituminous and arenaceous, with thin
ochreous layers and bands of septaria, may be seen.   A large

---

* Geol. Cambridge, p. 10.
† Jurassic Rocks of Cambridge, p. 42.
‡ Sedgwick, Rep. Brit. Assoc. for 1845, p. 44.
§ Jurassic Rocks of Cambridge, pp. 63–65.

number of Saurians and Fish-remains have been obtained by Mr. Marshall Fisher, of Ely, and these have been examined and named by Mr. E. T. Newton. A full list of fossils is elsewhere pub. lished. They show that both lower and upper portions of the Kimeridge Clay are present, for, among the Ammonites, there are *Ammonites alternans, A. biplex, A. eudoxus, A. longispinus,* and *A. mutabilis.* Among other fossils we may find *Aptychus, Astarte supracorallina, Cardium striatulum, Exogyra virgula, E. nana, Lucina minuscula, Ostrea deltoidea, Discina latissima, Lingula ovalis,* and *Rhynchonella inconstans.*[*]

The strata of this district have been studied in detail by T. Roberts, who notes in descending order the following local zones or fossil-beds :—[†]

	Ft. In.
Discina latissima - - -⎫	
Exogyra virgula - - -⎬ 15 0	
Ammonites alternans - -⎫	
Astarte supracorallina - -⎬ 80 0	
Ostrea deltoidea - - -⎭	

The full thickness of the Kimeridge Clay can nowhere be ascertained in the district, as we have no indication of Port. landian beds. The strata that can be assigned to the Upper Kimeridge Clay do not appear to exceed 15 feet, and the Lower Kimeridge Clay may be about 80 or 90 feet. Roberts has approximately estimated the full local thickness of the clay at 142 feet, but this estimate appears excessive when compared with his detailed observations and correlations.[‡]

Fenland and Norfolk.

Chatteris and March are situated on "islands" of Kimeridge Clay, with coverings of Drift.

At March the occurrence of *Ostrea deltoidea* points to the lower part of the Kimeridge Clay, and, as mentioned elsewhere, it is probable that Corallian Clay has been reached in the brickyards at this locality, where clay has been dug to a depth of 40 feet.[§]

At Littleport there is another "island" of Kimeridge Clay, and Prof. Hughes has drawn attention to the occurrence in the clay of subordinate bands of limestone.[‖] The beds are exposed in several brickyards near the village. Summarizing the information which they furnish, T. Roberts gives the following section :—[¶]

---

* Geology of parts of Cambridgeshire and Suffolk, p. 18, see also J. F. Blake, Quart. Journ. Geol. Soc., vol. xxxi. p. 211, and T. Roberts, *Ibid.*, vol. xlv. p. 558.
† Jurassic Rocks of Cambridge, p. 69.
‡ *Ibid.*, p. 74.
§ Skertchly, Geol. Fenland, pp. 192, 268 ; Geol. S.W. Norfolk and N. Cambridge-shire, p. 8; See also Hailstone., Ann. Phil., vol. v. 1815, p. 390.
‖ Proc. Geol. Assoc. vol. viii. p. 401.
¶ Jurassic Rocks of Cambridge, p. 73.

							FT.	IN.
Lower Kimeridge Clay	Black clay	-	-	-	-	-	6	0
	Septarian nodule-bed	-	-	-	-	0	8
	Clays with *Astarte supracorallina*, &c.		-	-	11	0
	Argillaceous limestone	-	-	-	-	1	0
	Black clays	-	-	-	-	-	4	0
	Grey limestone	-	-	-	-	-	1	6
	Black and bluish clays	-	-	-	about	9	0
	Grey limestone	-	-	-	-	0	9
	Grey and black clays	-	-	-	-	4	0
	Calcareous nodules	-	-	-	-	0	6
	Black clay	-	-	-	-	about	3	0
	Grey compact limestone	-	-	-	-	1	0
	Black clay	-	-	-	-	-	16	0
	Limestone	-	-	-	-	-	0	7

$$\overline{59\quad 0}$$

The following fossils are among those recorded by Roberts from the Kimeridge Clay of Littleport:—

Ammonites biplex.
—— Callisto.
—— mutabilis.
—— trifidus.
Belemnites abbreviatus.
Alaria trifida.
Arca reticulata.
Astarte supracorallina.
Exogyra nana.
—— virgula.
Lima pectiniformis.

Lucina minuscula.
Myoconcha.
Nucula Menkei.
Ostrea lœviuscula.
Pecten Grenieri.
Perna Flambarti.
Thracia depressa.
Trigonia elongata.
Lingula ovalis.
Rhynchonella inconstans.
Serpula intestinalis.

The occurrence of limestone-bands in the Kimeridge Clay of this district is interesting; their presence in the Oxford Clay of Huntingdonshire and Cambridgeshire has been noticed, so that in the great clay formations of this eastern-midland area we find calcareous bands to be somewhat markedly developed. Here, as in the case of the Lias, limestones occur in proximity to the old land-tracts.

Proceeding towards Downham Market we find an "island" of Kimeridge Clay at Southrey. Fitton recorded *Ostrea deltoidea,* *Aptychus* and Saurian bones from this locality; he also noted *Ammonites Lamberti* (an Oxford Clay fossil). This ammonite may be a wrong identification, or it may have been derived from the Boulder Clay which occurs on the Lower Greensand and Kimeridge Clay of this insulated tract.* The occurrence of bituminous clay in the same neighbourhood, mentioned by C. B. Rose, is suggestive of the presence of higher portions of the Kimeridge Clay.†

There are brickyards at Fordham, and again at Downham Market to the north and south-east of the railway-station. There we see about 5 feet of brown and grey clay resting on 8 feet of black shaly and bituminous clay, with here and there a gravelly clay at the surface. Saurian bones, *Ammonites biplex, Lucina*

* Trans. Geol. Soc. ser. 2, vol. iv. p. 316.
† Phil. Mag. ser. 3, vol. vii. p. 175.

minuscula, and other fossils may be obtained. Formerly clay was opened to a depth of 30 feet in one brickyard and a band of septaria formed the floor. A well-boring at Downham Market, after passing through 29 feet of Lower Greensand, was carried to a depth of 187 feet in beds regarded as Kimeridge Clay.[*] Here again, we have no means of telling the full thickness of the formation.

Kimeridge Clay has been worked in a brickyard south-east of Watlington and again at West Winch, south of Lynn. An interesting mass of the Clay has been worked at Fodderstone Gap between Shouldham and South Runcton. The brickyard showed 15 feet of dark blue shaly Kimeridge Clay and this was proved in a well to be 50 feet thick and to overlie Lower Greensand. Hence, Mr. C. Reid, who has described the section, remarks that " Clearly the mass of Kimeridge Clay is a huge boulder transported from the westward."[†]

The most northerly exposure of Kimeridge Clay in Norfolk is between North and South Wootton, south-west of Castle Rising. The occurrence of Kimeridge Clay on the foreshore at Hunstanton was mentioned by R. C. Taylor and others, but it is now well known that the clay here is a band in the Lower Greensand, which comes to the surface and is worked at Heacham brickyard, as pointed out by Mr. Teall.[‡]

In a well at Holkham, clay presumed to be Kimeridge Clay, was touched at a depth of 743 feet, beneath the Cretaceous and newer deposits.[§]

Lincolnshire.

Across the Fenland from Wisbech to Holbeach we have no definite evidence of the course of the underground Jurassic strata. At Boston a deep well (572 feet) after penetrating 24 feet of Alluvium, and 166 feet of Boulder Clay, passed into the Kimeridge Clay. This clay was penetrated for a depth, estimated by Mr. Jukes-Browne at about 294 feet, beneath which there was proved a further thickness of 88 feet of beds in or above the Oxford Clay.[‖]

We have no fossil evidence from the district of Tattershall, Coningsby, New Bolingbroke, &c. In this region the Jurassic clay is for the most part covered with Alluvium or Boulder Clay. On Revesby Bank, south of Revesby, blue clay was proved to a depth of 124 feet in an unsuccessful boring for water. The Kimeridge Clay has been worked to a depth of about 50 feet in a brickyard west of Stickney church. It contains a layer of septaria and " large

---

[*] Whitaker, Geol. S. W. Norfolk, &c., pp. 8, 157.
[†] *Ibid.*, p. 63.
[‡] Potton and Wicken Phosphatic Deposits, 1875, p. 17 ; see also Rose, Phil. Mag. ser. iii. vol. vii. p. 176, and Fitton, Trans. Geol. Soc. ser. 2. vol. iv. p. 316.
[§] H.B.W., Geol. Fakenham, &c., p. 51.
[‖] Quart. Journ. Geol. Soc. vol. xxxv. p. 420 ; see also G. Naylor, Phil. Trans. vol. lxxvii. p. 50.

smooth Ammonites." The clay has also been dug south of Hagnaby Corner, to the west of Stickford.*

Northwards in the area between Bolingbroke, Spilsby, and Fulletby, there is evidence of dark shaly and bituminous clay and paper-shales with occasional thin calcareous bands. Near Spilsby wells have been sunk into the clay to depths of about 90 feet in search of water, but without success—the thickness of the Kimeridge Clay being estimated at 300 feet by Mr. Jukes-Browne.

Phosphatic nodules have been observed sparingly in the clay at Halton Holegate and Raithby, and their occurrence is noteworthy.

Saurian remains have been found near West Keal, and other fossils have been obtained by Mr. J. Rhodes from the paper-shales exposed in the banks of the river Steeping, south-west of Ashby near Spilsby, and in a pit north-west of Salmonby. These include *Ammonites, Astarte, Discina latissima*, and *Lingula ovalis*.†

In a brickyard west of Fulletby, about 15 feet of paper-shales, with a band of cement-stone, were exposed, and there Prof. J. F. Blake obtained the following fossils :—‡

Ammonites biplex.	Cardium striatulum.
Aptychus biplex.	Gervillia tetragona.
Belemnoteuthis.	Lucina minuscula.
Trochus retrorsus.	Ostrea gibbosa.
Dentalium Quenstedti.	Pecten lens.
Astarte lineata.	Discina latissima.
Avicula vellicata.	

In these localities we have the Upper beds of the Kimeridge Clay.

A well sunk at Burgh reached Kimeridge Clay at a depth of 58 feet, and another at Driby reached it at 321 feet, and that at Skegness proved it at a depth of 363 feet.§

A brickyard (now abandoned) between Stixwould and Woodhall Spa showed about 14 feet of dark blue clay with septaria, and *Ostrea deltoidea*.|| The bore-hole at Woodhall Spa, after passing through 10 feet of gravel and Boulder Clay, penetrated 350 feet of clay assigned by Mr. Jukes-Browne to Kimeridge and Oxford Clays, beneath which there were Kellaways Beds and other strata.¶

From Campney, south-west of Bucknall, Mr. Jukes-Browne records *Ammonites mutabilis, A. plicatilis*, and *A. rotundus*.**

In these localities we have the Lower Kimeridge Clay which rests on the local equivalent of the Ampthill Clay. (See p. 148.)

---

* Jukes-Browne, Geol. S.W. Lincolnshire, p. 73.
† Jukes-Browne, Geol. E. Lincolnshire, pp. 9–11.
‡ Quart. Journ. Geol. Soc., vol. xxxi. p. 201 ; Jukes-Browne, Geol. E. Linc., p. 12.
§ A. Strahan, Geol. E. Lincolnshire, p. 168.
|| Geol. Lincoln, p. 79 ; J. F. Blake, Quart. Journ. Geol. Soc., vol. xxxi. p. 310.
¶ Geol. Lincoln, p 208 ; and Memoir on Lower Oolitic Rocks of England, p. 515.
** Geol. Lincoln, p. 79.

North of Horncastle, brickyards have been opened on both sides of the river Bain, that on the west is now abandoned. Clays with fine septaria occur here, and among the fossils that have been recorded are *Ammonites biplex, A. mutabilis, Astarte supracorallina, Cardium striatulum, Thracia depressa,* and *Lingula ovalis.** These beds belong to the Lower Kimeridge Clay.

The brickyard north-east of West Ashby showed shaly clay with septaria. Here Saurian bones occur, and I obtained *Ammonites biplex* and *Lucina minuscula.* The same species together with *Discina latissima* are recorded by Prof. Blake from Goulsby. In the brickyard south-east of South Willingham railway-station the shales are bituminous (see p. 328), and contain large septaria 5 or 6 feet across; here also Saurian bones and Ammonites may be found. These beds all belong to the Upper Kimeridge Clay, and we have a record of 309 feet of Kimeridge Clay in the boring at Donnington-upon Bain. (See p. 328.) This passed through clays, cement-stones, and bituminous shales.†

At Baumber, Hatton, and near Wragby the Lower Kimeridge Clay has yielded the following among other fossils :—‡

Ammonites alternans	Ammonites rotundus.
—— Berryeri.	—— triplicatus.
—— biplex.	Belemnites abbreviatus.
—— cordatus var. excavatus.	Pleurotomaria reticulata.
—— mutabilis.	Cardium striatulum.
—— decipiens.	Thracia depressa.

Market Rasen is a locality famous for the many fossils obtained, including beautiful iridescent and pyritic Ammonites. The brickyard east of the town, on the road to North Willingham, showed the following section :—§

		FT. IN.	FT. IN.
Drift.	False-bedded sand - - - -	6 0 to	8 0
Lower Kimeridge Clay.	Bluish-grey slightly micaceous clay with small nodules of argillaceous limestone, pyrites, &c. - - - - -	12 0 to	16 0
	Layer of slightly calcareous grit, in impersistent masses - - - -	0 4 to	0 5
	Dark clay.		

The following species have been recorded from Market Rasen:— ‖

×	Ammonites alternans.	× Gryphæa.
	—— Berryeri.	× Inoceramus rasenensis.
×	—— cymodoce.	Lucina minuscula.
	—— decipiens.	× Nucula Menkei.
	—— mutabilis.	Ostrea deltoidea (rare).
×	—— rotundus.	—— gibbosa.
	Belemnites nitidus.	Pecten demissus.
	Arca longipunctata.	Pholadomya æqualis.
	Astarte ovata.	× Thracia depressa.
	—— supracorallina.	× Trigonia juddiana.
	Cardium striatulum.	Lingula ovalis (rare).
	Exogyra nana.	

---

* J. F. Blake, Quart. Journ. Geol. Soc., vol. xxxi. p. 205. See also Judd, *Ibid.* vol. xxiv. p. 240.

† E. Bogg, Trans. Geol. Soc., vol. iii. p. 396; Conybeare and Phillips, Geol. England and Wales, p. 194.

‡ Strahan in Geology of Lincoln, p. 80; Blake, *op. cit.* p. 208; Roberts, Quart. Journ. Geol. Soc., vol. xlv. p. 553.

§ See also A. Strahan, Geol Lincoln, p. 80.

‖ J. F. Blake, Quart. Journ. Geol. Soc., vol. xxxi. p. 207; those marked × were obtained by myself.

A number of Gasteropods have been recorded. The occurrence here and elsewhere of *Ammonites Berryeri* is interesting, as that species occurs in the Corallian iron-ore of Westbury. (See p. 109.)

The occurrence of dark shales with *Ammonites biplex* and *Ostrea deltvidea* was noted by Mr. Strahan in a pit north of Claxby.[*]

From a brickyard on Claxby Moor, Mr. Ussher records *Ammonites rotundus*, and from another brickyard at Holton-le-Moor, he obtained the same species, together with *A. plicatilis, Thracia depressa*, &c.[†] Roberts notes *Ammonites decipiens* and other fossils from Holton-le-Moor,[‡] showing that along this line occurs the junction of Corallian and Kimeridge Clays.

At a clay-pit on Moortown Hill the following species were obtained by Mr. Ussher :—

Ammonites rotundus.	Avicula inæquivalvis.
—— (near to) cymodoce.	Ostrea deltoidea.
Alaria trifida.	Pecten lens.
Arca longipunctata.	Serpula tetragona.

A similar assemblage, including *Corbula, Modiola bipartita,* and *Opis,* was obtained at North Kelsey brickyard ; and Roberts records other species, *Ammonites plicatilis, A. achilles, Belemnites abbreviatus, Cardium striatulum, Thracia depressa, Astarte supracorallina,* &c.

At Nettleton, south of Caistor *Ammonites biplex, A. plicatilis, Lucina,* and some other fossils were obtained by Mr. Rhodes.[§]

In the railway-cuttings north-east of Wrawby, clays of Corallian age, and overlying Kimeridge Clay, have been observed. *Ammonites alternans, A. rotundus, Ostrea deltoidea, Exogyra virgula* and other fossils indicative of the lower beds are recorded by Mr. Ussher ; he also mentions *A. biplex* ; and Prof. Blake records this species, *Astarte supracorallina,* and other fossils.[‖]

In the brickyard south of Worlaby *Ammonites rotundus* and *Thracia depressa* were found by Mr. Ussher, and to these Roberts has added *Am. decipiens, Cerithium forticostatum, Pholadomya acuticosta, Ostrea gibbosa,* and *Serpula intestinalis.*[¶]

Further north the Kimeridge Clay is but little exposed, although it has been mapped along the base of the Cretaceous escarpment to South Ferriby on the shore of the Humber. It is not only in part concealed beneath the Cretaceous covering, but obscured by various Drifts, and along the escarpment itself by debris that has accumulated on the slope. Hence it is by no means certain that Corallian clay, or even Oxford Clay may not be present along the base of the cliffs between Saxby and South Ferriby.

The general evidence obtained by the fossils points to considerable unconformity between the Kimeridge Clay and overlying

---

[*] Geol. Lincoln, p. 81 ; see also H. Keeping, Quart. Journ. Geol. Soc., vol. xxxviii. p 240.
[†] Ussher, Geol. N. Lincolnshire, pp. 100–105.
[‡] Quart. Journ. Geol. Soc., vol. xlv. p. 552.
[§] See also Dikes and Lee, Mag. Nat. Hist., ser. 2, vol. i. 1887, p. 566.
[‖] Quart. Journ. Geol. Soc., vol. xxxi. p. 207.
[¶] *Ibid.,* vol. xlv. p. 552.

strata. Judging by this evidence we might infer that the Spilsby Sandstone, and overlying Cretaceous Beds extend directly on to the Lower Kimeridge Clay in the neighbourhood of Caistor. The dark bituminous clays proved in the shaft sunk at the Acre House Mine, north of Claxby, probably belong to the Upper Kimeridge Beds, while some of the fossils obtained at Nettleton and certainly those of Worlaby are Lower Kimeridge. The unconformity of the Upper Cretaceous Beds over the Lower Cretaceous or Neocomian Beds has been shown by Mr. Strahan; and as representatives of the Spilsby Sandstone occur as far north as Elsham, there can be no doubt of its unconformable relations with the Kimeridge Clay, a view to which he has been led from the evidence furnished by the nodule-band at the base of the Spilsby Sandstone.* Further reference is made to this subject in a later Chapter (p. 290).

Sussex and Kent.

In the Sub-Wealden Boring a considerable thickness of the Kimeridge Clay was penetrated; and this has been estimated at from 1,273 feet to as much as 1,512 feet. There being a passage from the Corallian Beds below, upwards through the Kimeridge Clay into the Portlandian strata above, we are by no means sure of the limits, but the lesser estimate appears better to coincide with the palæontological evidence. (See detailed section, p. 346.)

The fossils were identified by Mr. R. Etheridge, and the following may be mentioned with the depths at which they were found :—†

					FT. FT.
Ammonites biplex	-	-	-	-	275 to 1,083
—— Callisto	-	-	-	-	668 to 1,565
—— mutabilis	-	-	-	-	379 to 1,562
—— orthocera	-	-	-	-	1,039
Astarte supraccrallina	-	-	-	-	409 to 926
Cardium striatulum -	-	-	-	-	409 to 1,634
Exogyra nana	-	-	-	-	278 to 938
—— virgula -	-	-	-	-	934 to 1,634
Lucina minuscula	-	-	-	-	810 to 960
Ostrea deltoidea	-	-	-	-	280 to 1,423
Pholadidea compressa	-	-	-	-	526
Thracia depressa	-	-	-	-	284 to 1,634
Discina latissima	-	-	-	-	287 to 1,411
Lingula ovalis	-	-	-	-	267 to 1,390

The range here assigned to *Ostrea deltoidea* is remarkable since elsewhere it is found mainly in the lower portions of the Kimeridge Clay and the Upper Corallian Beds.

Moreover, the occurrence of beds of sandstone in the Kimeridge Clay is unusual. That obtained from depths of 1,121, 1,139, 1,140, and 1,360 feet, is hard and quartzose, and occasionally concretionary sandstone, unknown elsewhere in this country to form such thick masses in the Kimeridge Clay. Sands, grits, and

---

* Quart. Journ. Geol. Soc., vol. xlii. pp. 486, 488.
† Dixon's Geol. Sussex, Ed. 2, p. 159.

limestones however occur in the Kimeridgian Beds of Sutherland and Cromarty, of Dover, and of the Boulonnais.*   Much of the shale in Sussex, down to depths of 1,000 feet, is shelly and presents the ordinary characters of Kimeridge Clay.

In the Dover boring (see p. 343) a thickness of only 73 feet is assigned to the Kimeridge Beds.   It is clear that the limit cannot be taken *lower*, for undoubted Corallian fossils are met with in the oolites below.   Moreover, a bed of brown oolitic ironstone 12 feet thick, and resembling that of Abbotsbury and Westbury, is placed in the Kimeridge group (see p 150.).   Oolitic limestones, however, are known to occur in the Kimeridge Beds of the north-west of Germany,† and hard bands of rock similarly occur in the Boulonnais as already mentioned.

---

* See Barrois and others, Proc. Geol. Assoc., vol. vi. p. 1.
† H. Credner, Ueber die Gliederung der oberen Juraformation und der Wealden-Bildung, 1863 ; and J. F. Blake, Quart. Journ. Geol. Soc., vol. xxxvii. p. 497.

CHAPTER X.

PORTLAND BEDS.

(PORTLANDIAN.)

GENERAL ACCOUNT OF THE STRATA.

These beds take their name from the Island of Portland, where the stone has been extensively worked during the past two hundred years. The term "Portland Lime" was used by Michell in his table of strata 1788, while the name Portland Oolite was more definitely employed by Webster in 1812.[*]

The Portland Beds, like the Corallian rock-beds, comprise a very variable set of strata. In the upper part they include shelly limestones, oolite, chalky and compact limestones; with locally layers and nodules of chert, and some beds of sand and calcareous sandstone. The chert, which is well developed in Dorsetshire, becomes more prominent towards the base of the stone-beds, but nodules of black chert, resembling chalk-flints, occur in higher beds of chalky limestone.

The lower beds comprise alternations of yellow and greenish-grey sands, sometimes glauconitic, with loamy beds and clays. Here and there layers of sandy and clayey limestone occur, as well as nodules and large doggers of calcareous sandstone. The interbedded and associated clays are of considerable local interest as at Swindon and Hartwell. To the lower division, which is well seen in Portland and the Island of Purbeck, Fitton in 1827 gave the name of Portland Sand[†]; and at the same time he recognized its intimate connexion with the beds above and below. Thus the junction of the Kimeridge Clay and Portland Sands is transitional, the blue colour of the clay being repeated in the dark shaly sandstones and loamy sands of the beds above, as is the case with the Upper Lias clay and Midford Sands.

As remarked by Mr. Whitaker, the division into Portland Sand and Portland Stone does not hold good everywhere.[‡] Some confusion arose at Swindon where the Portland Stone consists very largely of sands, but here and also at Aylesbury the base of the Portland Stone is marked by a conglomeratic layer that contains small pebbles of lydite, and also phosphatized fossils. Thus locally we have indications of a pause in deposition accompanied perhaps by some reconstruction and irregular overlap. The proximity of land suggested by pebbly layers, as noticed in the Corallian Beds of Berkshire, &c., is therefore again indicated.

---

[*] Englefield's Isle of Wight, p. 187.
[†] Trans. Geol. Soc., ser. 2, vol. iv. pp. 206, 210.
[‡] Green, Geology of Banbury, p. 48.

The Portland Sands show that conditions of shallower water succeeded to those of the Kimeridge Clay, land probably appearing in the east or south-east. The Portland Stone, made up of calcareous mud with sandy beds, rolled fragments of organic remains and oolite grains, is again a comparatively shallow-water marine deposit.

Estimates of the thickness of the Lower Portland Beds vary for it is impossible to fix a definite plane of demarcation between them and the Kimeridge Clay. In Dorsetshire the estimates by different geologists vary from 80 feet to nearly 280 feet. In my own opinion an estimate of from 130 to 170 feet would be sufficient. The Upper Portland Beds in the Isle of Purbeck attain a thickness of about 120 feet, and at Portland about 100 feet. The entire formation in the Vale of Wardour and at Swindon is about 100 feet.

Doubtless there is some inconvenience in a term like Portland Sands, when it includes prominent beds of clay, like those of Swindon and Hartwell, and especially when, as at Swindon, the Portland Stone division consists largely of sand.

With a view to dispose of this incongruity Prof. Blake in 1881 suggested a new grouping of the beds, based essentially on that adopted in the Boulonnais.* There the Portlandian beds have been separated into three divisions, of which the Middle Portlandian is in the main equivalent to our Portland Sands, while the Lower Portlandian takes in a portion of our Kimeridge Clay— beds with *Discina latissima* and *Astarte suprajurensis*. Prof. Blake introduces the term Bolonian for the Lower and Middle Portlandian Beds, while he applies the name Portlandian not only to Portland Stone but also to the Purbeck Beds.

Against this rearrangement I have ventured, from a stratigraphical point of view, to protest.† We must not forget that the names Kimeridge Clay, Portland and Purbeck Beds, are derived from strata in this country, and it therefore seems hardly fair or right to modify their limits to suit the stratigraphical changes met with elsewhere.

Prof. Blake admits that the lower limit of his Bolonian must in this country be drawn in the midst of clays, where some introduction of new species commences; but he adds: "This takes place at no very well-defined line; so that the limit must remain open." In commenting on this subject, Mr. Hudleston said he could not agree with Prof. Blake in absorbing this upper part of the Kimeridge Clay into his "Bolonian."‡

As the term Bolonian is no universal remedy for the difficulties of Portlandian classification, it seems best to adhere to the stratigraphical divisions found in this country, and to adopt the following grouping :—

* Quart. Journ. Geol. Soc., vol. xxxvii. pp. 567, 580, 584. See also vol. xxxvi. p. 196, where the name "Boulognian" is introduced (1880).
† Geol. Mag., 1888, p. 469.
‡ Quart. Journ. Geol. Soc., vol. xxxvii. p. 587. *Bononia* is the Latin name of Boulogne.

Purbeck Beds.

| Upper Portland Beds | -{ | Portland, Tisbury, and Swindon Stones. |
| Lower Portland Beds | -{ | Portland Sand, Swindon and Hartwell Clays. |

Kimeridge Clay.

By employing the terms Upper and Lower Portland Beds we get over the awkwardness of calling the Swindon Clay and Hartwell Clay " Portland Sands."

With regard to the Lower Portland Beds, they shade downwards into the Kimeridge Clay, and in many respects their fossils are closely connected. In fact, the Portland Sands bear much the same relation to the Kimeridge Clay below, as do the Kellaways Rock to the Oxford Clay, and the Midford Sand or Midford Beds to the Upper Lias Clay. If Portlandian conditions commenced earlier in France than they did in this country, that seems no reason for modifying our nomenclature, for calling part of our Kimeridge Clay " Portlandian," or for burdening our Tables of Strata with another name, " Bolonian." In fact I quite agree with Prof. Blake, when in speaking of the Portland Sand, he says it may represent " a period which may have varied in *absolute* time from spot to spot."[*]

The Portland Beds are essentially marine accumulations. Indications of estuarine conditions have been considered to occur in the so-called " *Cyrena*-beds " of the Vale of Wardour; but the nature of the genus is not very clear, for it has been referred not only to *Cyrena*, but also to *Cytherea* and *Astarte*.[†] Mr. Hudleston observes that the " shell being frequently associated with *Cerithium*, represents a peculiar estuarine condition, which was the precursor of the Purbecks."[‡] *Neritoma* and *Corbula* which also occur may be taken to afford similar testimony.

The Portland Beds are exposed only over limited areas in this country, owing to the great Cretaceous overlaps. They are seen in portions of South Dorsetshire, Wiltshire, Oxfordshire, and Buckinghamshire, and they have been proved below ground in Sussex and Kent. Over the northern parts of Dorsetshire their outcrop is concealed, but the beds are probably continuous underground beneath the Dorsetshire and Wiltshire downs, except where shifted as near Upway by pre-Cretaceous faults. The outcrop is seen in the Vale of Wardour and again near Devizes. At Swindon the beds exposed may form part of an outlier; and further on in Oxfordshire we find straggling outliers, and perhaps part of the main outcrop extending onwards to Aylesbury. The beds may continue underground through parts of Bedfordshire and Cambridgeshire, for their northerly extension is indicated by derived fossils in the Lower Greensand of Brickhill, Wicken, &c. At the base of the Spilsby Sandstone of Lincolnshire we note the occurrence of Portlandian fossils, but the relation of the Sandstone to the Kimeridge Clay is one of unconformity, suggesting the derivation

---

[*] Quart. Journ. Geol. Soc., vol. xxxvi. p. 224.
[†] See Supp. to Damon's Geology of Weymouth, plates 7 and 19.
[‡] Proc. Geol. Assoc., vol. vii. p. 170.

of the fossils. Still further north, in Yorkshire, the Portland
Beds may be similarly represented, but these and the succeeding
Purbeck strata were formed under conditions different from those
in the area to the south, and more akin to the Volgian Beds of
Russia.

Zones.

With regard to zones Oppel (1856–58) included the Portland
Stone in his Kimeridge group, and marked it as the zone of
Ammonites giganteus and *Trigonia gibbosa,* associating with doubt
the Portland Sand. Renevier (1874) included the "Kimmeridgian,"
Portlandian, and Purbeckian in his Portlandian group, a compre-
hensive group equivalent to the Tithonic of some authorities. The
Tithonic formation of Oppel (1865) included the probable equivalents of the uppermost part of our Kimeridge Clay and higher
beds that shaded up into the Neocomian; in mass, however, it mainly
represented strata on the horizon of our Portland and Purbeck
Beds.* The Tithonic formation is therefore practically equivalent
to the Volgian Beds of M. Nikitin. These divisions both in the
Sub-Alpine (Rhone) region and in Russia represent the deeper-
water marine conditions of Portland and Purbeck times.

The following are the zones of the Portland Beds that are
more generally adopted, together with a few of the leading fossils
which are sometimes taken to indicate sub-zones :—

Ammonites giganteus	Cytherea (Cyrena ?) rugosa.
	Trigonia gibbosa.
	Ammonites boloniensis.
	Trigonia incurva.
Ammonites gigas	Trigonia Pellati.
	Cyprina Brongniarti.
	Exogyra bruntrutana.
	Astarte Sæmanni.

Ammonites gigas has not been very definitely recorded from this
country, and abroad the zone is sometimes separated and put
below the zone of *Cyprina Brongniarti.* Still the ranges of
species in different parts of the Continent seem so variable, it is
impossible to mark minor zones that will hold good over any
extensive region.† All I can attempt to do is to note the broader
general zones.

Organic Remains.

Among the fossils of the Portland Beds there are occasionally
found remains of Saurians, including species of *Cimoliosaurus,*
Pliosaurus, and *Metriorhynchus;* and of the Chelonians, *Steyochelys*
and *Pleurosternum.* The Fishes include *Ischyodus, Mesodon,* and
Lepidotus.

Ammonites are fairly abundant, especially the large *Ammonites
giganteus,* of which examples are obtained in considerable abundance
at Portland. Closely related to this species, if not a variety of it,
is *A. boloniensis.* Belemnites are extremely rare.

---

* Die Juraformation. pp. 725, 727, 807.
† See J. F. Blake, Quart. Journ. Geol. Soc., vol. xxxvii. p. 497 ; see also Table by
Fox-Strangways, Memoir on the Jurassic Rocks of Yorkshire, vol. i. p. 25.

PORTLAND FOSSILS

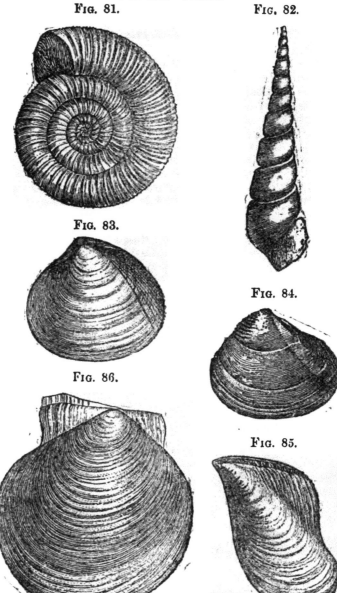

FIG. 81.

FIG. 82.

FIG. 83.

FIG. 84.

FIG. 86.

FIG. 85.

FIG. 81. Ammonites giganteus, *Sow.* ⅛.
„ 82. Cerithium portlandicum, *Sow.* Nat. size (cast).
„ 83. Cardium dissimile, *Sow.* ⅔.
„ 84. Trigonia gibbosa, *Sow.* ⅔.
„ 85. Perna mytiloides, *Lam.* ¼.
„ 86. Pecten lamellosus, *Sow.* ⅔.

PORTLAND FOSSILS.

FIG. 87.

FIG. 88.

FIG. 91.

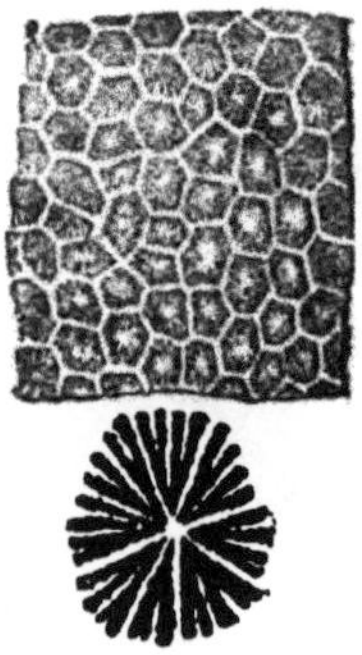

FIG. 89.

FIG. 90.

FIG. 87. Lucina portlandica, *Sow.* ½.
  ,,   88. Cytherea (Cyrena?) rugosa, *Sow.* Nat. size.
  ,,   89. Cyprina elongata, *Blake.* Nat. size.
  ,,   90. Exogyra bruntrutana, *Thurm.* 2.
  .,   91. Isastræa oblonga, *Flem.* ¾.

Many of the Portland Stone fossils have lost their shells and only casts or moulds are preserved, as in the Portland "Roach." Layers of this character occur on different horizons in the Upper Portland Beds. Thus the shells of *Cerithium, Neritoma, Trigonia,* &c., are removed, but *Pecten* and *Ostrea,* where present, retain their shelly matter.

The more abundant fossils are noted in the accompanying list. The Portland Sands do not as a rule afford a rich field for the collector, but fossil-beds occasionally occur, as at Swindon and Brill, from which *Ammonites, Pecten, Mytilus, Exogyra bruntrutana, Thracia, Trigonia,* &c., may be obtained. Brachiopoda and Crustacea are rare in the Portland Beds: of the latter, remains of *Eryma* and *Glyphea* have been found. No species of Polyzoa are recorded; Echinoderms are occasionally obtained.

One Coral, the Tisbury Star-coral, *Isastræa oblonga,* has been found in some abundance in the Vale of Wardour; and rarely in the Isle of Purbeck. It occurs in the Chert. Calcareous examples are not met with in England, but they occur in the Upper Oolites of Sutherlandshire. Sponge-remains are also found. (See p. 187.)

Remains of Plants (*Araucarites*) are but seldom obtained.

The more abundant and noteworthy fossils of the Portland Beds are as follows :—

	Lower Beds.	Upper Beds.
Ammonites biplex	1	2
—— boloniensis	—	2
—— giganteus (Fig. 81)		2
Buccinum ? angulatum		2
Cerithium portlandicum (Fig. 82)	—	2
Natica elegans	1	2
—— incisa	1	2
Neritoma sinuosa	—	2
Pleurotomaria rugata	1	2
Cardium dissimile (Fig 83)	1	2
Cyprina elongata (Fig. 89)	1	2
—— implicata	1	2
Cytherea rugosa (Fig. 88)	—	2
Exogyra bruntrutana (Fig. 90)	1	2
Lucina portlandica (Fig. 87)		2
Ostrea expansa	—	2
—— solitaria	1	2
Pecten lamellosus (Fig. 86)	1	2
Perna Bouchardi	1	2
—— mytiloides (Fig. 85)	—	2
Pleuromya tellina	1	2
—— Voltzi	1	2
Thracia tenera	1	2
Trigonia gibbosa (Fig. 84)	1	2
—— incurva	1	2
—— Pellati	1	2
Isastræa oblonga (Fig. 91)	—	2

Microscopic Structure of Portland Rocks.

Mr. Teall examined microscopic sections of the oolitic freestone (Whit and Base Beds, Portland; and the Top and Under Freestones, Winspit, St. Alban's Head). These showed oolitic grains, pellets, and organic fragments, with some quartz grains (excepting in Whit Bed). The matrix was clear crystalline calcite (Whit and Top beds), or of a granular or mixed nature (Base and Under beds). The oolitic grains on the whole were not very clearly defined, the concentric structure being apparently obliterated in some cases (Top bed) and obscure in others (Base bed). There were no indications of radiate structure.

The Upper Building Stone of Chilmark, near Tisbury, contains well-marked oolitic grains, which often contain quartz grains, and have no cementing material.* (See also p. 312.)

The chalky limestones of Upway, Portisham, and Brill consist of fine-grained granular calcareous matter; and in the Portisham rock there were small and often round patches of colourless crystalline calcite. Other earthy limestones, from Great Milton and Whitchurch, showed fine-grained granular and crystalline matrix, with organic fragments. Small quartz grains occur in the Whitchurch rock, and oolite grains at Great Milton.

A green marly rock from Brill showed round (but not rolled) grains of glauconite, together with organic fragments and quartz grains, embedded in a fine-grained granular calcareous matrix.

A specimen of silicified oolite, which I obtained from the Dancing Ledge quarry, St. Alban's Head, is a fine white oolitic rock, closely resembling an ordinary oolitic limestone, but not effervescing with acid. It consists of obscure oolitic grains or pseudomorphs after oolitic grains, and also of sponge-spicules of a coarse branching type now formed of a colourless chalcedony, and having traces of their axial canals sometimes preserved. The matrix appears to be largely formed of deeply coloured opal or chalcedony. This forms zones around the oolitic grains, thus leaving small irregular areas with cusp-like projections, which are now occupied by colourless chalcedony; sometimes it fills up the interspaces entirely.

The concretionary sandstone of Shotover, a grey calcareous and shelly sandstone, showed rounded and more or less angular grains of quartz cemented by calcareous matter.

Chert

The Chert, as seen in the Portland Beds of Portland, occurs sometimes in even, more commonly in irregular, bands; occasionally in smooth lenticular masses; and often in isolated nodules. The nodules as a rule are not so distinct as the flints of the Chalk, but end off in jagged outlines. The chert is often shattered. Cases are met with where the Chert appears as inclined bands or ribs, in parallel bands cutting across the bedding, as in cliffs below the Lighthouse on Portland, and again at the western end of Gad Cliff, near Worbarrow.

In an example of chert from the freestone-beds at Winspit, St. Alban's Head, Mr. Teall observed a few angular grains of quartz, but the main mass consisted of chert containing sponge-spicules with which some calcareous matter was associated.

A specimen of chert from Tisbury, examined by Mr. Teall, showed irregular patches of chalcedonic silica with radial and

---

* See also Memoir on the Lower Oolitic Rocks of England, p. 12; and Sorby, Quart. Journ. Geol. Soc., vol. xxxv. (Proc.), p. 80.

concentric (agate) structure. Another example showed sponge-spicules and quartz grains. Spicules of *Pachastrella* were identified by Dr. Hinde, both from this rock and from that of Winspit.

Examples of chert or "Portland flint" from the Chalky series of Chilmark Quarry were described some few years ago by Mr. Hudleston. He found shell fragments, sponge-spicules, and occasional oolitic granules in an opalescent ground-mass—in one case there was an inlet of agate silica and a "lake" of quartz. He concluded that the "Portland flints" were formed by the replacement of limestone by silica.* This view is supported by the silicified oolite before mentioned. At Tisbury the Portland Beds yield chert that is oolitic.

Dr. Hinde has remarked that, "From the Portland beds no entire fossil sponges are as yet known, but in some of the chert nodules in the limestones on the Isle of Portland and at Upway, near Weymouth, there are numerous detached spicules of *Pachastrella antiqua*, Moore, sp., and of *Geodites*, sp., thus showing that, in part at least, the chert of these rocks is derived from sponge remains."†

---

* Proc. Geol. Assoc. vol. vii. pp. 183, 184, and Plate 1. figs. 2–6.
† British Jurassic Sponges, p. 193.

CHAPTER XI.

PORTLAND BEDS.

LOCAL DETAILS.

Isle of Purbeck.

The Portland Beds are well shown in the cliffs between Durlston Head and St. Albans (St. Aldhelm's) Head, on the south-east coast of Dorsetshire. The sea all along beats directly against the cliffs, so that the beds can be best studied with the aid of a boat. Here and there we find natural caverns and fissures in the rocks, and there are a number of quarries where the " Purbeck-Portland stone" has been worked. The lower beds exhibited are dark-brown ragged-looking cherty rocks, above these come the freestones, which have a white appearance in the upper part, and these are surmounted by irregular and undulating layers of grey limestone that form the base of the Purbeck Beds, or a "cap" to the Portland Beds.

The general section of the Portland Beds may be stated as follows :—

					FT.	FT.
Upper Beds.	{ Freestone Series	-	-	-	40 to	50
	{ Cherty Beds -	-	-	-	60 to	75
Lower Beds.	Sands, &c. -	-	-	-	120 to	170

The most easterly exposure of the Portland Beds is at Durlston Head, on the eastern side of which the Portland Stone appears beneath the Lower Purbeck Beds. The junction is somewhat obscure, owing to the disturbed character of the beds; moreover, the Portland Rocks themselves are much weathered and encrusted with marine organisms. The beds are faulted, with a downthrow to the north, so that at the headland, the chert-beds below the Portland freestones appear. This fault is again seen west of Durlston Head, where the Lower Purbeck "caps" are let down on the west against the Portland freestones.

Passing Tilly Whim and Anvil Point, we find another fault below the lighthouse with a downthrow of 10 or 15 feet on the west, and again a slight fault further west, with also a western down-throw. Many other faults may be seen in the cliffs as we proceed westwards, and these are noted in a section of the cliffs which has been drawn by Mr. Strahan.

The beds may be examined at Tilly Whim, where the old "caves" show the extent to which the freestones were formerly quarried, up to about the year 1811. The name "Whim" is said to denote the crane or other contrivance used for raising and lowering the· blocks of stone.* The strata exposed are as follows:—

	Ft.	In.
Brown oolitic limestone; with Oyster-bed, yielding *Perna Bouchardi*, &c. - - - - -	8	0
Oolitic limestones, much sub-divided, with occasional lenticular patches of chert; and with shelly bed near base - - - - - - - -	21	0
Thick bed of oolitic freestone - - - -	11	0
Cherty beds: limestones with irregular bands and nodules of chert - - - - - - -	26	0
Limestones with occasional nodules of chert: *Ammonites giganteus, Pecten, Ostrea, Serpula* - -	16	0
Limestones with irregular bands and nodules of chert -	21	0

The thicknesses were measured by H. W. Bristow. It was evidently the lower or under freestone of other parts of this coast that was here worked.

The upper Portland beds were well shown in the quarry at Dancing Ledge, so called on account of a platform of cherty limestones, over which the breakers dash. Above these cherty beds about 30 feet of freestone is exposed in the quarry, overlaid by a few feet of irregular Purbeck strata. The freestones are separated roughly into an upper and lower division by an irregular cherty layer. The upper beds exhibit a good deal of false-bedding, and they are much broken up near the surface. Irregular sparry veins traverse the rock. The stone consists of oolitic limestone, and there are layers of compact white limestone. *Ammonites giganteus* 22 ins. in diameter occurs.

Passing Headbury (or Edboro) Quarry and Cliff Field Quarry east of Seacombe, we come to the Halsewell Quarry on the western side of Seacombe, so named after a vessel that was wrecked there in 1786. Here much stone is quarried and mined. The section showed the following beds:—

	Ft. In.	Ft. In.
Pale rubbly oolite, much weathered and broken up.		
Oolite with irregular cherty layers - -	12 0 to 15 0	
Bands of hard white cherty limestone - ⎫		
Massive oolite in 5 or 6 beds of varying thickness, with fossils here and there: *Ammonites giganteus, Ostrea, Trigonia* in shoals; near base lenticular masses of cherty oolite ⎬		18 0
Cherty limestones - - - - - -		40 0

The chert occurs mostly in irregular nodules in the mass of the beds, but here and there it is found in more or less continuous layers along the partings of the rock-beds.

The occurrence of a white cherty oolitic layer is of interest. (See p. 186.) Bands of oolitic chert occur also at Tisbury.

---

* See T. Webster, in Englefield's Isle of Wight, p. 174; and Robinson, A Royal Warren or Picturesque Rambles in the Isle of Purbeck, 1882, pp. 94, 95.

Fig. 92.

Section at Winspit Quarry, St. Alban's Head.

(From a photograph by A. STRAHAN).

The quarry at Winspit (or Windspit) further west affords the best section of the strata along this part of the coast. It was as follows :—

		Ft.	In.
Purbeck Beds.	Irregular limestones and clays. (Burr. &c.) - - - - -	4	0
Upper Portland Beds.	Fine-grained white limestone, formerly burnt for lime : SHRIMP STONE -	8	0
	Hard grey limestone. with *Perna Bouchardi*, &c. Durable stone, used for gate-posts, &c. The shells stand out in relief after long exposure of blocks : BLUE STONE - - -	”	0
	Oolitic limestone. Best stone : TOP. UPPER OR POND FREESTONE - -	7	0
	Bed with white flints : FLINT STONE -	4	0
	Soft freestone. NIST BED 2 0 to	4	0
	Coarse Limestone with *Ammonites giganteus*. Stone used for breakwaters. not so workable as other beds : HOUSE CAP - - 5 0 to	6	0
	Hard stone with lenticular white cherty stone : has to be blasted out : UNDER PICKING CAP - - 2 0 to	3	0
	Oolitic limestone, good stone, used for sinks. curbstone, &c. : UNDER FREESTONE - - - - -	6	0
	Cherty stone. known as the "CLIFF BEDS" in distinction from the Freestone series termed locally the "INLAND BEDS."		

The " Shrimp Bed," according to Mr. Hudleston, is so called from the quantity of remains of a small Crustacean.[*] It is sometimes 9 or 10 feet thick. The " Blue Stone" below is evidently the representative of the Oyster Bed of Tilly Whim. *Trigonia gibbosa* occurs at various horizons here and there in bands in the freestone beds.

In the higher part of Renscombe the Upper or Pond freestone is obtained by open work and the Under freestone by galleries. The Upper freestone has also been worked at the London Doors Quarry, north of Encombe House : there the top bed of Portland stone is a white limestone; the beds are much shattered. The Lower Portland Beds were shown in a small quarry above Kimeridge village, where the cherty rocks rest on sands with doggers and beds of sandy limestone.

The general section of the beds at St. Alban's Head, given by Prof. Blake,[†] may be compared with that of Winspit :—

		Ft.	In.	
Purbeck.	Botryoidal limestone and shale.			
	Creamy limestone, with *Trigonia gibbosa,* *Pecten lamellosus, Waldheimia boloniensis*? -	9	6	=Shrimp bed.
	Roach, with casts of *Trigonia,* and *Perna mytiloides* - -	4	8	}=Blue stone.
	Oolite with *Perna* - -	5	0	
	Sub-oolitic rock with chert, and *Trigonia* - - -	8	0	}=Top or Pond freestone.
	Band of cherty rock - 2 6 to 4	0		=Flint stone.
Upper Portland Beds.	Coarse-grained sub-oolitic rock, with *Trigonia* and broken shells - - - -	8	6	Nist bed = and House Cap.
	Fine-grained sub-oolitic rock, with few fossils, rather cherty towards base - - -	10	0	UnderPicking Cap and = Under free-stone.
	Brown sandstone - -	1	6	
	Cherty beds, with fucoids, *Serpula gordialis,* and *Ammonites boloniensis* - - -	16	0	}=Cliff beds.
	Cherty beds - - -	50	0	
Lower Portland Beds.	Sandy marl with indurated blocks of sandy cement-stone: *Ammonites biplex, Mytilus autissiodorensis, Pecten solidus, Trigonia incurva, T. Pellati,* &c. - - - -	39	0	
	Hardy sandy marl with small cement-stones; *Exogyra bruntrutana* - - -	42	0	
	Cement-stone, *Thracia tenera,* &c. - - -	2	0	
	Indurated sandy marl, *Rhynchonella portlandica* - -	30	0	
Passage-beds and Kimeridge Clay.	Cement stone - - -	5	0	
	Sand and marly beds, with indurated bands passing downwards into the Kimeridge Clay, with *Lingula ovalis* about 40 feet from base, about	126	0	

At Encombe Point, above Chapman's Pool (West Hill), the cliffs show the following general section :—

Upper Portland Beds.	} Hard beds of cherty stone.
Lower Portland Beds.	⎰ Sands (thin). ⎟ Soft marly sandstone. ⎟ Grey and greenish-grey sandy and marly ⎟ beds, with occasional indurated bands. ⎟ Thick band of shaly calcareous sandstone. ⎟ Shales and marls, mottled blue and brown, ⎟ with thin indurated bands and nodules ⎱ (passage beds).
Kimeridge Clay.	Blue clays with occasional cement-stones.

The thickness of the Lower Portland Beds appears to be about 170 feet. Fitton in his account of Emmit Hill gave a lesser thickness of 120 to 140 feet for these beds.[*]

FIG. 93.

Section at Gad Cliff, Isle of Purbeck.

The Portland Beds, while maintaining a general northerly dip, are bent into a gentle anticline between Emmit Hill and Gad Cliff. Swyre Head, to the south-east of Encombe House, is one of the highest points in Purbeck Isle, rising to an elevation of over 660 feet. It consists of Portland Sands, surmounted by Portland Stone.

Proceeding westwards we come to the grand exposure of Portland Beds in Gad Cliff. The junction-beds with the Kimeridge Clay are to be seen in the cliffs above Brandy Bay, and here, at the eastern end of Gad Cliff, huge doggers of calcareous grit are scattered with the talus at the foot of the cliffs. The indurated bands in the Portland Sands become thicker and more prominent as we trace the beds westwards. (See Fig. 93.)

The Portland Stone-beds are not very accessible, but on the under surfaces of some of the layers large Ammonites may be discerned. The beds break away in huge rectangular masses. The cherty series which forms the most prominent portion of the rugged cliffs, contains two bands of pale limestone, the lower of which, like certain layers in Portland, contains oblique bands or ribs of chert. Beneath the cherty rocks (6) we find in succession :—

Lower Portland Beds. 130 feet or more.	5. Buff sandstone, much honeycombed, and very dark in places.
	4. Blue and yellow marly sands.
	3. Blue and buff sandy marl with hard nodular beds and occasional persistent bands of calcareous sandstone.
	2. Bluish-grey calcareous sandy shales (Passage-beds).
Kimeridge Clay.	1. Dark blue clay.

It is interesting to note that specimens of the Tisbury Star-coral, *Isastræa oblonga*, have been found in blocks of chert at this locality, although it is far from abundant. In the sands we find *Ammonites biplex, Exogyra*, &c.

The pedestrian may be warned not to proceed westwards along this coast in the hopes of reaching Worbarrow Bay. Not many would be tempted to proceed very far, for there is no track-way, and progress has to be made over and between huge tumbled blocks of rough rock ; finally, there is no means, except by boat, of reaching the cove by Worbarrow Knob, as the sea dashes up against the cliffs for some distance east.

The top beds of the Portland Stone are seen at Worbarrow Knob, and, as at Durlston Bay, they are compact limestones. Bristow has recorded *Ostrea* and *Cardium dissimile* ; and from lower beds of oolitic limestone large forms of *Ammonites giganteus*, &c.

The Portland Beds are again met with east and west of Lulworth Cove, fringing the cliffs. At Mupe Bay the top bed of Portland Stone is a compact, cream-coloured shelly and oolitic limestone, with *Trigonia*, something like the " Roach " of Portland. The freestone below contains *Ammonites giganteus*, and in the limestone with bands of flint *Pecten lamellosus* occurs. Lower down limestones with 8 or 9 bands of chert occur. Altogether some 45 or 50 feet of Portland Beds are exposed.

To the west of Stare Cove or Stair Hole traces of Portland Sands were observed by Fitton,[*] and their occurrence was noted by Bristow in the headland known as Dungy (Horsewalls or Dungeon) Crags, which form the eastern horn of the bay in which Man of War Cove is situated. (See p. 258.)

Ringstead Bay, Upway, and Portisham.

` In the cliffs below South Holworth, on the eastern side of Ringstead Bay, we have the following section of the Lower Purbeck and Portland Beds, which here dip at an angle of 37° N. 5° E. :—

---

[*] Trans. Geol. Soc., ser. 2, vol. iv. p. 211.

		Ft.	In.
Lower Purbeck Beds.	Creamy limestones and shales - Hard marls, sandy and oolitic beds, with one or two bands of chert -	6	0
	Fissile limestones (resembling the Slatt or Slate of Portland) - -	6	0
	Hard cream-coloured marl - -	1	2
	Thin irregular dirt layer - -	0	1
	Hard irregular bituminous limestone (Purbeck Cap) - - -	2	6
	Dirt-bed with irregular calcareous seams - - - -	0	2
	Irregular decomposed sandy lime-stone - - - - -	1	0
	Laminated clay and dirt. Fish-remains - - - -	0	2
	Compact bituminous limestone, sandy at top, with casts of *Cyrena* -	0	10
Upper Portland Beds.	Irregular bed of limestone with casts and moulds of shells " Roach." Chalky limestone with *Pecten lamel-losus, Cardium dissimile,* &c. Oolitic limestone and Roach.	12 to 15	0 0
	Cherty beds, white sandy oolite and Roach ; with *Ammonites* - about	28	0
Lower Portland Beds.	Nodular creamy marls with *Ostrea* -	3	0
	Blue and creamy marls - -	3	4
	Earthy and sandy limestones, nodular in places - - - -	8	0
	Greenish-grey marly beds - -	5	0
	Calcareous sandstones with *Exogyra, Ostrea,* and *Trigonia* - -	10	0
	Blue and yellow mottled sandy clays - - - - -	40	0
Kimeridge Clay.	Shales, in upper part, with con-spicuous " white bed " as at Gad Cliff, and lower down with septaria and cement-stones.		

This section is of considerable interest as affording evidence of the change undergone by the Portland Rocks when we trace them westwards to Upway and Portisham. They differ considerably from the beds at St. Alban's Head and Portland, in the absence of the workable beds of freestone, which are replaced by beds of a very chalky character. In this area the beds are subject to rapid changes in thickness, as well as in the character of the individual layers.

Very chalky Portland rock, white or creamy in colour, occurs between Preston and Osmington. About 12 feet of the stone is exposed in a quarry south of the road. The top Portland beds, although disturbed and irregular, show bedding ; the lower beds are much broken up by vertical fissures. Casts of *Trigonia, Lucina portlandica,* &c. are most abundant in the upper beds.

At Upway the Portland beds are worked in a succession of quarries. They comprise beds so like Chalk-with-flints that only an appeal to the fossils would convince the geologist he was not in a Chalk-pit.*

---

* See also Fitton, Trans. Geol. Soc., ser. 2, vol. iv. p. 235; C. H. Weston, Quart. Journ. Geol. Soc., vol. iv. p. 252 ; and J. F. Blake, Quart. Journ. Geol. Soc., vol. xxxvi. p. 198.

Fig. 94.

Section at Upway, near Weymouth.

The general succession at Upway is as follows (see Fig. 94) —

		Ft.	In.
Lower Purbeck Beds.	9. Earthy marl and limestones.		
	8. Marl with seams of chert - -	2	6
	7. White fissile and somewhat sandy limestone, irregular - -	4	6
	6. Dirt Bed, with stools and trunks of trees - - - 0 2 to 1 0		
	Thin bands of limestone with Fish-remains, *Paludina*, and *Archæoniscus* - - - -	1	0
Upper Portland Beds.	5. Sandy limestone, with black specks, slightly oolitic, and very shelly here and there like Roach - -	3	0
	4. White chalky limestone with chert-bands; *Ostrea* and *Perna* abundant in upper bed, also *Trigonia*, *Lucina portlandica*, &c. - -	8	0
	Chalky limestone with small nodules of chert scattered through the middle and lower part - -	6	0
	3. White chalky rock, with many black and white flint or chert layers, *Ammonites* - - -	1	0
	2. Hard grey and yellow sandy lime-stone - - - -	0	10
	1. White chalky rock, with shelly fragments in places and a few irregular flints - - -	10	0

The beds are subject to many changes in thickness and character. Bed No. 7 in the Purbeck Series showed very irregular structure, and other Purbeck Beds were bent in places.

The general section at Portisham may be noted as follows :—

		FT.	IN.
Purbeck Beds.	Clays, shell-limestone, sands - - about	24	0
	Granular limestones, &c. - - about	20	0
	Marls, shales, and Cap - - - about	22	0
Upper Portland Beds.	Hard slightly oolitic limestone - about	5	0
	Chalky series, with bands and nodules of chert - - - - about	23	0
	Earthy and shelly limestone : Serpulite-bed - - - - - about	6	0
Lower Portland Beds.	Soft beds (not seen) - - - ⎫ Bands, more or less concretionary, of hard grey speckly limestones with small *Exogyra* - - - ⎬ about Rubbly beds of soft earthy limestone with casts of *Cardium*, &c. - Brown and greenish-grey calcareous sands - - - - - ⎭	40	0

The basement-bed of the Portland Stone yielded *Ammonites, Natica, Pleurotomaria, Cardium dissimile,* and *Trigonia.*

A well by the farm north-east of the quarry-ground was sunk to a depth of 130 feet, passing through about 60 feet of Purbeck Beds, and 34 feet of Portland Stone, into the Portland Sands.

The Sands with irregular beds of fossiliferous sandstone were shown also at Coryales, east of Portisham.

Isle of Portland.

The Portland Beds of the Isle of Portland can be studied in detail in many quarries, and in other exposures natural and artificial, while their structure and appearance may well be seen by boat.

The general dip is towards the south. The highest ground on which the Verne Fort is placed, rises to an elevation of 495 feet. This is formed of the Portland limestones,* further south these are overlaid by Purbeck Beds which extend southwards to Portland Bill.

Beneath the stone-beds at the northern end of the Island come the Lower Portland Beds, and then the Kimeridge Clay which forms the slopes, and whose slippery surfaces, combined with the action of springs and the great joints in the Portland stone-beds, have caused occasional landslips. (See Fig. 95.)

As the Lower Purbeck Beds are so well shown in connexion with the Portland stone-beds it will be best to consider them together.

---

* In their Map of the Island, Buckland and De la Beche marked Purbeck Beds in this northern part. Trans. Geol. Soc., ser. 2, vol. iv. Plate 1. Great portions of the Purbeck covering in the north have been removed in the process of quarrying.

FIG. 95.

Section through the Isle of Portland.

Longitudinal Scale 2 inches to a mile. Vertical Scale an inch to 1,200 feet.

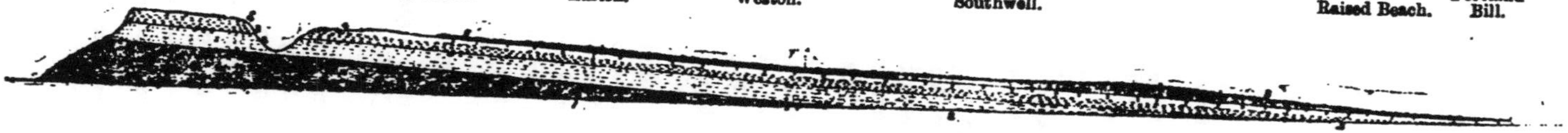

N.

S.

Verne.

Fortune's Well.

Kingbarrow.

St. George's Church, Easton.

Weston.

Near Well-boring west of Southwell.

Raised Beach.

Portland Bill.

5. Purbeck Beds.

4. Freestones.
3. Cherty limestones.   } Upper Portland Beds.

2. Lower Portland Beds.
1. Kimeridge Clay.

The numerous quarries and the economic applications of various beds has led to the introduction of certain local names well known to every worker in the Island. Thus the general section of the beds worked is as follows :—*

		Ft. In.	Ft. In.
Purbeck Beds, 10 to 30 feet.	Rubble, &c. - - - - - -	- 5	0
	Hard Slatt or Slate with Cyprides -	3 0 to 10	0
	Clay band or Dirt - - -	- 0	1
	Bacon Tier - - - -	1 6 to 2	0
	Clay band - - - -	- 0	3
	Aish - - - -	2 0 to 3	0
	Soft Burr - - -	0 9 to 2	0
	Dirt Bed or Black Dirt - - -	0 9 to 1	4
	Top Cap (including Top Rising about 1 ft. 6 ins.) - - - -	2 0 to 10	0
	Occasional dirt seam with Cycadeæ -	- 0	4
	Skull Cap, with occasional seam of clay at base - - - -	1 6 to 9	0
Upper Portland Beds.	Roach - - - -	1 6 to 4	0
	Whit Bed - - - -	4 0 to 9	0
	Curf and waste - - -	4 0 to 9	0
	Best (or Base) Bed - - -	5 0 to 9	0

All the beds are thus very variable in thickness. The strata above the Dirt-bed are known as " Rubbish " and " Head," although some of them have been found of local service.

The Portland Beds are much jointed, and the joints extend upwards to the Purbeck " Caps." Stalactitic deposits (known as " congealed water " and " sugar-candy ") are often found in the crevices. As a rule there are no surface indications of these fissures, some of which are very large, and no doubt they have assisted in producing landslips on the margin of the island.

The Roach forms the top stone of the Portland Series in the Island of Portland. It is an oolitic limestone, almost entirely composed of the cast and moulds of *Trigonia gibbosa* (known as "Horse-heads)† and *Cerithium portlandicum* (the " Portland Screw"). Casts of *Natica elegans, Neritoma sinuosa, Pleurotomaria, Lucina portlandica,* and *Sowerbya Dukei* have also been obtained. On the surface of the bed there are sometimes found irregular cherty patches, and shells of *Ostrea expansa* and *Pecten lamellosus* which have not suffered dissolution. Mr. A. M. Wallis has obtained some tiny forms of Mollusca in this situation. (See p. 200.)

South of Weston there is an upper brown and lower white Roach, and north-east of Rufus Castle an oblique band of chert traverses the beds.

The Roach appears to have been formed by masses or shoals of shells, for it merges irregularly downwards into the Whit Bed, while other layers presenting the character of Roach occur sometimes at lower horizons. *Cerithium portlandicum* appears to be confined to the uppermost or true Roach.

---

* See also Webster, Trans. Geol. Soc., ser. 2, vol. ii. p. 41 ; and W. Gray, Proc. Geol. Assoc., vol. i. p. 131.
† The name " Hippocephaloides " taken from the quarrymen's term, was applied by Plot to these casts of *Trigonia.*

The Whit Bed or Top Bed of Freestone is a fine-grained oolitic limestone from 5 to 15 feet thick, though seldom all good when of great thickness. It contains comminuted shells, and it has been suggested that its durability may be owing to the crystallized carbonate of lime derived from the contained shells.[*]

The Curf and flints form a bed which varies from 4 to 9 feet in thickness. Sometimes the flints are absent, at other times they are scattered. *Ostrea solitaria* occurs, and *Ammonites giganteus* is abundant. Where this band is absent, as at a quarry south of Weston, the Whit Bed lies directly on the Base Bed.

FIG. 96.

Section at Kingbarrow, Portland.

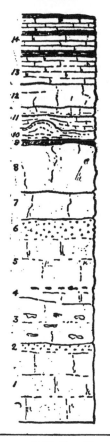

The Best Bed (Base Bed, Lower Tier or Bottom Bed) is a fine-grained oolitic-limestone, from 5 to 9 feet thick, and comparatively free from fossils. According to Prof. Blake this bed and the Curf above have yielded *Isastræa oblonga* and *Lithodomi*.

The beds below the Base Bed are not worked on acount of the flint.

Prof. Blake considers that the true *Ammonites giganteus* occurs in the upper beds, while large Ammonites referred to *A. boloniensis* occur at lower horizons.[†] Specimens of *A. giganteus* measuring as much as 38 inches in diameter have been obtained. These fossils are called "Conger Eels" by the quarrymen, and specimens, sometimes "whitewashed," may be seen in many cottage-gardens at Portland, and near Weymouth, Upway, and Portisham.

The Cherty series below the Portland Building-stones, consists of buff shelly and sandy limestone, with chert or "flints" in irregular bands, in smooth lenticular masses, and in scattered irregular nodules. Beneath the Lighthouse-cliff remarkable oblique bands traverse the rocks, appearing like ribs, and these are overlaid by approximately horizontal bands of chert.

---

[*] W. Gray, Proc. Geol. Soc. Assoc., vol. i. p. 131.
[†] Quart. Journ. Geol. Soc., vol. xxxvi. p. 191.

An excellent section of the Lower Purbeck and Upper Portland Beds is exposed in Steward's quarry, at Kingbarrow, where the local names of the beds were pointed out to me by Mr. A. M. Wallis (see Fig. 96, p. 199):—

		Ft.	In.
Lower Purbeck Beds.			
14. RUBBLE - -	Flaggy white limestone, rubble, and clay 6 0 or	7	0
13. HARD SLATT (Slate or Slattern)	Soft white flaggy limestone	4	0
	(Irregular band of marl and limestone.)		
12. BACON TIER -	Hard creamy and flaggy limestone, breaks irregularly - - 2 0 to	4	0
	(Thin shaly dirt-layer and sandy limestone.)		
11. AISH - -	Soft brown or white sandy limestone, laminated at base - - -	1	6
	Seam of brown, black, and white clay. Soft white banded limestone, somewhat sandy and with carbonaceous specks - -	2	0
10. SOFT BURR - -	Loose pale tufaceous limestone with hard cap 1 0 to	2	0
9. BLACK DIRT -	Dark clayey bed with carbonaceous matter, fragments of wood, and stones of grey earthy limestone and oolite - 0 9 to	1	0
8. TOP CAP - -	Buff and white limestone, hard and compact in places, in other parts full of drusy cavities in decomposed cherty rock running very irregularly through the stone; the rock splits up irregularly	7	0
	Irregular clay seam.		
7. SKULL CAP - -	Hard brown limestone with small cavities; irregular -	4	0
	(With at base nodules of chert or "hard rag" yielding small univalves.)		
Upper Portland Beds :— Freestone Series.			
6. ROACH - -	Oolitic limestone full of casts and moulds of *Trigonia, Cerithium*, &c.; merging into bed below - -	3	0
5. WHIT BED - -	Buff oolite, with a few layers containing casts of shells, and occasional flints	7	0
	Irregular cherty band with "sand holes" yielding small univalves.		
4. BOTTOM WHIT BED -	Buff oolite (local) 3 0 to	3	9
3. CURF and FLINTS -	Oolitic limestone with many nodules of chert: *Ammonites giganteus* - -	6	6
2. BASE BED ROACH - or LITTLE ROACH.	Oolitic limestone with *Ostrea* and casts of *Trigonia, Cerithium*, &c. - 1 0 to	1	6
1. BASE BED - -	Buff oolite - - 6 0 to	10	0

Sections by the Verne Fort showed the following beds, which appear to come directly beneath the Base-bed:—

		Ft.	In.
Upper Portland Beds:— Cherty Beds.	Buff limestones, shelly in places with *Ammonites boloniensis*, *Ostrea*, *Pecten lamellosus*, *Trigonia*, *Serpula*, and beds like "roach": with bands and nodules of chert throughout, about - - - -	65	0
Lower Portland Beds:— Portland Sands.	Blue buff and grey mottled marl. Dark blue clay. Greenish grey and blue mottled marl.	15	0

The cliffs near Rufus Castle afford a good section of the Portland Beds, with about 10 feet of Lower Purbeck Beds on top. The freestone-beds show shelly patches (Roach) at different horizons, and the cherty series below contains bands of pale and dark chert in even and irregular layers, in lenticular seams and nodules, with an occasional oblique vein. A sandy bed occurred about 8 feet from the base; below are shelly limestones with chert, and then a spring issues on top of a clayey band that marks the top of the Lower Portland Beds.

A section of the beds on the west side of the tramway leading to the breakwater was measured by H. W. Bristow. Beneath a thickness of 63 feet of stone-beds with chert, the following beds of the Portland Sands were seen:—

		Ft.	In.
Portland Sands.	Hard yellow sandy marl - - - -	4	6
	Grey marls with plates of fibrous gypsum -	12	6
	Hard grey sandy bed - - - - -	4	8
	Carbonaceous sand - - - - -	0	9
	Hard grey marl with large irregular cavities lined with carbonate of lime and containing sand and clay - - - - - -	6	3
	Soft grey marly sand, passing upwards into hard thick calcareous bed containing *Ammonites* and fibrous gypsum - -	8	0
	Hard grey calcareous sand with fossils - -	0	9
	Carbonaceous parting		
	Soft sand with fossils - - - - -	9	6
	Yellow sand with blue calcareous nodules: *Trigonia*	3	3
	Yellow sand - - - - - -	3	3
		53	5

We are indebted mainly to Prof. Blake for our knowledge of the palæontology of the Portland Beds. The fossils of the freestone-beds have been already mentioned. From the beds below, including the mass of the cherty series, he records the following species:—

Ammonites boloniensis.
—— pseudogigas.
—— trifidus (triplex).
Pleurotomaria Roseti.
—— rugata.
Cardium dissimile.
Cyprina elongata.
Lima rustica.
Ostrea multiformis.
Pecten lamellosus.
Perna mytiloides.
Pleuromya tellina.
Trigonia gibbosa.
—— incurva.
Serpula gordialis.

From the Lower Portland Beds he obtained the following fossils :—*

Ammonites biplex.	Exogyra bruntrutana.	Trigonia muricata.
Natica incisa.	Isocardia.	—— Pellati.
Arca.	Lima boloniensis.	Discina humphriesiana.
Avicula octavia.	Mytilus autissiodorensis.	Rhynchonella port-
Cyprina elongata.	Pecten solidus.	landica.
—— implicata.	Trigonia incurva.	Serpula.

The Lower Portland Beds are well exposed in the cliffs of Clay Hope on the western side of the Island. They consist of marly and sandy beds that contain indurated beds in the upper part and lower down doggers of calcareous sandstone ; they pass downwards into the Kimeridge Clay.

Mr. Whitaker has kindly given me the following particulars of a boring made by Mr. J. L. Webster at Southwell, Portland :—

		THICKNESS.	DEPTH.
Purbeck Beds.	Rubble, &c. - - - -	37 0	
	Cap - - - - - -	9 0	46 0
	Skull cap - - - -	4 6	50 6
Portland Stone.	Roach - - - - -	9 0	59 6
	Whit Bed, &c. - - -	16 0	75 6
	Limestone and flint - -	67 3	142 9
Portland Sand.	Clay - - - -	} 127 3	270 0
	Sands, &c. with water - -		

Kimeridge Clay.

A shaft was sunk 230 feet, and the rest bored. The boring was made rather less than a quarter of a mile W.N.W. of the old "South well." Galleries were driven N. 20 feet, S. 35 feet, E. 40 feet, and W. 35 feet. The yield of water has been equal to about 90,000 gallons a day.

Mr. Whitaker has remarked that a supply of water was to be expected at the base of the Portland Stone, where water would naturally be held up by the clayey-bed that forms the top of the Portland Sand. It was a matter of surprise, therefore, that the shaft was practically dry, for the Government water-works at Chene, N.E. of Southwell, got their supply from the Portland Stone above the clay-bed, a supply amounting to about 50,000 gallons a day. He thought there must be joints or other means of communication in the top part of the Portland Sand, by means of which water which in some places is stopped there, passes through and sinks to a lower level. Fissures having a direction nearly north and south were met with in the shaft and galleries.

Near Portland Bill we find in places that the entire mass of the building-stones presents a "roachy" character ; in other places, and at the Bill, the top bed of Portland stone is of very compact nature.

The full thickness of the Portland Beds, difficult to obtain from measuring the beds in the Cliffs, is now proved by the well-boring before noticed. I had estimated the Portland freestones at 30 feet at Kingbarrow, and 28 feet at Rufus Castle ; and the chert-beds at Kingbarrow at 65 feet, while those of Rufus Castle appeared to be no more than 45 feet. We may, however, take the general thickness as follows :—

	FT.
Freestones - - - - -	25
Cherty Beds - - -	67
Lower Portland Beds - - -	128
	220

No doubt the thicknesses of these subdivisions vary to some extent in different parts of the island.

---

* Quart. Journ. Geol. Soc., vol. xxxvi. p. 193.

CHAPTER XII.

PORTLAND BEDS.

LOCAL DETAILS—*continued.*

Vale of Wardour.

The Vale of Wardour is one of the classic regions of the geologist. Large quarries have been opened in the Portland and Purbeck beds, many fossils have been obtained, and the district has been described by many observers. Pyt House, a few miles west of Tisbury, was the home of Miss Etheldred Benett, one of the earliest of lady geologists, who gave especial attention to the fossils of Wiltshire, and published the first detailed account of the strata.[*] They have subsequently been studied by Fitton,[†] Mr. W. H. Hudleston,[‡] Prof. J. F. Blake,[§] and the Rev. W. R. Andrews[||] (formerly of Teffont Evias).

The general section of the Oolitic strata in the Vale of Wardour is as follows :—

			Ft.	In.
	Lower Purbeck Beds.	Flaggy limestones, dirt-beds, and peculiar oolitic beds. (See page 267.)		
Upper Portland Beds.	Upper Building Stones.	Buff sandy and oolitic limestones, compact limestone, and occasional chert-seams in lower part - 10 to	16	0
	Chalky Series.	Soft white chalky limestone, with nodules and veins of black chert - 4 to	24	0
	Ragstone.	Brown gritty and shelly limestone, divided in places by seam of rubbly marl - 4 6 to	5	6
		Pale shelly and oolitic limestones, with rubbly shelly marl at base - -	3	3
		Trough Bed: Hard buff sandy and oolitic limestone, the surface covered with bivalves (*Trigonia gibbosa*), the bed merging into that below - -	2	8
	Building Stones.	Glauconitic and sandy limestones: divided locally into :— Green Bed - 5 0 / Slant Bed - 1 0 / Pinney Bed - 2 0 / Cleaving or Hard Bed - 1 0 / Fretting Bed - 3 4 / Under Beds - 3 0	15	4
Lower Portland Beds. Kimeridge Clay.				

---

[*] A Catalogue of the Organic Remains of the County of Wilts, 1831; (privately issued).
[†] Trans. Geol. Soc., ser. 2, vol. iv. pp. 251, 254.
[‡] Proc. Geol. Assoc., vol. vii. pp. 167–170, and Geol. Mag. 1881, p. 387.
[§] Quart. Journ. Geol. Soc., vol. xxxvi. p. 200.
[||] Proc. Dorset Nat. Hist. Club, vol. v. p. 66.

The finest exposures of the Portland Beds are to be seen in the Chilmark ravine, about a mile south of the village, but there are a number of quarries near Tisbury.

The Oolitic series has a general inclination towards the E.S.E., but the beds are affected by undulations, which bring the Portland Beds to the surface at Chilmark. The Upper Cretaceous Rocks extend across the denuded surfaces of the older strata and form a margin to the vale both north and south. (See Fig. 97.)

On the west side of the Chilmark ravine only the lower building-stones are worked, on the east side both Upper and Lower Beds are quarried, for the most part in underground workings. The overlying Purbeck Beds at the Teffont (Chilmark) Quarry are noted on p. 267. The lower building-stones which I noted were as follows:—

		FT.	IN.
Upper Portland Beds.	Chalky Series.		
	WHITE BED: gritty limestone, used for hearthstone - - - 1 6 to 4		0
	Rubbly marl (Rag) - - - - 0		6
	Shelly limestones - - - - 3		6
	TROUGH BED: pale shelly oolitic limestone - 1		3
	Rubbly marl, passing into Roach - - 0		6
	GREEN BED: hard buff or pale greenish-grey oolite merging into bed below - 2 6 to 2		9
	PINNEY BED: brown glauconitic and oolitic sandy limestones in three or four layers - 12		0

A well at the base of the Chilmark (Teffont) Quarry was sunk to a depth of 39 feet through clays and calcareous sandy beds to very black clay (Kimeridge Clay), and water rose to within 19 feet of the surface.

The thicknesses at this locality may be thus summarized:—

		FT.	IN.
Upper Portland Beds.	Upper Building Stones - - -	16	0
	Chalky Series - - - -	24	0
	Ragstone - - - -	9	0
	Lower Building Stones - - -	18	0
Lower Portland Beds	- - - - -	38	0
		105	0

The Lower (or Chief) Building Stones have yielded but few fossils. Prof. Blake records *Ammonites boloniensis* and *A. biplex*. They contain occasional cherty masses, and a quantity of sponge-spicules may, according to Mr. Hudleston, be found in some of the beds. He observes that *Trigonia gibbosa* is found in a state of chalcedonic replacement, and states that the Pinney Bed is penetrated by a small Serpula, from the appearance of which the name is derived.[*]

The names applied to these building-stones vary in different parts of this district, and other names besides those mentioned have been used.[†]

---

[*] Proc. Geol. Assoc., vol. vii. pp. 171, 172.
[†] See Section of Chicksgrove Quarry by Miss. E. Benett, Sowerby's Mineral Conchol., vol. ii. 1818, p. 58.

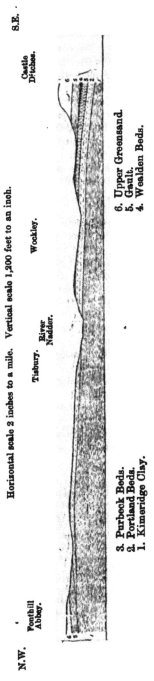

Fig. 97.

Section across the Vale of Wardour.

Horizontal scale 2 inches to a mile.   Vertical scale 1,200 feet to an inch.

N.W.

Fonthill Abbey.

Tisbury.   River Nadder.

Wockley.

Castle Ditches.

S.E.

3. Purbeck Beds.
2. Portland Beds.
1. Kimeridge Clay.

6. Upper Greensand.
5. Gault.
4. Wealden Beds.

The Ragstone Beds are characterized by *Cytherea* (*Cyrena*) *rugosa*, and they have been termed the "Lower Cyrena Beds." Gasteropods are fairly abundant, including the form known as *Cerithium concavum*, the small *Natica elegans*, *Pseudomelania teres*, *Neridomus transversus*, *Neritoma sinuosa* and *Actæonina signum.** The beds yield also *Corbula*, *Cardium dissimile*, *Lucina portlandica*, &c.

Mr. Hudleston considered there was evidence of a break between these beds and those below. He noted at the base an irregular *Trigonia*-bed, with *T. gibbosa* and *Mytilus jurensis*. *Trigonia Manseli* also occurs in these beds. The shells in this division are well preserved. The absence of Ammonites is noteworthy, and the general assemblage is considered by Mr. Hudleston as suggestive of fluvio-marine conditions.

The Chalky Series calls to mind the similar beds at Upway. The fossils are marine and include *Ammonites boloniensis*, known as "Horns," *Pleurotomaria rugata*, *Turbo apertus*, *Cardium dissimile*, *Lucina portlandica*, *Ostrea expansa*, *Pecten lamellosus*, *Pholadomya tumida*, *Pleuromya tellina*, and *Trigonia gibbosa*.

The Upper Building Stones, termed by Mr. Hudleston the "Upper Cyrena Beds," yield fossils for the most part in casts, and the beds have been compared to the Roach of Portland. *Cerithium portlandicum* is characteristic, and among other fossils there are *Neritoma sinuosa*, *Cytherea* (*Cyrena*) *rugosa*, *Trigonia gibbosa* (not uncommon), *Cardium dissimile*, *Lucina portlandica*, and *Pecten lamellosus*.

Portland Beds with usually some thickness of Purbeck Beds on top, were worked on the north side of River Nadder by Chicksgrove Mill, and there were other quarries (with underground workings) on the south side of the river. An account of one of these was published by Miss Benett in 1818.† The locality is that of Upper Chicksgrove, and pits were opened westwards in Quarry Copse, south of the railway. A cutting on the railway west of Chicksgrove Mill, and north of Wockley, showed the following section :—

		Ft.	In.
Upper Portland Beds.	Greenish sandy bed	1	0
	Hard grey sandy limestone, weathering white: *Trigonia, Ammonites*	1	3
	Greenish and grey beds of more or less calcareous sandstone or sandy limestone.	2	6
	Shelly limestone: *Serpula*	2	6
	Sandy and shelly limestone: *Serpula*	2	6
	Sandy marl: *Ostrea, Serpula*, Spine of Echinus	0	4
	Grey shelly limestones	4	0
Lower Portland Beds.	Brown and greenish-brown sand with clay seams and bands of indurated sand: casts of shells hereand there, thin beds of stone near top	15 0 to 20	0

---

* See Hudleston, Geol. Mag. 1881, p. 387; and Proc. Geol. Assoc., vol. vii. pp. 167, &c.

† Sowerby, Min. Con., vol. ii. 1818, p. 58.

The uppermost three beds merge one into the other and are much shattered at the top in places. The fissures that sometimes traverse the rocks in this district are known as "lets."[*]

At Wockley, south-east of Tisbury, the beds are much reduced in thickness, and their character, especially in the lower beds, is altered for we miss the Ragstones of Chilmark. (See Fig. 134, p. 268.) The individual layers of rock also vary much in thickness. On top there is from 18 to 20 feet of Lower Purbeck Strata, beneath which we find :—[†]

		FT.	FT.	IN.
Upper Portland Beds.	4. Bed of Roach, with lenticular mass of chert at top: *Trigonia gibbosa.*			
	3. Chalky limestones obliquely bedded, with *Ammonites biplex, Pleurotomaria rugata, Ostrea expansa, Pecten lamellosus.*		10 to 15	0
	2. Buff and greenish, glauconitic sandy limestone　-　-　-　-		2 to 4	0
	1. Compact and very shelly limestone, passing down into sandy limestone (quarried for freestone)　-　-		4 to 5	0

A quarry south of Tisbury Station afforded evidence of the variable nature of the beds. The section was as follows (see Fig. 98):—

		FT.	FT.	IN.
Upper Portland Beds.	5. Rubbly stone and marl with seam of clay　-　-　-　-			4 0
	4. Shelly limestone (Roach) with *Trigonia incurva.*			
	3. Impure shelly and tufaceous limestone　-　-　-　-		4 to 5	0
	2. Compact, but rotten chalky limestone much shattered, with Gasteropods -		3 to 4	0
	1. Greenish glauconitic sandy limestone with lenticular seams of oolitic chert : three layers seen　-　-		10 to 12	0

FIG. 98.

Quarry South of Tisbury, Wiltshire.

Beds of greenish sandstone, that become paler when dry, are dug to a depth of about 10 feet in a quarry between Tisbury and Newtown. Specimens of *Trigonia* are abundant on some of the blocks. Formerly much stone was obtained at Lower Lawn, and extensive old quarries are to be seen there.

Looking generally at the variations exhibited in the different quarries, to the attenuation and local absence of beds that may be classed with the Upper Building Stones, Mr. Hudleston remarked there was

---

[*] Fitton, Trans. Geol. Soc., ser. 2, vol. iv. p. 255.

[†] See also Fitton, Trans. Geol. Soc., ser. 2, vol. iv. p. 253 ; and Hudleston, Proc. Geol. Assoc. vol. vii. p. 173.

evidence of discordance between the Purbeck and Portland Beds
in the Vale of Wardour.*

No doubt there are abrupt changes here and there between
the formations, as there sometimes are between individual beds
in the Portland series. There is, however, no discordance such
as would imply upheaval and denudation of the strata. The
phenomena may be attributed in part to contemporaneous erosion,
in part to the attenuation and local deposition of certain sediments ;
while again the variations in the lithological characters of different
layers serve to render the results of minute correlation very
difficult and uncertain.

Tisbury has been long famous as the locality for the Star
Coral, *Isastræa oblonga*, which occurs in the Portland chert, and
polished specimens of which are to be found in most collections.

The exact position of the bed yielding this " Siliceous
Madrepore" has been a matter of some doubt. It was described
in 1729 by John Woodward as the "Starr'd Agate." He says
" This was found, amongst several others, lying on Floors, like the
common black Flints, amongst Chalk ; * * * Underneath
these Floors of starred Flints lay Strata of Sand-stone, in a
Quarry in *Tisbury* Parish."†

This position in the Chalky Series agrees with that assigned
to it by the Rev. W. R. Andrews, who has found the fossil above
the Ragstones at Newtown, Tisbury. Most of the specimens have,
however, been obtained from ploughed fields to the north-west of
Tisbury. The horizon also agrees with that noticed by Miss
Benett, who states that the Coral was found above the Portland
rock (*i.e.* building-stone), in a well sunk at Burton's Cottage,
near the Inn, at Fonthill Giffard ; the same authority recorded the
following section of a well at Butcher's Knap(field), Tisbury :—‡

					FT.	IN.
Rubble of Portland Beds	-	-	-	-	10	0
Siliceous Madrepore	-	-	-	-	1	0
Portland Beds	-	-	-	-	42	0
(Water.)						

Fitton mentions the finding of *Ammonites biplex* in chalcedonic
flint, and states that many specimens of the Tisbury Coral had
been obtained from "a continuous bed of flint, about 2 inches
thick," exposed in one of the quarries formerly worked to the
south of Fonthill Giffard.§ It has been found also at Chilmark.

In a cutting on the road-side between Tisbury and Wardour
near Hazelton, and in other sections, Mr. Hudleston noted the
following beds beneath the main building-stone :—||

---

* Hudleston, Proc. Geol. Assoc. vol. vii. pp. 170, 173, 174. See also J. F. Blake,
Quart. Journ. Geol. Soc., vol. xxxi. p. 191.
  † Nat. Hist. Fossils of England, Tome II. p. 77.
  ‡ Cat. Org. Rem. Wilts, p. iv ; see also Hudleston, Proc. Geol. Assoc. vol. vii. p.
167.
  § Trans. Geol. Soc., ser. 2, vol. iv. p. 255. For a reference to disintegrated Port-
land Stone from Fonthill Giffard, see p. 233 of this Memoir.
  || Proc. Geol. Assoc., vol. vii. p. 172.

		Ft.	In.
Upper Portland Beds.	3. Loose sands with doggers - -	7	0
	2. Greenish concretionary limestone grit, with occasional lydite; originally a Trigonia-bed -	3	0
Lower Portland Beds.	1. Loamy sands and clays - -	21	0

From the hard band(2) he records the following species :—

Natica elegans.	Mytilus jurensis.
Avicula credneriana.	Pecten lamellosus.
Cardium dissimile.	Perna Bouchardi.
Exogyra bruntrutana.	Trigonia gibbosa.

Prof. Blake, who first noted these beds, identifies the *Trigonia* as *T. Pellati*.* The fauna, as remarked by Mr. Hudleston, seems very little different from that of the building-stones,—"It is what one would call an average Portland stone fauna of the large type, somewhat modified."

The occurrence of the bed with lydites is interesting, as we find other such pebbly layers as we trace the beds over the exposures onwards to Buckinghamshire, and the horizon seems to be fairly constant.

Worton and Potterne, near Devizes.

A small area of Portland rocks is exposed between Coulston and Potterne, to the south of Devizes, at the western end of the Vale of Pewsey.

The passage-beds into the Kimeridge Clay have been opened up in the brickyards at Worton and between Potterne and Dues Water. At the brickyard south of Potterne the following section was exposed :—

	Ft.	In.
Brown and blue sandy loam with casts of *Trigonia* and *Exogyra* - - - - - - -	14	0
Hard irregular nodules of calcareous sandstone with *Thracia* and *Perna*.		
Blue loamy clay.		

Prof. Blake has noted that here we find the usual Swindon and Hartwell fossils.† On a higher horizon we find, in the road-cutting east of Worton, grey, brown and greenish-grey sands. The spur north of the Potterne brickyard is partially covered with pale quartzose sand (Lower Greensand); but, as I am informed by Mr. Jukes-Browne, there is fine-grained Portland sand by Larborough Farm. Eastwards by Potterne Park, he noticed a considerable area of fine sand containing glauconite, and he remarks that the Portland sand is thus locally green, whereas the Lower Greensand of the area is never so.

---

* Quart. Jourt Geol. Soc., vol. xxxvi. p. 202.
† Ibid., vol. xxxvi. p. 203.

Cherty Beds were formerly dug in the ridge between Hurst Farm and Greenlands Farm north of Great Cheverell. Mr. Jukes-Browne was informed by Mr. W. Cunnington that about 50 tons of stone were obtained.

The highest beds seen in this area are exposed at Crockwood. In the farmyard we find 2 or 3 feet of irregular and rubbly beds of gritty and oolitic limestone, with *Pecten lamellosus, Cardium dissimile,* and *Serpula.* To the north-west of the farm, a lane-cutting showed 3 feet of the same rock, here glauconitic and containing an occasional lydite pebble. Prof. Blake has recorded *Trigonia gibbosa* from this locality, and I obtained, in addition to the species before mentioned, some Gasteropods, also *Modiola, Pleuromya,* and *Ostrea.* The beds rested on greenish sand with hard concretionary masses of glauconitic sandstone, shown to a depth of 1 ft. 8 ins.; while on top of the stone-beds stiff brown clay (probably Gault) was shown.

The fossiliferous stone-beds are no doubt on the same horizon as the basement Portland Stone at Swindon.

Swindon and Bourton.

The next exposures of the Portland Beds are at Swindon, where the large quarries to the south-west of the old town afford fine sections of Purbeck and Portland beds, and many fossils may be obtained.

The sections have attracted a good deal of attention, and they have been described by Fitton,[*] Godwin-Austen,[†] the Rev. P. B. Brodie,[‡] J. Buckman,[§] and others, more particularly Prof. J. F. Blake, to whom reference will subsequently be made.

The Portland Beds at Swindon are partially concealed beneath the Cretaceous rocks to the south-east, and they were considered by Sir A. C. Ramsay to occur probably as an outlier. In confirmation of this, it may be mentioned that no trace of them was found in a boring carried through the Cretaceous rocks of Burdrop, between Wroughton and Chiseldon on the south.[‖]

The Portland Beds at Swindon comprise in their upper part two calcareous beds, and these are separated by a mass of sands, with layers of hard calcareous sandstone that furnish the Swindon building-stone.

When the sections were described by Ramsay,[¶] these sandy strata were classed as "Portland Sands," and he represented them at one place as overlaid irregularly by Portland Limestone; but it is now known, and mainly from the observations of Prof. Blake, that the Swindon stone is part of the Portland Stone division, and that the true Portland Sands are well developed beneath.

When Prof. Blake's observations were made the details of the Lower Portland beds could only be fitted together from somewhat

---

[*] Trans. Geol. Soc., ser. 2, vol. iv. p. 266.
[†] Quart. Journ. Geol. Soc., vol. vi. p. 467.
[‡] *Ibid.* vol. viii. p. 53.
[§] *Ibid.,* vol. xiv,, p. 128.
[‖] *Ibid.,* vol. xlii. p. 308.
[¶] Geol. parts of Wilts and Gloucestershire, pp. 25, &c.

scattered data. Since then sections, opened up during the construction of the new railway from Marlborough by Swindon to Cirencester, have furnished clear evidence of the sequence of the strata; and additional information has been afforded by cuttings along a road made from Old Swindon to the new town adjoining the Great Western Railway. The information thus obtained confirms the succession made out by Prof. Blake.*

The general section at Swindon is as follows :—

Purbeck Beds. (Resting irregularly on different members of the Portland Series.)

		Ft. In.	Ft. In.
	Chalky beds.		
Upper Portland Beds.	Oolitic limestone with casts of *Cerithium*, &c. - - - -	5 0 to	6 0
	Irregular grey clay.		
	Sands with beds of Swindon stone -	20 0 to	25 0
	Fossiliferous limestone, " Cockly Bed " -		4 0
	Bluish-grey limestone with lydite pebbles		3 6
Lower Portland Beds.	Blue clay.		
	Exogyra-bed - - - -	6 0 to	8 0
	Sands with doggers - - -	30 0 to	40 0

Kimeridge Clay.

The Kimeridge Clay passes up most gradually into the Lower Portland Beds, and these comprise the strata Nos. 1 to 5 in the Section (Fig. 99, p. 212). A more detailed description of these may be given.

Lower Portland Beds.

1-3. Grey and greenish or buff sands, and loamy beds, with hard concretionary masses or "doggers" of calcareous sandstone, some of which are as much as eight feet in diameter. The upper portions of this series were exposed in the railway-cutting (Fig. 99, p. 212), in the cuttings for the new road from Swindon town to the railway-station, and in an adjoining sand-pit where material was obtained for making mortar. Very fossiliferous layers occur at these localities in this higher portion. The full thickness of the beds is not less than 40 feet. A number of fossils were obtained by Mr. Rhodes and myself, but many more have been recorded by Prof. Blake. The following is a list of the species obtained from these beds :—†

Ammonites biplex.
—— pectinatus.
Aporrhais Thurmanni.
Cerithium Lamberti.
Delphinula globata.
Turbo Foncardi.
Arca velledæ.
Astarte polymorpha.
—— Sæmanni.
Cardium morinicum.
× Cyprina implicata.
—— pulchella.
—— swindonensis.
Exogyra bruntrutana.
Lima boloniensis.
Lithodomus.
Lucina fragosa.
Myoconcha portlandica.
—— Sæmanni.

Mytilus antissiodorensis.
—— boloniensis.
—— longævus.
Ostrea bononiæ.
—— multiformis.
Pecten lens var. Morini.
—— solidus.
—— suprajurensis.
Perna Bouchardi.
Pholadomya tumida.
Pinna suprajurensis.
Placunopsis Lycetti.
Pleuromya Voltzi.
Sowerbya longior.
× Trigonia irregularis.
—— Pellati.
—— swindonensis.
× Unicardium sulcatum.
Acrosalenia Kœnigi.

---

* Quart. Journ. Geol., Soc. vol. xxxvi. pp. 210-212, and Proc. Geol. Assoc., vol. xii. p. 326; see also H.B.W., Geol. Mag., 1888, p. 469.

† The additional species named by Messrs. Sharman and Newton are marked × : the others were all recorded by Prof. Blake, Quart. Journ. Geol. Soc., vol. xxxvi. p. 212.

FIG. 99.

Section along the Midland and South-Western Junction Railway west of Swindon Town Station.

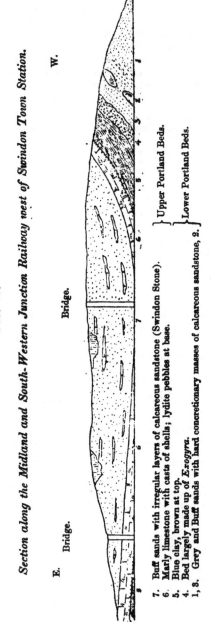

} Upper Portland Beds.

} Lower Portland Beds.

7.   Buff sands with irregular layers of calcareous sandstone (Swindon Stone).
6.   Marly limestone with casts of shells; lydite pebbles at base.
5.   Blue clay, brown at top.
4.   Bed largely made up of Exogyra.
1, 3.   Grey and Buff sands with hard concretionary masses of calcareous sandstone, 2.

4. *Exogyra-bed :* a marly sandstone that varies from three to eight feet in thickness. It was well shown in the railway-cutting, and less distinctly in the road-cuttings. As noted by Prof. Blake the most abundant species is *Exogyra bruntrutana.* Mr. Rhodes and myself obtained the following fossils :—

Goniomya v.-scripta.	Ostrea læviuscula.
Lithodomus.	Pholadomya tumida.
Mytilus autissiodorensis.	Serpula.

5. Swindon clay: a blue racy clay, weathering brown at the surface, and from 14 to about 20 feet thick. It was well shown in the railway-cutting (Fig. 99). Small lydite pebbles occur in the top part of the clay, together with small quartz pebbles and phosphatized fossils, that appear to have been worked into the surface of the clay after its deposition. Towards its base the clay becomes loamy, and passes into dark greenish sand (3 feet thick) that overlies the *Exogyra-*bed. From this clay I obtained *Ostrea læviuscula,* and Prof. Blake has recorded the following species :—

Arca Beaugrandi.	Mytilus autissiodorensis.
Corbula dammariensis.	Perna Bouchardi.
Cyprina elongata.	Trigonia.

Upper Portland Beds.

6. Resting on the Swindon Clay in the railway-cutting, there was a bed of marly limestone (like Roach), with abundant casts and moulds of shells, underlaid by fossiliferous sandy and glauconitic limestone with lydite pebbles. These beds were little more than 3 feet thick in the railway-cutting. In the great quarries they are exposed beneath the Swindon stone. There the lower bed, which is burnt for lime, and used for road-metal, is a bluish-grey limestone, with lydite pebbles, and is about 3 feet 6 inches thick.* The upper bed is a hard marly limestone, oolitic and sandy in places, and about 4 feet thick. It merges upwards into calcareous sandstone, belonging to the division above. It is very fossiliferous, and is known as the "Cockly Bed," but specimens are most abundant in the lower portion of it. These include the following fossils :—

Ammonites biplex.	Lucina portlandica.
—— giganteus.	Modiola unguiculata.
Buccinum ? angulatum.	Ostrea læviuscula.
Natica elegans.	Pecten lamellosus.
Pleurotomaria rugata.	—— suprajurensis.
Arca.	Pleuromya tellina.
Cardium dissimile.	Trigonia Carrei.
Cyprina elongata.	—— gibbosa.
Exogyra.	—— incurva.

7. Buff and white false-bedded sands, shelly and calcareous in places, and with bands, irregular lenticular layers, and some smooth masses of calcareous sandstone (Swindon Stone). Lignite occurs, and the sands are variable, some portions being composed of comminuted shells of *Ostrea* and *Pecten.* These beds attain a thickness of 25 feet, and are surmounted in places directly by Purbeck Beds. The top layer is a hard sandy limestone 2 to 3 feet thick with casts of shells. From these beds Prof. Blake has recorded the following fossils :—†

Corbula dammariensis.	Ostrea solitaria.
Cyprina pulchella.	Perna.
Lima rustica.	Pleuromya tellina.
Modiola unguiculata.	Trigonia gibbosa.

Ramsay called attention to the evidence of local erosion between these sands and the overlying Portland beds ;‡ and Prof. Blake has stated that

---

* This lydite-bed was noticed by Fitton, who included it with the "Portland Sand," Trans. Geol. Soc., ser. 2, vol. iv. p. 267.

† Quart. Journ. Geol. Soc., vol. xxxvi. pp. 204, 208. See also Fitton, Trans. Geol. Soc., ser. 2, vol. iv. p. 266.

‡ Geology of parts of Wiltshire and Gloucestershire, p. 25.

these "basal sands" may be considered to be the equivalents of the Tisbury stone (lower building-stone), and of the cherty series of Portland.

8. Irregular grey clay. This is a thin seam which may owe its irregularity to dissolution of calcareous matter from the underlying strata.

9. Oolitic limestone, 4 or 5 feet thick, with casts of *Cerithium* and *Trigonia*, a bed like the Roach of Portland. It yields also *Cytherea rugosa* (the "Cyrena" of the Vale of Wardour). The following are the fossils recorded from this bed, and most of them have been noted by Prof. Blake :—

Buccinum ? angulatum.	Cytherea rugosa.
Cerithium portlandicum.	Pecten lamellosus.
Neritoma sinuosa.	Pleuromya tellina.
Cardium dissimile.	Trigonia gibbosa.
Corbicella moraeana.	

10. Pale marly limestone, described by Ramsay as hard cream-coloured limestone from 1 to 6 feet thick, with *Trigonia*, &c. I obtained no fossils from this bed, which, but for the occurrence of *Trigonia*, might have been regarded as Purbeck.

The higher Portland Beds are of a somewhat variable nature, and the sections were not sufficiently continuous to display their relations very clearly ; so that I saw no evidence of discordance between them and the underlying Swindon Stone-beds. As remarked by Ramsay, nearly 40 years ago, " The section varies so rapidly in different parts of the quarry that, as the works proceed, any details now given may probably not apply to the section as it may exist in a few years."[*]

Nor did I find any freshwater fossils in the Portland Beds. The records of these discoveries require confirmation. W. Keeping described these upper beds under the name of " Swindon Series," remarking that some of the beds were of Purbeck type, and noting the occurrence of *Valvata ?* in a cherty band just below bed No. 8.[†]

It may be useful now to give a few notes on the particular sections observed in and around Swindon.

The section in the road-cutting (below Victoria Street) was noted as follows in 1887 :—

	Ft. In.
Loamy sand passing down into dark greenish sand	3 0
Very shelly layer, with *Ostrea læviuscula* &c.	1 0
Bed with *Exogyra*, in clayey matrix, with concretionary marly stone	3 3
Lower Portland Beds. Green, brown, and red sands, false-bedded and shelly in places (particularly at the base) with large concretionary masses of marly sandstone ; *Ammonites, Ostrea, Pecten, Trigonia*, &c.	3 6
Grey or buff sand with hard concretions in places (doggers), some at top	30 0 to 40 0

There was a marked plane of division on top of this lower bed of sand. The beds are much disturbed and slipped at the outcrop, and small local faults are thus shown as well as undulations, both due entirely to the slipping.

* Geol. parts of Wiltshire and Gloucestershire, p. 25.
† Fossils of Upware, &c., 1883, p. 41 ; see also J. F. Blake, Quart. Journ. Geol. Soc., vol. xxxvi. p. 207.

An old cutting made for an extension of the Great Western Railway east of Victoria Street showed the lower beds of sand with large lenticular doggers near the base. These lower beds of Portland Sand were noted by Fitton.[*] A brickyard further east shows the junction of these sands with the Kimeridge Clay, as follows :—

		FT.	IN.
Lower Portland Beds.	{ Greenish and yellow sands (even line) - 8	0 or 9	0
	{ Dark bluish-grey sand becoming loamy lower		
	{ down - - - - - -	6	0
Kimeridge Clay.	{ Blue clay with cement-stones :		
	{ *Thracia depressa, Perna mytiloides, Corbula,*		
	{ *Ammonites biplex* (abundant).		

About 20 feet of clay was shown. Bricks, tiles, drain-pipes, and flower-pots are manufactured.

Below King's Hill, west of Swindon, the *Exogyra*-bed, and Portland Sand with doggers below, have been exposed in the road and bank-cuttings. The Kimeridge Clay is worked further on at a brickyard, where the upper loamy beds are seen.

A section showing the irregular junction of the Purbeck Beds where they rest on the Sands with the Swindon Stone, is given in Fig. 138, p. 276.

The highest portion of the Portland Beds was shown at the eastern side of the great stone-pits at Swindon. There the following section was to be seen :—

		FT.	IN.
Purbeck Beds	{ 9. Rubble, &c.		
	{ 8. Pale marl.		
	{ 7. Grey clay with pebbles.		
	{ 6. Marl and greenish clay with boulders.		
Upper Portland Beds.	{ 5. Chalky beds.		
	{ 4. Oolitic limestone (Roach) with casts of *Cerithium, Neritoma sinuosa, Cytherea rugosa, Pleuromya tellina, Trigonia gibbosa,* &c. - - - 5	0 to 6	0
	{ 3. Irregular grey clay.		
	{ 2. Hard sandy limestone with casts of shells near top - - 2	0 to 3	0
	{ 1. Sands with indurated bands, false bedded (Swindon Stone).		

Mr. Hudleston suggests to me that Bed No. 4 may be on the horizon of the Lower Cyrena-beds of the Vale of Wardour. Its position beneath chalky beds strengthens the indications of the fauna.

A specimen of *Neritoma*, described by Morris, showed old colour-markings distinctly on its surface.[†]

At Coate, to the south-east of Swindon, a quarry to the north of the road showed the following beds :—

		FT.	IN.
Upper Portland Beds.	{ Fissile sandy beds - - - 1	0
	{ Calcareous sands - - - 3	0
	{ Sandstone - - - - 1	0
	{ Roach-beds and sandy limestone - 5	0

---

[*] Trans. Geol. Soc., ser. 2, vol. iv. p. 267.

[†] Quart. Journ. Geol. Soc., vol. v. p. 333.

The section here has also been described by Ramsay* and by
Prof. Blake;† and the following are the more abundant fossils
from the Roach beds :—

Pleurotomaria rugata.	Pecten lamellosus.
Cardium dissimile.	Pleuromya tellina.
Lima rustica.	Trigonia gibbosa.
Modiola.	—— incurva.

Prof. Blake records *Ammonites Boisdini, A. boloniensis, A.
pectinatus,* and other fossils ; as he remarks, the beds correspond
with those seen in the lower part of the Swindon quarry, the
sandy beds of Coate representing the Swindon stone series that
overlies the fossiliferous beds (No. 6, p. 213).

At Bourton, to the south of Shrivenham railway-station, there
is an outlier of Portland Beds, and stone has been quarried in places.
The section was described by Ramsay, and is as follows :—‡

		Ft.	In.
Upper Portland Beds.	Soft thin-bedded chalky oolite, with grains of sand - - -	8	0
	Hard bluish limestone, with pebbles of lydian stone and white quartz: *Ammonites* "*giganteus,*" *Cardium dissimile, Ostrea, Pleuromya, Trigonia gibbosa,* &c.		
	Yellow sands.		

Here the basement-limestone (No. 6, p. 213, of Swindon), rests
on sands, but I was unable to observe their thickness. They
probably belong to the Lower Portland Beds, and in this case
indicate conditions somewhat different from those of Swindon,
where clay directly underlies the fossiliferous limestone.

Culham, Garsington, Shotover, and Great Milton.

In the area to the south and south-east of Oxford we find
evidence of several outlying masses of Portland Beds, and it
appears probable that the rocks exposed at Great Milton are
continuous underground with those of Thame and Aylesbury.
That these beds do not, however, extend below the Cretaceous
rocks near Wallingford seems evident from the record of the
well-boring at Shillingford, to which attention has already been
drawn (p. 127).

In the general section taken from Faringdon across country to
the south-east of England (Fig. 145, p. 299) the structure indicates
a syncline in the Jurassic rocks that may extend in a north-
easterly direction beneath the range of the Chiltern Hills. Hence
the Portland Beds from Great Milton to Aylesbury may form
an irregular basin of strata, and the Kimeridge Clay and underlying Jurassic rocks may in succession occur directly beneath the
Cretaceous covering as we proceed towards the south-east from
the neighbourhoods of Thame and Aylesbury.

---

* Geol. parts of Wilts, &c. p. 25.
† Quart. Journ Geol. Soc., vol. xxxvi. p. 209 ; see also Etheridge, Stratigraphical
Geology and Palæontology, 1885, p. 480.
‡ Geol. parts of Wilts, &c. p. 27 ; see also Godwin-Austen, Quart. Journ. Geol.
Soc., vol. vi. p. 468.

At Culham, to the south-east of Abingdon, there is evidence of the presence of the passage-beds from Kimeridge Clay to Portland Beds, for John Phillips in 1860 noted the occurrence there of fossils like those of Hartwell, near Aylesbury. (See p. 167.) Moreover at Toot Baldon, five miles north-east of Culham, he noted the occurrence of beds of sand and sandstone, with some black pebbles, and obtained from them an *Ammonite* (belonging to the group of *Ammonites polyplocus*), also *Pecten, Cardium,* and other fossils.* The occurrence of this pebbly bed is interesting, especially when we note the presence of lydite-beds in the Portland rocks of Swindon, Brill, and Aylesbury. Further investigation of these beds near Toot Baldon is therefore desirable.

The Portland rocks form a large outlier extending from Shotover Hill to Garsington and Cuddesden; and the lower beds have been exposed in excavations at brickyards that are opened in the Kimeridge Clay.

As remarked by Prof. Hull, "Over the northern part of the hill it is seldom that the upper bed of limestone is visible, as the iron-sands of the Lower Greensand rest generally upon the yellow sand of the Portland; but at the southern-part of the range the upper limestones occur in force." He notes a section, one mile north of Garsington, showing white oolitic limestone (Portland Beds) with *Ostrea, Trigonia,* &c. overlaid irregularly by Lower Greensand.† (See Fig. 100.) The presence of Purbeck Beds at Garsington was observed by Fitton (see p. 278), and he also noted a section near Langcombe, where the lower sandy Portland Beds were seen.‡

FIG. 100.

Section near Garsington, south-east of Oxford.

(Prof. E. Hull.)

A. ‖ Soil.
B. Lower Greensand: Beds of variegated sands and clays, with lenticular bands of iron-ore.
C. Portland Beds: White oolitic limestone, much eroded at the upper surface, containing *Trigonia, Pinna, Ostrea expansa,* &c.

Shotover Hill, which lies to the east of Oxford, includes the commons of Cowley and Horsepath, and affords some famous sections of the Portland Beds and Lower Greensand. (See p. 278.)

---

* Quart. Journ. Geol. Soc., vol. xvi. pp. 310, 311. See also Hull, Explan. of Hor. Sec., Sheets 71 and 72, p. 2; and Green, Rep. Brit. Assoc. for 1894, p. 644.
† Geol. parts of Oxfordshire and Berkshire, p. 11.
‡ Trans. Geol. Soc., ser. 2, vol. iv. p. 278.

The general section above Headington may be noted as follows :—

		Ft. In.
Lower Greensand	{ White and buff sands, cherty sandstones, ferruginous sands with ironstone and ochre, clay and fuller's earth ; total (thickness according to Phillips) - - - - - -	80 0
Portland Beds.	7. Hard brown sandstone with *Trigonia* - -	1
	6. Rubbly limestones with casts of *Trigonia*, *Cerithium portlandicum*, &c.	
	5. Greenish (glauconitic) sand with shelly bands, *Cardium dissimile*.	
	4. White sand with large spheroidal masses of calciferous sandstone known as "sand ballers" or "giant's marbles," some being 3 to 6 feet in diameter. Shelly in places, with bored shells of *Ostrea* and *Perna*.	
	3. Sandy loam and pipe-clay - - -	2 0
	2. Layer of ironstone.	
	1. Brown and greenish sands with some hard masses and doggers of sandstone ; shelly at top, with *Perna*.	

Kimeridge Clay.

I was unable to measure the thicknesses of the beds, but Phillips has estimated that the Portland Beds at this locality are from 70 to 80 feet thick.

Bed 7 is noted on the authority of Strickland (as quoted by Fitton); it may be the bed from which Mr. Teall obtained a specimen of *Trigonia gibbosa* var. *damoniana*.[*]

Beds 1–3 may doubtless be grouped with the Lower Portland Beds. From these "Shotover Sands" Prof. Blake records *Pholadomya tumida*.[†] From some of the large doggers Mr. James Parker has obtained *Perna*, *Lima*, and *Trigonia*; and he has found *Astarte hartwellensis* in the top Kimeridge Clay at Wheatley.

The upper shelly beds of Shotover yield the following fossils :—

Cerithium portlandicum.	Ostrea expansa.
Natica elegans.	Pecten lamellosus.
Cardium dissimile.	Trigonia gibbosa.
Lucina portlandica.	

Among other fossils from the Portland Beds, recorded by Phillips, are the following :—[‡]

Ammonites pectinatus.	Modiola unguiculatus.
—— trifidus (triplex).	Perna mytiloides.
Littorina paucisulcata.	Pholadomya inæqualis.
Pleurotomaria rugata.	Pinna.

L. Sæmann in a section of Shotover Hill noted Purbeck Beds ? above the strata with *Trigonia gibbosa*;[§] this may have been

---

[*] Fitton, Trans. Geol. Soc., ser. 2, vol. iv. pp. 275, 278 ; Teall, Potton and Wicken Phosphatic Deposits, pp. 32, &c.

[†] Quart. Journ. Geol. Soc., vol. xxxvi. p. 213, and Plate VIII.

[‡] Ibid., vol. xiv. p. 238, and Geology of Oxford, &c. pp, 326, 413, 421; see also Hull, Geol. parts of Oxfordshire and Berkshire, p. 12.

[§] De Loriol and Pellat, Mem. Soc. Phys. Hist. Nat. Genève, vol. xlx. 1866, Plate 1, Fig. 6

but a general reference, as no Purbeck Beds have been observed nearer than Combe Wood, south of Wheatley. (See p. 278.)

A mass of Portland strata appears from beneath the covering of Cretaceous rocks at Great and Little Milton, and the stone-beds have been worked beneath the Lower Greensand at Great Hazeley and near Great Milton.

The section which I saw at Great Milton was on the eastern side of the village, and there the following beds were to be seen :—

		Ft.	In.
	Brown loamy soil.		
Lower Greensand.	Sand with bands of white and ochreous clay, with lignite - - -	3 0 to 6	0
	Buff and white false-bedded sand with ferruginous layers and concretions; with at base lydite pebbles and ironstone - - -	3 0 to 6	0
Portland Beds.	Rubbly bed of sand with hard calcareous sandstone and chert ; *Ostrea expansa.* This had the appearance of a remanié bed, like that described by Prof. Blake at Great Hazeley* -	3 0 to 4	0
	Clayey and shelly layer with hard nodules - - - -	0	4
	Hard calcareous sandstone, cherty in places, passing down into grey and greenish sandy oolitic limestone (burnt for lime) ; with *Ostrea, Trigonia*: seen to depth of - -	6	0

A section at Great Hazeley showed the following beds :—

		Ft.	In.
Lower Greensand.	Ironstone - - - -	4	0
	Ochreous and white clays with bands of ironstone and ferruginous sand : more sandy towards the base and false-bedded - - -	15	0
	Laminated clay and ironstone.		
Portland Beds.	Fine mealy calcareous sand.		
	Shelly, earthy, and sandy, oolitic rock, with *Trigonia*, &c. - - -	3	0

The beds were much disturbed at the surface, and the Lower Greensand rests so irregularly on the Portland Beds that the details shown in the sections vary considerably from time to time. Prof. Hull noted the presence of small pebbles of lydian stone in sandy oolite, which rested on 2 feet of white sand. Above the oolite there was then exposed 5 feet of chalky limestone, with *Ostrea expansa*, &c. This bed he observed was sometimes eroded.† It is the layer called " Curl " by Fitton.

Prof. Phillips mentions that " At Great Hazeley the Portland stone has been quarried from ancient time, and it there furnishes a limited supply of better quality than usual for building, being of good colour and firm and equal texture, except for the shells,

* Quart. Journ. Geol. Soc., vol. xxxvi. p. 214.
† Geol. parts of Oxfordshire and Berks, p. 11 ; See also Fitton, Trans. Geol. Soc., ser. 2, vol. iv. p. 276 ; and Blake, Quart. Journ. Geol. Soc., vol. xxxvi. p. 213.

which, however, mostly lie in bands. A thick grey or greenish sand is at the bottom ; over this the stony series, the lower part workable freestone, the top hard splintery limestone, 2 feet thick, much jointed, fit for roads and rough walling, called ' Curl,' which suggests the idea of Purbeck beds. This is immediately followed by the iron-sand and clay series."*

Thame, Brill, and Aylesbury.

Reference has already been made to the possible underground course of the Portland Beds beneath the Cretaceous rocks in this neighbourhood (p. 216). At Thame there are now but few sections of the Portland Beds to be seen, but fortunately here as elsewhere we have the careful records of Fitton.

The following section of a pit at Barley Hill, north-east of Thame, was recorded by Fitton :—†

		Ft.	In.
	Soil and loam - - - - -	2	6
Upper Portland Beds.	Irregular concretional lumps of calcareous stone, with green particles, interspersed in sand: *Perna*, *Lithodomus*, *Ammonites*, &c. -	6	0
	Greenish sand, with concretions of stone near the top - - - - -	1	0
	Coarse stone, interspersed in sand: with *Ammonites*, *Trigonia gibbosa*, &c. - -	1	3
	More compact and uniform greenish or bluish rock, with green particles ; like soft Kentish rag. In some of the quarries it is fit for building - - - - -	1	0
Lower Portland Beds.	Sand ; with nodules of great size - about	30	0

The Lower Portland Beds were exposed to the west of Thame, south-west of Priestend, where a pit was opened in false-bedded greenish-grey and buff and white sand. Streaks of clay occur here and there, and towards the top, where the sand becomes ochreous, there were fissile and impersistent beds of glauconitic sandstone.

At Long Crendon the following sequence of beds may be observed in brickyards and stone-pits :—

		Ft.	In.
Gault	Clay - - - - - -	5	0
Lower Greensand.	Clay and sand, with ironstone-nodules ·} Laminated Clay - - - -	2	0
Purbeck Beds.	Variable beds of marly clay, soft marly limestone, &c , with Cyprides - - -	4	0
Upper Portland Beds.	Pale-grey and greenish marl, with *Trigonia gibbosa*, *Ostrea expansa*, passing down into more fossiliferous bed with *Ammonites*, casts of *Trigonia gibbosa*, *Cardium dissimile*, *Pleuromya tellina*, and *Perna*, with seam of shelly marl at base - - - - -	4	6
	Greenish and brown calcareous sandstone about	3	6

* Geol. Oxford, &c., p. 417.
† Trans. Geol. Soc., ser. 2, vol. iv. p. 282.

The beds below, described by Fitton as seen "on the descent towards Thame," were as follows:—[*]

		Ft.	In.
Upper Portland Beds.	⎰ Sand with green particles (glauconite) and concreted nodules - - - -	2	9
	Rubbly, sandy, and calcareous stone, with green particles: *Trigonia.* At its junction with the bed below, and for about 6 or 8 inches downwards, "are worn fragments of black flint" (lydite) - - - 12 0 to 18	0	
Lower Portland Beds and Hartwell Clay.	Greenish grey sand, with *Trigonia, Perna,* &c. - - - - - about 30	0	
	Clay - - - - - -		

At Brill brickyard, and in an adjoining quarry on the north-western side of the hill, the following beds were exposed:—

		Ft.	In.
Upper Portland Beds.	⎰ Oolitic and shelly rock - - - -⎱ Oolitic marl - - - - -⎰	2	0
	Marl, with *Ostrea expansa* - - -	1	0
	Chalky rock, with *Ammonites, Lucina portlandica, Trigonia* - - -	4	0
	Brown clay, with *Ostrea* - - -	0	6
	Oolitic earthy rock - - - -	1	0
	Greenish sandy and marly beds: *Pecten lamellosus* - - Glauconitic marly limestone, becoming sandy in lower part: *Cardium dissimile, Ammonites giganteus, Trigonia, Ostrea.* } about 7 0 or 8	0	
	Nests of green sand in places.		
	Lydite bed: brown clay with pebbles - -	1	0
Lower Portland Beds and Hartwell Clay, passing down into Kimeridge Clay.	Brown and greenish mealy sand - -	3	0
	Sandy clay, passing down into stiff blue clay, with iridescent Ammonites: *A. biplex, Thracia* - - - -	20	0
	Astarte hartwellensis.		

This section agrees in the main points with that recorded by Fitton,[†] who enumerates a number of fossils.

The pits were visited under the guidance of Prof. J. F. Blake in 1893,[‡] who remarked that the sandy clay at the base yielded *Perna,* and should be correlated with the Hartwell Clay. The fossiliferous glauconitic bed he regarded as Lower Portlandian (Bolonian), and identified the following fossils:—

± Ammonites biplex.
× —— boloniensis.
× Cardium Pellati.
× Lima rustica.
× Mytilus boloniensis.

× Pecten lamellosus.
× Perna Bouchardi.
× Pleuromya tellina.
Trigonia muricata.
± —— Pellati.

* Trans. Geol. Soc., ser. 2, vol. iv. p. 282; see also J.F. Blake, Proc. Geol. Assoc., vol. xiii. p. 74.

† Trans. Geol. Soc., ser. 2, vol. iv. pp. 280, 299; see also Whitaker, in Geol. Banbury, pp. 47-49; J. Mitchell, Proc. Geol. Soc., vol. ii. p. 6; Brodie, Quart. Journ. Geol. Soc., vol. xxiii. p. 197.

‡ Proc. Geol. Assoc., vol. xiii. p. 73; Quart. Journ. Geol. Soc., vol. xxxvi. pp. 209, 226, &c.

He remarked that the lydite-bed was not to be confounded with one which occurs at the base of the *Trigonia*-beds at a higher level; but in this I cannot concur, for all the species above noted are recorded by Prof. Blake himself from the higher stages of the Portland Beds. those marked × occurring in the "*Trigonia*-beds" above the lydite-bed at Swindon, and the two species marked ± occurring elsewhere in equivalent or higher stages, as well as in lower stages.

Other outliers of Portland Beds occur at Muswell Hill, north-west of Brill, at Ashendon, Nether and Over Winchendon and Coney Hill, Waddesdon, Quainton Hill, &c. Mr. Whitaker noted the presence of limestone on Muswell Hill; it occurs above the glauconitic sands.[*]

Sections at Quainton Hill have been described by Fitton and others.[†] Fitton notes the occurrence of a large Ammonite with a Belemnite embedded in it.

Prof. Blake describes the upper layer of the Portland Beds of Buckinghamshire as a "compacted shell-brash," while below come beds which he groups as "Creamy Limestones." These include the stone chiefly quarried for building-purposes and lime; and they are characterized by *Natica ceres.* Among other fossils he notes the following :—[‡]

× Ammonites boloniensis.	× Lima rustica.
× —— pseudogigas.	Lucina portlandica.
× —— trifidus (triplex).	Ostrea expansa.
Cerithium portlandicum.	× Pecten lamellosus.
Natica elegans.	× Perna Bouchardi.
× Pleurotomaria rugata.	× Pleuromya tellina.
× Cardium dissimile.	× Trigonia gibbosa.
Cytherea rugosa.	× —— incurva.
Exogyra bruntrutana.	× —— Pellati.

On a lower horizon are limestones which he terms the "Rubbly Beds." Of the species before mentioned from the "Creamy Limestones," those marked × occur also in these "Rubbly Beds." They include also the following species :—

Ammonites pectinatus.	Ostrea solitaria.
Cardium Pellati.	Trigonia muricata.
Myoconcha portlandica.	—— Voltzi.
Mytilus boloniensis.	

Below these strata come the Glauconitic Beds and Lydite-bed to which attention has already been drawn.

One of the most interesting features in connexion with these beds is the occurrence of the conglomeratic band at the base of the Portland Stone; because we find a similar bed, occupying apparently the same position, at Tisbury, near Devizes, and

---

[*] Geology of Banbury, &c., p. 47. See also Fitton, Trans. Geol. Soc., ser. 2, vol. iv. p. 283.

[†] *op. cit.* p. 290 ; J. Mitchell, Proc. Geol. Soc., vol. ii. p. 6.

[‡] Quart. Journ. Geol. Soc., vol. xxxvi. p. 216.

Swindon, and also at Aylesbury and Hartwell, where the strata have been so fully investigated by Mr. Hudleston,[*] and others. In the Aylesbury district, as pointed out by Mr. Hudleston, the lydite-bed rests on the Hartwell Clay ; at Swindon it rests on a bed termed the Swindon Clay.

Much has been written about the Hartwell Clay, because, as first pointed out by Sæmann, the fossils it has yielded differ from those of the true Kimeridge Clay ; and he therefore classed it as equivalent to the " Middle Portlandian " of Boulogne.[†]

There can be no question that the clay below the Portland Stone at Swindon is homotaxial with the Hartwell Clay, for although the Swindon Clay is not so fossiliferous as that at Hartwell, yet it has yielded some species, and the beds immediately below the Swindon Clay have yielded many fossils identical with those of the Hartwell Clay, as shown by the lists of Prof. Blake and Mr. Hudleston. Among the species thus found are the following :—

Ammonites biplex.	Mytilus boloniensis.
Astarte Sæmanni.	Ostrea bononiæ.
Cardium morinicum.	Pecten lens var. Morini.
Cyprina elongata.	Perna Bouchardi.
—— pulchella.	Pinna suprajurensis.
Exogyra bruntrutana.	Placunopsis Lycetti.
Myoconcha Sæmanni.	Trigonia Pellati.
Mytilus autissiodorensis.	

Hence we may admit that on stratigraphical and palæontological grounds the Hartwell Clay belongs to the Lower Portland Beds.

Where the lydite-bed occurs the Lower Portland Beds appear to be more distinctly separated than in other areas from the Portland Stone, but this is a local feature, and is not the case on the Dorset coast, where no conglomeratic bed has been observed at the base of the Upper Portland Beds. The lydite-bed, with its associated phosphatized fossils, thus indicates some local change accompanied probably by a pause in the deposition of sediment, but there is no palæontological break.

The Portland and Purbeck Beds of Hartwell and Aylesbury have been repeatedly described by Prof. Morris,[‡] Prof. J. F. Blake,[§] and Mr. W. H. Hudleston.[||] The Beds have been opened up at the Bugle Pit, near the Bugle Inn at Hartwell, in the brickyard on the road to Aylesbury, and in various places at Aylesbury and Bierton. The general section may be stated as follows :—[¶]

---

[*] Proc. Geol. Assoc., vol. vi. p. 344, vol. x. p. 167.

[†] Sæmann in De Loriol and Pellat, Etage Portlandien des environs de Boulogne. Mem. Soc. Phys. Hist. Nat. Genève, vol. xix. 1866, pp. 189, 196, and Plate 1, Fig. 7.

[‡] Proc. Geol. Assoc., vol. iii. p. 210.

[§] Quart. Journ. Geol. Soc., vol. xxxvi. p. 215 ; xxxi. pp. 211, 212.

[||] Proc. Geol. Assoc., vol. vi. p. 346, vol. x. p. 167. (A fuller account of one Excursion was printed in the " Bucks Herald," for June 18, 1887.)

[¶] See H B.W., Explanation of Horizontal Section, Sheet 140 (Geol. Survey).

FIG. 101.

Section at Aylesbury.

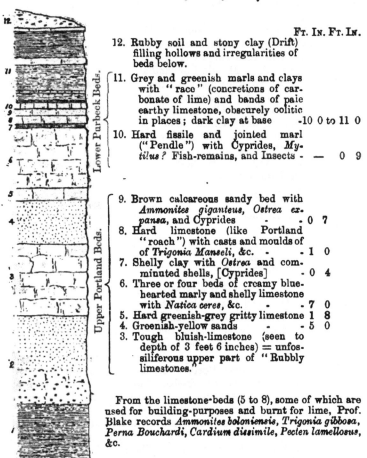

FT. IN. FT. IN.

12. Rubby soil and stony clay (Drift) filling hollows and irregularities of beds below.

Lower Purbeck Beds.

11. Grey and greenish marls and clays with "race" (concretions of carbonate of lime) and bands of pale earthy limestone, obscurely oolitic in places; dark clay at base   -10 0 to 11 0

10. Hard fissile and jointed marl ("Pendle") with Cyprides, *Mytilus?* Fish-remains, and Insects -   —   0 9

Upper Portland Beds.

9. Brown calcareous sandy bed with *Ammonites giganteus, Ostrea expansa,* and Cyprides   -   0 7

8. Hard limestone (like Portland "roach") with casts and moulds of of *Trigonia Manseli,* &c. -   1 0

7. Shelly clay with *Ostrea* and comminuted shells, [Cyprides] -   0 4

6. Three or four beds of creamy bluehearted marly and shelly limestone with *Natica ceres,* &c. -   7 0

5. Hard greenish-grey gritty limestone 1 8

4. Greenish-yellow sands   -   - 5 0

3. Tough bluish-limestone (seen to depth of 3 feet 6 inches) = unfossiliferous upper part of "Rubbly limestones."

From the limestone-beds (5 to 8), some of which are used for building-purposes and burnt for lime, Prof. Blake records *Ammonites boloniensis, Trigonia gibbosa, Perna Bouchardi, Cardium dissimile, Pecten lamellosus,* &c.

Examples of the large Ammonites were built in the walls bounding Hartwell Park, in the time of the former proprietor, Dr. John Lee.

In other spots at Aylesbury lower beds have been traced, the town itself, as remarked by Mr. Hudleston, being based to a large extent on the "Rubbly limestones" (of Prof. Blake), very fossiliferous beds, to which the name "Aylesbury Limestone" has been applied.[*] These beds were well shown during drainage-operations by the George Hotel in Aylesbury, and the lydite-bed was disclosed "about 12 feet below the surface in the lower part

---

* Name used by Farey in 1811.

of the Market Place, opposite the ' Green Man.' "   The succession
is continued thus :—

		Ft.	In.

Upper Portland Beds.
{
3. Rubbly white limestones (like Portland
"roach") with many fossils, including
some of the species mentioned above, together with *Pleuromya tellina, Myoconcha
portlandica, Unicardium circulare, Cardium Pellati, Trigonia Pellati,* &c.   -   8   0

2. Yellow and greenish sands (6 or 8 feet),
with conglomerate or lydite-bed at base,
yielding *Ammonites, Cardium dissimile,
Pleuromya tellina, Serpula,* &c. -   -   10   0

Hartwell Clay.
{
1. Clays with *Ammonites biplex, Belemnites
Oweni, B. Souichi, Arca longipunctata,
Astarte hartwellensis, A. Sœmanni, Cardium morinicum, Cyprina pulchella, C.
elongata, Ostrea læviuscula, Mytilus
boloniensis, Pecten lens* var. *Morini,
Perna Bouchardi, Pleuromya tellina,
Thracia tenera, Trigonia Pellati.*

The Portlandian conglomerate occurs in a disintegrated form
at a brick-pit on the Bierton Road, Aylesbury, and as a hard blue-
rock at the brickyard (Locke's pit) between Aylesbury and Hartwell.   It is composed of quartz and black siliceous pebbles known
as "lydites."   These, as Mr. Hudleston remarks, may be derived,
some from Palæozoic cherts, others from indurated siliceous beds
of mechanical origin, such as are found in these old rocks.   They
are extremely hard, but usually contain more alumina than true
flint.   This conglomerate band is 4 or 5 feet thick, and contains
glauconite; Mr. Jukes-Browne and I have also obtained phosphatic
nodules and some phosphatized fossils.

About 10 feet of grey micaceous clay was exposed at Locke's
brickyard, and as much as 23 feet, according to Mr. Jukes-
Browne, at the Bierton brickyard.   There the clay beneath the
pebbly layer is very sandy towards the top, where it contains
glauconite, and there the shells have been destroyed by atmospheric agencies.   At Aylesbury, as noted by Mr. Hudleston, the
clay opened up during drainage-excavations was perfectly fresh
up to its junction with the overlying strata, and well-preserved
fossils were found in it.

No outcrop of sandy beds such as we find near Swindon has
been traced beneath the clay at Hartwell, for the "Portland
Sands," marked on the Geological Survey Maps, and on the
Sections between Stone and Coney Hill, belong to the beds above
the conglomerate.   The record of a well-boring at Stone is
given on p. 337.   This record, however, is not sufficiently precise
to enable us to fix the horizons of the beds with certainty.   It is,
moreover, evident that the sands of the Lower Portland Beds are
gradually replaced by clay as we proceed northwards.   At Brill,
beneath the lydite-bed, there is 3 feet of brown and greenish
sand, which passes down gradually into stiff blue clay, yielding
Ammonites biplex, Thracia, &c., and is not separable from the
Kimeridge Clay.

A section noted by Mr. A. Strahan on the new Metropolitan Extension Railway near Bishopstone, showed beneath the Gault, with its coprolite-bed, the following Upper Portland Beds :—

	Ft.	In.
Limestone with hard layer at base (6 to 8 ins) - -	3	6
Sand [Equivalent to bed 4 in the Section p. 224.]	5	0
Limestone - - - - - - -	2	6
Hard blue shaly stone seen to depth of · - -	3	0

The Portland Beds had a slight dip towards the south-east, so that the Gault overlapped the top limestone to the north and rested directly on the sand. Judging by this section the Portland Beds may extend some distance beneath the Cretaceous covering towards Wendover. (See p. 216.)

The following species from the Upper Portland Beds of Aylesbury and Hartwell, include those in the collection of the late Dr. John Lee (L.), determined by Messrs. Sharman and Newton (1888) ; and those recorded by Mr. Hudleston (H.). The numbers 1 refer to the Lydite-bed ; 2 to the Rubbly Beds, and 3 to the higher strata :—

Ammonites biplex. 1 (H.), 2 (H.).
—— boloniensis. 2 (H.).
—— giganteus. (L.)
—— pseudogigas.* (L.)
—— triplicatus. (L.)
Natica ceres. 3 (Blake.)
—— elegans. 2 (H. L.).
—— incisa. (L.)
Neritoma, n. sp. (L.)
Pleurotomaria rugata. 1 (H. ?), 2 (L.).
Pseudomelania collisa. (L.)
Trochus. 2 (H.).
Arca. 2 (H. L.).
Cardium dissimile. 1 (H.). 2 (H. L.).
—— Pellati. 1, 2 (H.).
Cyprina elongata. (L.)
Exogyra bruntrutana. 2 (H.).
Lima boloniensis. 2 (H.).
—— rustica. 2 (H. L.).
Lucina. 2 (H.).
Modiola unguiculata. 1 (L.).
Myoconcha portlaudica. 2 (H.L.).
Mytilus boloniensis. 2 (H.).
—— jurensis. 1, 2 (H.).

Ostrea expansa. (O. falcata of Morris), 3 (L.).
Pecten lamellosus. 1 (L.), 2 (H.).
—— solidus. 1, 2 (H.).
Perna Bouchardi. 2 (H. L.).
Pholadomya tumida. (L.)
Pleuromya tellina. 1 (H. L.), 2 (H.).
—— Voltzi. 1 (H.).
Quenstedtia. 2 (H.).
Sowerbya longior. 2 (Blake).
Thracia depressa. (L.)
Trigonia Carrei. 1 (H.).
—— gibbosa. 1 (H.), 2 (H. L.).
—— incurva. 1 (H.).
—— Manseli. 3 (Geol. Surv.).
—— muricata. (L.)
—— Pellati. 1 (H.).
—— tenuitexta. (L.)
Unicardium circulare. 1 (H. L.).
Eryma. (L.)
Vermilia quinquangularis. 1 (H.).

The following is a list of fossils from the Hartwell Clay, including species in the collection of the late Dr. Lee (L.), determined by Messrs. Sharman and Newton, and those recorded by Prof. J. F. Blake (B.) and Mr. Hudleston (H.) :—

---

* Phillips recorded *A. gigas* from Aylesbury, Geol. Oxford, &c., p. 331.

Crocodilian tooth. (L.)
Ammonites biplex. (L.)
—— rotundus. (L.)
Belemnites nitidus (explanatus).
  (J. Phillips.)
—— Oweni. (L.)
—— Souichi. (L.)
Alaria. (L.)
Cerithium crebrum? (L.)
—— multiplicatum. (L.)
Chemnitzia Beaugrandi. (L.)
Delphinula, n. sp. (L.)
Littorina. (L.)
Neritopsis delphinula. (L.)
Pleurotomaria reticulata. (L.)
Anatina, n. sp. (L.)
Anomia Dollfusi. (B.)
Arca longipunctata. (L.)
—— æmula (mosensis.) (B.)
—— rhomboidalis. (B.)
Astarte hartwellensis. (L.)
—— (Lucina) lineata. (H.)
—— mysis. (L.)
—— Sæmanni. (L.H.)
Avicula. (L.)
Cardium morinicum. (H.)
—— striatulum. (L.)
Corbula Deshayesea (L.B.)
Cyprina elongata. (L.)
—— implicata. (H.)
—— pulchella. (L.)
Cytherea, n. sp. (L.)
Exogyra bruntrutana. (H.)
Lima duplicata. (L.)
—— rustica. (L.)
Myoconcha Sæmanni. (L.)

Myoconcha texta. (L.)
Mytilus autissiodorensis. (H.)
—— boloniensis. (L.)
—— pectinatus. (L.)
Nucula Menkei. (B.)
Ostrea bononiæ. (H.)
—— duriuscula. (L.)
—— læviuscula.
  (= O. deltoidea of Morris.) (L.)
Pecten lamellosus. (L.)
—— lens var. Morini. (L.)
—— nitescens. (L.)
—— Thurmanni. (B.)
Perna Bouchardi. (H.)
—— Flambarti. (L.)
—— mytiloides. (L.)
Pholadidea. (? boring in lignite.)
  (L.)
Pholadomya Protei. (L.)
Pinna granulata. (L.)
—— suprajurensis. (H.)
Placunopsis Lycetti. (L.)
Pleuromya tellina. (H.)
Thracia depressa. (L.)
—— tenera. (H.)
Trigonia Pellati. (H.)
Lingula ovalis. (L.)
Rhynchonella subvariabilis. (H.)
Waldheimia boloniensis. (L.)
Eryma. (L.)
Glyphea. (L.)
Pentacrinus. (L.)
Cristellaria nummulitica. (L.)
—— rotulata. (L.)
Nodosaria (cf.) raphanistrum.
  (L.)

In the irregular outlier of Oving and at Whitchurch to the north of Aylesbury, there are several quarries in the Purbeck and Portland Beds, and in one between the two villages the following section was to be seen :—*

		Ft.	In.
	Thin grey soil - - - - - 0	6 to 1	0
Purbeck Beds.	Grey and white marls with impersistent indurated bands passing down into - Hard compact limestone and calcareous gritty rock - - - - -	5	0
	Brown calcareous sands with casts of shells -	2	0
	Grey clay - - - - -	0	3
.	Clay and comminuted shells, [Cyprides] -	1	6
	Shell-bed with *Trigonia, Cardium dissimile* 0 6 to 0		8
	Shelly clay - - - - -	0	6
Upper Portland Beds.	Very tough brown and grey limestone compact and blue-hearted - - 1 0 to 1		6
	Shell-bed with large Ammonites, some 3 feet across, *Pecten lamellosus, Pleurotomaria*, &c.		
	about 2		6

Cyprides were observed by Prof. Green in the lowest bed of clay here grouped with the Purbeck Beds. Their presence,

---

* See also section by Fitton, Trans. Geol. Soc., ser. 2, vol. iv. p. 291.

however, as shown by the evidence at Hartwell affords no criterion for fixing a plane of division.

The hard grey limestone near the base is quarried for building-stone and road-metal, and it is burnt for lime. Many old walls in Whitchurch have been constructed of the shell-beds, but they weather badly. The building-stone is grouped by Prof. Blake with the "Creamy Limestones," and he remarks that "a large part of the village of Whitchurch has this for its natural paving."[*]

Below these beds the following strata have been noted by Prof. Green :—

		Ft.	In.
Upper Portland and Hartwell Clay.	Soft light brown sand - - - -	8	0
	Hard white limestone with *Ammonites* "*giganteus*," &c. - - - -	6	0
	Yellowish sandy clay - - -	0	6
	Sandy and ferruginous clay with lydites	0	4
	Light blue and yellow clay with lydites -	4	0
	Dark bluish-black clay.		

He has remarked that to the south of Creslow, east of Whit-church, "quite low down in the valley is a mass of Portland Stone, from which a powerful spring breaks out." Whether this mass was slipped or faulted was uncertain.

The Lower Greensand at Oving and Whitchurch lies irregularly on the Purbeck and Portland Beds.

There is an outlier of Portland Beds at Weedon, south of Hardwick, described by Prof. Green as comprising limestones with intervening masses of sand as at Whitchurch ; and inlying portions of the beds have been traced beneath the Gault, between Cublington and Wing, and they outcrop for the last time about a mile south of South End, Stewkley.

Fitton has given a section of a pit at the Warren, south of Stewkley, which showed 2 or 3 feet of Purbeck Beds resting on Portland Beds, as follows :—

		Ft.	In.
Purbeck Beds.	Clay and rubbly limestone		
Upper Portland Beds.	Marly limestone, with *Ostrea expansa*, Pecten lamellosus, Trigonia, Pleuroto-maria, large *Ammonites*, &c. - -		
	Yellowish grey sand about - -	6	0
	Dark grey stone, nodular in character	7	0
		2	0

Mr. Jukes-Browne observes that Portland Beds appear to have been reached in a well at North Cottesloe at a depth of 20 feet, and Prof. Green notes that Portland Beds were formerly quarried east of Cublington.

Fitton has stated that at Dunton the lowest bed above noted, is represented by a rubbly calcareous stone, alternating with sand, the whole full of green particles, with small pebbles of black flint (lydite). Below these occurred a green rock, with large *Ammonites*, *Perna*, &c., about 2 feet thick, and still lower 3 feet of green stone had been opened up.[†]

* Quart. Journ. Geol. Soc., vol. xxxvi. p. 216.
† Trans. Geol. Soc., ser. 2, vol. iv. p. 292.

Prof. Green was unable to find any evidence of Portland Beds *in situ* at Dunton and Hogston. He observes that " The Drift about Hogston is very sandy, and may have been mistaken for Portland Sand, and in many places pebbles and even large blocks of Portland Limestone lie in the Drift and might have given rise to the idea of there being Portland Beds in place."

We have now come to the north-easterly termination of the Portland Beds that have been proved in the area under consideration. It is possible they may extend some way eastwards beneath the Dunstable and Luton Downs, but we have no evidence that this is the case. Some account of the strata that overlie the Kimeridge Clay in Lincolnshire will be given in the next Chapter; at present no definite evidence of Portlandian Beds *in situ* has been obtained in that region, although derived fossils occur in the nodule-bed at the base of the Spilsby Sandstone. Particulars of the Yorkshire beds have been given by Mr. Fox-Strangways.*

Sussex and Kent.

Beds of Portlandian age were probably passed through in the Sub-Wealden boring, but nothing distinctively characteristic of the beds was obtained. The beds evidently vary within short distances, for one record given of these beds describes them as follows:—†

	THICKNESS.		DEPTH.	
	FT.	IN.	FT.	IN.
Sandy shale with chert	23	0	200	0
Sand and soft whitish sandstone	52	0	252	0
Darker sandstone	5	0	257	0

It is possible that some portion of the underlying sandy shales may represent in part the Lower Portland Beds or the passage-beds from Portland Beds to Kimeridge Clay. Indeed, in Mr. Topley's account of the first boring (p. 346) a thickness of 110 feet was doubtfully assigned to the Portland Beds, and it is possible that a greater thickness might be included; elsewhere 250 feet of Portland Beds has been recorded.

At a depth of 206 feet the sandstone was streaked with shale. *Serpula* was found in some of the specimens.

In the Dover boring, the record of which is given on p. 342, Prof. Boyd Dawkins has assigned a thickness of 32 feet to Portlandian Beds (depth, 505 to 537 feet), though the evidence of a *Rhynchonella* and an *Exogyra* (species not given) affords no clue to the age of the strata. The attenuation of clayey Kimeridgian beds is remarkable.

* Jurassic Rocks of Yorkshire, vol. i. p. 379.
† Topley in Dixon's Geol. Sussex, Ed. 2, p. 154; see also pp. xxiii. and 6.

CHAPTER XIII.

PURBECK BEDS.

(PURBECKIAN.)

GENERAL ACCOUNT OF THE STRATA.

These strata take their name from the so-called Isle of Purbeck, in Dorsetshire, where they are well exposed in the cliff-sections of Durlston and Worbarrow Bays, and where inland the beds of Purbeck Marble and Swanage Stone have been worked for long ages.

As a stratigraphical term, the name Purbeck Beds was employed in 1811–12 by Thomas Webster, and the term Purbeck Stone was used by William Smith in 1812.

The Purbeck Beds comprise a series of clays and shales with "Beef," marls, marly, tufaceous, and shelly limestones, and occasionally of granular oolitic beds, and sandy strata. In thickness the series varies from about 80 or 90 feet in Wiltshire to nearly 400 feet in Dorsetshire.

The term "Beef" applied by workmen to the seams of fibrous carbonate of lime, was so given from "the resemblance of its small and parallel fibres to the fibres of animal muscle."[*] It often presents cone-like structures, similar to that known as "cone-in-cone."[†]

Pseudomorphous crystals of rock-salt were noticed by H. W. Bristow in the Lower Purbeck Beds of Durlston Bay and Lulworth Cove ; and similar pseudomorphs have been observed in the Vale of Wardour.[‡] Gypsum also occurs, and here and there we find nodules of Chert.

It is remarkable that in general lithological characters we find much similarity between the Rhætic Beds and the Purbeck Beds, and a glance at the cliffs in Durlston Bay calls to mind the Rhætic Beds and Lower Lias of Watchet, Blue Anchor, and Penarth. The grey marls with gypsum, the black shales with "beef," a band with rude arborescent markings (recalling the Cotham Marble), and occasional beds like White Lias, are all to be found in the Purbeck strata of Dorsetshire ; and some of the beds in Dorsetshire and Wiltshire, termed "Lias" by the quarry-men, closely resemble the argillaceous limestones that form the basement-beds of the Lower Lias.

The band with the rude arborescent markings occurs in the "Soft Cockle Beds" of Bristow's section,[§] and about the horizon marked by him as a "Hard brecciated Limestone," above the

---

[*] Buckland and De la Beche, Trans. Geol. Soc., ser. 2, vol. iv. p. 11.
[†] See Memoir on the Lias of England and Wales, p. 308.
[‡] Andrews and Jukes-Browne, Quart. Journ. Geol. Soc., vol. l. p. 52.
[§] Vertical Sections, Geol. Survey, sheet 22, No. 1.

"Soft blue shaly Marls" that contain the main masses of gypsum. My attention was first attracted by small slabs on the beach, for these showed the mammillated surface shown in Fig. 103, that is so characteristic of the Cotham or Landscape Marble. On breaking the stone, it presented the appearance shown in Fig. 102. This thin bed of stone occurs in impersistent layers, in the first exposure of the Lower Purbeck marls north of Durlston Head, and the features it presents are of interest when studied in connexion with the Landscape Marble of Cotham. They may serve to throw light on the formation of more or less nodular masses of stone. In the case of the Landscape Marble, I have been led to suggest that the arborescent markings were produced by the disarrangement of dark and pale bands of calcareous mud, during the solidification of the stone, and that the mammillated surface was rucked up in the course of the process, when individual portions of the calcareous mud shrunk into isolated nodular masses.[*]

Somewhat analogous features were produced in the Purbeck Limestone known as the Cutlet, found near Battle, an example of which is shown in Fig. 104, p. 232.

The irregular nodular masses of argillaceous limestone found in the Lower Lias are well known, but the rock being homogeneous, no particular structure is shown in the mass. Where, however, the original sediments were banded with layers of

FIG. 102.

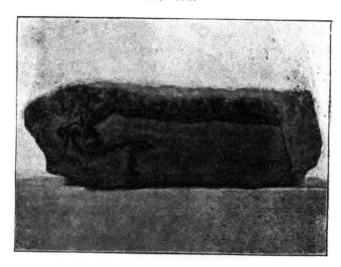

Limestone, Lower Purbeck Beds, Durlston Head.

---

[*] Geol. Mag., 1893, p. 110. See also Memoir on the Lias of England and Wales, p. 30, and B. Thompson, Quart. Journ. Geol. Soc., vol. l. pp. 393, 409, &c.

Fig. 103.

*Mammillated surface of Limestone, (Fig. 102), Lower Purbeck Beds,
Durlston Bay.*

Fig. 104.

*Bed known as " Outlet," Purbeck Beds, near Battle.**

---

* These three illustrations have been prepared from photographs taken by Mr.
W. W. Watts.

different tints, and afterwards solidified, then, in some cases, we find a simple banded limestone, and, in other cases, where the sediment was concreted in isolated masses, we find the mammillated surfaces and the irregular markings that in some instances assume an arborescent character. Even in the Cotham Stone, many obscure forms of arborescent marking may be found, but, as a rule, only the ornamental varieties are polished, and these only come under general notice.

Curious trifid, vermiform, and fucoidal markings occur on some of the beds, especially in Durlston Bay ;* and large ripple-marked slabs, of which an example is preserved in the Museum of Practical Geology, occur in the Middle Purbeck Beds of the same locality. Ripple-marks have also been noticed in the Upper Purbeck Cypris Shales, of Mupe (Mewps) Bay, and I have noticed similar markings in the rocks of the Vale of Wardour.

Granular beds, having an oolitic appearance, occur in the Lower Purbeck Beds at Ringstead, Portisham, and in the Vale of Wardour. Mr. Teall remarks that the Ringstead rock comprises mud-pellets, with or without thin coatings of radial (oolitic) carbonate of lime. Where two or more mud-pellets are in contact, there is no distinct line of demarcation, and there is no matrix.

Prof. Rupert Jones, who previously examined another specimen which I collected at Ringstead, found it to consist " of a mass of small roundish granules and Cyprides (*Cypris purbeckensis*) coated more or less with calc-sinter, including also small tubes (*Serpulæ ?*). This is comparable with a similar Purbeckian bed near Boulogne."† The disintegrated "Portland Stone." from Fonthill Giffard, described by J. A. Phillips, bears some resemblance to the granular Purbeck beds, but is much finer in grain. About a quarter in bulk of that material was found to consist of rounded quartzose sand, and this was associated with, but not enclosed by ovoid grains of calcite.‡

Microscopic Structure of Purbeck Rocks.

A granular limestone from Portisham, examined by Mr. Teall, was "Mainly composed of pellets which may contain traces of *Girvanella*. These pellets are often surrounded by a thin zone of oolitic character (concentric and radiate). The interspaces are mostly empty."

A granular limestone from the Lower Purbeck Beds, east of Lulworth Cove, examined by Mr. Teall, showed " Mud pellets often taking the form of rice-grains. Sometimes where two or more are in contact they appear to blend. At other times one appears as if pressed into the other. The matrix is fine-grained, clear crystalline calcite."

A brown shell-limestone from the "Hard Cockle Beds" of the Lower Purbeck Beds at Mupe Bay, showed "Sand-grains, organic fragments, oolitic grains, and more or less rounded pellets (showing *Girvanella.* structure?) in a fine-grained crystalline matrix. Some of the fragments

* T. R. Jones, Quart. Journ. Geol. Soc., vol. xli. p. 323.
† Quart. Journ. Geol. Soc., vol. xli. p. 326. See also Proc Geol. Assoc., vol. viii. p. 58; and Bull, Soc. Géol. France, ser. 3, vol. viii. p. 616.
‡ Quart. Journ. Geol. Soc., vol. xxxvii. p. 16. A sample of this material is preserved in the Museum of Practical Geology.

have a narrow coating of oolitic material, showing concentric and radiate structure."

An earthy limestone from Teffont Evias, showed "slender organic remains in a fine-grained granular calcareous matrix."[*]

The calcareous sandstone from near Battle showed "quartz grains cemented by calcareous matter."

Chert.

Nodules of chert occur in the Lower and Middle Purbeck Beds of the Isle of Purbeck, Portisham, the Vale of Wardour, Sussex, &c. Specimens from Portisham and near Battle, examined by Mr. Teall, showed chalcedonic silica; the rock from Sussex contained stems of *Chara* and an agate-like structure around cavities. A specimen of chert or flint from the lower part of the Purbeck Beds in Chilmark quarry, in the Vale of Wardour, was examined by Mr. W. H. Hudleston, who observed that it showed portions of marly limestone partially silicified, sealed together with purer chalcedony. The rock contained oolitic granules, a fragment of shell, and valves of *Cypris*: some of the oolitic granules of characteristic type were immersed in more or less pure chalcedony. Hence the flint resulted from the silicification of calcareous matter. Another specimen of oolitic chert from Chilmark examined by Mr. Teall, showed the oolitic grains to be partially silicified, the inner portions of the grains being calcareous.[†]

Spicules of a sponge named *Spongilla purbeckensis* by Mr. J. T. Young, were obtained from the Purbeck chert at Stair Cove.[‡] In reference to these Dr. Hinde remarks that "The spicules are exceedingly numerous, so as to constitute the main portion of the chert in which they occur." [§] It is interesting to know that in a MS note by Edward Forbes, he mentions the occurrence of *Spongilla?* from the Purbeck Beds of Dorsetshire.

Relations of the Purbeck Beds with Strata above and below.— Method of Formation.

The Purbeck Beds in most places rest conformably on the Portland Beds, and are overlaid conformably by the Wealden Strata.

The junction with the Wealden Beds shows a much more gradual passage than that between the Purbeck and Portland Beds, and yet on the whole, as remarked by Sedgwick, the general lithological characters of the Purbeck Beds seem to unite them more closely with the stony Portland Beds.[||]

Many authorities following Webster and Fitton, amongst whom Godwin-Austen, Oppel, and Ramsay, grouped the Purbeck Beds with the Wealden, for palæontologically by reason of their fresh-

---

[*] See also Sorby, Quart. Journ. Geol. Soc., vol. xxxv. (Proc.), p. 79.
[†] Proc. Geol. Assoc., vol. vii. p. 181, and Plate I., Fig. 1.
[‡] Geol. Mag., 1878, p. 220.
[§] British Jurassic Sponges, Pal. Soc., p. 212.
[||] Ann. Phil., vol. xxvii. p. 344.

water fauna they are more intimately connected with that formation.*

That some beds like the Purbeck Marble were of freshwater origin was pointed out long ago by John Woodward, and in 1812 by Webster, who later on observed that the beds contain a mixture of freshwater with marine shells.

Edward Forbes who made a study of the Purbeck Beds in 1849 and 1850 in company with Bristow, did not recognize any passage from the Portland into the Purbeck Beds in the Isle of Purbeck, and observed that the top beds of the Portland series were marine, the lowermost Purbeck Beds purely freshwater. At the same time he concluded that the Purbeck Beds were connected with the Oolitic group rather than with the Cretaceous, and he was evidently influenced by his discovery of the Echinoderm *Hemicidaris purbeckensis*, which he found in a layer above the Cinder Bed (with *Ostrea distorta*).† The same view is taken by De Loriol, and also by Coquand, Contejean, and others; while the more recent researches of Pavlow and others, on the marine equivalents of the Purbeck Beds, tend to show that their affinities are rather with the Jurassic than with the Cretaceous system.

On the whole it may be said that the marine fossils ally the Purbeck Beds with the Portland strata, and the freshwater fossils link them with the Wealden Beds. The vertebrate as well as the invertebrate remains have Jurassic as well as Wealden affinities.‡ The fish-fauna as remarked by Mr. Smith Woodward shows a close connection with the Upper Jurassic lithographic stones of France, Bavaria, and Wurtemburg.§

Hemicidaris purbeckensis occurs in the zone of *Ammonites gigas*, Lower Portlandian, of Boulogne-sur-Mer.

The Purbeck Beds mark changing conditions; freshwater limestones, botryoidal and tufaceous in character, like beds of travertine, are succeeded in places by evidences of land vegetation in the now silicified remains of Cycads and Conifers. The Lower Purbeck Beds indicate freshwater and terrestrial conditions, which may have been marked by a lake or series of lagoons, for they were attended locally by the deposition of gypsum. Subsequently an irruption of the sea in Middle Purbeck times allowed the incursion of marine forms like the *Pecten*, *Thracia*, *Trigonia*, *Avicula*, and *Hemicidaris*. These were succeeded by a gradual change from brackish to freshwater conditions in Upper Purbeck times, when the *Unio*, *Limnæa*, *Physa*, *Valvata*, *Paludina*, and *Planorbis* flourished. The freshwater Purbeck Beds were

---

* Fitton, Trans. Geol. Soc., ser. 2, vol. iv. p. 159. See also Webster, *Ibid.*, vol. ii. pp. 40, 44, and Ann. Phil., vol. xxv. p. 47; Mantell, Geol. I. of Wight, ed. 3, 1854, Table, p. 42; Marcou, Geologist, 1859, p. 1; Dr. C. Struckmann, Geol. Mag., 1881, p. 556.

† Forbes, Decade III., Geol. Survey, Plate V.; and Rep. Brit. Assoc. for 1850, p. 80; P. de Loriol and A. Jaccard, Mem. Soc. Phys. Hist. Nat. Genève, vol. xviii. p. 112.

‡ Lydekker, Quart. Journ. Geol. Soc., vol. xlvi. p. 49. See also Judd, Rep. Brit. Assoc. for 1870, p. 77.

§ Proc. Zool. Soc., 1890, p. 346.

succeeded gradually by the great accumulations of Wealden strata formed by river agency, but whether distributed over the bottom of a huge lake, or as a delta deposit, is a somewhat debatable subject.

Godwin-Austen remarked that " the changes in the Purbeck series are readily accounted for by reference to areas of water such as occur on the American coast at present, and which may be salt or brackish, according to the extent to which the sea-waters are excluded by sand-bars from mixing with the fresh waters flowing from the land."* Indeed Forbes in his early account of the Purbeck Beds, indicates nine or ten alternations of freshwater, brackish water, and marine conditions in the Purbeck Beds ; and remarks that they are not marked by any striking physical characters or mineral changes.† Brackish-water conditions were indicated by the occurrence of *Corbula, Cyrena, Cardium, Melanopsis,* and *Rissoa,* and perhaps by the *Ostrea distorta.*

No zones are recognized in our Purbeck Strata, though judging by the position in the series, they may be equivalent to beds on the Continent with a marine fauna, and grouped in the zone of *Ammonites transitorius ;* while in Lincolnshire and Yorkshire they may be in part equivalent to the zone of *Belemnites lateralis* of the Spilsby Sandstone and Speeton Beds.

Organic Remains.

The Purbeck Beds have yielded an exceedingly varied series of fossils. Perhaps of the highest interest are the Mammals, which at present have been obtained only from the base of the Middle Purbeck Beds of Durlston Bay—in a thin earthy or " Dirt " layer—in the very bed in which Forbes suggested they might be found.‡

We are indebted to W. R. Brodie, of Swanage, and Charles Willcox, for the discovery of *Spalacotherium* in 1854, and two years later S. H. Beckles obtained a number of additional forms, including *Plagiaulax.* These fossils are far from abundant, and long and patient search may be unrewarded. In 1880, after 10 days' search, Mr. E. W. Willett obtained one specimen of *Triconodon mordax.*

At least 12 genera and 22 species of Mammals have been described. They are Marsupials, insectivorous and herbivorous in habits, and are of diminutive size. The principal genera, such as *Spalacotherium, Triconodon,* and *Plagiaulax,* have been described by Sir Richard Owen and Dr. H. Falconer.

Until recently no such remains had been found in the Wealden formation, and it is therefore interesting to note that a specimen described by Mr. A. Smith Woodward as *Plagiaulax Dawsoni,*

---

* Rep. Brit. Assoc. for 1872, Sections, pp. 92, 93. See also Meyer, Quart. Journ. Geol. Soc., vol. xxviii. p. 244 ; and Ramsay, Address to Geol. Soc., 1864 p. 32 ; and Physical Geology and Geography of Great Britain, ed. 6, 1894.
† Rep. Brit. Assoc. for 1850, p. 81.
‡ Geikie, Memoir of Edward Forbes, 1861, pp. 460, 461.

PURBECK MAMMALS.

FIG. 105.

FIG. 106.

FIG. 1 7.

FIG. 105. Plagiaulax Becklesi, *Falc.*, 2.
Middle Purbeck, Swanage.

,, 106. Triconodon mordax, *Owen.*
Middle Purbeck, Swanage.
Lower jaw and teeth.

,, 107. Spalacotherium tricuspidens, *Owen.*
Middle Purbeck, Swanage.
a, b. Imperfect left ramus of mandible.
c, d. Lateral and upper views of a molar tooth.

and another, described by Mr. Lydekker as *Bolodon,* have been
obtained from the Wadhurst Clay, near Hastings.*

No traces of Birds have as yet been found. Remains of Frogs
were recorded from the Purbeck Beds of Swindon, by Charles
Moore,† but their occurrence is considered exceedingly doubtful
by Mr. E. T. Newton.

Among the fossils from the Purbeck Beds the Saurian remains
are important. The Dinosaurs include *Iguanodon* and *Nuthetes;*
then we have the "Swanage Crocodile," *Goniopholis crassidens,*
and the dwarf Crocodiles known as *Nannosuchus* and *Theriosuchus.*
Owen estimates the average length of a mature *Theriosuchus* at
eighteen inches. Turtles have been obtained mostly from the
quarrymen, as they occur in the stone-beds : they include *Treto-*
sternum, Pleurosternum concinnum, P. Bullocki, and *Chelone*
obovata, some fine specimens of which are preserved in the
museum at Corfe Castle. The Fishes include *Asteracanthus*
verrucosus, Caturus, Coccolepis, Hybodus, Lepidotus, Macrosemius,
Ophiopsis, and *Pleuropholis,* and they are mostly found in the
stone-beds worked at the quarries. Specimens occur at various
horizons in the Purbeck Beds, and some of the best preserved
examples have been obtained from the neighbourhood of Teffont
Evias, in the Vale of Wardour.

It was the intention of Edward Forbes to publish an account
of the Invertebrata of the Purbeck Beds of Dorsetshire; and he
had assigned names to a number of new species of Mollusca,
which he was the first to discover.‡ Some of these have been
figured in Lowry's Chart of Characteristic British Fossils, and
others in works published on the Continent. They all belong to
living genera, and taken by themselves, possess a Tertiary or
even recent aspect.

Reference has already been made to the principal genera of
Mollusca found in the Purbeck Beds, but it may be mentioned
that Mr. Carruthers has described, under the name *Teudopsis*
Brodiei, a cuttle-bone from these strata in Dorset.§

Insects are represented very fully by remains of Coleoptera
Orthoptera, Diptera, Neuroptera, and Hemiptera. To the various
remains very numerous names were applied by Prof. Westwood,||
and for our knowledge of these we are chiefly indebted to the
Rev. P. B. Brodie. There were Butterflies, Beetles, Dragon-flies,
Locusts, Grasshoppers, Ants, and the earliest known Aphides;
forms that are considered to indicate temperate conditions.

At Durlston the Insect-remains occur most abundantly in the
Lower Purbeck Beds, elsewhere they are found in Middle Purbeck strata.

---

* Proc. Zool. Soc., 1891, p. 585; and Quart. Journ. Geol. Soc., vol. xlix. p. 281.
† Proc. Cotteswold Club, vol. ii. p. 192; and Proc. Geol. Assoc., vol. iv. p. 544.
‡ Some figures of Purbeck fossils will be found in the works of Fitton, Trans.
Geol. Soc., ser. 2, vol. iv. Plates XXI. and XXII.; Mantell's Geological Excursions
round the Isle of Wight, ed. 3; and Damon's Supp. to the Geology of Weymouth,
&c., ed. 3, 1888.
§ Quart. Journ. Geol. Soc., vol. xxvii. p. 448.
|| *Ibid.,* vol. x. p. 391.

PURBECK MOLLUSCA.

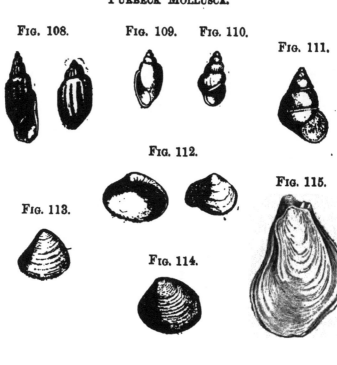

FIG. 108.   FIG. 109.   FIG. 110.   FIG. 111.   FIG. 112.   FIG. 115.

FIG. 113.

FIG. 114.

FIG. 116.   FIG. 117.

FIG. 108. Melanopsis harpæformis, *Dunk.* Nat. size.
  ,, 109. Physa Bristovii, *Forbes.* Nat. size.
  ,, 110. Paludina elongata, *Sow.* Nat. size.
  ,, 111. —— carinifera, *Sow.* 2.
  ,, 112. Corbula alata, *J. Sow.* Nat. size.
  ,, 113. —— sp. 2.
  ,, 114. Cyrena media, *J. Sow.* Nat. size.
  ,, 115. Ostrea distorta, *J. Sow.* Nat. size.
  ,, 116. Unio sp. Nat. size.
  ,, 117. —— valdensis, *Mant.* ⅓.

PURBECK FOSSILS.

FIG. 118.   FIG. 119.

FIG. 120.   FIG. 121.

FIG. 122.   FIG. 123.

FIG. 118. Cypridea punctata, *Forbes.*
  „ 119. —— granulosa, *Sow.*
  „ 120. Cypris purbeckensis, *Forbes.*
  „ 121. Archæoniscus Brodiei, *M. Edw.* 1½.
  „ 122. Hemicidaris purbeckensis, *Forbes.* ½.
  „ 123. Mantellia (Cycadeoidea) microphylla. *Buckl.*¼.

Among Crustacea, the Isopod, *Archæoniscus* occurs in profusion in the Middle Purbeck Beds in the Vale of Wardour. It has been found, however, in the Lower Purbeck Beds of Wiltshire and Dorset.

In 1850 Edward Forbes named eight species of Ostracoda ("Cyprides") from the Purbeck Beds of Dorsetshire, but unfortunately he never described them.[*] Figures of these species, reduced from diagrams used by Forbes in his lectures at the Royal School of Mines, were afterwards published by Lyell.[†] It has not, in every case, been possible to identify the specimens upon which the names of these species were based, while some doubts have arisen concerning the particular horizons from which they were obtained. These difficulties have been pointed out by Prof. T. Rupert Jones, and more recently he has given a full account of the subject, with a revision of the names and descriptions of new species.[‡] These are for the most part freshwater forms, but the genus *Cythere* is marine, and *Candona* is said to be estuarine.

Among the Ostracods it is found that *Cypris purbeckensis, Candona ansata*, and *C. bononiensis* are most abundant and characteristic in the Lower Purbeck Beds; *Cypridea granulosa* in the Middle Purbeck; and *C. punctata* in the Upper Purbeck. Their particular range was not, however, restricted, and it is interesting to note that Prof. Jones has recorded *Cypris purbeckensis*, and a variety of *Cypridea tuberculata* from the Upper Cretaceous strata of Wyoming, in North America.[§] (See also p. 280.)

Here and there at various horizons in the Middle and Lower Purbeck Beds a number of "Dirt Beds" occur. The name has been applied to layers of carbonaceous clay or shale, but it originated in Portland Island, where the Great Dirt Bed which occurs near the base of the Lower Purbeck Beds is especially noted for the silicified remains of Cycads and Coniferous trees which it has yielded. There great "Burrs" of siliceous and calcareous (tufaceous) material have been accumulated around the old tree stumps, an example of which seen by the cliffs east of Lulworth Cove is shown in Fig. 124, p. 242, and another at Portisham is shown in Fig. 133, p. 263.

A bed yielding similar more or less silicified plant-remains has been observed near Lulworth and eastwards to Gad Cliff; [||] and there are traces of Burrs to the east of Dancing Ledge as observed by Mr. Strahan and myself in 1893.

Chara was found in the cherty freshwater band of the Middle Purbeck Beds by Edward Forbes.[¶] (See p. 244.)

---

[*] Rep. Brit. Assoc. for 1850, pp. 79–81.
[†] Manual of Geology, edit. 3, p. 231 ; and Elements of Geology, edit. 6, 1865.
[‡] Geol. Mag., 1878, p. 105 ; and Quart. Journ. Geol. Soc., vol. xli. p. 311.
[§] Ibid., 1893, p. 386.
[||] See Fitton, Trans. Geol. Soc., ser. 2, vol. iv. pp. 220, 223 ; see also Buckland and De la Beche, Ibid., p 14.
[¶] See also E. Wethered, Proc. Cotteswold Club, vol. x. p. 101.

FIG. 124.

Purbeck " Burrs," seen on cliffs east of Lulworth Cove.
(From a photograph by A. Strahan.)

The following may be included as among the more common
and characteristic fossils of the Purbeck Strata :—

　　Melanopsis harpæformis (Fig. 108).
　　Paludina carinifera (Fig. 111).
　　——elongata (Fig. 110).
　　Physa Bristovii (Fig. 109).
　　Corbula alata (Fig. 112).
　　Cyrena (Cyclas) media (Fig. 114).
　　—— (——) parva.
　　Modiola.
　　Ostrea distorta (Fig. 115).
　　Unio compressus.
　　—— valdensis (Fig. 117).
　　Archæoniscus Brodiei (Fig. 121).
　　Candona ansata.
　　—— bononiensis.
　　Cypridea granulosa (Fig. 119).
　　—— punctata (Fig. 118).
　　Cypris purbeckensis (Fig. 120).
　　Hemicidaris purbeckensis (Fig. 122).
　　Mantellia microphylla (Fig. 123).
　　—— nidiformis.

CHAPTER XIV.

PURBECK BEDS.

LOCAL DETAILS.

Dorset Coast.—Isle of Purbeck.

The Purbeck Beds are well exposed in the cliffs at Durlston Bay, Worbarrow Bay, Mupe (Mewps) Bay, Lulworth Cove, and in the quarries and railway-cutting at Ridgeway or Upway. These sections have been measured in great detail by H. W. Bristow, assisted at Ridgeway by the Rev. Osmond Fisher, and at Lulworth by Mr. W. Whitaker.* The most complete section is that at Bacon Hole, Mupe Bay, where the gradual passage into the Wealden Beds is well shown.

The beds were locally sub-divided by Edward Forbes and H. W. Bristow, and the principal beds recognized in the abovementioned sections, may be summarized as follows, the thickness being stated in feet and inches :—

	PURBECK BEDS OF DORSET.	Durlston Bay.	Worbarrow Bay.	Mupe Bay.	Lulworth Cove.	Ridgeway.
Upper Purbeck Beds.	PALUDINA CLAYS. Blue, green, grey, and purple marls, and sandy shale. Fish-remains, *Paludina*, Cyprides -	—	—	11·0	1·0	5·10
	UPPER CYPRIS CLAYS AND SHALES. Clays, shales, sands, red and green marls, and alum shale, with occasional layers of "beef" (fibrous carbonate of lime), selenite, &c. Bands of shelly limestone with *Paludina carinifera* (Purbeck Marble); green sandy limestone with Unio (Unio Bed). Fish-remains (*Pycnodus*), Coprolites, *Cyrena, Cypridea punctata, C. ventrosa, Darwinula leguminella* -	45·10	35·9	29·2	17·0	23·4
	UNIO BEDS. Clays with bands of greenish limestone and sandy layers, occasional "beef" and alum shale. Carbonaceous limestone (Crocodile Bed) at Durlston Bay. Turtle and Fish-remains, Coprolites, *Paludina, Cyrena, Unio*, Cyprides -	4·9	6·6	6·0	5·9	9·
	UPPER BROKEN SHELL LIMESTONE (Soft Burr Stone). Limestone made up largely of comminuted shells (*Cyrena*). Turtle and Fish-remains (*Pycnodus*), *Limnæa, Paludina, Unio* -	10·0	7·0	4·3	3·9	10·6
	CHIEF "BEEF" BEDS. Dark (alum) shales with "beef" and selenite, beds of limestone, and layers of perished shells. *Cyrena* and Cyprides -	29·11	13·10	6·6	8·2	4·6
	CORBULA BEDS. Layers of shelly limestone, shale, alum shale, and marl, with "beef" and selenite. Turtle and Fish-remains (Asteracanthus), *Melanopsis harpæformis, Cardium, Corbula alata, Cyrena, Modiola, Ostrea, Pecten, Perna, Thracia.* Insects and Cyprides -	34·1	40·2	22·4	15·11	18·6

---

* Vertical Sections (Geol. Survey), sheet 22 ; Horizontal Sections, sheet 56 ; see also Damon, Geology of Weymouth, ed. 2, 1884, p. 201.

	PURBECK BEDS OF DORSET.	Durlston Bay.	Worbarrow Bay.	Mupe Bay.	Lulworth Cove.	Ridgeway.
Middle Purbeck Beds.	SCALLOP BEDS. White (Roach) shelly limestones and occasional shale. Fish-remains, *Corbula alata, Ostrea, Pecten,* (Pecten-bed of Forbes) - - -	4·6	2·9	2·4	1·10	—
	INTERMARINE BEDS. Upper Building Stones. More or less shelly limestones, with shale partings. Saurian, Turtle and Fish-remains (*Hybodus, Lepidotus*), *Hydrobia, Limnæa, Melanopsis harpæformis. Paludina, Corbula alata, Cyrena. Modiola, Ostrea, Serpula, Cyprida punctata,* Plant-remains - -	46·0	8·0	8·7	7·5	7·5
	CINDER BED. Earthy limestone made up chiefly of *Ostrea distorta.* Fish-remains, *Cardium, Perna, Trigonia, Serpula, Hemicidaris purbeckensis*	8·6	5·6	4·0	4·0	5·0
	CHERTY AND MARLY FRESHWATER BEDS. Lower Building Stones. Shelly limestones with partings of shale and marl. Marly limestone with nodules of black chert (Flint), bands of clay, limestone, and carbonaceous shale (Dirt Bed or Mammal Bed of Durlston Bay). Mammals, Saurian, Turtle and Fish-remains, also *Hydrobia, Limnæa, Melanopsis, Paludina, Physa, Planorbis, Valvata, Rissoa, Corbula, Cyrena, Pinna,* Insects, *Cyprida fasciculata, Chara* - -	43·9	19·3	5·7	5·9	12·10
Lower Purbeck Beds.	MARLY FRESHWATER BEDS. Marls and limestones. Fish-remains (*Lepidotus*), *Hydrobia, Limnæa, Planorbis, Physa,* Cyprides, Lignite - - - -	7·4	12·3	8·0	12·3	3·9
	UPPER INSECT BEDS. Marls and clays, Insects and Cyprides -	—	—	—	—	5·4
	SOFT COCKLE BEDS. Marls and marly limestones, granular oolitic limestones, occasional bands of brecciated limestone, and band showing arborescent markings. Ripple-marks. Gypsum in veins and nodules; pseudomorphous crystals of rock salt. Fish-remains, *Cardium, Corbula, Leda, Modiola, Serpula, Archæoniscus,* Cyprides -	82·1	62·7	58·8	39·0	25·10
	HARD COCKLE BEDS. Marls and marly limestone, with chert. Pseudomorphous crystals of rock-salt. *Rissoa, Cardium, Cyrena media,* Insects and Cyprides. *Cypris purbeckensis* - -	10·0	24·6	24·4	13·7	8·11
	LOWER INSECT BEDS. Clays and Sands with marly layers. Insects -	—	—	—	—	5·9
	CYPRIS FREESTONE. Soft shales and marly and fissile limestones. Gypsum in places. Fish-remains, *Planorbis, Rissoa, Cyrena, Cypris purbeckensis* - -	36·3	26·5	19·10	21·10	26·7
	BROKEN BANDS. Bituminous slaty and sandy limestones, cherty and earthy matter; much broken and disturbed, excepting at Ridgeway. *Limnæa, Valvata,* Cyprides -	15·0	14·0	15·0	10·0	9·3
	SOFT CAP. Impure bituminous and botryoidal limestone with chert in places. Cyprides - - -	7·6	5·0	6·11	} 4·6	}
	DIRT BED, with trunks and stools of trees (Ridgeway, &c.) - -	—	1·0	0·9		} 6·8
	HARD CAP. Bituminous, sandy, or cherty and botryoidal limestone. Fish-remains (*Histionotus*), *Paludina, Archæoniscus,* Cyprides -	11·0	} 6·3	17·5	4·6	
	DIRT BED - - - - -	0·3				—
		396·9	290·9	250·8	176·3	189·5

¶ (The thicknesses of the Purbeck sub-divisions, measured by Bristow, are as follows:—

Purbeck Beds.	Durlston.	Worbarrow.	Mupe.	Lulworth.	Ridgeway.
Upper · · ·	60·7	49·3	50·5	27·6	49·1
Middle · · ·	166·9	89·6	57·4	55·4	52·0
Lower · · ·	169·5	152·0	142·11	93·5	88·4
	396·9	290·9	250·8	176·3	189·5

Thus there appears to be a general attenuation of the beds from east to west.

The value of the three divisions, except as a matter of local convenience, has been questioned, for with the exception of the Cinder Bed there are no marked lithological characters whereby they can be traced from point to point, and the palæontological changes are not of a very distinctive character, as remarked by Morris.[*]

Broadly speaking the Upper Purbeck Beds comprise the clays with Marble Rag and the *Unio*-Bed; the Middle Purbeck Beds include the chief building-stone of Swanage, the Cinder Bed, the band with "flints," and the Mammal Bed of Durlston Bay; the Lower Purbeck beds include the marls with gypsum at Durlston Bay, the building-stone of Upway and Ridgeway, and the cap-beds and Great Dirt Bed of Portland and elsewhere.

Durlston Bay, Swanage.

In the cliffs and quarries along the coast between St. Alban's (St. Aldhelm's) Head and Durlston Head, the basement portions of the Lower Purbeck Beds may be seen. For the most part they comprise thin and somewhat irregular and undulating layers of grey limestones with clay partings, and the beds are more or less botryoidal in character like the Portland Cap-beds which they represent. The higher portions of the Lower Purbeck Beds, for the most part of a soft character, are concealed along this southern coast beneath the pleasant grassy slopes that rise above the cliffs.

Rounding Durlston Head we have a good view of the cliffs that extend northwards to Peverel Point. (See Fig. 125, p. 246.) There Durlston Bay is formed by the denudation of the softer Lower Purbeck Beds, the hard Portland Stone at Durlston Head, and the limestones of the Upper Purbeck at Peverel Point forming headlands and reefs. Smaller projections along the coast are produced where the stone-beds of the Middle Purbeck descend to the sea-level and run out in ledges. The destruction of cliffs has no doubt been hastened by the quarrying of the stone and the removal of blocks from the beach.

---

[*] Proc. Geol. Assoc., vol. vii. p. 396.

FIG. 125.

Section from Durlston Head northwards to Peverel Point, Swanage.   (A. Strahan.)

Distance one mile.

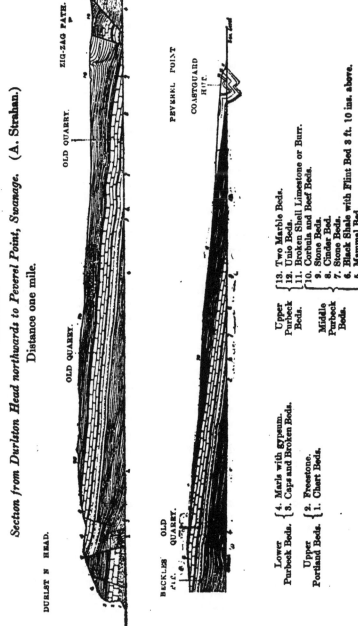

Lower   { 4. Marls with gypsum.
Purbeck Beds. { 3. Caps and Broken Beds.

Upper   { 2. Freestone.
Portland Beds. { 1. Chert Beds.

Upper   { 13. Two Marble Beds.
Purbeck   { 12. Unio Beds.
Beds.   { 11. Broken Shell Limestone or Burr.
             { 10. Corbula and Beef Beds.

Middle   { 9. Stone Beds.
Purbeck   { 8. Cinder Bed.
Beds.   { 7. Stone Beds.
             { 6. Black Shale with Flint Bed 3 ft. 10 ins. above.
             { 5. Mammal Bed.

The cliffs do not present a clear unbroken series, for the strata, while dipping generally towards Swanage, are dislocated and repeated in several places. The Portland Stone occurs at the base of Durlston Head, overlaid by somewhat shattered Purbeck strata; but the junction is not easy to determine, owing to the tumbled masses of rock, and to the fact that the Portland stone presents a rough scoriaceous appearance, and is much covered by limpets, periwinkles, and barnacles. A fault that may be noticed on the other side of Durlston Head is again intersected by the cliffs, and the tract adjoining the headland is a good deal obscured by slips. (See Fig. 126.) Hence although a nearly complete

FIG. 126

Section at Durlston Head.

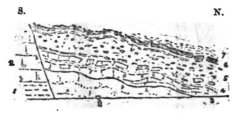

7. Cypris Shales.	4. Soft Cap.
6. Limestones and Shales.	3. Hard Cap.
5. Broken Beds.	2. Portland freestone.
	1. Portland cherty limestone.

section of the Purbeck Beds may be measured, it has to be made up from different portions of the cliffs.

The top bed of the Portland Stone seems to be compact in character, and there are indications of slight. irregularity at the junction with the overlying Purbeck strata. (See Fig. 126.) The basement " Dirt parting " of the Purbeck beds is much indurated in some places, and has been partially eroded by the sea. The Hard and Soft Caps above, of a somewhat fissile nature, are exposed just north of Durlston Head. Surmounting these strata we find the remarkable division marked by Bristow as the " Broken Bands," which he describes as "thin slaty beds of bituminous sandy limestone much broken and squeezed together, with a few broken fragments of chert." Strata of this description, from 10 to 15 feet or more in thickness, occur about the same horizon at Worbarrow Bay, Mupe Bay, and Lulworth Cove. So dislocated and tumbled are the strata that they were familiarly termed the "Grand Smashery" by Edward Forbes.* Below, the beds are for the most part undisturbed, and above, they gradually assume an undisturbed aspect, though the overlying layers of " Cypris Freestone " are occasionally disrupted.

---

* See Geikie, Life of E. Forbes, p. 465.

Broken Beds of this character may be due to various causes Similiar appearances are presented in many places in the Isle of Portland where Purbeck Beds immediately overlie fissures or "gullies" in the Portland Rocks. There they are simply in a tumbled condition, and the effect is local. In the sandy limestones or calcareous sandstones belonging to the Northampton Sand, somewhat similar appearances have been locally produced by the removal of calcareous matter by carbonated water. In this way some of the overlying strata have lost support and become "Broken Beds," and the mass in places has been partially re-cemented by carbonate of lime.

On the coast from Durdle Door and Lulworth to Durlston the effect appears to be fairly uniform along an extensive tract; and the Broken Beds cannot be attributed to faults (in the ordinary sense), nor to springs washing material into fissures or cavities of the Portland Rocks beneath, although such causes, together with frost, have no doubt in places exerted some influence. Indeed on the west side of Bacon Hole, Mupe Bay, and on the east side of Lulworth Cove, the beds are not only contorted, but slightly faulted, and the same is the case at Durlston Head.

The appearances in question being confined practically to one general horizon, it was suggested by the Rev. O. Fisher that the beds, originally deposited on the debris of the old Purbeck forest, subsequently fell into vacuities caused by the decay of the vegetable matter.[*] Where, however, the great dirt-bed is so well shown on Portland we do not see any " Broken Beds " that can be compared with those between Lulworth and Durlston Head : although I noticed that the "Bacon Tier" in places at Kingbarrow and near St. George's Church becomes a broken-bed, and so do the upper strata. These appearances, however, are exceptional, and it must be borne in mind that Portland is on the south side of the Weymouth anticline, where the dip is comparatively gentle. (See p. 265.)

It seems to me clear that the beds have for the most part been broken up and disturbed since their deposition and consolidation. The disturbances are by no means confined to one exact horizon, for there is a gradation from the broken and shattered beds into undisturbed strata both above and laterally; at any rate the amount of dislocation varies at different points, so that the disturbances affect strata higher in some places than in others. This proves that the main features of the disturbances could not have been produced in Purbeck times. (See Figs. 127, 130, and 131, on pp. 249, 256, and 258.)

Webster attributed the phenomena to disturbance,[†] and when it is borne in mind that the strata dip at angles of 20° to 40° in a N.N.W. direction, and repose on the hard Purbeck " Caps " and underlying Portland Stone, it seems to me that much shattering

---

[*] Trans. Cambridge Phil. Soc., vol. ix. p. 566.
[†] Englefield's Isle of Wight, pp. xix. and 173 ; see also Sedgwick and Murchison, Trans. Geol. Soc., ser. 2, vol. v. p. 655 ; H. B. Woodward, Geol. England and Wales, Ed. 2, p. 346 ; and Strahan, Quart. Journ. Geol. Soc., vol. lit., p. 549.

must have been caused during the tilting of the strata. Indeed the Rev. W. B. Clarke, in a lengthy account of the great disturbances in the Chalk and other rocks of the Isle of Purbeck, speaks of "the whole range of strata from Tilly Whim to Ballard Down being drawn out like a pack of cards " *—presumably as when higher layers are shifted slightly over lower ones.

FIG. 127.

Section at Bacon Hole, Mupe Bay, Dorset.

4. Shales and bands of limestone.
3. Broken Beds.
2. Caps.
1. Portland Stone.
About 20 feet shown.

That some sliding of individual beds has taken place is shown by the striated pavements noticed by Bristow, on the Upper Tombstone Bed, above the Cinder Bed, in Durlston Bay.† I thought at first that these appearances were due to shifting of loosened slabs of the rock and that the scratching might have been produced by stones and grit ; but on a subsequent visit with Mr. Strahan we found that the striæ passed beneath the superincumbent limestones, and that they were evidently slickensides.

The weakness produced by the shifting of the strata has doubtless been subsequently acted upon by weathering-agents, for a certain amount of dissolution of calcareous matter has taken place from the sandy limestones that form the mass of the strata.

We are therefore justified in concluding that the horizon now known as the " Broken Beds," formed one of weakness during the disturbances to which the beds have been subjected, and this may have been due to some extent to the character of the strata, and to the fact that they were somewhat irregularly accumulated over the terrestrial surfaces and hummocky " burrs."

We find no evidence of Broken Beds at Ringstead Bay, nor at Ridgeway Hill (Upway), further west, although the strata are highly tilted ; but here the beds are changing in character : the mass of the Portland Stone is softer than in the Isle of Purbeck, the " Caps " are thinner ; and we find above the " Caps," beds of fissile limestone and marls. In other places we see Broken Beds produced locally, as in the Soft Cockle Beds north of Durlston Head where they are disturbed by faulting and permeated by springs ; and further instances might be given.

---

* Mag. Nat. Hist., ser. 2, vol. i. p. 465 ; vol. ii. pp. 79, &c.
† Vertical Sections, Geol. Survey, sheet 22, No. 1.

Not until we pass the zigzag path below the Refreshment House in Durlston Bay, can we clearly trace the upward succession of the Purbeck Beds from the lower, through the middle, to the upper division of these strata.

A considerable fault has thrown the Middle Purbeck Cinder Bed and other strata against the Lower Purbeck Beds on the north side of the zigzag; these latter consist for the most part of soft grey and yellow marls, with occasional hard bands, and with masses of gypsum or alabaster. At their base traces of the Broken Beds were just visible, overlaid by the impure limestone of the Cypris Freestone. The soft grey Cypris shales contain thin layers of gypsum, but larger masses of this mineral occur higher up, in the " Soft Cockle Beds." (See Fig. 125, p. 246.)

FIG. 128.

Junction of Lower and Middle Purbeck Beds, Durlston Bay.

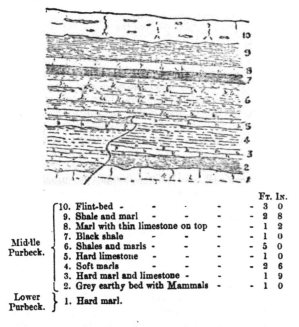

		FT.	IN.
	10. Flint-bed - - - - -	3	0
	9. Shale and marl - - - -	2	8
	8. Marl with thin limestone on top -	1	2
Middle	7. Black shale - - -	1	0
Purbeck.	6. Shales and marls - - -	5	0
	5. Hard limestone - - -	1	0
	4. Soft marls - - -	2	6
	3. Hard marl and limestone - -	1	9
	2. Grey earthy bed with Mammals -	1	0
Lower Purbeck.	1. Hard marl.		

Undulations and slight dislocations affect the beds here and there as we trace them along, but two dark bands are conspicuous among the strata, above the soft grey and yellow marls of the Lower Purbeck Beds. The lower and rather lighter band is that known as the Dirt Bed or Mammal Bed, which must not be confused with the Great Dirt Bed of Portland, that comes on a much lower horizon. It may be mentioned that where this Dirt Bed comes to the surface in the cliff to the north of the Refreshment House, an excavation, known as Beckles' Cutting, was made many years ago by S. H. Beckles, in search of the fossils for which the layer is noted. Two Dirt-beds were mentioned by Mr. H. Willett, but they belong to the same horizon, being shifted by faults. This bed is a dark grey and brown carbonaceous shale about 1 foot in thickness, resting on an irregular and sometimes ruptured surface of the Lower Purbeck marl beneath. (See Fig. 128.) In places, however, the beds appear to merge one into the other, and the irregularity is seen to be partly due to slight faults or joggles, which die out upwards, and are evidently produced by disturbance, that led to slight shifting of the lower strata. A Dirt-bed occurs on the same horizon at Worbarrow Bay, but no Mammals have there been discovered.

In addition to the Mammals found in this thin band, Reptilian remains, freshwater shells and Cyprides have been obtained. The Reptilian remains include *Macellodus Brodiei, Theriosuchus pusillus, Nannosuchus gracilidens,* and *Nuthetes destructor.* The Mollusca, recorded by Bristow, are *Valvata, Paludina, Rissoa, Physa Bristovii, Limnæa,* and *Planorbis.*

The Dirt-bed marks the base of the Middle Purbeck strata, which comprise a series of blue and grey stone-beds, including the famous Swanage Stone and the Cinder Bed, overlaid by dark clays.

The black carbonaceous shale that occurs about 10 feet above the Mammal Bed, forms a conspicuous band in the cliff; it is not known to have yielded any vertebrate remains.

The next bed of particular interest in ascending the series, is the "Flint" layer, a pale marly limestone about 3 feet thick, with irregular nodules of black chert. Freshwater shells occur in both chert and limestone. The nodules do not stand out in relief as they are readily broken away with the rock by the action of the sea. (See Fig. 129.)

Higher up we come to the Lower Building-stones (including the Feather Bed), which were formerly worked in the cliff.

FIG. 129.

Flint-bed, Middle Purbeck Beds, Durlston Bay.

These stone-beds are surmounted by the Cinder Bed, one of the most easily recognized beds in the Purbeck Series. It forms a rugged cliff of rock, and was so named on account of its rough cindery appearance. Such a well-marked layer, as remarked by Fitton,* is very useful as a guide to fix the position of the strata

* Trans. Geol. Soc., ser. 2, vol. iv. p. 209.

where they are disturbed. It attains a thickness of 8 or 9 feet, and is somewhat thicker inland ; it is almost entirely made up of the small oyster *Ostrea distorta,* although it is not easy to procure any fine examples of this shell. Bristow records from this bed Fish-remains, *Cardium Gibbsi, Perna, Trigonia, Serpula,* and *Hemicidaris purbeckensis.*

Above the Cinder Bed, come the Upper Building-stones, which jut out at the base of the cliffs, and form hard ledges on the shore, where the shaly partings that separate the beds of lime-stone are cleared away in many places by the sea.

Turtle and Fish-remains occur in these building-stones together with Mollusca and Plant-remains that suggest estuarine con-ditions. False-bedding is shown in one of the Freestone Veins.

Some of the shelly limestones, such as the White Roach and Laning (Leaning or Lane End) Bed, show cavernous weathering. There is evidence of a slight fault along the face of the cliff where these beds are exposed, and perhaps, owing to this cause, and to the fact that sea-water produces some superficial alterations in the strata, it is difficult to recognize all the individual beds of limestone recorded in Bristow's section.

Near the top of the Corbula Beds, there is a prominent band of Broken-shell Limestone with yellow sandy seams, and over-lying this is a layer marked the Toad's Eye Limestone, that con-tains irregular seams of indurated marl, and has curious tridactyl and vermiform markings on its surface.*

The Upper Building-stones like those at a lower horizon, have, in places, been quarried in the cliffs, and old excavations and tunnels remain to mark the spots. The beds were exposed along the cliff-road leading from the Refreshment House to Durlston Head. The higher portions of the Middle Purbeck Beds, con-sist, for the most part, of dark clays and paper-shales with much fibrous carbonate of lime (chief " Beef" Beds), with selenite, and occasionally gypsum, and ochreous matter.

The shales here and those intercalated with the stone-beds below, frequently show efflorescences of alum, and many of the beds are described as alum-shales, by Bristow. The Mollusca found in the shales above the stone-beds are freshwater and estuarine forms, but, in many cases, we find the shelly layers in a very decayed and rotten condition.

On the whole, the Middle Purbeck Beds have a bluish-grey appearance, weathering brown only in places : the Upper Purbeck Beds have a generally brown appearance in these cliff-sections, and they occur on top of the cliffs about half way between the Refreshment House and Peverel Flag-staff, descending gradually to the sea-level, and forming the cliffs to Peverel Point. (See Fig. 125, p. 246.)

The base of the Upper Purbeck beds is here marked by the " Soft Burr" or Upper Broken-shell Limestone, consisting of about 10 feet of sandy limestone, largely made up of comminuted

---

* See T. R. Jones, Quart. Journ. Geol. Soc., vol. xli. p. 313.

shells, but with the fragments dissolved out in many cases, so that the rock is full of tiny cavities. Formerly it was much used for building-purposes. The shell-fragments, according to Bristow, belong chiefly to *Cyrena*, and he also records *Unio, Paludina*, and remains of Fishes and Turtles.

Overlying these beds are the "Unio Beds," bands of irregular limestone and shale, with here and there many examples of *Unio*. One of the limestone-bands is termed the Crocodile Bed by Bristow, and he records from it *Unio, Paludina*, Fish and Turtle-remains, Coprolites, and the Crocodile, *Goniopholis*.

The highest beds exposed in the cliffs are grouped as the Upper Cypris Clays and Shales, and among these, we find one or two bands of sandy and shaly limestone with *Unio*. These are usually characterized by much greenish colouring matter, and one band in particular constitutes the Unio Bed.

In these upper beds we find also three bands of Purbeck-Marble, hard shelly limestone largely made up of *Paludina carinifera*, and therefore known also as the *Paludina*-Marble. *Unio* and teeth and scales of Fishes are occasionally met with in these beds. They are well exposed beneath the Peverel Flag-staff, and they form the ledges of Peverel Point, that stretch far into the sea, and form a protection to this part of the coast. The beds, however, are much disturbed. (See Fig. 125, p. 246.)

The marble is much broken up by joints, and the lines of weathering on the surfaces exhibit false-bedding. The marble is mostly grey or bluish-grey : red and bluish marble occur in the same mass. The grey stone according to Mr. G. Maw, contains carbonate of iron.[*] The colour of the red marble appears super-ficial, and due to staining by iron-oxide along the joints, the colour penetrating irregularly into the mass of the stone. The green colouring-matter fills the cavities of the shells of Paludina, and it has been found by Mr. Teall to resemble glauconite in many respects. This green tint which is so common in the Unio Beds, occurs also occasionally in lower beds, and the "Lias Rag," belonging to the Upper Building-stones, contains argillaceous kernels of a green colour.

On the southern margin of Swanage Bay, blue shelly and crystalline limestones are seen dipping northwards at an angle of about 15°. Beneath them and running along the line of the High Street, Swanage, the Purbeck Marble was formerly traced. The junction of Purbeck and Wealden Beds is, however, con-cealed by the town and harbour of Swanage, and northwards on the shores of Swanage Bay we come to the mass of the Wealden Strata, which dipping northward consist of many alternations of yellow sands and coloured clays, with occasional bands of grit and much lignite, and in the upper part of shale, with Cyprides.

The Purbeck Marble is now occasionally quarried at Easton near Langton Maltravers, south-east of Woody Hyde, south-west of Afflington Farm, and south-west of West Orchard.

---

[*] Quart. Journ. Geol. Soc., vol. xxiv. pp. 356, 366, and Plate XIII. Fig. 22.

The pit at Easton (open in 1884) was situated just south-east of the farm-buildings. The section was as follows, the beds dipping E. 35° N. at angle of about 25° :—

		Ft.	In.
	Two bands of marble—the upper bed 2 to 3 ins. thick : the lower 4 to 5 ins. The beds are broken up and only remnants occur in the soil.		
	Flaggy marble, much weathered : *Paludina* on joints - - -	0	8
	Shales with "race," and thin flaggy limestone - - -	4	0
Upper Purbeck Beds.	Paludina-marble, much broken by joints, and occurring in interrupted masses - - - - -	0	3
	Clays and calcareous shales with "race" - - -	5	6
	Brown limestone, blue hearted - -	0	4
	Paludina-marble - - -	0	6
	Shaly parting - - -	0	1
	Paludina-marble, irregular bed -	0	3
	Calcareous shales with "race" and thin limestone bands - - -	1	6
	Paludina-marble - - -	0	7
	Paludina-marble - - -	1	4
	Paludina-marble (another bed, just exposed).		

The Marble at this quarry is variable, and green, bluish-grey, and occasionally red in colour—where iron-stained : it generally weathers brown outside.

The clays are sometimes locally disturbed, a feature produced by the washing down of material through cracks or joints of the underlying bands of limestone. The surfaces of the hard beds are sometimes covered with curious markings, due, perhaps, to shrinkage and cracking of the sediment, and the subsequent infilling of the cracks.

Worbarrow Bay.

The Purbeck Beds are well shown in the cliff at the south-eastern sides of Worbarrow (or Worthbarrow) Bay, on both sides of the headland known as Worbarrow Knob or Tout, and in the little cove to the east of it. The details of the strata have been recorded by Bristow (Vertical Sections, Sheet 22, No. 2).

Above the somewhat argillaceous and compact limestones that form the upper part of the Portland Stone, we find the Hard and Soft Caps of the Lower Purbeck Beds, which appear closely united with the Portland Stone. The Hard Cap exhibits cavernous weathering, and is separated from the Soft Cap by an impersistent Dirt Bed. Near the base of the "Hard Cockle Beds" there is a band of irregular rubbly limestone, with white cherty seams here and there; and white cherty particles occur in a higher band of "Hard slaty sandy limestone." Above this is a band of "Hard limestone," which forms a ledge in the Cove, it exhibits false-bedding, and contains thin layers of sandy limestone, and pale cherty seams. Near the base of the "Soft Cockle Beds," the "Hard bituminous limestone" exhibits ripple-marks; higher up the bands of brecciated limestone-conglomerate are well shown.

A "Dirt Bed" occurs at the base of the Middle Purbeck division, but no Mammals have been found in it. The hard limestones with bands of black chert form a series of prominent beds above. The Cinder Bed is well shown, and the Scallop Beds stand out as a band of hard limestone, beneath the signal staff on the Knob.

In the Upper Purbeck Beds we find the Unio Beds on the northern side of the Knob, forming a face of rock, containing much green matter, with *Unio*, and dipping northwards at an angle of about 40°. The Upper Cypris Clays and Shales form the soft beds under the Coast-guard station. Near this point there is a ravine, which occurs at about the junction of the Purbeck and Wealden Beds, though the strata are much obscured : but on the north side we find the coloured clays and sands of the newer division.

Mupe Bay.

The most complete section of the Purbeck Beds is seen in the cliffs and rocks of Mupe (Mewps) Bay and Bacon Hole, between Worbarrow Bay and Lulworth Cove.

The Portland Beds are shown at the base of the cliffs, and in the islets known as Arish Rock, Slip Rock, and Wreckneck Rock, where the highly inclined strata are surmounted conform-ably by Lower Purbeck Beds. The strata may most readily be examined in Bacon Hole. A cave has been formed at the junction of the Portland and Purbeck Beds, by removal of slabs of the Hard Cap. The lower Purbeck beds here include brecciated and banded layers, and the Hard Cap is separated from the Soft Cap by a Dirt Bed containing pebbles of limestone. The Soft Cap is much broken in places, and indeed merges up-wards into the "Broken Bands." These again are overlaid by the Cypris Freestone, also much shattered and broken in places. (See Fig. 127, p. 249.) The upper portions of the Hard Cockle Beds stand out in prominent bands, overlaid by the Soft Cockle Beds, and higher up the Cherty limestone again forms a hard band, and, together with the overlying Cinder Bed, runs out in ledges on the shore. Higher up we may trace the Chief Beef Beds, the Upper Broken Shell Limestone, and the Unio Beds, which also form ledges running eastwards along the shore.

The junction with the Wealden Beds is shown in the cliff to the north of the Flag Staff and again a little further east, in Bacon Hole.

The Paludina-clays which form the top of the Purbeck Beds, consist of purple, green, blue, and grey marls full of *Paludina*. The top layer of Paludina Marble contains also *Unio*. The layer of Marble below contains veins of beef in the stone, and beneath are ferruginous layers with *Unio.* One bed of red and green Marble, of which I saw a loose block, was 1 ft. 10 ins. in thickness.

Fig. 130.

Section of Cliff at Bacon Hole, Mupe Bay.

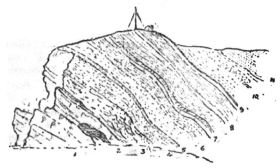

11. Wealden Beds.
10. Shales, &c., with Purbeck Marble.   } Upper
9. Unio-beds.   } Purbeck.
     Broken-shell Limestone.
8. Corbula Beds, &c.; Shales and limestones.   } Middle
7. Cinder Bed.   } Purbeck.
     Hard limestone ; flint-bed.
6. Marls, with conspicuous band of harder white marl.   }
5. Sandy and honeycombed bands ; top of Hard Cockle Beds.   }
4. Cypris Freestone, &c., the lower beds much shattered   } Lower
     in places.   } Purbeck.
3. Broken Beds, merging downwards into Soft Cap.   }
2. Caps with intermediate Dirt-bed.   }
1. Portland Stone ; with *Ammonites giganteus, Trigonia gibbosa, Cardium dissimile,* and *Pecten lamellosus.*

The coloured clays are naturally like the red and variegated Wealden Clays ; but unlike the mass of those clays, the beds here grouped as Upper Purbeck, contain *Paludina* in abundance

The lowest beds of the Wealden series, dipping at about $40°$, comprise red and grey sandy and laminated clay, white sand and yellow carbonaceous sandstone. Small crystals of selenite are abundant : rusty springs are thrown out, at or near the base.

The following section was noted in 1893, in company with Mr. Strahan :—

			FT.	IN.
Wealden Beds.	{ White sands, &c.			
	{ Laminated clay and sands	- -	6	0
Purbeck Beds.	{ Band of clay-ironstone nodules with *Cypridea punctata,* and *C. tuberculata* var. *adjuncta* -	- - -	0	1
	{ Shales, &c.	- - -	17	0
	{ Greenish shelly limestone	- -	0	4
	{ Shales with lignite	- -	4	0
	{ Purbeck marble -	- - -	0	9

The Ostracods were kindly identified by Prof. T. R. Jones.

Lulworth Cove.

In the cliffs west of Mupe Bay and a short distance east of Lulworth Cove we find the first clear traces of the Great Dirt Bed of Portland, and the stools of trees, although evidences of it as before noted occur further east. (See p. 241.)

Here near Lulworth, huge "burrs" are shown in the Dirt Bed where it is exposed on the highly inclined surfaces of strata projecting from the cliffs. (See Fig. 124, p. 242.) This Dirt Bed occurs between the Hard and Soft Caps and these are overlaid by the Broken Bands. Another Dirt Bed occurs at the base of the Hard Cap and directly on the Portland Stone. Here as near Durlston Head, these lower Purbeck Beds, form conspicuous irregular and undulating bands of rock above the Portland Stone, and in places the beds split up so as to form three bands of Cap limestone.

The Purbeck Beds are exhibited on both sides of Lulworth Cove. On the east the "Broken Beds" exhibit ruptures that do not affect the overlying strata. Marked ledges are formed by the "Hard Cockle Beds," and one of the upper layers, a "Hard pale grey Limestone" is somewhat cherty, and weathers in a cavernous form. The Cinder Bed forms ledges, and the lower portion of it contains Fish-remains. Here as in Stair Cove the lower portion contains few specimens of *Ostrea distorta*. The Scallop Bed, hard-shell limestones with bivalves and broken shells, and the Chief Beef Beds, are well shown in ledges.

Towards the northern end of the section of Purbeck Beds, the higher strata are much folded, as well as slightly displaced by slips, and a false appearance of unconformity is presented.

Where the Cinder Bed descends to the shore the mass of it and other beds are slightly inverted, and there is evidence of disturbance by a fault having a very low hade, a "thrust-plane" in fact which causes a slight repetition of the Cinder Bed.

Lulworth Cove has been formed by the erosion of the soft Wealden Beds, after the sea had broken through the barrier of Portland Rocks.

At Stair Hole or Cove there are two or three caves or tunnels through the Portland Rocks, the sea perhaps having enlarged fissures in these rocks. Overlying and overhanging masses of Purbeck Caps and Broken Beds occur in places on the eastern side of the barrier that contains these caves.

As the Upper Purbeck beds are denuded, so the sea may in process of time come upon the soft Wealden Beds and then uniting with Lulworth Cove, a bay like that of Mupe and Worbarrow will be formed, with islets of Portland rocks capped by Lower Purbeck Beds.

The well-known section at Stair Hole shows how much the strata are disturbed. Here the beds are much broken and cemented by veins of calc-spar. The shales squeezed up in the central part of the Cove open out towards their outcrop at the edge of the cliff, and are in some instances slightly inverted. This is also seen to be the case on the western side of Lulworth Cove.

Fig. 131.

Diagram-section of part of Stair Hole, Lulworth, Dorset.

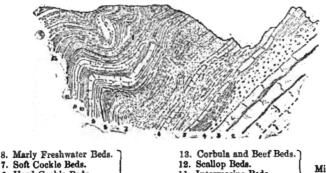

8. Marly Freshwater Beds.	⎫	13. Corbula and Beef Beds.	⎫
7. Soft Cockle Beds.	⎪	12. Scallop Beds.	⎪ Middle
6. Hard Cockle Beds.	⎪	11. Intermarine Beds.	⎬ Purbeck.
5. Cypris Freestones.	⎬ Lower	10. Cinder Bed.	⎪
4. Broken Beds.	⎪ Purbeck.	9. Cherty Freshwater Beds.	⎭
3. Soft Cap.	⎪		
2. Hard Cap.	⎪		
1. Portland Stone.	⎭		

There are many slight disturbances and faults, on the west end of Lulworth ridge, and by Durdle Door or Barn Door. Here minor repetitions of beds are caused by what may be called step-faults, that traverse the beds in a very oblique direction, and sometimes almost at right angles to the highly inclined strata : so that at first I was disposed to think that faulting had taken place prior to the great tilting of the beds.

On the east side of the Durdle promontory the Purbeck Beds appear to be faulted against the Wealden Strata, and in no case west of Mupe Bay can we see the junction of these beds. As Forbes remarked in 1849 "some of the sections are imperfect, the top-beds being squeezed out in a line of fault."[*] The character of these disturbances has been investigated in detail by Mr. Strahan, and they will be illustrated in a Memoir that deals more particularly with the Isle of Purbeck.

Ringstead Bay, Upway, and Portisham.

The Lower Purbeck Beds are exposed on the eastern side of Ringstead Bay and to the west of the high Chalk cliffs of White Nore. The Cretaceous Beds as pointed out by the Rev. O. Fisher rest unconformably on the inclined Purbeck and Portland Beds, and Mr. A. Strahan informs me that there is no fault along the line of junction of the Cretaceous and Oolitic Strata as marked on the Geological Survey Map: it is simply a case of the unconformable overlap (or overstep) of the Cretaceous on to the Purbeck and Portland Strata, and on to the Kimeridge Clay.

---

[*] Geikie, Memoir of E. Forbes, p. 466.

The section of Purbeck Beds, which dip N. 5° E at an angle of about 37°, was as follows :—

		Ft.	In.
Lower Purbeck Beds.	Creamy limestones and shales.		
	Hard marls, sandy and granular oolitic beds, with one or two chert bands: *Cypris purbeckensis.*		
	Fissile limestones (Purbeck " slates ") 5 0 to	6	0
	Hard creamy marl - - -	1	2
	Thin irregular dirt layer -	0	1
	Hard irregular bituminous limestone ·	2	6
	Dirt bed, with irregular calcareous seams - - - - -	0	1
	Irregular decomposed sandy limestone	1	0
	Laminated clay and dirt with fish-remains - - - -	0	2
	Pale compact bituminous and cherty limestone with casts of *Cyrena;* sandy at top - - - -	1	5
Portland Stone	Irregular beds of " Roach," chalky limestone and oolite.		

Mr. Fisher has observed that the junction of the Purbeck and Portland Beds is "often so intimate, that the same block of stone, as it comes out of the quarry, contains the marine fossils of the Portland series in its mass, and about 4 inches of freshwater limestone on its upper face full of Cyprides and *Cyclas* (*Cyrena*); "[*] a fact that coincides with what has been observed in the Vale of Wardour (p. 269).

At Ridgeway Hill, north of Upway, the beds have been opened up in several quarries, and also in the cuttings of the Great Western Railway on the south side of the Bincombe Down or Ridgeway tunnel. The railway-cuttings were described by Mr C. H. Weston and Captain L. L. B. Ibbetson,[†] who gave detailed measurements of the strata, but included with the Wealden division some beds that were subsequently grouped with the Purbeck Beds by Messrs. Bristow and Fisher, while they included with the Portland Beds strata shown by Forbes to be Purbeck.[‡] (See Fig. 132, p. 260.)

The lowest beds are seen above the Portland Stone in the quarries east of Upway, and they have been previously noted (p. 195).

The building-stones which belong to the Lower Purbeck Beds are opened up in quarries on the east side of the old high road, north of Upway. There the " Cypris Freestones " are worked, and they yield *Cypris purbeckensis.* Fish-remains occur also in the "Top" or "Little Rock." The best building-stone is obtained from the "Hard Rock," a shell-limestone composed largely of bivalves. The Soft Cockle Beds are also quarried, and a " Hard marly Rock," known as the "List Bed" is used for " Blue

---

[*] Trans. Cambridge Phil. Soc., vol. ix. p. 555.
[†] Quart. Journ. Geol. Soc., vol. viii. p. 116; see also Weston, vol. iv. pp. 252-254.
[‡] Geikie, Memoir of Edward Forbes, p. 463; see also Fisher, Trans. Cambridge Phil. Soc., vol. ix. p. 574; and Vertical Sections, Geol. Survey, sheet 22, No. 4.

Fig. 132.

Section along the Great Western and South-Western Junction Railway on the south side of Ridgeway Hill, Dorset.

(L. L. B. Ibbetson and C. H. Weston.)   Scale, 400 feet to 1 inch.

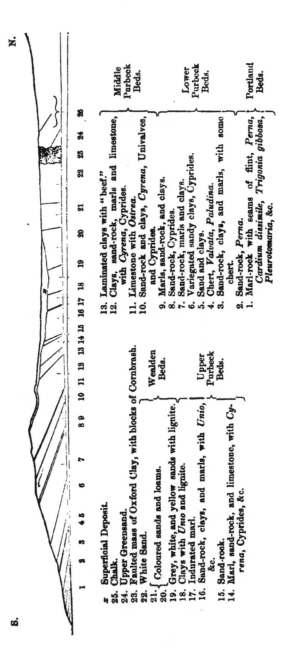

s   Superficial Deposit.
25.   Chalk.
24.   Upper Greensand.
23.   Faulted mass of Oxford Clay, with blocks of Cornbrash.
22.   White Sand.
21.   Coloured sands and loams.   } Wealden
20.   Grey, white, and yellow sands with lignite. } Beds.
19.   Clays with *Unio* and lignite.
18.   Indurated marl.
17.   Sand-rock, clays, and marls, with *Unio*, } Upper
       &c.   } Purbeck
16.   Sand-rock.   } Beds.
15.   Marl, sand-rock, and limestone, with *Cy-
       rena, Cyprides*, &c.

13.   Laminated clays with "beef."
12.   Clays, sand-rock, marls and limestone, } Middle
       with *Cyrena, Cyprides*.   } Purbeck
11.   Limestone with *Ostrea*.   } Beds.
10.   Sand-rock and clays, *Cyrena,* Univalves,
       and Cyprides.
9.   Marls, sand-rock, and clays.
8.   Sand-rock, Cyprides.
7.   Sand-rock, marls and clays.
6.   Variegated sandy clays, Cyprides.   } Lower
5.   Sand and clays.   } Purbeck
4.   Chert, *Valvata, Paludina.*   } Beds.
3.   Sand-rock, clays, and marls, with some
       chert.
2.   Sand-rock, *Perna.*   } Portland
1.   Marl-rock with seams of flint, *Perna,*   } Beds.
       Cardium dissimile, Trigonia gibbosa,
       Pleurotomaria, &c.

Lias lime ; " higher beds of marly and granular oolitic limestone, are used for inside work in building, and also for lime-burning.

The Middle Purbeck Beds have been exposed in the railway-cutting south of the short tunnel.

The Cherty Freshwater Beds, which include a band of irregular gritty limestone, have yielded Saurian bones, many Mollusca, Cyprides, and seed-vessels of *Chara*. The Cinder Bed and overlying strata may be traced as we proceed northwards. The Upper Purbeck Beds and the junction with the Wealden strata are for the most part obscured. Layers with *Unio, Paludina*, &c., have been observed, including a representative of the Purbeck Marble.

It seems probable that the Purbeck and Wealden strata were faulted against the Oxford Clay and denuded prior to the Upper Cretaceous overlap. (See p. 20.)

The Purbeck Beds at Portisham occupy a somewhat undulating trough, faulted on the north against the Chalk. Some of the granular oolitic beds have been extensively quarried for building-stone.

The following Lower Purbeck Beds were exposed in a quarry on the hill west of the Farm-buildings, and at the Lime-works :—*

		Ft.	In.
	Brown soil with water-worn layer of limestone - - - -	2	6
Soft Cockle Beds.	Laminated clays, marls, and yellowish sands, with hard bands - - -	5	0
Hard Cockle Beds.	Pale earthy limestone, with *Cyprides* -	0	3
	Laminated marl - - - -	0	2
	Compact banded and somewhat crystalline limestone - - -	1	1
	Shelly beds (not well shown) - - -	2	3
	Hard pale shell-limestone, with *Corbula*: like the " Hard Rock " of Upway - - - 2 0 to	4	0
Lower Insect Beds.	Sands with streaks of clay - - -	1	9
	Pale compact limestone, marls and clays - - - - -	2	3
	White banded marly limestone - - -	2	3
	Soft shales and pale limestone - - -	4	0
Cypris Freestones.	Granular limestones, passing down into variable limestones that for the most part split up into thin tiles; with *Cyprides* - - - -	8	0
	Earthy and granular limestones, wavy-bedded; occasional slabs 3 ft. and more square, obtained, and used for tombstones - - -	4	2
	Fissile sandy marlstone, passing up into granular oolitic limestone, like bed at Ringstead Bay - - -	0	6
	Tough grey banded limestone, with irregular bedding and sparry joints, as if disturbed and re-cemented— the top portion granular and oolitic. Cavernous weathering shown along lines of bedding - - 4 0 to	5	0

* The subdivisions are noted in accordance with the grouping of the strata at Ridgeway, by H. W. Bristow and the Rev. O. Fisher.

		Ft.	In.
Strata equivalent to the " Broken Beds " of other localities ; and to the Hard and Soft Caps.	Rubbly marl and clay with irregular band of tough grey limestone and decomposed chert - 1 0 to	1	6
	Irregular earthy and granular limestone - - - - -	1	6
	Soft white marly limestone - -	1	0
	Irregular sandy marls and dark shales, about - - - - -	1	0
	Soft chalky limestones - - -	1	10
	Hard and soft limestones and marls, irregular and fissile beds, with clayey seams near top. Fish-scales -	6	0
	Thin irregular dirt layer with lignite.		
	Impure marls with seams of chert -	3	0
	Hard beds of variable limestone, some massive and irregular, with occasional chert - - - -	3	3
	Irregular fissile banded marls, hard and soft beds - - - -	2	0
	Impure clays and marls with two black earthy bands with lignite and stumps of trees (disturbed in places) - -	1	7
	Thin and variable beds of compact and earthy limestone, fissile in places 1 8 to	2	8
Portland Beds.	Hard white shelly and oolitic limestone with fragments of *Perna:* merging in places into more compact limestone, and passing down into -	1	6
	Hard white chalky limestone with occasional nodules of chert - -	3	6
	White chalky limestone, with bands and nodules of chert.		

A well by the farm-buildings was sunk to a depth of 130 feet for water, and passed through about 64 feet of Purbeck Strata.

The granular Purbeck Beds were opened up in a small (now disused) quarry, north-east of the farm.

On the lower Dirt-bed an immense cylindrical and ribbed mass of rock was found about the year 1888. It was 15 feet long, tapering, and about 3 feet in diameter : in transverse section the form was oval, with a cavity containing fragments of silicified wood. This was at the time announced as a fossil Elephant, in the local newspaper, but the " trunk " of which portions were preserved in this huge calcareo-siliceous earthy envelope, belonged to one of the Purbeck trees. (See Fig. 133.)

Isle of Portland.

The Lower Purbeck Beds cover the greater part of Portland Isle, and they may be seen in the various stone-quarries, and in the cliff-sections south of the Verne, and as far as Portland Bill. Traces of Middle Purbeck strata have been recognized only in fragments obtained from superficial deposits on the northern slopes of Portland.*

---

* W. Gray, Proc. Geol. Assoc., vol. i. p. 138.

FIG. 133.

Silicified mass of Purbeck Wood, Portisham.

The Lower Purbeck Beds attain a thickness of about 40 feet or even more in places; the greatest thickness seen is in the cliff section south of Black Nore, where nearly 30 feet of hard white marls and grey marly clays overlie the Hard Slatt and other beds exposed in the quarries. The general succession has been previously noted (p. 198), and details of the beds seen in a quarry at Kingbarrow, have also been given (p. 200). It remains, therefore, only to note more generally the beds that occur.

Immediately overlying the Portland Roach we find an impersistent layer of carbonaceous clay or *"Dirt,"* but no plant-remains have been obtained from it. It was first noticed in 1830 by Prof. Henslow.[*] Prof. Blake has spoken of it as a constant band, and as containing fragments of chert derived from the Portland Beds. Such evidences of erosion are however local, and although the Lower Purbeck Beds are somewhat irregular and undulating in appearance, there is a general conformity between the stratification of Portland and Purbeck Beds, and we find no evidence of any general upheaval of the former strata, previous to the deposition of the Cap-beds.

The Purbeck CAP BEDS are hard dense grey limestones, often cavernous and somewhat botryoidal in character. They are extremely variable in thickness: the Skull (or School) Cap, varies from 1 ft. 6 ins. to 9 feet, and contains occasional cherty nodules. The overlying Top Cap, which includes sometimes a "Top rising" layer is from 2 to 10 feet thick. Both beds are much shattered in places.

---

* See Buckland and De la Beche, Trans. Geol. Soc., ser. 2, vol. iv. p. 13.

It has been observed by Fitton and others that the Portland Caps were partially of subäerial origin, for they contain minute ramifying cavities lined with botryoidal carbonate of lime, such as characterize beds of calcareous tufa, like the travertine of Italy.* These features, as noted by H. W. Bristow, are met with in the Tertiary limestones of Sconce and Bembridge in the Isle of Wight. Hence it is likely, as Prof. Blake remarks, that the Caps have been "derived from the denudation of the Portland rocks which had elsewhere emerged from the sea at an earlier date, though here the emergence had been so recent that very little atmospheric action had occurred."†

A second *Dirt* layer, that is found between the two beds of Cap, was also noticed by Prof. Henslow. This is a loamy and carbonaceous bed that sometimes contains small fragments of stone. Fitton in 1835 noticed the occurrence of Cycadeæ of large dimensions in the bed between the two layers of Cap. They "were in the upright position, and apparently in the places where they originally grew."‡

The term "chaff-holes" is applied by workmen to the hollows or cavities which extend from this lower dirt-bed into the Top Cap, and sometimes through it : they contain remains of decayed silicified wood. At this same horizon the so-called "Bird's Nests" or "Crow's Nests" are more commonly found : these are the silicified Cycadean stools, of which the finer examples are those of *Mantellia nidiformis (Cycadeoidea megalophylla).*§ (See also Fig. 123, p. 240.) They are found from 8 ins. to 1 ft. in diameter, and from 5 to 10 ins. in height; but Mantell says the largest specimens are about 2 feet high and 3 feet in circumference.

Judging from the general stratigraphical aspect of the beds, as seen in the cliffs south of Black Nore and other places, the Cap beds might be taken to belong rather to the Portland than to the Purbeck series, for the Great Dirt-bed appears to mark the more important plane of demarcation. Buckland and De la Beche,‖ took it as the base of the Purbeck Beds, although Webster had previously suggested that the Cap has more analogy with the lower part of the Purbeck Beds than with the Oolite below; while Fitton,¶ finding *Cyprides* in all the beds above the Portland "roach," took that bed as the top of the Portland Stone, and this division is now adopted. The observations of Forbes tended to confirm this, for writing to Ramsay in 1849, he remarks "When you say that the line of division between the Purbecks

* Fitton, Trans. Geol. Soc., ser. 2, vol. iv. pp. 222, 224.
† Quart. Journ. Geol. Soc., vol. xxxvi. p. 190; see also Andrews and Jukes-Browne, *Ibid.*, vol. l. p. 51.
‡ Proc. Geol. Soc., vol. ii. pp. 185, 186 ; Trans. Geol. Soc., ser. 2, vol. iv. pp. 218, 223.
§ Buckland, Trans. Geol. Soc., ser. 2, vol. ii. Pls. 47, 48; Mantell, Geol. Excursions round the I. of Wight, &c., Ed. 3, p. 289 ; see also Carruthers, Trans. Linn. Soc., vol. xxvi. Pls. 61, 63.
‖ Trans. Geol. Soc., ser. 2, vol. iv. p. 15; Webster, *Ibid.*, vol. ii. p. 44 ; and Buckland, Bridgewater Treatise, Plate 57.
¶ Trans. Geol. Soc., ser. 2, vol. iv. p. 219.

and Portlands is exceedingly well marked in the Isle of Portland, you speak of the supposed line (I thought the same here [Lulworth] for the first two months), as marked by the dirt-beds and physical features. One of my points is proving how far *exactly* the Purbecks extend downwards, and the result is, that all the so-called "*caps*" are Purbeck without a question, the freshwater beds with Purbeck fossils lying directly upon the Portland beds with marine fossils, though strangely alike in mineral character and consolidation."[*]

Below the Light House the Purbeck Caps become thinner and appear to merge into one bed. Towards the Bill both beds are seen, the Top Cap being much paler in colour than the underlying Skull Cap. It was near the Bill that remains of *Valvata* and *Cyrena* were somewhat doubtfully recorded from the Cap Beds, on the authority of Edward Forbes.[†] Mr. A. M. Wallis has obtained Cyprides and some Fish-remains in a seam of soft stone at the base of the Top Cap. (See p. 280.)

Above the Top Cap we come to the celebrated GREAT DIRT BED (or Black Dirt) of Portland. This is an irregular layer of clay and carbonaceous earth, with numerous partially rounded stones ; these are composed of more or less oolitic limestone, and appear to have been derived chiefly from Portland rocks, although I believe some pieces may be derived from the Cap beds. Mr. Horace T. Brown indeed suggested to me in 1884 that superficial weathering of the underlying beds in Purbeck times, might account for the stones. Analyses which he has made of some of the fragments and of the Top Cap support this view. No stones that could not have been obtained from local strata have been noticed. This Dirt Bed is very irregular, seldom exceeding 1 foot in thickness, and almost thinning out in places.

The BURR BED (Burr of the Aish or Soft Burr), above the Dirt Bed, although practically forming part of it, is a calcareo-siliceous layer from 9 inches to 1 foot thick, having an irregular mammillated appearance, well shown in a drawing made in 1832 by Henslow.[‡] Remains of silicified trees are found in hollows of these mounds, which are sometimes of circular form and as much as 6 feet in diameter, and sometimes oval and as much as 9 feet in length. Occasionally these oval masses are indented like the figure 8 with one or two rings of corrugated masses of earthy limestone. (See Figs. 124 and 133, pp. 242 and 263.)

Gentle undulations in the overlying Purbeck Beds that affect the beds to a height of 12 or 15 feet, appear to correspond with these mounds, but the beds are but rarely " broken " as in the Isle of Purbeck, excepting where they occur above the great fissures or "gullies " that traverse the Portland stone.

Six or seven tree-stumps are sometimes to be seen on the bared surface of the Great Dirt Bed over an area of about

---

* Geikie, Memoir of E. Forbes, p. 465.
† See Weston, Quart. Journ. Geol. Soc., vol. viii. pp. 116, 118.
‡ Buckland, Bridgewater Treatise, Plate 57.

400 square yards. They rise from a few inches to about 3 feet, and rarely to as much as 6 feet, occasionally penetrating the over- lying Aish Bed. Trunks of Conifers are sometimes found lying prostrate, and partially imbedded in the Dirt Bed. Specimens from 2 to 4 feet in diameter have been obtained, and as much as 23 feet in length. One example, 18 feet in length, is placed outside a house in Fortune's Well. The "Bird's Nests" (Cycads) are also occasionally found at this horizon. All the plant-remains are more or less silicified, but they contain coatings of carbonate of lime, and some ferruginous matter. Roots have never been observed to penetrate the Top Cap.

What is called the cap of the Burr is a hard brown cherty limestone, that decomposes in places into a sandy calcareous earth. As remarked by Fitton, the Burr Bed "has obviously been deposited around the lower part of the petrified trunks."[*]

The silicification of the beds and plant-remains has been attributed to the action of siliceous springs.[†] Webster remarked: " The woody part is siliceous, and the longitudinal vessels are filled by and surrounded with radiated quartz; numerous veins of chalcedony and quartz also pass through these stems, but always following the direction of the concentric and radial structure."[‡]

The AISH (or Ash) is a soft earthy and sometimes fissile limestone, closely connected with the underlying bed, and having often a banded appearance. Mr. Lydekker has obtained remains of *Cimoliosaurus* from this bed.[§]

The BACON TIER consists of earthy limestone with layers of sand, presenting sometimes the fissile character of the " slate" above.

The HARD SLATT (or slate) is a fissile limestone from which Fitton obtained Cyprides and a small *Modiola*. Fish-remains have also been found in this bed by W. Gray. The Cyprides include *Candona ansata, C. bononiensis,* and *Cypris purbeckensis.*[‖]

Occasional Dirt-layers are found at the base of the Bacon Tier and of the Hard Slatt.

The higher strata include bands of clay and fissile limestone, usually presenting a rubbly appearance. In these upper beds at Steward's quarry, Kingbarrow, there were found several Bee- hive shaped pits, 7 to 9 feet deep, built of thin flaggy stone and covered with slabs. One of the pits extended through the Hard Slatt. Attention was called to them by Mr. A. M. Wallis. They had evidently been used for storing grain.[¶]

* Trans. Geol. Soc., ser. 2, vol. iv. p. 220.
† On this subject see W. H. Weed, 9th Annual Report, U.S. Geol. Survey, p. 619; H. Graf zu Solms-Laubach, Fossil Botany (English Translation, Oxford), 1891, p. 29; and H. G. Lyons, Quart. Journ. Geol. Soc., vol. l. p. 545.
‡ Trans. Geol. Soc., ser. 2, vol. ii. p. 42.
§ Quart. Journ. Geol. Soc., vol. xlvi. p. 47. The specimen was described as from the Cinder Bed, which is not known in Portland.
‖ T. R. Jones, Quart. Journ. Geol. Soc., vol. xli. p. 325.
¶ Damon, Geol. Weymouth, 1884, p. 165; and Holmes, Proc. Geol. Assoc., vol. viii. p. 404.

From Portisham and Upway we find no exposures of Purbeck Beds until we come to the Vale of Wardour, the rocks probably stretching beneath the Cretaceous covering in the direction of Dorchester and Blandford. It must be remembered however that the concealed outcrop.is probably shifted a few miles eastward by the Ridgeway fault.

Vale of Wardour.

In this district the Purbeck Beds are well shown in many quarries and cuttings, but we have no continuous section to enable us to estimate the full thickness with accuracy. The beds have been described by Fitton, Brodie, O. Fisher, J. F. Blake, Hudleston, W. R. Andrews and A. J. Jukes-Browne.

The general succession of strata appears to be as follows:—

			Ft.	In.
Purbeck Beds.	Middle. Upper.	Shell-marls, clays and marls, with "beef," and sandy layers with bands of calcareous sandstone -	20	0
		Marls and sandy rocks, with "beef," limestone with *Archœoniscus*, Cinder Bed, and other calcareous bands -	12	0
	Lower.	Limestones and shales ("Lias," &c.) -	18	0
		Limestone, clay, and oolitic beds -	15	0
		Fissile limestones, tufaceous beds and dirt-beds, with chert nodules, &c. -	20	0
			85	0

It is difficult to mark any divisional planes to correspond with the Lower, Middle, and Upper Purbeck divisions of the Dorset Coast. The occurrence of *Ostrea distorta* was noticed by Fitton, but the presence of a bed equivalent to the Cinder Bed of Durlston Bay, was first recognized by the Rev. O. Fisher.[*] The presence of Upper Purbeck Beds was generally questioned until, in 1894, Messrs. Andrews and Jukes-Browne described beds that belonged to this division;[†] but they included as Purbeck, strata that I was led to regard as Wealden, during my examination of the sections in 1885. My observations then led me believe that the Wealden Beds rested quite conformably on Upper Purbeck Beds. I have been able to confirm this view by a subsequent examination of the ground in 1894, in company with Mr. A. Strahan.

The lowest Purbeck Beds are exposed in the great quarries of the Chilmark ravine.[‡] A pit on the north-east side showed the following beds:—

[*] Quart. Journ. Geol. Soc., vol. x. p. 477.
[†] *Ibid.*, vol. l. p. 59.
[‡] See also J. F. Blake, Quart. Journ. Geol. Soc., vol. xxxvi. p. 200; and Andrews and Jukes-Browne, *Ibid.*, vol. l. p. 48.

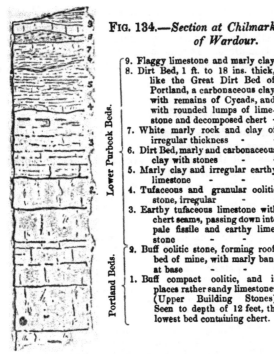

FIG. 134.—*Section at Chilmark, in the Vale of Wardour.*

Lower Purbeck Beds.

9. Flaggy limestone and marly clay
8. Dirt Bed, 1 ft. to 18 ins. thick, like the Great Dirt Bed of Portland, a carbonaceous clay with remains of Cycads, and with rounded lumps of limestone and decomposed chert -
7. White marly rock and clay of irregular thickness -
6. Dirt Bed, marly and carbonaceous clay with stones - -
5. Marly clay and irregular earthy limestone - - -
4. Tufaceous and granular oolitic stone, irregular - -

         FT. IN.

    8 0 to 10 0

3. Earthy tufaceous limestone with chert seams, passing down into pale fissile and earthy limestone - - -

Portland Beds.

2. Buff oolitic stone, forming roof-bed of mine, with marly band at base - - -
1. Buff compact oolitic, and in places rather sandy limestones. (Upper Building Stones). Seen to depth of 12 feet, the lowest bed containing chert.

         3 0

Messrs. Andrews and Jukes-Browne record *Mantellia (Cycadeoidea) microphylla;* and in the layer above noticed as the upper Dirt Bed, they noticed an upright and rooted stump of a tree, the stem standing about 6 feet high.

The Lower Purbeck Beds were observed by Fitton in some of the old quarries at Upper Chicksgrove. The details of the strata vary considerably from place to place, even in one quarry, as at Wockley. There the general section which I noted was as follows :—

		FT.	IN.
	14. Loamy soil - - 0 6 to	1	0
	13. Fissile limestones, some oolitic; and marls and clays with layers of sandy limestone and sand: about - - - -	8	0
	12. Hard marly limestone - -	0	8
Lower	11. Banded limestones and marls -	2	0
Purbeck	10. Dark clays - - - -	1	0
Beds.	9. Sandy limestone - - -	0	9
	8. Earthy marl with irregular (? concretionary) masses of stone -	3	0
	7. Fissile limestones - -2 0 to	3	0
	6. Dark shaly clay, much squeezed up in places.		
	5. Compact limestones - -	2	0
Upper Portland Beds. (See p. 204.)	4. Roachy bed with chert at top. 3. Chalky limestones. 2. Sandy limestone. 1. Shelly and sandy limestone.		

The Purbeck Beds resemble in some respects the lower beds of Lulworth Cove and Worbarrow. They undulate and are much broken up in places. Bed No. 6, which was best shown in the northern part of the pit, may represent one of the Dirt Beds, while Bed 8 reminded me of the Soft Burr and Bacon Tier of Portland.

Messrs. Andrews and Jukes-Browne* figure a curious disturbance in the beds, but I saw nothing so striking as they have represented (in a slightly diagrammatic manner), either in 1885, or on a subsequent visit, in 1894, in company with Mr. Strahan.

FIG. 135.

Section at Wockley, near Tisbury.

In the Museum of Practical Geology there is a " Large Block of Limestone, showing the junction between the Portland and Purbeck formations from Oakley [Wockley] Quarry, near Tisbury." It was thus described by H. W. Bristow: " In the quarry from which the specimen was taken, the uppermost bed of Portland Stone is harder than the chalky limestone upon which it reposes, and is crowded with marine shells common to the formation, viz., *Trigonia, Cardium dissimile, Ostrea,* &c. Immediately above this stratum is a bed of hard, grey, bituminous limestone, the upper foot of which is fissile and used for flagstones. In the specimen, as in the quarry, the exact line of junction between the shelly bed and the fissile limestone is scarcely distinguishable to the eye, but when broken by a heavy blow the Portland Stone and the Purbeck split off from each other at the junction, along a smooth and even surface. The line of demarcation between the two strata is crowded with fish."†

In their section at Wockley Messrs. Andrews and Jukes-Browne group with the Portland Stone the hard flaggy limestone that I have included at the base of the Purbeck Beds. They state that the flaggy and shelly portions " are firmly welded together, and would yield a slab like that at the Museum of Practical Geology, in which Portland shells are visible in the lower and Cyprids in the upper part, but these Cyprids are not freshwater species, being in fact *Candona ansata* and *C. bononiensis* (which are estuarine forms). From the flaggy portion two species of fish have been obtained (*Ophiopsis breviceps* and *O. penicillatus*) and also a large species of *Archæoniscus.*" Furthermore, they obtained from the beds above, *Candona ansata, Cypridea, Cypris, Cardium,* and *Corbula alata.*‡

* Quart. Journ. Geol. Soc., vol. l. pp. 49, 52.
† Catalogue of Rock Specimens, ed. 3, p. 139.
‡ Quart. Journ. Geol. Soc., vol. l. p. 51. See also Fitton, Trans. Geol. Soc., ser. 2, vol. iv. p. 253.

To be consistent, however, we must continue to regard the old plane of division as the best, and going again to the district with Mr. Strahan I found no difficulty in determining this junction in the quarries near Tisbury and Chilmark. The evidence of the fossils shows that the change of conditions was not so marked as in other localities where freshwater beds overlie those of a marine character. Here we have estuarine beds overlying marine beds, while in other places in the Vale of Wardour the "Cyrena-beds" of the Upper Portland Beds have been regarded as of a semi-estuarine character. (See p. 181.)

An interesting section to the south-east of Ridge has been noted by Messrs. Andrews and Jukes-Browne as follows :—*

		Ft.	In.
	Dark brown soil -	1	0
Lower Purbeck Beds.	Weathered marlstone or "lias"	1	0
	Buff-coloured marl, with seams of grey clay -	0	6
	Soft fine-grained, marly oolite, a mixture of oolitic particles with triturated shells, cyprids, &c.; with thin layers of harder compact marlstone in the lower part -	2	3
	Soft yellowish calcareous oolitic sand -	0	9
	Very hard limestone, consisting of shelly layers alternating with seams of compact marlstone -	0	10
	Soft marl with yellowish oolitic stone -	3	2
	Hard grey shelly limestone -	2	4
	Oolitic stone, with layers of marl -	3	3
	Soft calcareous stone passing down into hard limestone with pseudomorphous crystals of rock-salt: *Corbula alata, Perna, Cardium, Leda, Serpula*, &c. -	3	3
	Grey laminated marl -	1	0
	Buff marlstone -	3	0
		22	4

To the north of this quarry and just east and also south of the letter e of Ridge on the 6-inch map—by the road-side—two large trees that had been blown down, had torn up blocks of Purbeck stone—shelly and sandy limestone. These were observed by Mr. Strahan and myself, and served to indicate that hereabouts the Gault rests directly on the Purbeck Beds.

To the south-west of Teffont Evias Church there is a long excavation in the Purbeck Stone-beds which present a general resemblance to Lower Lias limestones, and many beds are known to the quarrymen as "Lias." The best section is near the Lime-kiln at the northern end of the workings; this I noted as follows :—

---

* Quart. Journ. Geol. Soc., vol. l. p. 52.

		FT.	IN.
	Brashy soil, brown sandy loam.		
Middle Purbeck Beds.	Cherty layer with many bivalves [Cyclas].		
	CINDER BED : hard greyish brown limestone, much broken up; Ostrea distorta, [Trigonia gibbosa, and spine of Hemicidaris] - - -	1	0
	Clay and rubble - - - -	0	8
	Hard grey limestone with dendritic markings - - - - -	1	2
	Grey shelly limestones, splitting up irregularly; the bottom bed called WHITE BED (6 ins. to 1 ft.) [Chelonian bones, Hybodus, Cyprides] - -	2	3
	Shaly limestone, with curious concretionary projections from base of White Bed, which disturb this stratum [Modiola] - - -	0	6
Lower Purbeck Beds.	Pale grey rubbly marls - - -	0	4
	White limestones [Lias No. 1] - -	1	6
	Sandy marl and clay [Mesodon, Estheria, and Cyprides] - -	0	4
	Sandy shell-limestone BLUE ROCK [or FLAGSTONE]: blue-hearted stone, weathering buff, with brown ferruginous base called SCALE, showing tridactyl markings on under surface. Cyrena [Fish-remains] - -	2	3
	Clays and shales with Cypridea granulosa, [Cypris purbeckensis] - -	1	0
	Hard white marl - ⎫	1	3
	Soft marl - ⎬ [Lias No. 2] ⎰	1	0
	Hard marly limestone - ⎭	1	3
	Soft white marl - - - -	1	3
	Hard white marly limestone [Lias No. 3] - - - - -	1	3

The above section has been described in more detail by Messrs. Andrews and Jukes-Browne, and I have added in square brackets some of the fossils recorded by them. They had an excavation made below the floor of the quarry, and their observations showed that the lowest bed of " Lias " (No. 3) above noted, was 3 ft. 6 ins. thick, and beneath were nearly 8 feet of marls and marly limestones.

From the Lower Purbeck Beds, and especially from the bands of " Lias," Mr. Andrews has obtained many fish-remains, including Caturus, Coccolepis, Leptolepis, and Pleuropholis. Many of these are very beautifully preserved, but all are diminutive when compared with the Purbeck fishes of Dorsetshire.[*]

The finding of Estheria subquadrata was recorded by Prof. T. R. Jones.[†]

---

[*] Quart. Journ. Geol. Soc., vol. l. p. 53. ; and A. Smith Woodward, Geol. Mag., 1895, p. 145.
[†] Geol. Mag., 1890, p. 389.

The occurrence of the Cinder Bed *in situ* in the Vale of Wardour was first observed by the Rev. O. Fisher.[*] It has since been proved in a well at Teffont Rectory, as I am informed by Mr. Andrews, who "saw a rough section of the well, about 12 feet below the floor of the cellar, showing the beds below the Cinder." It has yielded *Ostrea distorta, Cardium, Trigonia, Corbula,* &c.

The Purbeck stone-beds, comparable with those of Teffont Evias, have also been quarried for road-metal and building-stone on the south of Lower Chicksgrove. There is a band of hard grey limestone, like the Swanage stone, and compact smooth-grained limestone termed "Lias" (2 feet thick). These beds overlie shelly limestones and marls, with decomposed shelly layers and "beef;" with *Paludina* and *Modiola.* The Cinder Bed, as noticed by Messrs. Andrews and Jukes-Browne occurs above these beds, and is surmounted by a marly oolitic limestone, and by clays with "beef," &c.[†]

Higher up occurs the Isopod Limestone discovered by the Rev. P. B. Brodie,[‡] a band containing *Archæoniscus Brodiei* in multitudes here and there, although the stone may in places be split up without any specimens being observed. This fossil occurs also at other horizons, but the particular bed above-mentioned is a smooth-grained limestone that may be readily identified in the neighbourhood of Dinton.

About a mile west of Dinton Station and extending north-westwards along the scarp into Teffont Park there are traces of old stone-pits. Some of these must be at or near the spot where Fitton noted his section at Dallard's Farm. This showed about 12 feet of slaty stone and clay with *Ostrea distorta, Modiola, Corbula alata,* and Cyprides. Others are nearer the present line of railway, and are those described by the Rev. P. B. Brodie and the Rev. O. Fisher. Fitton mentions that on Ladydown, quarries have long been worked for the sake of tilestone—a fissile stone, yielding *Cyrena* and remains of Fishes.[§]

Somewhat higher beds probably were opened up at Dashlet, on the south side of the Nadder, to the north of Fovant, for there Fitton noted oolitic particles in the top layer of stone, and found *Ostrea distorta,* Fish-remains, &c., in a compact limestone at the base of the quarry. In these quarries no doubt the stone was obtained below the Cinder Bed.

Clear sections of the strata from the Cinder Bed up to the junction with the Wealden Beds have been exposed in the two railway-cuttings to the west of Dinton Station. The second cutting west was described in detail by the Rev. W. R. Andrews

---

[*] Quart. Journ. Geol. Soc., vol. x. p. 477.
[†] *Ibid.*, vol. l. p. 54.
[‡] Proc. Geol. Soc., vol. iii. pp. 134, 780 ; History of Fossil Insects, pp. 3, 18, 19 ; and Quart. Journ. Geol. Soc., vol. x. p. 474.
[§] Trans. Geol. Soc., ser. 2, vol. iv. pp. 249–251 ; see also Brodie, Proc. Geol. Soc., vol. iii. p. 780.

FIG. 136.—*Section in the Railway-cutting west of Dinton Station, Wiltshire.*

Distance about 165 yards.

Near Dinton Station.

The numbers indicate the same beds as those shown in Fig. 137, p. 274.

in 1881, who then took Bed 19 as the top of the Middle Purbeck Beds, believing that they were overlaid unconformably by the Wealden Beds 20.* The clear connexion with the overlying beds was not manifest, and when I examined the sections in 1885 I noted, as well as I could, the various strata that occurred above the Isopod Limestone onwards to the white and coloured clays on top of the series near Dinton Station. These white clays I took to be Wealden and thought they were the beds so described by Mr. Andrews. The bed of hard eroded marl (19) described by him, is identical in character with Bed 27, and this caused my misinterpretation of his section. In the meanwhile Messrs. Andrews and Jukes-Browne have given particulars of the strata, and have included with the Upper Purbeck Beds, strata that I regard as Wealden.† In the spring of 1894 Mr. Strahan and I examined the area and were fortunate in finding a fresh cutting near Dinton Station, the railway having been widened for the extension of the siding. We were thus enabled to measure all the strata from the base of the white clays, which we regard as Wealden, down to the blue clay taken by Messrs. Andrews and Jukes-Browne as the base of the Upper Purbeck. Examining also the second cutting where the Isopod Limestone is well shown, we were led to believe that that band would be met with a foot or two below the lowest bed exposed in the first cutting. Getting assistance in digging a hole we were successful in finding this well-marked band of limestone No. 13, and thereby confirmed our previous inference that there was no discordance and no evidence of faulting between the two cuttings.

* Proc. Dorset Nat. Hist. Club, vol. v. p. 68 ; Quart. Journ. Geol. Soc., vol. xxxviii. p. 251.

The following is the section exposed in the railway-cuttings west of Dinton railway-station (see also Fig. 136, p. 273) :—

FIG. 137.

Section at Dinton.

		Ft. In.
Wealden Beds.	34. Irregular gravel passing down into whitish stony clay - - - -	5 0
	33. White, grey, and mottled clay, passing down into white and ochreous clay with seam of greenish sand - -	3 0
	32. Laminated yellow ochreous clay and sandy seams - - - -	2 3
	31. Brown, black, and white sand, and thin layer of laminated clay - -	0 6
Upper Purbeck Beds.	30. White marl passing down into clay : with Cyprides - - -	1 4
	29. Shelly calcareous grit - - 0 1½ to 0	3
	28. Gritty marl - - - -	0 3
	27. White marl with black (carbonaceous ?) matter on top - - 0 4 to 0	8
	26. Blue clay - - - -	0 5
	25. Bluish-grey calcareous sandstone -	0 3
	24. Marls and clays, with thin bands of "beef," and thin impersistent layers of sandstone - - - 1 9 to 2	6
	23. White shell-marl, with thicker bands of "beef" - - - -	3 0
	22. Dark blue clays, with shell-marl, "beef," and ferruginous matter - 2 6 to 3	3
	21. Blue-hearted shelly and sandy limestone, with greenish earth in places, lignite, *Unio, Paludina.* Brown calcareous sandstone. The whole passing into sand with ferruginous layers - - - -	2 8
	20. Yellowish sands and laminated sands and clays, passing downwards, and laterally into stiff blue clay - - 4 0 to 6	0
Middle Purbeck Beds.	19. Hard-jointed white marl, the surface eroded and the hollows filled with clay (like bed 27) - - - 0 1½ to 1	3
	18. Thin laminated marl, with layers of clay and sand, shelly bands and "beef" - 1	6
	17. Calcareous sandstone passing into sand 1 0 to 1	6
	16. Clay with shelly bands - - - 1	3
	15. Brown sandy rock with *Cyrena* - 0	6
	14. Shell-marl with greenish tinges - 1	0
	13. Smooth-grained grey limestone with *Archæoniscus* - - - 0	3
	12. Sandy shell-marl - - - 0	3
	11. Grey marly and ferruginous limestones - 0	9
	10. White limestones - - 1	3
	9. Shelly and sandy limestones. Fish-remains - - - - 0	5
	8. Shell-marl with "beef" - - 0	5
	7. Cinder Bed : earthy limestone with *Ostrea distorta* - - - 1	3
	6. Marly and sandy layer with "beef" - 0	5
	5. Grey sandy and shelly limestone with marly seam - - - 1	0
Lower Purbeck Beds.	4. White limestone - - - 0	8
	3. White limestone - - - 0	6
	2. Marly bed - - - 0	7
	1. Brown sandy limestone - - 1	0

Still lower beds were noted as follows, in the adjacent quarries, by the Rev. O. Fisher in 1853 :—*

		FT.	IN.
Lower Purbeck Beds.	Ribbon clays and sands, with compressed shells	1	0
	Hard crystalline limestone, with comminuted shells, *Cypris* and *Cyrena*	1	6
	Brown sand, full of crushed bivalves and Serpulæ	0	9
	Blue and grey laminated clay with limestone nodules, thin "beef," and crushed bivalves	1	0
	Hard grey marly limestone	3	6
	Dirt-bed	0	3
	Laminated clay and soft and hard marls	1	6
	Hard marl with conchoidal fracture	0	8

These lower beds represent the "Lias" beds, &c. of Teffont; from the Middle Purbeck Beds Mr. Rhodes and I obtained the following fossils :—

Lepidotus.	Cyrena? gibbosa.
Avicula dorsetensis.	—— media.
Cardium.	Modiola.
Corbula.	Mytilus.

The "Upper Marls" noted by Messrs. Andrews and Jukes-Browne were proved in the well at the cottages north of Dinton Station. This well was probably carried down through Wealden and Purbeck clays to the fossiliferous bed No. 21. The thin bands of calcareous sandstone above are, like the thicker beds, readily decomposed, and their presence would not have attracted the notice of the well-sinker. The water would have been obtained in the sandy beds on this horizon and below. Indeed the same beds yielded water in a shallow well sunk by the old railway-siding, and the supply failed when the siding was extended.

In the material thrown out from this well Mr. Andrews obtained some Fish-scales, also the following fossils :—

Paludina carinifera.	Cypridea (cf.) valdensis.
Unio.	Cyprione Bristovii.
Cypridea punctata.	Darwinula leguminella.

The evidence clearly establishes the contention of Messrs. Andrews and Jukes-Browne, that the Upper Purbeck Beds are represented, though I believe their "Upper Marls" are for the most part Wealden. In his original notes on the Vale of Wardour, the Rev. P. B. Brodie appears to have observed strata about as high as No. 23 in the Dinton section.

The hard marl (No. 19) and also the similar bed (No. 27) show evidence of local dissolution. The former layer, which sometimes is 15 inches thick, has almost disappeared in places, being reduced to 1½ inches. These facts illustrate some of the points to which

---

* Quart. Journ. Geol. Soc., vol. x. p. 476; see also !Andrews, *Ibid.*, vol. xxxviii. p. 351; Andrews and Jukes-Browne, *Ibid.*, vol. l. p. 55; and Geol. Mag., 1891, p. 292.

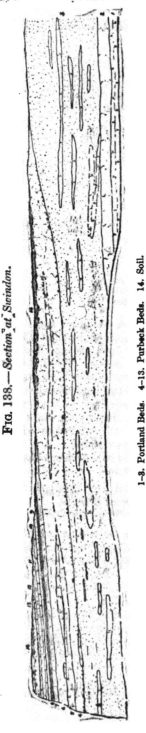

FIG. 138.—*Section at Swindon.*

1-3. Portland Beds. 4-13. Purbeck Beds. 14. Soil.

attention has been drawn by Mr. F. Rutley in a paper "On the Dwindling and Disappearance of Limestones."[*]

Swindon.

From the Vale of Wardour northwards we see no traces of Purbeck Beds until we reach the outlier at Swindon, for in the intermediate area the beds are concealed beneath the Chalk of Salisbury Plain. Their underground course might be anticipated to lie beneath the Cretaceous covering at Market Lavington, near Patney in the Vale of Pewsey and beneath the Marlborough Downs; possibly the Swindon outlier is not far from a concealed mass of the strata, but we have no evidence of their continuity, though we may infer that Purbeck Beds would be present where both Portland and Wealden Beds occur.

At Swindon we have but the lower portion of the Purbeck Beds exposed, and the accompanying section (Fig. 138.) shows all the beds I was enabled to observe.

Section on the western side of the great quarries at Swindon :—

		Fr.	Ix.
	14. Pipes of reddish-brown clay -		
	13. Rubble of greenish-clay and chalky marl - -	3	0
	12. Chalky sandstone and fissile sandy limestone, with intervening bed of marl - -	0	8
Purbeck Beds.	11. Marl, and hard marly and cavernous limestone -	1	4
	10. Hard, white-banded, marly, fissile limestone, with Cyprides -	0	7
	9. White marly bed [*Paludina*] -	1	0
	8. Dark, bluish-grey, marly and earthy clay -	1	0

		Ft.	In.
Purbeck Beds.	7. Light marly and racy clay, and ferruginous marl - - -	0	6
	Yellow, ochreous, and calcareous bed - - - - -	1	4
	6. Pale marl, ochreous and sandy in places - - - - Bluish-grey marl, with lenticular blocks, masses of stone and pebbles at base - - -	3	6
	5. Hard limestone, ochreous and nodular in places - - -	0	9
	Marl and hard white marly limestone - - - -	2	0
	4. Rubbly marls, with rolled lumps of limestone at base - - -	3	0
Upper Portland Beds.	3. Calcareous shelly sands, with Ostrea, Pecten, and lignite - Hard calcareous sandstone - -	6	0
	Sands, with beds of hard calcareous sandstone - - - about	10	0
	2. Hard marly limestone.		
	1. Bluish-grey limestone, with lydite pebbles.		

The Purbeck Beds here rest in a markedly unconformable manner on the Portland Beds, for they occur on different members of that series, eroding into the sands with the layers of Swindon stone, and containing blocks derived from them. The higher bands of the Purbeck Beds appear to overlap the lower portions in places, but many of the latter are impersistent and variable in character.

Traces of Lower Greensand were recorded by Fitton as resting irregularly on the Portland Beds,[*] but the evidence was not satisfactory, and the only indications since observed appear to be those of ferruginous earth resting in pipes and furrows of Purbeck and Portland Beds.

The unconformity of the Purbeck and Portland Beds was noticed by the Rev. P. B. Brodie, in 1846,[†] and he first observed the occurrence of fresh-water Mollusca, Cyprides, Insects, Fish and Saurian remains and Plant-remains. In 1880 the Swindon sections were described in detail by Prof. Blake,[‡] who showed that there was no evidence of an inosculation of beds representing 'Purbeck' and 'Portland' conditions, as had been suggested, for the evidence of unconformity was clearly shown. He noted the occurrence in the Purbeck beds of blocks or pebbles containing Cerithium portlandicum, and also of rolled fragments of Kimeridge (?) Clay.[§] At the same time he expressed the opinion that the Purbeck Beds of Swindon were formed at an earlier period than any of the Purbeck Beds of Dorset, having been laid down contemporaneously with the later Portland Beds of that region.

* Trans. Geol. Soc., ser. 2, vol. iv. p. 265.
† Quart. Journ. Geol. Soc., vol. iii. p. 53.
‡ Godwin-Austen, Ibid., vol. vi. p. 466; and Moore, Proc. Geol. Assoc., vol. iv. p. 544.
§ Quart. Journ. Geol. Soc., vol. xxxvi. pp. 203, 207, 222.

Such a view seems to me purely hypothetical, although it may be freely admitted that the conditions attending the deposition of our Portland and Purbeck Beds may have changed somewhat irregularly over the area.

The fossils from the Purbeck Beds of Swindon include the following : the Ostracoda having been determined by Prof. T. R. Jones :—*

Paludina.	Cypridea Dunkeri.
Planorbis.	—— punctata.
Unio.	Cypris purbeckensis.
Candona ansata.	Cythere retirugata, and var.
—— bononiensis.	rugulata.

Shotover Hill and Garsington.

Shotover Hill is situated between two and four miles east of Oxford. The old coach road to London passed over the summit, and the several groups of strata opened up from near Headington at the western foot of the hill to the " Iron Sands " on the top, early attracted the attention of geologists. These Iron Sands which have yielded deposits of ochre and fuller's earth, are usually assigned to the Lower Greensand, but certain fresh-water shells having been found in the lower strata, Prof. Phillips and Prof. Prestwich have considered that Wealden Beds also are represented.† The Ostracoda would lend support to this view.‡ These Iron Sands, however, appear to rest unconformably on the Portland Beds below, whereas in all other localities where we find Wealden Strata they rest conformably on Purbeck and Portland Strata, and they are in the south-west of England overlaid with marked unconformity by the Lower Greensand. There is no evidence of Purbeck Beds on Shotover Hill ; but the Portland Beds form a straggling outlier stretching in a south-easterly direction to Garsington and Cuddesden, and covered irregularly by two other outliers of the Iron Sands, which in one place near Littleworth overlap the Portland Beds, and rest directly on Kimeridge Clay.

In this south-eastern portion of the Portlandian outlier, traces of Purbeck Beds were noted here and there by Fitton.§ Thus above the Portland Stone at Combe Wood, south of Wheatley, he observed beds like the "Malm" of Garsington. They consisted of compact and oolitic limestone and rubble, and yielded remains of *Cypris, Mytilus, Modiola, Paludina elongata* and *Planorbis*? They were overlaid by strata referred to the Lower Greensand. (See p. 218.)

The following section of a stone-pit at Garsington was also recorded by Fitton :— ‖

* Quart. Journ. Geol. Soc., vol. xli. p. 330.
† See also Morris, Proc. Geol. Assoc., vol. ii. p. 86.
‡ T. R. Jones, Quart. Journ. Geol. Soc., vol. xli. p. 320.
§ Trans. Geol. Soc., ser. 2, vol. iv. pp. 272, 275.
‖ *Ibid.*, ser. 2, vol. iv. p. 277.

		FT.	IN.
	Loamy soil	2	0
Lower Greensand.	Ferruginous brown sand and greenish sand, with yellow ochre and fuller's earth - - - - -	8	0
Purbeck Beds.	Malm, soft limestone, and softer marl, much decomposed; comprising:—		
	a. Light greenish-grey marl with, in the upper part, detached fragments of silicified coniferous wood, like that of Portland, and portions of bone;		
	b. Limestone, with oolitic grains; Paludina, Planorbis?, Mytilus, and Cypris;		
	c. Limestone, in some places like the "Pendle" of the pits at Whitchurch, oolitic and botryoidal in places, with Paludina, &c. - - -	4	0
Portland Stone.			

Long Crendon, Brill, Aylesbury, and Whitchurch.

Purbeck Beds, attaining at most a thickness of 30 feet, have been observed on several of the Portlandian outliers near Aylesbury. Their presence at Long Crendon, Dinton, and Bishopstone was noted by Fitton;* and their occurrence at Brill was pointed out by the Rev. P. B. Brodie.† Purbeck Beds have also been observed near Cuddington, &c. On the whole there are few sections now to be seen, and the strata do not present many features of general interest. The occurrence of silicified wood at Garsington is of interest, for Buckland said he found traces of a dirt-bed above the Portland Stone about two miles north of Thame—probably at Long Crendon. (See p. 220.)

In the neighbourhood of Aylesbury and Whitchurch, as remarked by Mr. Jukes-Browne, the Purbeck Beds form a very variable series, and although in some places he thought there was evidence of contemporaneous erosion between them and the underlying Portland Beds, yet in other sections it "is very difficult to say where one group ends and the other begins."

The Lower Purbeck Beds were well shown at the Bugle Pit, Hartwell. On top there was an accumulation of rubble and clay with stones, that rested irregularly in hollows of the underlying beds. This appeared to me to belong to the Drift. The strata have been previously noted in the description of the Portland Beds. (See p. 224.) They have yielded remains of Plants, Cyprides, Cyrena, Mytilus, Insects, &c. From the collection of the late Dr. Lee at Hartwell, Messrs. Sharman and Newton identified Turtle Bones, and remains of the Fishes, Aspidorhynchus, Lepidotus minor, Pleuropholis serrata, Mesodon (Pycnodus), and Strophodus.

* Trans. Geol. Soc., ser. 2, vol. iv. pp. 282–287.
† Quart. Journ. Geol. Soc., vol. xxiii. p. 197.

Prof. T. R. Jones obtained the following species of marine and freshwater Ostracoda from .beneath the junction of Purbeck and Portland Beds; the material having been forwarded by Mr. Jukes-Browne :—

Candona ansata.	Cythere retirugata.
—— bononiensis.	—— —— var. rugulata.
Cypris purbeckensis.	

This evidence, in Mr. Jukes-Browne's opinion, points to a gradual transition between the formations; as all the above-mentioned species were also obtained from the Purbeck Beds.* (See p. 227.)

The occurrence of Purbeck Beds on Quainton Hill was noted by Fitton. Beneath the Lower Greensand, consisting of red sand and ferruginous conglomerate with ochre, clay, fuller's earth, and sands, altogether about 17 feet thick, the following beds were observed :—

	Ft.	In.
Clays with fragments of shells (obscure)		
Hard sandstone or siliciferous grit, with *Paludina* and *Cyrena?*	3	0
White sand and clay, with decomposed shells; *Mytilus, Modiola,* and *Cypris*	3	0
Fissile oolitic stone (Pendle), with *Cypris, Cyrena (Cyclas) parva, Modiola,* and *Mytilus*	0	6

The Pendle rested on the Portland Stone, but there was evidently some unconformity with the overlying beds, for at one point the Purbeck Beds were covered directly by the hard ferruginous conglomerate which elsewhere occurred above the main mass of sandy strata grouped with the Lower Greensand. Fitton thought that the siliciferous grit might belong to the Wealden Beds; but all the layers may be grouped with the Purbeck formation. Similar beds were noted by him in the outlier at Oving and Whitchurch, to which reference has previously been made; and he recorded thin layers of Purbeck limestone and clay in a pit at the Warren, south of Stewkley.† Further to the north-east we have no evidence of any Purbeck Beds of the type of the freshwater and estuarine beds already described.

Sussex.

The lowest beds exposed in the Wealden area emerge from beneath the Ashdown Sand to the north and north-west of Battle, in Sussex. The summit of the anticlinal which is much broken by faults, is to a certain extent repeated at Ashdown Forest, but we have no exposures there of beds beneath the Ashdown Sand. These older strata extend from near Whatlington on the east, to near Little Tattingworth, north-east of Heathfield on the west, a distance of nearly 10 miles. The breadth of the exposures is nowhere so much as one mile, and they are separated into three

---

* Quart. Journ. Geol. Soc., vol. xli. p. 329.
† Trans. Geol. Soc., ser. 2, vol. iv. pp. 272, 288, 290, 291.

p)rtions, the two easterly exposures being faulted. The western exposure includes the beds at Poundsford and Rounden Wood ; the central exposure, that of Darvel and Limekiln Woods, with the site of the Sub-Wealden Boring north of Netherfield; and the eastern exposure includes the beds in Archer Wood.

Conybeare and Phillips, in 1822, remarked that the " argillo-calcareous formation " near Battle " will probably be found, on an attentive examination of all its beds, especially the lowest, to coincide with that of Purbeck." *

In 1824 Thomas Webster remarked on the very close resemblance between the beds quarried near Battle Abbey and the Purbeck strata ; † and in the same year Fitton noted the occurrence of " *Cyrena* " and other fossils " in slaty clay between the beds of limestone, near Darvel Wood," strata which he stated to be " below the Hastings' Sands."‡ Later on he commented on the similarity between the clays with " *Cyclas* " found in association with the limestone north-west of Battle, and certain Purbeck strata near Swanage.§ In the meantime Mantell || also referred to these strata, but the fossils from Poundsford, or Pounceford, were noted as Wealden.

Fitton, and afterwards Mantell, described the strata under the name of ASHBURNHAM BEDS, although the village of Ashburnham is situated more than three miles to the south-west of the nearest exposure near Netherfield. Under this name the strata were mapped by Mr. C. Gould, and they were so designated on the Geological Survey Map, sheet 5 ; but in 1875, when Mr. Topley described the Wealden area in detail, he decided to group the Ashburnham Beds with the Purbeck Beds, because they were known to contain some Purbeck fossils, "and also because of the striking lithological difference between these beds as a whole and the Wealden proper." ¶

No sections showing the junction of the Purbeck Beds and Ashdown Sand have been observed, but Mr. Topley remarks that " the two divisions are well marked off from each other, although in the lower part of the Ashdown Sand there is a good deal of loam and clay, the representative of the Fairlight Clays of the coast." He considered it probable that the Purbeck Beds " occur at no great depth below the base of Fairlight Cliff." ** The Fairlight Clays comprise coloured clays, with many beds of sand and sandstone, and lignite, and present characters that clearly associate them with the Wealden rather than with the Purbeck formation.

The Purbeck Beds consist of a group of shales with important bands of limestone, occasional beds of hard calcareous sandstone, and in the lower part with bands of gypsum.

The succession of the strata has mainly to be made out from the records of old shafts, sunk to obtain the beds of limestone ; the Sub-Wealden Boring added further particulars, and there are a few exposures of the strata along the banks of streams and in quarries.

---

* Outlines of Geol. Eng. and Wales, p. 148.
† Trans. Geol. Soc., ser. 2, vol. ii. p. 44.
‡ Ann. Phil., 1824, vol. xxiv. p. 379.
§ Geological Sketch of Hastings, 1833, p. 50, and Plate Fig. 1 ; Trans. Geol. Soc., ser. 2, vol. iv. pp. 176, 209, 216.
|| *Ibid.*, vol. iii. p. 215 ; Geol. South-east of England, 1833 ; Geol. Isle of Wight, ed. 3, 1854, Table p. 42.
¶ Geol. Weald, p. 31.
** Geology of the Weald, pp. 30, 33, and 45.

The limestones, as remarked by Mr. Topley, occur chiefly on two horizons ; the higher series is termed the "Greys," and the lower the "Blues."

The total thickness of the Purbeck Beds was estimated at about 400 feet by Mr. Topley, and the following appear to be the main sub-divisions :—

	FT.	IN.
Shales, &c. - - - - - - - about	30	0
"Greys," Limestones (about 12 or 14 beds, 2 to 12 inches thick), calcareous sandstones and shales, with selenite, &c. In the upper part of the "Greys" there are two or three layers of shelly limestone known as "Ragged Bulls," which pass into hard calcareous sandstone -	50	0
Black, grey, and greenish shales and clays, with ironstone nodules, occasional bands of limestone, and sandy beds - - - - - - -	130	0
"Blues," Limestones (about 10 beds, 3 to 20 in. thick), and shales, *Ostrea* - - - - -	25	0
"Rounden Greys," Limestones and shales - -	50	0
Shales, with gypsum - - - - -	130	0

The following fossils have been obtained from the Purbeck Beds of Sussex, and most of them have been recorded by Mr. Topley :—

Goniopholis.
Megalosaurus.
Lepidotus Mantelli.
Melanopis attenuata.
—— harpæformis.
—— Popei.
—— rugosa.
—— tricarinata.
Paludina.
Corbula alata.
Cyrena angulata.
—— elongata.
—— media.
—— —— var. gibbosa.
—— membranacea.
—— parva.

Modiola.
Mytilus Lyelli.
Ostrea distorta.
Psammobia tellinoides.
Unio.
Insect-remains.
Estheria elliptica var. subquadrata.
Cypridea granulosa.
—— punctata.
—— valdensis.
Darwinula leguminella.
Chara.
Equisetites Lyelli.
Mastonidium Göpperti.
Onychiopsis Mantelli.

At Archer Wood according to the observations of Mr. C. Gould, both the "Greys" and "Blues" had been worked, but only the latter to any considerable extent. The works have been abandoned for nearly 60 years. Shafts were sunk to depths of 100 feet and more, through shales, down to the "Main Blue," and they proved that the "Greys" occur only over portions of the tract.

The road-cutting north of Copyhold Bridge showed papershales with "beef" (fibrous carbonate of lime), like beds in the upper portions of the Middle Purbeck Beds of Swanage. Mr. Gould mentions that the "Greys" were also exposed in this road-cutting.

At Limekiln Wood, the information obtained by Mr. Gould showed that both "Greys" and "Blues" were formerly worked by means of quarries and bell-pits. (See p. 321.)

Several pits had also been opened east of Darvel Beach Farm "for the purpose of extracting the hard calciferous sandstone which occurs in the upper part of the Greys." Two layers of this rock intercalated with beds of limestone, were worked. They were exposed in quarries to the south-east, and west of Snep's Wood: where the following section was noted when I visited the district in company with Mr. Topley :—

	Ft.	In.
Shale and clay, with bands of shelly limestone and shale.		
Top Rock - - - - - - - -	0	8
Shale, &c. - - - - - - -	17	0
Bottom Rock - - - - - -	1	8

The rock-beds consist of blue gritty limestone or calcareous sandstone with plant-remains. They are used for road-metal. Other limestone-beds occurred lower down, and dark shales were exposed by the stream.[*] The beds in Darvel Wood occur beneath the main mass of the "Blues."

At the Sub-Wealden boring the site chosen was in Councillor's Wood, just south-east of the stream which divides that wood from Limekiln Wood, and north of Netherfield.

The boring was commenced in the strata locally known as the "Bastard Blues," below the "Main Blue," which is the top bed of limestone in the series of "Blues."

Some of these beds are exposed in the banks of the adjoining stream—there as described by Mr. Topley, "they (the 'Bastard Blues') consist of calcareous shales, with bands of argillaceous limestone or cement-stone, closely resembling in appearance the lowest beds of the Lower Lias."

Limestone with *Ostrea* is found south-east of the present Gypsum-works, and some of the beds of limestone break into rhomboidal fragments like the Lias of Aberthaw and other places.[†]

The following detailed account of the Purbeck strata passed through in the first Sub-Wealden boring (Sept. 1874) was drawn up by Mr. Topley :—[‡]

BORING AT NETHERFIELD.	THICKNESS.		DEPTH FROM SURFACE.	
	Ft.	In.	Ft.	In.
Shales - - -	16	6		
Blue limestone (Spring tapped here) - - - -	2	6	19	0
Shale - - - -	5	0	24	0
Blue limestone - - -	2	0	26	0
Shale - - - -	4	0	30	0
Limestone - - - -	1	6	31	6

[*] See also C. Gould, in Topley's Geol. Weald, p. 37.

[†] See Memoir on the Lias of England and Wales, pp. 116, &c.; also Conybeare, Proc. Geol. Soc., vol. ii. p. 172.

[‡] Geol. Weald, p. 43. Details of the upper part of the second boring are given in the 10th Quarterly Report of the Sub-Wealden Exploration, 1875.

BORING AT NETHERFIELD.	THICKNESS.		DEPTH FROM SURFACE.	
	FT.	IN.	FT.	IN.
Shale - - - -	4	0	35	6
Limestone - - - -	3	0	38	6
Shale (Spring tapped here) -	4	0	42	6
Limestone - - - -	4	0	46	6
Hard blue shale - - -	15	6	62	0
Hard grey shale - - -	3	0	65	0
Hard shale - - - -	14	6	79	6
Shales with crystals of carbonate of lime - - - -	9	0	88	6
Grey shale - - - -	13	0	101	6
Greenish shales with gypseous veins - - - -	20	0	121	6
Impure gypsum - - -	8	6	130	0
Pure white gypsum - -	4	0	134	0
Impure gypsum - - -	5	6	139	6
Pure white gypsum - -	3	0	142	6
Gypsum, more or less pure, hard and dark - - - -	14	6	157	0
Black shale (very sulphurous) -	3	6	160	6
Gypsum in nodules and veins -	12	0	172	6
Gypseous marl - - -	6	6	179	0
Sandy marl (water level lowered here) - - - -	0	6	179	6
Black sulphureous shale - -	0	6	180	0

Purbeck Beds, 180 feet.

At the Gypsum works the first seam (6 feet thick) of gypsum was met with at a depth of 119 feet; below this 25 feet of shales and laminated marly beds with a 4 feet seam of gypsum (5 feet down) were proved; and beneath, another bed of gypsum 7 feet thick is worked.

The beds in many respects resemble those of Durlston Bay near Swanage, although the gypsum here occurs in a more massive and persistent form than it does in Dorsetshire.

In the Sub-Wealden boring Cyprides were found at various depths down to 136 feet, including *Cypridea granulosa, C. punctata,* and *C. valdensis. Estheria elliptica* var. *subquadrata* was found in calcareous shale by Mr. J. E. H. Peyton at a depth of 86 feet. Seed vessels and stems of *Chara* were found at a depth of 85 feet, and shale with *Unio* at a depth of 78 feet.[*]

At Rounden Wood, a series of limestones known as the "Rounden Greys" has been worked beneath the "Blues." Mr. Topley mentions that attached to the "Main Blue" and extracted with it in the workings is a band of oysters known as the "bottom grey."

There are shallow pits on the north-east side of the wood, where hard buff and grey limestones like the cap-beds of Portland are still to be seen. In other portions of the wood we find banded limestones and beds exhibiting obscure arborescent markings, such as I have noticed in the Lower Purbeck Beds of Durlston Bay.

[*] T. R. Jones, Geol. Mag., 1878, p. 110; and Quart. Journ. Geol. Soc., vol. xli. p. 880; Dixon's Geol. Sussex, ed. 2, pp. 153, &c.

A specimen of a limestone termed "cutlets," now placed in the Rock-collection at the Museum of Practical Geology, was obtained during the course of the Geological Survey. Mr. Topley mentions that the term "Cutlet" seems to have been applied to any bed of inferior limestone, at whatever horizon it may have occurred.[*] An illustration of this rock has been previously given (Fig. 104, p. 232).

Mr. Gould mentions that the beds above the Greys were to be seen in Little Poundsford Lane, just south of the stream, where he noted the following section :—

		Ft.	In.
	Sand and loam.		
	Shale, &c. - -	30 0 or 35	0
Greys.	Thin bed of limestone (Ragged Bull?).		
	Shales, &c. - -	10 0 or 15	0
	Limestone (Main Bull?).		

The Greys Beds are well developed, and some of the strata were to be seen in the banks of the stream south of Poundsford. *Cyrena* occurs abundantly in some of the layers of limestone and shale, and in a bed of calcareous sandstone beneath the "Ragged Bull," *Melanopsis*, *Hybodus*, and Reptilian remains, were found by Mantell, and later on by C. Gould. This bed is on about the same horizon in the Greys as the calciferous sandstone noted near Darvel Beach Farm. The old quarry described by Mantell [†] is situated near Poundsford Farm. Good sections of the strata are exposed by the cascades in Poundsford Gill. There we find beds of dark shale with thick irregular and interrupted bands of stone, the general appearance being like a section of Coal-measures. The "Greys" were formerly worked above the cascades by means of a shaft. Some of the shales associated with the Greys are described by Mr. Topley as "peculiar tough, dark brown and blackish leathery shales, which when dry somewhat resemble stiff brown paper in character, and occasionally also in colour."[‡] They contain crushed specimens of *Cyrena*. At a lower horizon, and separating the main mass of Greys and Blues, we find a series of green, grey, and black shales with ironstone-nodules, together with sandy beds.

Dark grey clays and shales with shelly layers and occasional bands of shelly limestone, were exposed along the banks of the Dudwell stream south-west of Poundsford.

From the beds exhibited in this neighbourhood which I visited under the guidance of Mr. Topley, and in company with Mr. R. H. Tiddeman and Mr. J. E. H. Peyton, we obtained a few fossils, as follows :—

Goniopholis.	Cyrena parva ?
Cardium.	Ostrea.
Cyrena media.	Unio.

Ostrea distorta has been recognized from Poundsford.

---

[*] Geol. Weald, p. 36.
[†] Geol. South-east of England, pp. 219, 222, &c.
[‡] Geol. Weald, p. 31.

At Perch Hill, south-east of Burwash Wheel, the Greys beds are worked by means of a shaft 30 or 40 feet deep. Here also beds of hard calcareous gritty sandstone are obtained. The rock is too hard to be dressed for building-purposes, and it is employed, like the beds at Darvel Beach Farm, for road-metal. Bivalve Mollusca and plant-remains occur abundantly in the stone.

The general characters and the organic remains of these Sussex Purbeck Beds thus accord well with those of the Purbeck Beds of Dorsetshire and Wiltshire.

Lincolnshire.

Although in passing to the north-east of the Aylesbury area we have no further evidence of Portland strata *in situ* in the district under consideration, yet we find relics of these beds in the Lower Greensand in various localities.[*]

In describing the Lower Greensand of Wicken, Potton, and Brickhill, near Woburn, Mr. J. F. Walker,[†] Mr. J. J. H. Teall[‡] and subsequently W. Keeping,[§] have noted the occurrence of many fossils derived from the Oxford Clay, Corallian Rocks, and Kimeridge Clay, and the following forms which belong to Portlandian species :—

Ammonites giganteus ?	Cytherea rugosa.
Natica elegans.	Lucina portlandica.
Neritoma sinuosa ?	Myoconcha portlandica.
Arca.	Sowerbya.
Astarte cuneata.	Trigonia gibbosa.
—— hartwellensis.	—— incurva.
Cardium dissimile.	

In addition to these there have been recorded from Potton, *Endogenites erosa,* and remains of *Iguanodon,* that may have been derived from Wealden strata ; and silicified wood has been found at Brickhill.

Mr. Teall has remarked that "the phosphatic deposit which has been identified at Potton, Ampthill, Brickhill, and Leighton, and which is probably far more persistent than was at first sight supposed, occupies different positions in the different localities. Thus at Potton it is found towards the top of the sands, at Brickhill and Ampthill at a much lower horizon, while at Rushmore Pond, near Leighton, it rests directly on the [Oxford] clay."[||]

There is thus locally a considerable mass of sands, including the fuller's earth of Woburn, that is older than the Potton nodule-bed ; and although the nodule-beds of the Lower Greensand of Cambridgeshire, Bedfordshire, and Buckinghamshire have been compared with the Folkestone Beds, including the Bargate Stone,

* In a list of Fossils from the Fenland area (Skertchly, Geol. Fenland, p. 317) a column is given for Portland Beds; but there is no evidence to support the statement.
  † Rep. Brit. Assoc., for 1866, Sections, p. 67.
  ‡ Potton and Wicken Deposits, 1875.
  § Fossils of Upware, &c., pp. 40, 45, 155; Geol. Mag., Dec. II., vol. ii. p. 373.
  || Potton and Wicken Phosphatic Deposits, p. 29.

yet in the midland area there may be earlier stages of the Lower Greensand in the Shotover Iron Sands and in the fuller's-earth beds of Woburn, the precise relations of which to other beds in the south of England and in Lincolnshire have yet to be determined.[*]

In Yorkshire the researches of Prof. Judd,[†] followed by those of Mr. G. W. Lamplugh,[‡] have shown that overlying the Upper Kimeridge Clay there is a thin "Coprolite-bed" and a series of clays and shales, forming part of the Speeton Clay, that yield a mixture of Jurassic and Cretaceous forms. In the bed with phosphatic "pebbles," *Lucina portlandica,?* *Arca,* and some other fossils were found, and the overlying clays have yielded *Lingula ovalis?* *Pecten lens* var. *Morini, Thracia,* and *Avicula inæquivalvis.* There also occur a variety of *Exogyra sinuata, Ammonites gravesianus* (Fig. 139, p. 288), together with *Belemnites lateralis* (Fig. 140, p. 288), a form taken to characterize the zone.

In Lincolnshire the beds have been studied more particularly by Prof. Judd, Mr. Henry Keeping, and subsequently by Mr. Jukes-Browne and Mr. Strahan. It should not be forgotten, however, that a very good account of the strata and their fossils at Nettleton Hill, by W. H. Dikes and J. E. Lee, was published in 1839.[§]

Mr. Lamplugh remarks that in Lincolnshire the zone of *B. lateralis* comprises both the Spilsby Sandstone and great part of the Claxby ironstone. There the base of the Spilsby Sandstone is marked by a nodule-bed which yields *Lucina portlandica* and other fossils noted p. 293. In reference to the Yorkshire "coprolite-bed," Mr. Lamplugh remarks: "It is difficult to say whether the fossils which occur in this seam are indigenous to it or have been derived," but he adds, "it seems to me that this coprolitic band may have been formed during a period when, either through the increased strength of the current, or through lack of material, or from some other cause, the deposition of the clay ceased and allowed time for the heavier nodular matters dropped over the sea-bottom to accumulate as a band." Prof. Blake however regards the fossils as *remanié.*[||]

The observations of M. Serge Nikitin,[¶] and more especially those of Prof. A. Pavlow,[**] tend to show that our Purbeck and uppermost Portland Beds, and also the Spilsby Sandstone, may be correlated with the Upper Volga Beds of Russia, while the Lower Volga Beds include the upper portions of our Kimeridge

[*] Keeping, Fossils of Upware, &c., p. 47 ; see also Strahan, in Geol. Lincoln, p. 88 ; and Green, Rep. Brit. Assoc. for 1894, p. 644.
[†] Quart. Journ. Geol. Soc., vol. xxiv. p. 218 ; vol. xxvi. p. 326.
[‡] *Ibid.,* vol. x.v. pp. 583, 608, 609 ; Rep. Brit. Assoc. for 1890, p. 808 ; see also Strahan, Geol. Lincoln, p. 88.
[§] Mag. Nat. Hist., ser. 2, vol. i. p. 561.
[||] Proc. Geol. Assoc., vol. xii. p. 140.
[¶] Bull. Soc. Belge de Géol., vol. iii. 1889, p. 29.
[**] Jurassique superieur et Crétacée inferieur de la Russie et de l'Angleterre, Bull. Soc. Imp. Nat. Moscow, 1889 ; and Pavlow and Lamplugh, Argiles de Speeton et leurs équivalents, Bull. Soc. Imp. Nat. Moscow, 1892.

Clay, and a great part of our Portland Beds. These Upper
Volga Beds are thus characterized :—

Zone of Belemnites lateralis { Ammonites gravesianus (and gravesiformis). A. subditus.

It is mentioned by Pavlow that *Belemnites lateralis* (*B. boloniensis*,
Sauvage and Rigaux) has been met with in "Portlandian Beds" at
Boulogne. The Claxby Ironstone which is said to contain
Ammonites noricus, *A. regalis*, and *Belemnites jaculum* in its upper
part, is referred partly to the Neocomian; and as it is also said to
yield in its lower part *Ammonites Blakei*, *Belemnites lateralis*, &c.,
that portion is grouped as Jurassic.

Fig. 140.

Fig. 139.

Ammonites gravesianus, *d'Orb*, ½.

Belemnites lateralis, *Phil.*
Nat. size.

In these studies the horizons assigned to particular species of
Ammonites will be found to vary to some extent in different
parts of the Continent. The general sequence is maintained; but
as might be expected the assemblages of fossils and the ranges of
the individual species vary in different areas. The subject is
unfortunately rendered more complex by the differences of
nomenclature and the differences with regard to species that
prevail among palæontologists.

The Kimeridge Clay in East Lincolnshire from Spilsby to the
neighbourhood of Caistor is directly overlaid by strata that have
been regarded as the lower division of the Neocomian Series, the

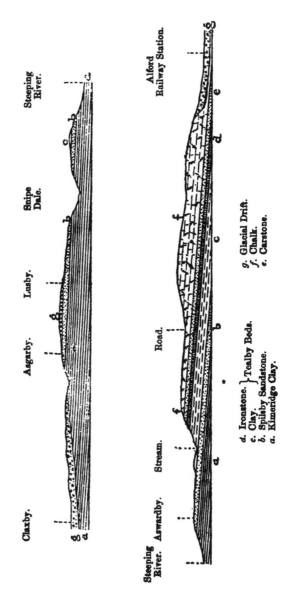

FIG. 141.—Section from Claxby to Alford, Lincolnshire.   (A. J. Juker-Browne.)

Longitudinal scale, one inch to a mile.   Vertical scale, one inch to 800 feet.

g. Glacial Drift.
f. Chalk.
e. Carstone.

d. Ironstone. } Tealby Beds.
c. Clay.
b. Spilsby Sandstone.
a. Kimeridge Clay.

Lower Sand and Sandstone of Prof. Judd, to which the name
Spilsby Sandstone has been applied by Messrs. Jukes-Browne
and Strahan.*

At the base of this division and resting directly on the Kimeridge
Clay, there is always a thin seam of derived phosphatic nodules
and casts of fossils, similar to those of Potton and Sandy in
Bedfordshire. The mass of the rock above consists of brown,
white, or greenish-yellow sand, in many places compacted into
irregular bands of sandstone or hard grit, and often into boulder-
like concretions of hard calcareous sandstone. It contains small
pebbles of quartz, lydian stone, &c. It attains a thickness of from
25 to 45 feet. There is evidence of erosion between the Kimeridge
Clay and Spilsby Sandstone, and this erosion is accompanied by
unconformable overlap as the beds are traced northwards. Hence
there appears no reason to believe that any Portlandian strata
occur *in situ* in Lincolnshire. (See p 177.)

The Spilsby Sandstone is exposed over a considerable tract
bordering the river Steeping, from Spilsby to Salmonby : and
there are many quarries where the stone has been obtained for
building-purposes. Its thickness is about 40 feet, but at
Skegness 26 feet. The actual junction with the Kimeridge Clay
is rarely to be seen, but the phosphatic nodules and phosphatized
fossils, as remarked by Mr. Strahan, "may be picked up in
almost any freshly-ploughed field or in the soil along and just
below the junction." In such situations I have seen them, under
his guidance, to the east of Eresby Hall, south of Spilsby, and to
the north of Hundleby.

FIG. 142.

Section across Hundleby Brickyard.   (A. J. Jukes-Browne.)

c. Claxby Ironstone.   e. Boulder Clay.
b. Spilsby Sandstone.   d. Tealby Clay.
a. Kimeridge Clay.

The relations of the Spilsby Sandstone with the overlying
strata were well shown in the Hundleby brickyard described by
Mr. Jukes-Browne.† He remarks that above the Spilsby Sandstone there is " a hard calcareo-ferruginous rock, full of brown
oolitic grains like the Tealby [and Claxby] ironstone, and of a
bluish-grey colour inside, but weathering to a yellowish brown ;"
its thickness was over 14 feet. He adds that among the fossils
obtained only one species, *Belemnites lateralis,* is found also in

---

* Geol. E. Lincolnshire, 1887, p. 14.
† *Ibid*, pp. 22, 141.

the Spilsby Sandstone.   The facies of the fauna, as I am informed by Mr. Lamplugh, is characteristically that of the upper part of the zone of *Belemnites lateralis* of Speeton.

Pecten cinctus also occurs in the ironstone, and this species has been recorded from the Spilsby Sandstone.   *Pecten orbicularis* occurs in the Sandstone, and this species and *Belemnites lateralis* are also recorded from the Tealby Clay of East Keal.

The Spilsby Sandstone has been traced by Mr. Strahan from Scamblesby and Donnington-upon-Bain and Tealby to Audleby north of Caistor, and again from the Barnetby Gorse Hills to Elsham.   This further outcrop may belong to a concealed outlier, or it may be part of the main mass.   In the southern part of this tract the thickness is estimated at 50 feet, at Tealby 42, Claxby 30, Nettleton 35, Audleby 15, and Elsham 10 feet.

The Spilsby Sandstone is well shown in the Benniworth cutting of the Louth and Bardney Railway, south-west of Donnington-upon-Bain.   The following beds have been described by Mr. Strahan in whose company I examined the sections :—*

		Ft.	In.
Claxby Ironstone.	{ Ferruginous clay, with oolitic grains of iron-oxide - - - -	3	0
Spilsby Sandstone.	Greenish and brown sand and sandstone; in places fine white sand, cemented into a hard rock here and there by iron-oxide - - -	19	0
	Pebbly band with pebbles of quartzite, lydian-stone, &c.; also *Belemnites lateralis*, and fragments of silicified wood - - - - -	0	0
	Hard ferruginous sandstone, with scattered pebbles; impressions of *Pecten* and *Ammonites* seen to depth of	14	0

In drawing attention to the irregular cementation of the Spilsby Sandstone, Mr. Strahan has expressed the opinion that in places it had originally been a hard calcareous sandstone, and that subsequently portions of the calcareous matter had been replaced by carbonate of iron, which would be converted into the oxide wherever the rock was exposed.   In places near Donnington he found the concretionary structure to take "the form of vertical, oval, or round cylinders 3 to 6 feet in diameter, and 4 to 5 feet high.   The cylinders are formed of six or eight concentric rings, and stick up through the grass like the stumps of hollow trees.   The sandstone is pale yellow or white."   Elsewhere a kind of botryoidal weathering has been noticed, and this tendency to irregular weathering has led to a great number of blocks of the Spilsby Sandstone being incorporated in the Glacial Drift, in Norfolk as well as in Lincolnshire.†

A boring at Skegness was carried through Drift and Cretaceous Beds, &c., to a depth of 321 feet.   Then followed the Spilsby

* Geol. Lincoln, p. 91; see also H. Keeping, Quart. Journ. Geol. Soc., vol. xxxvii. p. 239.

† Geol. Lincoln, pp. 88, 91–93.

Fig. 143.

Section near Donnington-upon-Bain, Lincolnshire. (A. Strahan.)

Longitudinal scale, 6 inches to 1 mile.   Vertical scale twice exaggerated.

6. Tealby Clay.
7. Claxby Ironstone.
8. Spilsby Sandstone.
9. Kimeridge Clay.

1. Lower Chalk.
2. Red Chalk.
3. Carstone.
4. Upper Ironstone.
5. Tealby Limestone.

Sandstone, 26 feet, and Kimeridge Clay, 78 feet; the boring having been carried to a depth of 425 feet.[*]

In addition to phosphatic nodules, the nodule-bed at the base of the Spilsby Sandstone, contains phosphatized fossils, many of them internal casts, much waterworn and scarcely recognizable, but some show a nacreous lustre. The following have been recorded :—[†]

Ammonites biplex.	Cyprina.
—— plicatilis.	Isocardia ?
—— spectonensis.	Lima.
Belemnites.	Lucina portlandica.
Natica.	Myacites or Panopæa.
Pleurotomaria.	Pectunculus.
Arca.	Thracia.
Astarte.	Trigonia.
Cardium.	Terebratula ovoides.
Cucullæa.	Waldheimia Woodwardi.

Some of these fossils have been derived from the Kimeridge Clay; others, like *Lucina portlandica*, have been derived from the Portland Beds.

The following species have been recorded as indigenous to the Spilsby Sandstone, many of the specimens having been obtained from the large concretionary masses of calcareous sandstone :—[‡]

Ammonites stenomphalus.	Inoceramus.
—— plicomphalus.	Lima tombeckiana.
—— multiplicatus.	Lucina crassa.
—— cf. rotundus.	—— lirata.
—— subditus.	Panopæa or Myacites.
Belemnites lateralis.	Pecten cinctus (? Pavlow).
—— russiensis.	Pecten orbicularis.
Pleurotomaria.	Pinna.
Trochus.	Trigonia cf. alina.
Astarte.	—— Keepingi.
Cardium subhillanum.	—— cf. Moretoni.
Cucullæa donningtonensis.	—— robinaldina.
—— errans.	—— tealbiensis.

Mr. G. Sharman, who identified the species collected by the Geological Survey, says: "Judging from the Survey specimens alone, there is a preponderance of Oolitic forms, but Mr. Keeping gives a greater proportion of Neocomian forms. The stratigraphical evidence, it seems, points to these beds as being of Lower Neocomian age, but the fossils they have yielded form a very distinct group with strong Oolitic affinities. It is tolerably evident, therefore, that these 'Calcareous concretions' occupy a lower horizon than any Neocomian beds hitherto described, and in so far as Palæontological evidence goes, seem to occupy an intermediate position between the Lowest Neocomian and the upper-

---

[*] Strahan, Geol. East Lincolnshire, p. 169; and Jukes-Browne, Quart. Journ. Geol. Soc., vol. xlix. p. 472.

[†] Geol. E. Lincolnshire, p. 139; Geol. Lincoln, p. 93.

[‡] W. Keeping, Fossils of Upware and Brickhill, p. 64; H. Keeping, Quart. Journ. Geol. Soc., vol. xxxviii. p. 241; Jukes-Browne, Geol. E. Lincolnshire, pp. 140, 141; A. Strahan, Geol. Lincoln, p. 102; and A. Pavlow, Bull. Soc. Imp. Nat. Moscow, 1891, p. 159.

most Oolites."* Prof. A. Pavlow has since published some remarks on the species that have been recorded from Spilsby. That given by W. Keeping as *Amm. Kœnigi* he believes to be *A. subditus*, Trautsch. The evidence on the whole tends to justify the grouping of the Spilsby Sandstone as a marine equivalent of the Purbeck Beds.

The Claxby Ironstone is described by Mr. Strahan as "a yellow ferruginous clay packed with minute spherical oolitic grains of iron-oxide, and very fossiliferous. It rests with a generally sharp base on the Spilsby Sandstone, but passes quite gradually up into the less ferruginous mass of the Tealby Clay . . . . . The grains are almost perfectly spherical, polished, and about the size of millet seed. They are made up of alternating concentric layers of clear silica and opaque hydrated sesquioxide of iron."† The grains are sometimes cemented by a calcareous sandy matrix; and in this form the bed has been worked as an iron-ore. According to Prof. Judd, the iron-ore may have been known to the Romans, as slag has been found in association with Roman pottery, near Claxby.

This bed is represented at Hundleby, and is well shown at Donnington-upon-Bain, both in the Benniworth railway-cutting and in a pit west of Donnington railway-station. At the Acre House Mine its thickness was 13 or 14 feet as noted by Prof. Judd. It thins out to the north and disappears "a short distance beyond Caistor."‡ It has been proved to the south-east in the borings at Willoughby, near Alford, and at Skegness.

The fossils from the Claxby Ironstone show an admixture of forms recognized as Neocomian or Lower Cretaceous and Jurassic, but that any distinction can be made in the bed itself seems utterly impossible, though the attempt has been made to separate some of the zonal fossils by Prof. Pavlow. The majority of the species are distinctly Neocomian (or Lower Cretaceous) in character.

List of Fossils from the Claxby Ironstone.

⊕ Ammonites Beani (lower bed).
⊕ —— Blakei (lower bed).
×⊕ —— noricus (upper bed), Pavlow.
× —— nutfieldensis.
⊕ —— regalis (upper bed).
‡ —— plicomphalus.

⊕ Belemnites explanatoides (upper bed).
⊕ —— jaculum (upper bed).
× ‡ —— lateralis.
‡ —— quadratus.
⊕ —— russiensis (lower bed.)
‡ Emarginula.

× Geol. Survey: Geol. North Lincolnshire, pp. 108, 109; Geol. East Lincolnshire, p. 141.
‡ Il. Keeping, Quart. Journ. Geol. Soc., vol. xxxviii. p. 241.
⊕ Authority of A. Pavlow and G. W. Lamplugh, Argiles de Speeton, 1892.

---

* Geol. East Lincolnshire, p. 141; Études sur les couches Jurassiques et Crétacées de la Russie, I., Bull. Soc. Imp. Nat. Moscow, 1889.
† Geology of Lincoln, pp. 94, 95; see also Judd, Quart. Journ. Geol. Soc., vol. xxvi. p. 329.
‡ Fox-Strangways, Geol. N. Lincolnshire, p. 108.

‡ Neritopsis.
‡ Pileopsis neocomiensis.
‡ Pleurotomaria neocomiensis.
‡ Trochus.
‡ Turbo.
× Arca Raulini.
‡ Astarte robusta.
‡ Avicula macroptera.
× Cucullæa gabrielis.
‡ Cyprina.
× Exogyra conica.
× ‡ —— sinuata.
‡ —— tombeckiana.
× Lima.
× Lucina.
× ‡ Modiola.
‡ Myacites.
‡ Ostrea frons var. macroptera.
× Panopæa.
× ‡ Pecten cinctus.
× —— orbicularis.
× ‡ —— striato-punctatus.
× Pholadomya Martini.
× Pholas.
‡ Sphæra.

‡ Sowerbya.
‡ Trigonia ingens.
× —— nodosa.
‡ Rhynchonella lineolata ?
—— multiformis.
‡ —— speetonensis.
‡ —— Walkeri.
‡ Terebratula depressa var. cyrta.
‡ —— prælonga.
‡ —— sella.
‡ Waldheimia faba.
× —— hippopus.
‡ —— var. tealbyensis.
× —— Juddi.
× ‡ —— tamarindus.
‡ —— Walkeri.
× Serpula antiquata.
× —— filiformis.
× —— gordialis.
‡ —— lophoides.
× —— plexus.
× Vermicularia.
‡ Nucleolites.

× Geol. Survey : Geol. North Lincolnshire, pp. 108, 109 ; Geol. East Lincolnshire, p. 141.

‡ H. Keeping, Quart. Journ. Geol. Soc., vol. xxxviii. p. 241.

CHAPTER XV.

GEOLOGICAL STRUCTURE, SCENERY, AND AGRICULTURE.

General Structure of the Jurassic Area.

The physical features of the country occupied by the Middle and Upper Oolitic rocks are very much modified by the irregular overlaps of the Lower and Upper Cretaceous strata. The main mass of these newer formations now and again entirely conceals these Oolitic rocks or stretches across their outcrops in irregular spurs with occasional outliers.

We have evidence that prior to these overlaps, the Jurassic rocks were in many places folded and faulted, and over wide areas denuded. The evidence of great folding is shown when we draw sections to show the underground structure from Battle to Chatham (Fig. 144, p. 298) or from Faringdon to Dover (Fig. 145, p. 299). Though diagrammatic, these sections are based on deep borings and therefore show the general structure. Evidence of folding and faulting prior to the overlap is shown in the country near Beaminster and Weymouth, and again near Sherborne. Owing to these disturbances it is difficult to depict the underground geology beneath the Chalk Downs of Dorchester, although the evidence obtained in making the railway at Ridgeway showed the remarkable double system of faulting to which attention has been directed. (See Fig. 132, p. 260.) The faults and disturbances in this southern part of Dorset have since been worked out in detail by Mr. Strahan, who has found evidence of pre-Cretaceous folding along the line of the Weymouth anticline, which is rucked up from south to north.*

In the Vale of Wardour the Upper Cretaceous strata are seen to extend across the Wealden, Purbeck, and Portland Beds on to the Kimeridge Clay, and there is evidence of pre-Cretaceous disturbance in the gentle anticline of the Chilmark ravine, rucked up from east to west.

The general easterly or south-easterly dip of the Jurassic strata is thus locally modified by flexures of considerable importance, especially when we consider the underground structure in the south-east of England. At present, however, we know nothing of the extent of the strata beneath the Cretaceous rocks of Hampshire.

The evidence shows that during or prior to the accumulation of the Lower Greensand, the Jurassic rocks and even the Wealden Strata in the west of England were subjected to wide denudation ; and that afterwards during the Upper Cretaceous period a still

---

* Six-inch maps of the Dorset area, geologically coloured by Mr. Strahan, are mounted as a wall-map in the Museum of Practical Geology : See also his paper in Quart. Journ. Geol. Soc., vol. li., p. 549.

more extensive plain of denudation was formed. The Lower Greensand was then partially removed, while the Gault and Upper Greensand, and the Chalk doubtless spread across the entire Jurassic area, and far beyond ; and in due course the slightly tilted Oolites were further planed off—so that in the south of England these Upper Cretaceous rocks stretch across the denuded outcrops of all the members of the Jurassic and New Red Series on to the borders of the Palæozoic regions of Devonshire. The present features have thus resulted after a series of disturbances and great denudations.

In areas where the Cretaceous rocks have been removed, the ages of the faulting and disturbance cannot be fixed, but attention has been drawn to the subject by Prof. Judd in his remarks on the Jurassic rocks of Northamptonshire and Rutlandshire.*

Escarpments and Outliers.

Reference has been made to the varying character of the escarpments, and to the varying elevations attained by different formations.† In some areas we have a succession of escarpments, separated by vales, as in passing westwards from Dinton to Crewkerne, or from Swindon to Bath. In other areas several escarpments may be locally concentrated as it were, as near Bath and Minchinhampton, where the summits are formed of Great Oolite and the vales of Lower Lias.

Of all the present escarpments there seems to be little doubt that the Chalk escarpment was the first formed, for although the previous plain of marine denudation was by no means a strictly level tract, yet there were no decided escarpments in the Liassic and Oolitic rocks formed or retained previous to the deposition of the Chalk.‡ The present relative nearness or distance of the successive scarps of the Jurassic rocks, has resulted from the way in which the beds were locally tilted, and abraded in the formation of the pre-Cretaceous plain.

How far the Tertiary plains of marine denudation may have extended, there is no evidence to say, though it seems likely that in Eocene and Pliocene times much of the Chalk was removed by marine action, while during the intervening Miocene period great subaërial denudation must have taken place.

From the elevated region of Lower Oolites in the country between Daventry and Naseby there flow in diverse directions some of the chief rivers that intersect the Jurassic rocks—the Warwickshire Avon that runs into the Severn Valley ; the Nene, the Welland, and the Great Ouse that flow into the Wash ; and the Cherwell, one of the tributaries of the Thames. As Prof. Judd has pointed out, the confluence of a number of very

---

* Geol. Rutland, pp. 176, 177, 255, 259. See also Hull, Quart. Journ. Geol. Soc., vol. xi. p. 480.
† Memoir on the Lower Oolitic Rocks of England, p. 459 ; See also Judd, Geol. Rutland, p. 264.
‡ H. B. W., Geol. East Somerset, &c. 1876, p. 201.

FIG. 144.

Diagram-section from near Battle northwards to Chatham.

Distance about 80 miles.

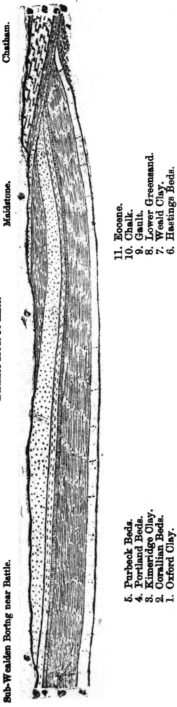

Sub-Wealden Boring near Battle.

Maidstone.

Chatham.

5. Purbeck Beds.
4. Portland Beds.
3. Kimeridge Clay.
2. Corallian Beds.
1. Oxford Clay.

11. Eocene.
10. Chalk.
9. Gault.
8. Lower Greensand.
7. Weald Clay.
6. Hastings Beds.

This diagram differs from one drawn by Mr. Whitaker, in which the Corallian Beds are shown to taper away to the north, and the Weald Clay to overlap the Upper Oolites and rest directly on Oxford Clay.*

* Quart. Journ. Geol. Soc., vol. xliii. p. 47.

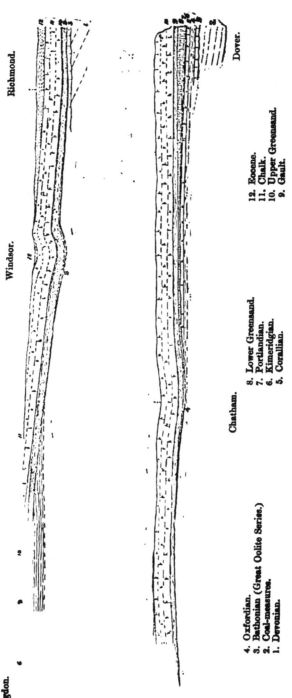

FIG. 145.

Diagram-section from Faringdon in Berkshire to Dover.

Distance about 32 miles.

12. Eocene.
11. Chalk.
10. Upper Greensand.
9. Gault.

8. Lower Greensand.
7. Portlandian.
6. Kimeridgian.
5. Corallian.

4. Oxfordian.
3. Bathonian (Great Oolite Series.)
2. Coal-measures.
1. Devonian.

considerable rivers, namely, the Witham, Welland, Nene, and Great Ouse, which, with their tributaries, drain a very large extent of country, has effected a breach in the great mass of Chalk strata, and thus the sea has been able to find admission to, and to operate on, the soft clays of the Middle and Upper Oolites. While the harder Chalk rocks have been cut back over a breadth of only 20 miles, at the mouth of the Wash, the soft clays over the area known as the Fenland or Bedford level, have been overspread by the sea for a far greater distance.*

The Cotteswold Hills may be described as a land of combes, as testified by the names of villages. In seeking to account for some of the dry valleys, E. Witchell pointed out that formerly the Fuller's Earth extended further westward, that the springs then thrown off from it would have helped to excavate the valleys, and as the Fuller's Earth was denuded, the combes became dry.† Certain valleys may have been started in this way; but a more probable explanation seems to be that of denudation on frozen surfaces, in the manner suggested by Mr. Clement Reid to account for some of the dry valleys on the South Downs. There the result has been to form the Coombe Rock, a gravelly deposit made up of chalk and flint. In the Cotteswold area, we have the Oolitic rubble, to which attention was previously directed.‡

The main features of all our escarpments were formed prior to the Glacial period, as shown by the fact that Boulder Clay and other Glacial Drifts occur both on the uplands and in the vales in the area from Buckinghamshire to Lincolnshire. At the same time, as Professor Judd has pointed out, in reference to Rutlandshire and Northamptonshire, many of the minor valleys on the plateaus and some in the lower grounds have been formed in Post-Glacial times;§ and Mr. Jukes-Browne is of opinion that the Steeping valley in Lincolnshire is for the most part of similarly recent age.‖

Various outliers have been left as monuments of the former extent of the strata. In few cases, however, are these at a great distance from the main mass of the formations. The exceptions being with those Liassic outliers in Shropshire and Cumberland, and these owe their preservation to causes unconnected with the particular formation of the main escarpments.

The general inclination of the strata towards the east and south-east, even if we estimate a dip of 1 degree or 1 in 57, would in most cases have carried the original formations above the present elevations of the land, and consequently the outliers that do occur are not far removed from the escarpments. The outlier of Inferior Oolite which gives rise to Bredon Hill has been preserved by means of a fault that lowered the level of the strata.

---

* Judd. Geology of Rutland, p. 262.
† Proc. Cotteswold Club, vol. iv. p. 214.
‡ Memoir on the Lower Oolitic Rocks of England, p. 462.
§ Geol. Rutland, pp. 262, 263.
‖ Geol. East Lincolnshire, p. 118.

In the vale below the Cliff of Lincoln the absence of outliers is noteworthy; and it is by no means improbable that some may have been removed by ice-action during the Glacial period. To such destruction indeed we may attribute the large masses of Marlstone and Lincolnshire Oolite that have sometimes been found incorporated in the Boulder Clay. (See Map.)

Brent Knoll that lies to the south-west of the Mendip range probably owes its origin in part to subaërial and in part to estuarine agencies.

Knolls like Brent Knoll and Glastonbury Tor are met with in many other parts of the south-west and west of England, but nowhere else are they so conspicuous, from the fact that at these localities they are further removed from the main mass of Oolites in the escarpment. Near Bridport many other outliers of the Sand and Inferior Oolite are met with. Further north again similar Knolls occur near Montacute and other places between Ilminster and Castle Cary. Those near South Cadbury indicate how the severance of outliers from the main mass may take place. Subterranean drainage and erosion in the first instance may lead to the formation of underground channels in the impervious Lias Clays beneath the porous Midford Sands and the limestones of the Inferior Oolite. The limestones themselves may be in part wasted by chemical dissolution, and if channels are formed in subjacent strata, slight subsidences must take place here and there, and pave the way for the disconnection of portions of the main limestones to form outliers. The subsequent more complete isolation of the severed masses is due to the superficial and subterranean denudation by rain and streams. Thus in the case of Brent Knoll, as well as in that of Glastonbury Tor (to some extent), the influence of the estuarine waters that once spread over the Somersetshire levels must have helped to complete the denudation. The preservation of both Brent Knoll and Glastonbury Tor, however, appears to be due to the basin-shaped arrangement of the strata, and this, although slight, has exercised some influence on the agents of subaërial denudation.[*]

It has been observed that the Kimeridge, Oxford, Lias, and other clays present, when their surface is exposed, a "corrugated character," and "the cutting through one of these subterranean stanks or ridges will often lay a large tract immediately dry. This was the case in the City of Oxford, when a system of deep sewer-drainage was attempted, by which the house-wells were laid dry through a considerable district, till the stank of clay cut through was restored by artificial means, when the water was restored also."[†] C. Moore observed similar deep furrows at Bath on the surface of the stiff Lias Clays where the gravels rest upon them.[‡] These features appear to me to be due to subterranean erosion.

---

[*] H. B. W., Proc. Bath Nat. Hist. Club, vol. vi. p. 125.
[†] Rev. J. C. Clutterbuck, Journ. Roy. Agric. Soc., ser. 2, vol. i. p. 277 ; Rep. Brit. Assoc. for 1860, p. 75.
[‡] Proc. Bath Nat. Hist. and Antiq. Field Club, vol. ii. p. 87.

Landslips.

Landslips have occurred here and there on the borders of the Corallian and Portland escarpments, but as a rule they are not of great extent, nor so important as those that have affected the Lias and Lower Oolites.

At Hermitage, in the Vale of Blackmore, a landslip of Corallian rocks over Oxford Clay occurred in 1585. Near Claxby, west of Normanby-on-the-Wolds, in Lincolnshire, landslips have taken place, owing to the superincumbent strata slipping over the Kimeridge Clay.

On Portland isle many landslips have occurred from time to time, and there the natural fissures in the Portland Stone have facilitated the breaking away of great masses of strata over the slippery foundation of Kimeridge Clay.*

Over the same Clay considerable founders have taken place of Portland Beds at St. Alban's Head, and of Cretaceous rocks at White Nore.

Coast Scenery.

So far as the coast scenery is concerned, the features of the Jurassic rocks are much modified by the coverings of Cretaceous strata, excepting as regards the lower beds of Lias in Glamorgan-shire, and those of Watchet on the Somerset coast, where we can well study the formation on a small scale of a plain of marine denudation.

East of Burton Bradstock the coast onwards to Portland Bill is protected by the great Chesil Beach.†

Along the coast east of Ringstead Bay the sea has worn back the hard Portland rocks that formerly stood out in cliffs like those of Gad Cliff and St. Alban's Head. It has breached the narrow fringe of these rocks and hollowed out the bays of Lulworth Cove and Worbarrow in the soft Purbeck and Wealden strata. At Stair Hole it has made its entry through enlarged fissures, which have been widened into arches, and produced a sea-loch not unlike the " Buller of Buchan " figured by Sir Archibald Geikie.‡

Kimeridge Bay has been formed by the enlargement of an inlet eroded originally by streams. An anticlinal arrangement of the beds is here apparent, and this structure may have influenced the agents of subaërial denudation, to which the lateral valleys were due. The cliffs that slope inland were no doubt protected by cappings of Portland Beds, now removed by the encroachment of the sea. The base of the cliffs is strengthened by layers of cement-stones which form the well-known Kimeridge ledges ; in fact, the irregularities along the coast-line of the Kimeridge Clay are due to ledges formed by these cement-stones, which serve to arrest marine denudation.

Scenery of the Middle and Upper Oolites.

The country occupied by the Middle and Upper Oolites con-sists in great part of low lands. In the south-west of England as

---

* Conybeare and Dawson, Memoir on Landslips ; Fitton, Trans. Geol. Soc., ser. 2, vol. iv. p. 218 ; and O. Fisher, Geologist, vol. vi. p. 250.

† See Reports on Coast Erosion, Rep. Brit. Assoc. for 1885, pp. 422–427, and 1888, p. 900.

‡ Class-Book of Geology, 1891, p. 71.

far as Oxford, the rocks extend in two vales separated by the low escarpment of the Corallian Rocks—the continuity being obscured only by the Cretaceous rocks of the Dorsetshire and Wiltshire Downs.

In the Oxford Clay, in addition to the minor vales near Weymouth, we have Vale of Blackmore in Dorset, the valley of the Upper Thames, Otmoor above Oxford, and the Vale of Bicester.

The Corallian escarpment is shown between Weymouth and Abbotsbury on either side of the Weymouth anticline. Northwards it is fairly well-marked throughout its course, though interrupted near Penzlewood and again near Devizes and Calne. It never attains any very great elevation, perhaps the outlier of Wytham Hill, 583 feet, is the highest point on the Corallian rocks, south of Yorkshire.

The Kimeridge Clay is well shown in the Vale of Kimeridge and in portions of the area between Osmington and Abbotsbury. Northwards it forms the Vale of Shaftesbury extending partly into the Vale of Wardour; and further on in conjunction with the Gault forms the vales north of Westbury in Wiltshire, and near Swindon, the Vale of White Horse* that lies between Faringdon, Abingdon, and Wantage, and part of the Vale of Aylesbury. Many of these vales, like those in the Lias of the midland counties, form well-known fox-hunting districts.

Further north the Oxford Clay and Kimeridge Clay are parted chiefly by clays of Corallian age, covered in places with accumulations of Drift. So that we have a wide tract of low land extending from Winslow and Stewkley through Bedfordshire and Cambridgeshire to the Fenland.

Northwards from Bourn through great part of Lincolnshire by Horncastle and Market Rasen, beneath the Lincolnshire Wolds and the more diversified tracts of Spilsby Sandstone, there extends a vale formed of the same series of clay formations.

Over parts of the area extensive tracts of woodland formerly existed. Thus Gillingham Forest, which included the White Hart Forest of the Vale of Blackmore, extended in Leland's time as far south as Yeovil, and northwards joined the Forest of Selwood, which extended from near Wincanton over the Greensand heights of Penzlewood (Pen Selwood), and along the vale to Frome.

In North Wilts there was the Forest of Bradon or Braydon, known also as Orwaldes Wood, which was disafforested in the reign of Charles I., and which extended over the vale from near Melksham and Chippenham to Malmesbury, Purton, Minety, Cricklade, and Wootton Bassett. Bernwode, or Brentwood, near Brill, Whaddon Chase north-east of Winslow, and St. Ives Heath,† were other noteworthy tracts of woodland.

The Portland Beds form a pleasant escarpment around the vale of Kimeridge Clay in the Isle of Purbeck. Encombe,

---

* From the White Horse cut in the Chalk south of Uffington.
† J. K. Watts, Rep. Brit. Assoc. for 1852, p. 63.

known as the Golden Bowl, owes its shape to subaërial denudation effected in part by landslips, which have given an amphitheatre-like form.[*] It is remarkable, however, that but little detritus of Portland Beds occurs over the area, except along the deeper ravines, as in Renscombe. The Portland Beds form steep slopes bordering the vales of Kimeridge Clay as near Portisham, and again in the Vale of Wardour.

Superficial Accumulations and Drift.

No superficial accumulations of any extent are found over the Middle and Upper Oolites in the country to the south-west of Aylesbury and Buckingham.

Occasional valley-drifts are found as at Radipole, near Wey-mouth, whence the Mammoth has been obtained. Gravels formed largely of limestone débris are met with here and there in the vales of Dorsetshire; in the Avon Valley near Trowbridge, near Melksham, and again in the country from Chippenham north-eastwards, by Tytherton, Christian Malford, Dauntsey, and Somerford, and more particularly in the Upper Thames Valley, near Cricklade, and Lechlade.

To the north-east of Aylesbury and Bicester the country is largely covered with Boulder Clay and Glacial gravels.

The clay-country includes much permanent pasture and meadow land, although the value for dairy and grazing-purposes varies considerably in different areas. Corn and beans are cultivated in places; and as a rule the land is divided by good hedgerows. Some of the richest grazing and dairy lands in the vales are those areas where, as in North Wiltshire, the Oxford Clay is covered with gravel, or the Kimeridge Clay is lightened by the downwash of material from the Greensand; while the Vale of Aylesbury, said to be the richest meadow and pasture-land in England, and dividing its honours between the Gault and Kimeridge Clay, yet owes much to material derived from the waste of Upper Greensand and Portland Beds, as well as to coverings of rich vegetable mould and scattered accumulations of Drift.

The North-Wiltshire cheese is made on the dairy-lands chiefly of Kimeridge but also of Oxford Clay. Near Cambridge the Cottenham cheese is a product of the clay-land pastures. Stilton cheese, most largely made in Leicestershire, took its name from Stilton, on the Oxford Clay, in Huntingdonshire—cheeses being formerly sold to passengers who travelled that way by coach.

SOILS.

Oxford Clay.

The area occupied by Oxford Clay is, much of it, gently undulating ground; and here and there we find a low escarpment, above

---

[*] See Buckland and De la Beche, Trans. Geol. Soc., ser. 2, vol. iv. p. 34.

'the Cornbrash, where the sandy strata of the Kellaways Beds are developed. A good deal of the ground, however, is very low-lying and almost Alluvial in aspect; the soil is a stiff and heavy clay,* and the slopes are damp and rushy in places. The clay-flats near Oxford, including the district of Otmoor, are often much flooded in the autumn and winter. The hedgerows, as a rule, are well-timbered, with oak and elm, ash, willow, and poplar.

The soil in general is considered expensive to cultivate; it is improved by lime-manure, and much draining is required. In dry weather the cracks and fissures formed in the ground extend to some depth, as I noticed near Witham Friary in 1885.†

The country is thinly populated, and there are many green lanes in Dorsetshire and Wiltshire.

Through Buckinghamshire, Bedfordshire, Huntingdonshire, and further north, the Oxford Clay is largely covered with Boulder Clay, &c., so that the soils are not so largely influenced by it, and there is more land under the plough. The features are thus partly due to the Drift, as at Alconbury and Kimbolton Hills. Numerous spinneys or thorn woods occur in this region of Huntingdonshire.

Prof. Judd remarks that the Oxfordian strata constitute a band of country, in the western part of the Fenland, which rises into numerous swelling hills, usually of no great elevation. The land is in the main devoted to grazing purposes, but some considerable areas of it have been brought under the plough, while others remain as woodland. The clay in places is dug out in trenches for the purpose of getting material to clay the land. He further states that the rapidity with which the surface-water soaks away over the areas throughout which the sandy Kellaways Beds outcrop, causes their soil to present a remarkable contrast with that of the districts occupied by the stiff and impervious portions of the Oxford Clay. The light-coloured sandy soils thus formed, constitute what is locally known as "drummy land," and can easily be traced, over many miles of country, near the limits of the Oxford Clay and Cornbrash formations.‡

In Lincolnshire the Oxford Clay forms a stiff and cold soil, but this varies from the mild friable loam, characteristic of the Kellaways Beds, to very tenacious clay. Much of the land is under grass, but beans, wheat, oats, and clover are cultivated in places.

Corallian Beds.

The Corallian Beds yield a very variable soil, though for the most part light and brashy. Much of it is a rich, friable, sandy loam, generally brown, but sometimes of red colour. The ground is mostly under cultivation for corn and roots, and to some extent

---

* For analyses of soils on Oxford Clay in Wiltshire, see Dr. A. Voelcker, Journ. R. Agric. Soc., ser. 2, vol. ii. pp. 382, 383, 386.
† Memoir on the Lias of England and Wales, p. 319.
‡ Judd, Geol. Rutland, pp. 237, &c.

as meadow-land, but there are occasional rabbit-warrens on the sandy areas of the Lower Calcareous Grit.

In the Weymouth district, where there are few trees, except in sheltered hollows near the farmsteads, the land has a bare aspect, and the fields are divided by stone-walls and hedgerows.

Near Abbotsbury the ironstone-beds yield a bright red ferruginous soil, on which potatoes are much cultivated.

Kimeridge Clay.

The soil is as a rule heavy, requiring to be well-drained. Like that on the Oxford Clay, it is improved in places by lime-manure. Much of it is, however, rich, and in places phosphatic, while it is improved by downwashes from the escarpments formed by overlying strata.

The land is comparatively flat or gently undulating, and the greater part is pasture and meadow. Extensive dairy-farms occur near Seend, Potterne, and Westbury; and in this neighbourhood the by-roads consist chiefly of green lanes (known as bridle ways in Wiltshire and halter paths in Dorsetshire), very muddy, and almost impassable in wet weather. This clay in Wiltshire is considered a warm clay as compared with the Oxford Clay, so that the land, as a rule, is of more value.

Near the junction with the Portland Sands, or with the Lower or Upper Greensand where they rest immediately on the Kimeridge Clay, or again where thin gravelly patches occur, the soil is often more or less loamy, and there are occasional gorse-covered tracts, and commons. Wheat, beans, and potatoes are grown in places. West of Kimeridge Bay there is a small rabbit-warren in the dry shaly surface-soil on the top of the cliff.

Good hedgerows are generally found, and from the prevalence of the oak, the term "Oak Tree Clay" was applied to the Kimeridge Clay by William Smith in 1816, though in those old days it was not strictly confined to the formation.

The village of Kimeridge lies near the junction of the Kimeridge Clay and Portland Beds, and the rich pastures on the former tract contrast pleasantly with the scanty herbage on the latter. Further east, Encombe Valley, known as the "Golden Bowl," is marked by wooded and grassy slopes, which, like the Wealden tract further north, form a relief to the bleak Purbeck hills.

In the vale below Shaftesbury, in that near Westbury in Wiltshire, in the Vale of White Horse, and onwards by Nuneham Courtney to near Thame and the Vale of Aylesbury,[*] we have broad tracts of Kimeridge Clay, where the stiff soil is ameliorated in places by down washes from the Portland Sands, the Lower or Upper Greensand. Further on the Clay is concealed in places by patches of Boulder Clay, so that in its course through Cambridgeshire and Lincolnshire the character of the land is much modified

---

[*] An analysis of soil from Putlowes (? Putlers) north-west of Aylesbury, was given by C. S. Read, Journ. R. Agric. Soc., vol. xvi. p. 281.

by Glacial Drift. In Lincolnshire there are tracts of Drift sand of a comparatively sterile character; and there in former times the Kimeridge Clay was dug to spread over lands that needed clayey material.*

Portland Beds.

The Portland Beds of Dorsetshire form high-rounded and grass-covered hills with here and there bare rocks jutting out. The slopes are steep and the ground is not often cultivated, though supporting a herbage suitable for sheep-pastures. The steep slopes of Portland Sand are marked by slipped ground and sheep-terraces.

The land is divided mostly by stone fences. These remarks apply to the escarpment from Portisham to Upway, to the marked spur of Swyre Head formed of cherty Portland Beds (667 feet high), and to the combes of the Isle of Purbeck, at Renscombe and West Hill. The elder and the teazel grow abundantly on the débris that covers the slopes of Kimeridge Clay.

On both Portland and Purbeck Beds near St. Alban's Head and Worth, there are many old terraces of cultivation, known as lynchets.†

In the Vale of Wardour the ground is mostly under cultivation, roots and corn being grown; but there are few trees, so that the country occupied by the Portland Rocks looks bare, except along the wooded escarpment near Pyt House on the north, and near Wardour Castle on the south side of the vale.

In Buckinghamshire the Portland Beds give rise to a dry sandy and fertile loam, that forms useful arable land.

Purbeck Beds.

Portland Isle, which is covered almost entirely by Purbeck Beds, is a bleak and dreary tract owing to the absence of trees— there being a few only near Pennsylvania Castle and at Fortune's Well. The ground is cultivated in long and narrow strips in places, known as "Lawns," where corn, grass, and potatoes are grown. Near the higher lighthouse I counted 35 such strips within a space of 530 yards. Over great part of the Island there is good pasture for sheep.

In the Isle of Purbeck the soil is a brown and somewhat brashy clay, a foot to 18 inches deep. Much of it is heavy and requires draining, being in this respect like that of the Forest Marble.

Crops of corn, beans, potatoes and peas are grown; but much of the land is in pasture, partly owing to the broken nature of the ground, where the hollows remain from old workings for stone. Thus the Lower Purbeck Beds form a tract that is more cultivated

---

* Strahan, Geol. Lincoln, p. 80.
† See also F. Seebohm, The English Village Community, 1883, p. 5; and H.B.W., Memoir on the Lias of England and Wales, p. 313.

than that of the higher beds.   Near the edge of the cliffs, where the land is naturally drained, the gorse grows, and near Durlston Head there are fir plantations, but over great part of this Purbeck area there are few trees excepting near the habitations.   The fields are sometimes divided by hedgerows, but more often by stone walls.

In the Vales of Wardour the Purbeck Beds are mostly under cultivation.   On Lady Down the ground rises high above the Portland Beds, dipping thence generally towards the east.

In the Sussex area, as remarked by Mr. Topley, the soil formed by the Purbeck Beds "is mostly stiff, as clays and shales predominate in their composition ; a great part is woodland, only a small proportion being under the plough."*   Hops are cultivated in places.

In Lincolnshire the Spilsby Sandstone "everywhere forms a tract of light sandy soil between the heavy clays [Kimeridge and Tealby] below and above."†   Much of the area is under cultivation, but there is a pleasing mixture of arable land and woodland.

---

* Geology of the Weald, p. 33.
† Jukes-Browne, Geol. East Lincolnshire, p. 16.

CHAPTER XVI.

ECONOMIC PRODUCTS.

BUILDING STONES.

General remarks on the Building Stones of the Oolitic Series have been made in a previous volume.* In the Middle and Upper Oolites, freestones are obtained from the Corallian Beds, the Portland Beds, and occasionally from the Purbeck Beds. These include limestones more or less oolitic, and calcareous sandstones.

Corallian Rocks.

Among the Corallian freestones those of Calne, and of Marnhull and Todbere near Sturminster Newton, are the principal.

The Osmington oolite near Weymouth, the Goatacre freestone, north of Hillmarton, and other beds have locally been quarried, but never to any large extent.

The Marnhull stone was used in the churches of Gillingham, Hinton St. Mary, Sutton Waldron, &c. About 10 feet of freestone is worked, and blocks up to 20 cubic feet have been obtained. (See p. 103.)

The Headington Stone ("Shotover Limestone") east of Oxford, though it has been employed as a freestone, is for the most part exceedingly poor, many of the buildings in Oxford where it was used, having come to a deplorable state of decay. The upper beds of impure comminuted shell-limestone were those so much used; but all the beds are very variable and shelly, false-bedded shelly limestone passes laterally into rubbly coral-rock; and there are layers of shelly sand and hard lenticular bands of grey oolitic limestone. In the village of Headington Quarry most of the better stone has been removed, but here and there old quarries have been re-opened. A layer of the "best stone" (of its kind) occurs near the base of the limestones, about 5 feet above the top of the Sands of the Lower Calcareous Grit.

According to Plot, stone from the Wheatley quarries was used for some of the older buildings in Oxford;† it is a variable stone, but on the whole more durable than that of Headington.

Of other building-stones there are certain hard sandstones, shelly limestones, &c., which, although not freestones, have been found serviceable for building-purposes.

In the Lower Calcareous Grit the layers of hard calcareous sandstone furnish building-blocks that are quarried in many parts of Wiltshire.

At Westbury, in Wiltshire, the slag from the iron-furnaces is prepared with cement to form building-blocks, which have been used in some

---

* Memoir on the Lower Oolitic Rocks of England, p. 470.
† Plot, Nat. Hist. Oxfordshire, 1677, p. 76.

railway-bridges near Reading. A kind of paving-material is also made from the slag.

North-west of Highworth an oolitic and pisolitic rock belonging to the Upper Corallian Beds has been employed for building-purposes.

Portland Beds.

The merchantable stone of Portland comprises the Whit Bed and the Best Bed, the latter, notwithstanding its name, being of inferior quality, the name it is believed being a corruption of Base Bed. It has, however, been remarked by C. H. Smith that, " Formerly, when none but east cliff quarries were worked, the lowest bed was considered the best stone [the Whit Bed there being coarse and shelly], hence the name amongst the quarrymen of 'best bed;' but in the north, north-east, and west cliff quarries, the top bed, though not always the finest grained, is certainly the best stone for standing the weather."* Both beds thus vary sometimes even in the same quarry, as might be expected, and both are somewhat harder where the stone is unprotected by coverings of Purbeck Beds, as near the Bill and around the cliffs generally.

Towards Portland Bill, and at the Bill, the Whit Bed, becomes very flinty, and partly on that account it is not now worked there. In that neighbourhood the Curf or Middle Bed has yielded some good stone.

Both Whit and Best Beds are used for Scantlings, Landings, Sills, Steps, Cornices, Piers, Columns, &c., and also for paving and for monumental purposes. The Best Bed is considered as a rule the more suitable for fine carving, and for inside work generally ; the Whit Bed being as a rule harder and more durable as a weather-stone.

Blocks weighing 10 or even 15 tons are obtained, about 16 cubic feet of the stone weighing one ton; much larger blocks are often detached during the working of the quarries.

The refuse of the stone-beds is burnt for lime in a kiln on the top of the hill above Fortunes Well.

The top Roach, of which blocks weighing 20 tons have been procured, is employed, when wanted, for sea-walls, harbours, and docks, but is otherwise rejected among the waste-stone or " ridding." It stands great pressure, and is not affected by frost. It is considered to be much more durable than other beds of Roach, that occur at lower horizons, and consequently they are always rejected.

The thickness of the Building-stone series is from 20 to 28 feet ; and is on the whole rather thicker towards the west than at the eastern side of the island.

The Portland Stone was not much known beyond the neighbourhood of Portland before the beginning of the seventeenth century.

Among the old quarries were those known as the Weycroft quarries, where the stone was shipped from the King's Pier below. They were situated a little north or north-west of the Admiralty quarry, while the Grove quarries (Redcroft) were situated a little north-west of the Convict Prison. Important quarries are now worked at Weston, Kingbarrow, and near the Prison.

* Lithology, Journ. R. Inst. Brit. Architects, 1840, p. 158; and W. Gray, Proc. Geol. Assoc., vol. i. p. 142.

The Kingbarrow quarries lie to the west of the Weycroft quarries, near Yeales, and the Maggot quarry was a little further south. Other quarries were known as the Trade, Vern Street, Castles, and Goslings quarries.* During the construction of St. Paul's Cathedral (1675-1717) Sir Christopher Wren had control of the Portland quarries, the stone was carefully selected and seasoned, and some blocks said to have been quarried at the time are still lying about.†

Prof. T. R. Jones has remarked in reference to buildings constructed of Portland Stone in London, that sulphate of lime is sometimes formed by sulphuric acid derived from the London atmosphere ; and he points out that the black portions are due not merely to smoke, but probably also to a black *Protococcus*. "The degradation of this limestone by rain and carbonic acid, however, is considerable, for the cement stands up in edges, an inch high, on the parapet of Somerset House, and the component fragments of oyster shells almost as high."‡

On the whole, however, it is considered that Portland Stone is less affected by the London atmosphere than other Oolitic limestones. St. Paul's has stood well, but, as already remarked, great care was used in the selection of the stone, and many blocks were subject to long seasoning before being employed. The chief symptoms of decay are said to be in those parts of the edifice that are most exposed to the south-west winds and rains.

The Portland Stone was used in the construction of the Monument on Tower Hill (1671-7), the Custom House (1815), General Post Office (1829), Somerset House (1776-92), British Museum at Bloomsbury (1753), St Pancras Church (1819-22), &c § It was also used for the Horseguards, the India House and Foreign Offices in Downing Street, the Post Office Savings Bank, the Reform Club, the National Liberal Club, the National Provincial Bank in Piccadilly (1893), the new Record Office in Chancery Lane (1895), &c.

On Portland itself, the stone was locally used for the old Bow and Arrow Castle, and, on the mainland, for the later Castle of Sandsfoot.

The "Purbeck-Portland stone" is said to excel the true Portland stone in the qualities of closeness, slight absorption, hardness, and durability, characters which improve in an easterly direction from St. Alban's Head, so that the old quarries of Tilly Whim yielded the most durable stone. However, the stone there is so much harder that it is thought on this account the old quarries were disused.

With few exceptions all the quarries of the Purbeck-Portland area are in the face of the sea-cliffs, "the first portion being cut down perpendicularly from the crest, which when effected at once by blasting is termed *ridding*, so as to form a platform level with the base of the merchantable stone, which in most of them is afterwards extracted by driving galleries into the rock, forming deep caverns, and leaving pillars for the support of the superincumbent mass. From the position of the quarries the produce can only be shipped in very calm weather, so that the greater part of the year they are unapproachable."‖

---

* Report on the Selection of Stone for the New Houses of Parliament, 1839, p. 4.

† Memoir of Sir Christopher Wren, by L. Phillimore, 1893, p. 221.

‡ Proc. Geol. Assoc., vol. vi. p. 250. See also Memoir on the Lower Oolitic Rocks of England, p. 474.

§ Builder, September 18, 1858.

‖ P. Braunon, Guide to Swanage, 1872, p. 30.

e 82428.

The stone from Tilly Whim was used in the construction of Corfe Castle; elsewhere the stone obtained from near St. Alban's Head, at Seacombe, Winspit, &c., and known as "Cliff Stone," is employed for building-purposes, for sinks, troughs, steps, landings, curb-stones, columns, rollers, &c.

At Seacombe the stone is mostly obtained by tunnelling. Where excavated it has a pale bluish-grey appearance, but when cut, dressed, and dried it has a yellow appearance. The stone varies, some beds being too shelly to be of service as freestone.

At the Winspit Quarry the Under Freestone, on top of the chert-beds, is about 6 feet thick. It is used for sinks, curb-stones, &c., and is sent to the Isle of Wight, Southampton, Portsmouth, Brighton, &c. The top or Pond Freestone is considered the best bed; this is 7 feet thick. The House Cap above the Under Freestone, is a coarse-grained limestone, used for breakwaters, &c., and the "Blue Stone" above the Top or Pond Freestone is a hard, durable stone, used for gate-posts, &c.

Inland the Portland Freestone has been quarried at London Doors and other places on the Eldon estate, but not westwards towards Kimeridge. This may partly arise from the fact that at London Doors the beds are much shattered, and they are much weathered where exposed at the surface.

In the Vale of Wardour the Portland Beds, known as Wardour Stone or Tisbury Stone, are extensively quarried and mined for building-stone near Tisbury, at Chilmark, Chicksgrove, and Wockley; and formerly there were old quarries at Lower Lawn to the north of Tisbury.

The lower beds of the Portland Stone are those chiefly worked as freestone—they consist generally of greenish sandy limestones or calcareous sandstones, which become paler when dry. They are used not only for building-purposes, but for troughs, tombstones, &c. The most extensive quarries with galleries, are those in the Chilmark ravine, where beds of variable character are present, passing from sandy and glauconitic limestones into calcareous sandstone. Mr. Hudleston notices how full of quartz grains these beds are, but they appear to be cemented to a certain extent by opaline silica.*

The principal beds are, the *Trough Bed*, a sandy limestone about 2 ft. thick, considered the best weather-bed, and used for building-stone, steps, paviours, &c.; the *Green Bed*, a sandy and glauconitic limestone sometimes shelly, and about 5 feet thick; the *Pinney Bed*, a sandy and glauconitic limestone, a good weather stone, about 2 feet thick; and the *Fretting Bed*, a very sandy and partially calcareous rock, a little over 3 feet thick; the thicknesses above given being those of the merchantable stone.

The Portland Stone of the Vale of Wardour was employed in the Cathedrals of Salisbury, Rochester, and Chichester, in Wardour Castle, Longford Castle, Fonthill Abbey, Wilton Abbey, Romsey Abbey, Westminster Abbey (Chapter House), Christchurch Priory, Balliol College at Oxford, &c.

An upper and oolitic freestone has been obtained by means of galleries at Chilmark; it was employed in the west front of Salisbury Cathedral, and has lately been again worked.

---

* Proc. Geol. Assoc., vol. vii. p. 171.

The following statistics show the density and absorbent powers of some of the Portland and Purbeck rocks:—

Building Stones.	Specific Gravity of Dry Specimens.	Bulk of Water absorbed.	Weight : Per Cubic Foot.
		Per cent.	lbs.
Purbeck (Swanage) Stone -	—	—	169
Seacombe - - -	—	—	150
Portland Stone - - -	2·14	—	133 to 147
„ Roach - - -	—	—	155
„ Whit Bed - -	—	13·5	145
„ Base Bed - -	—	—	137
Chilmark Stone - - -	2·48	—	—
Tisbury Stone - - -	—	8·6	153

The following analyses show the general composition of some Portland rocks :—

————	Portland Stone.	Trough Bed, Chilmark Stone.
Carbonate of lime - - - -	95·16	79·0
„ „ magnesia - - - -	1·20	3·7
Iron, Alumina - - - - -	0·50	·2·0
Silica - - - - - -	1·20	10·4
Water and loss - - - -	1·94	4·2
Bitumen - - - - -	trace	trace
	100·00	99·3

Analyses of the Whit and Base Beds of Portland are said to show practically the same composition, although the former stone is as a rule found to be the more durable.[*]

The hard calcareous sandstone in the Upper Portland Beds at Swindon, known as Swindon Stone, has been quarried for building-purposes.

In the Portland Beds of Oxfordshire and Buckinghamshire there are beds of hard, compact, and sometimes sandy limestone that are employed locally for building-purposes. At Great Hazeley the top layer of the Portland Stone, called "Curl stone," was formerly used for chimney-pieces. It is a grey and somewhat gritty limestone with fossils.

Purbeck Beds.

At Swanage all the stone-quarries are in the Middle Purbeck Beds, while at Ridgeway the Lower Beds yield the best stone.

The stone-quarries or mines are situated on the ridge or hill above Swanage (Swanwich), and to the south of the town, but they extend westwards to Langton Maltravers, and there are a few on towards Kingston. There is said to be no good workable stone west of Kingston, and none is now worked, except in an occasional open quarry. The ridge at Swanage is indeed riddled

---

[*] Report with reference to the Selection of Stone for building the New Houses of Parliament, 1839. Analyses by Prof. Daniell, p. 30 ; see also Mem. Geol. Survey, vol. ii. Part 2, p. 691.

with pits. some deserted, many now worked (upwards of 90), and most of them marked by heaps of refuse, or "ridding," called "Scar-heaps," some of which are 20 feet high.

The stone is really mined, being worked underground, because the layers of merchantable stone are thin in comparison with the worthless strata associated with them. Moreover, the beds dip at a considerable angle, and are therefore not exposed near the surface over a very large area; they are worked along the dip, and the pits are from 40 to 60 feet deep, or even deeper.

Sloping shafts, about 5 feet square, are made, so that the stone can be dragged up in trucks by means of a capstan and chain, with the aid sometimes of a donkey. The stone is extracted by means of crowbars and wedges. The shafts are paved with stone for the trucks, and rude steps are cut alongside for the use of the workmen. As the stone is worked away the galleries are supported sometimes by timber, but more usually by blocks of useless rock. The stone is dressed at the surface, and then sent to merchants at Swanage, where the blocks or "bankers" are piled ready to be shipped or to be sent away by rail.

The quarries are distinct, and generally owned by the workers, and their area is marked out on the surface by stone-walls. Many curious customs still prevail in connexion with the labour, and it has been suggested that the quarrymen may be descendants of a colony of Norman stone-workers.* They form a distinct class of people, owing to their curious customs and intermarriages. The men constitute the Company of Marblers and Stone-cutters of the Isle of Purbeck, and maintain the privilege of confining the stone-trade to themselves and sons, rigourously excluding strangers.

The Purbeck Stone is shaped into curb-stone of various widths, gutter-stones, square slabs for paving, for steps, gate-posts, window-sills, and landings, sinks, horse-troughs, and granary rick-leg and cap stones, &c.

Tha walls are often built with gently undulating layers of stone, or layers inclined now in one direction now in another, with occasional horizontal layers, larger irregular slabs are piled on the top, either vertically or in a sloping direction.

The stone is much used in the south-east of England, being sent to Brighton, Portsmouth, Winchester, Salisbury, &c. Formerly the stone was largely used for paving in London, and Hutchins states that during the years 1764–1771, about 14,000 tons of stone were annually obtained. At the present day the output is from 15 to 20 thousand tons.†

In the beginning of this century much stone was obtained from the cliffs in Durlston Bay. Webster remarked that the mode of quarrying the Purbeck Stone "is by making excavations and inserting slight props; in the course of time, these props giving way, a part of the cliff falls down, and the fragments are worked by the masons into the forms proper for several purposes."‡

Formerly much stone was quarried on the hills above the Lighthouse near Durlston Head, as indicated by the heaps of rubbish, and there are shallow pits further north where portions of the stone-beds are repeated by a fault.

The Purbeck Stone Beds of marketable value, excepting the Marble beds, occur beneath the Corbula Beds, in the Middle Purbeck division. They were noted in detail by Thomas

---

* A good account of the Quarries is given by Mr. C. E. Robinson, in "A Royal Warren or Picturesque Rambles in the Isle of Purbeck," 4to, Etching Co. 1882.

† Report on the Dorsetshire Purbeck Stone for 1880 by C. Le Neve Foster and R. J. Frecheville.

‡ Englefield, Isle of Wight, p. xviii; Trans. Geol. Soc., ser. 2, vol. ii. p. 37.

Webster,* and later on by H. W. Bristow.† Their records differ in detail and in the grouping of the "veins," but the principal bands of stone are given in the same sequence. They may be stated as follows :—

		Ft.	In.
White Roach	· · · · · ·	4	6
Laning, Lane-ond or Leaning Vein	· · ·	5	6
Royal	· · · · · · ·	5	0
Red Rag	· · · · · ·	2	2
Rag	· · · · · · ·	1	0
Under Rag	· · · · · ·	1	6

		Ft. In.
Freestone Vein.	Top Shingle · · · · Shingle · · · · Under Picking · · · Lower Shingle · · · Grub · · · · Roach · · · · Pink Bed · · · · Grey Bed · · · · Thornback · · · · Freestone Bed · · · Blue Bed · · · ·	16 0 (including shales.)
Downs Vein.	Lias · · · · Lias Rag · · · · Laper · · · · Under Picking · · · Upper Tombstone Bed · · Brassy Bed · · · · Lower Tombstone Bed · · ·	13 0 (including shales.)

Below the Cinder Bed the following beds occur :—

		Ft. In.
New Vein	Button · · · · Feather · · · · Under Feather · · · Cap · · · ·	6 3 (including shales).
	Sly Bed (Waste) · · ·	2 0
	Upper 5-foot Bed · · ·	5 0
	Under 5-foot Bed · · ·	5 0
	Tombstone Bed · · ·	4 1
	Shear Bed · · · ·	0 4

Different beds are worked in the many quarries. As an example a section may be given of Mr. Squib's quarry, Swanage, which I visited in 1884. It was situated on the eastern part of the hill, and the following beds were worked :—

			Ft.	In.
Used for road-metal and paving.	Grub · · · · ·	0	5
	Roach · · · · ·	1	0
	Thick (or Pink) Bed · · ·	0	6
Used for curbs and steps.	Grey Bed	· · · · ·	0	7
Used for curbs and paving.	Thornback · · · · ·	0	10
Used for curbs, steps, and building.	Freestone · · · · ·	2	0

The rubbish here was used for lime-burning, road-metal, and for ballast on the railway. Very little roofing-material was obtained.

---

* Trans. Geol. Soc., ser. 2, vol. ii. p. 88.
† Vertical Sections, Geol. Survey, Sheet 22.

It has been remarked that "all the useful beds are broken up, by natural partings, into blocks and slabs of various sizes, generally irregular rectangles, varying in size from ten to twelve feet long by five to eight feet wide, down to eight or fifteen inches long by six or ten inches wide, and three to eight inches thick. The latter class are termed pitchers, the term 'horse pitchers' being applied to the larger sizes."*

The Laning vein is used for tombstones, paving, walling, and marine works.

The Freestone veins yield curb,' step, and tile-stone, and freestone used for all kinds of architectural work ; but none of the beds can be considered as durable as the better kinds of Portland freestone.

The Downs veins yield good paving-stone and tombstones.

The Feather bed, a hard bluish-grey shelly limestone, is considered the best stone for curbs and paving: associated beds are used for walls and marine works.

The New veins supply slabs and blocks for steps, tombstones, &c.

The beds at the base of the Upper Purbeck of Swanage, known as the Soft Burr, rough decomposed shell-limestones, are not now much worked, for the stone can only be obtained in small sizes. It is, however, considered a durable stone, and was employed in the construction of Corfe Castle, in the old tower of Swanage Church, and in the restoration of Wimborne Minster.

A quarry north of Langton Maltravers showed blue and grey shelly limestones (mostly blue hearted) and arenaceous limestone with shells and comminuted shells, chiefly bivalves, dissolved away. This latter rock is the "Soft Burr." A slight anticlinal was shown in the pit—the beds dipping to the north-east and to the west.

At Upway, near Weymouth, the Lower Purbeck Beds are quarried for building-stone, as previously mentioned (p. 259). The spire of All Saints' Church, Dorchester, was built of the "Cypris freestone."

In the Purbeck Beds of Portland the Cap beds are too hard and uncertain in fracture to be of use, for they cannot be "squared." They are blasted in the quarries. The Soft Burr is sometimes used for local building-purposes, and it is laid according to the bedding, though it is but a rough stone. It is employed for coigns of chimneys as it stands fire. The Bacon Tier is occasionally used for walling in Portland. The Hard Slatt is sometimes employed for paving, as at the Brewery at Wyke, but it is usually too hard to square ; it is used also for walling.

Purbeck Marble.

In his "Picturesque Rambles in the Isle of Purbeck," Mr. C. E. Robinson gives an interesting sketch of the history of the workings of Purbeck Marble or "Marble Rag." In early times it was one of the best marbles readily obtainable in England, and consequently it was much used in our principal Abbeys, Cathedrals, and Churches. The numerous columns, the fonts, and the tombs, were often chiefly formed of Purbeck Marble. The stone for the pillars in the interior of Salisbury Cathedral was obtained at Langton. In those early days the district was in a flourishing state, for Afflington (or Affington)—a name perpetuated in some farm-buildings near Kingston—was a market-town in the time of Henry III.

About the beginning of the sixteenth century the introduction of foreign marbles worked a change in the industry, and in course of time the quarries for marble were abandoned. Webster observed

* P. Brannon, Guide to Swanage, 1872, p. 28.

in 1811 that it "is now out of use, and the quarries are filled up, and scarcely known."[*]

Of late years occasional demands have arisen for the stone in the restoration of churches, and a considerable amount was required for the new church at Kingston designed by Street and built for Lord Eldon.[†] Here many columns and supports both inside and outside were made of greenish Purbeck Marble. Some of the columns were 7 feet in length.

Good specimens of Purbeck Marble may readily be procured near Peverel Point, and it is occasionally quarried inland, as at Easton, near Langton Maltravers, at West Orchard, and Woody Hyde, south of Corfe Castle.

The solid rock is grey, being coloured by carbonate of iron, and this is often rusted on the outside. The interiors of the *Paludina*-shells, which form the mass of the rock, are often filled with green matter, like that which tinges the *Unio*-beds. (See p. 253.)

South-east of Woody Hyde the Marble, chiefly of a grey colour, has been obtained by mining in a pit situated in a small plantation.[‡] Slabs were lying about of the following dimensions, 5 ft. × 16 in. × 2 ft.; 6 ft. 7 in. × 1 ft. 11 in. × 1 ft. 5 in.; and 2 ft. 3 in. × 3ft. 6 in. × 1 ft. 5 in. The thickness of the principal bed varies from 1 ft. 4 in. to 1 ft. 5 in.

South-west of West Orchard there were (in 1884) two small quarries opened in the Upper Purbeck beds, from which marble had been obtained. The beds dip N. 5° W. at an angle of 18° to 20°. The marble here occurs at the top of the pit, and is stripped off: the underlying beds comprising bluish-grey mottled sandy limestone with *Paludina* (2 inches), and a thin flaggy and sandy limestone (1 in. or more). Slabs of marble 4 ft. × 2 ft. × 11 in.; 3 or 4 ft. × 1½ ft. and 3 to 4 in. thick; and 2 ft. × 2 ft. × 8 in. are obtained. The shells are weathered on the surfaces and joints of the rock, and some of the beds are quite rotten, so that the Paludinas may be picked out. The marble is rather flaggy and splits up in places.

As the marble gets exhausted along its outcrop, it may in future be necessary to go deeper, even in places under the covering of Wealden beds where the dip is not too high. Indications of old workings are met with along the outcrop of the Upper Purbeck beds for some distance west of Swanage.

In the Vale of Wardour at Teffont Evias, the beds of hard smooth-grained limestone are worked for building-purposes. Fitton long ago suggested that some of these beds of uniform whitish limestone (locally known as "Lias") at Chicksgrove, might possibly be of service for lithographic purposes.[§]

Hard limestone belonging to the Purbeck Beds of Battle (a bed known as the Grizzle in the Bastard Blues) was formerly used for paving-stone.[||]

A band of hard limestone above the "Pendle," in the Purbeck Beds of Hartwell, has locally been used as pitching-stone.

The Spilsby Sandstone is described by Mr. Jukes-Browne as "a fine greenish sandstone, weathering brown, which can be easily quarried as a freestone." It has been largely used for building-purposes, in East Lincolnshire, and "the principal quarries seem to have been at Salmonby and at Holbeck, near Ashby Puerorum," also in Harrington Carrs.[¶] In

[*] Englefield's Isle of Wight, p. 172.
[†] The pavement outside is of Purbeck stone, and the main portion of the building of Purbeck shell-limestone (Burr). See Hudleston, Proc. Geol. Assoc., vol. vii. p. 381.
[‡] See also Rev. J. H. Austen, Guide to Geology of the Isle of Purbeck, 1852.
[§] Trans. Geol. Soc., ser. 2, vol. iv. p. 251.
[||] Topley, Geol. Weald, p. 34.
[¶] Geol. E. Lincolnshire, p. 132.

quality the stone is variable, in many old churches it has lasted well, but in some places it has decayed and crumbled. Much of the stone is soft and useless.

Miscellaneous uses of Fossils.

The palatal teeth of fishes, more especially those of *Lepidotus maximus* (*Sphœrodus gigas*) from the Kimeridge Clay, as well as similar teeth from other formations, have been utilized in old times as ornaments, or employed as charms, being supposed to possess extraordinary virtues.

Conybeare and Phillips thus state (1822) that " these little fossil productions were a century since, in common use with the ladies, as ornaments ; and, what is a still more singular circumstance connected with their history, they seem to have been applied to the same purpose by our ancient British ancestors : as several strings of them were discovered in the Wiltshire Barrows, among other ornaments, opened by the late Mr. Cunnington, of Heytesbury."* Later on Mr. W. Cunnington described a " Crapaudine Locket " formed of two circular teeth of *Lepidotus*, found at Devizes.†

Dr. Henry Woodward has recently given a further account of these articles,‡ mentioning' that the old Naturalists (such as Lhwyd) described the circular enamelled palatal teeth of fishes as Bufonites or Toad-stones.

Mr. Cameron informs me that Belemnites obtained from the Oxford Clay, south of Bedford, have been ground up and the material has been used for sore eyes ; and specimens of the same fossil were formerly collected in Buckinghamshire and Bedfordshire " by the villagers who consider them, when pounded, an excellent cure for rheumatism."§

At Fletton near Peterborough, the very abundant specimens of *Belemnites Oweni* are collected by the workmen, who receive 2s. a scuttle-full for them ; but I am informed by Mr. A. N. Leeds that the men are paid simply to pick them out from the clay as they would spoil the bricks.

Stone-tiles.

Purbeck stone-tiles have been employed for roofing-purposes from Roman times.

Thin limestones in the Lower Purbeck Beds on Portland, known as the Slatt Beds and some of the overlying fissile limestones, have been employed for roofing. The same is the case at Swanage.

Stone tiles were formerly obtained from the hill above Tilly Whim‖ near Swanage, and used in the town, also at Corfe, &c.

At Swindon the Lower Purbeck Beds have yielded tilestones used in some of the old buildings in North Wilts and at Devizes.

None of these stone-tiles can, however, compare with those obtained from the Stonesfield and Collyweston Beds or from the Forest Marble.

Hearth Stone.

Hearth-stone has been obtained from some of the soft Lower Purbeck Beds at Portland.

The Aish is used for holystone, and for whitening stones and passages. The holystone sent away is obtained from the Base Bed of the Portland freestone.

Hearth stone has been also obtained from a pale gritty limestone, White Bed, 18 ins. to 4 feet thick, on top of the Rag Beds belonging to Upper Portland Beds in the Vale of Wardour.

---

* Outlines of the Geol. England and Wales, p. 208.
† Wilts Arch. Mag., 1870, p. 249.
‡ Geol. Mag., 1893, p. 246.
§ J. H. Macalister, Geologist, vol. iv. p. 215 ; see also H. B. W., Memoir on the Lias of England and Wales, p. 297,
‖ Fitton, Trans. Geol. Soc., ser. 2, vol. iv. p. 209.

Road Metal.

The Kellaways Rock of Wiltshire was in old times used locally for road-mending.

Prof. Judd remarks that the Oxford Clay, as at Ramsey, has been dug and burnt as a substitute for gravel, a use to which the clay is frequently applied. It has been so used in Wiltshire, but when the surface is covered by only 5 or 6 inches of the "metal," it is found unsuitable for heavy carting.

The Corallian Beds, more especially the hard calcareous sandstones of the Lower Calcareous Grit, have furnished good material for roads, as at Derry Hill, west of Calne ("Black stone"), near Conygre Farm, Calne (Conygre stone), Bullingdon, &c. Sometimes, as at Purton and Headington, the Coral Rag is used.

In many places in the neighbourhood, the Westbury Slag from the iron-furnaces has been employed for road-metal, though the smell of it is sometimes unpleasant.

South-east of Encombe House, in the Isle of Purbeck, a rubble of Portland and Purbeck Beds is worked for gravel.

On Portland the "Hard Slatt" of the Purbeck Beds has been used for road-mending, and in the Isle of Purbeck some of the roads are mended with the waste material from the quarries: such roads are white and dusty or very muddy.

Hard calciferous sandstones in the "Greys" of the Purbeck Beds of Sussex have been used for road-metal.

As before-mentioned, the principal roads are now mended with material that is often obtained from a distance.[*]

Sands.

At Casewick, north-east of Stamford, sand for the use of the glass-maker and potter has been obtained.[†] Prof. Judd mentions that in this district irregular beds of whitish sand occur in the Kellaways Beds, so that probably this was the deposit formerly worked.

Sands belonging to the Corallian Beds have been used for mortar, at Purton, Highworth, and other places; and elsewhere they have been utilized in brickyards.

At Shotover Hill in the Portland Beds, the white gritty sands (with "sand ballers") are used for moulding bricks, while a lower bed just above the Kimeridge Clay yields a soft mealy sand used for moulding in iron-furnaces as "foundry loam." Dr. Percy remarks that casting sands for moulds in foundries may contain 93 per cent. of fine quartzose sand, 2 of ochre, and 5 of aluminous earth (non-calcareous).[‡]

The sands belonging to the Spilsby Sandstone are used for mortar-making, and also in brickyards. At Acre House near Claxby the sand, "being exceedingly sharp and clean, is quarried and sent away by rail."[§]

Septaria and Cement Stones.

Septaria from the Oxford Clay have occasionally been cut and polished for ornamental purposes (slabs and tables) as at Radipole and Melbury Osmund in Dorsetshire, where the stone is known as Melbury Marble, or Turtle Stone, the waxen veins of calc-spar forming its chief beauty. The original name applied to septaria was *Ludus Helmonti*,[||] as when cut they resemble surfaces of dice. (See p. 17.)

---

[*] Memoir on the Lias of England and Wales, p. 12.
[†] Hunt, Mineral Statistics for 1858, Part 2, p. 375.
[‡] See Metallurgy, Refractory Materials, Ed. 2, p. 152.
[§] Fox-Strangways, in Geol. N. Lincolnshire, p. 108.
[||] After J. B. Van Helmont.

Septaria from the Kimeridge Clay of Portland have been cut and polished, and Fitton mentioned the occurrence of "Turtle Stones" in the Kimeridge Clay west of Shaftesbury.*

Fine Septaria occur in the Kimeridge Clay near Horncastle, where they are sold for rockeries. Sometimes a gill or two of water was found when the stones were broken. This may have originated when the concretion was formed, the outer portion of which having solidified, the cavity resulted from the shrinkage of the inner portion. A band of very large septaria occurs in the lower part of the Oxford Clay west of Bourn.

Referring to a brickyard in the Kimeridge Clay at South Willingham, Mr. Strahan says that at its base it is full of very large septaria, averaging from 3 to 4 feet in diameter, by about 1½ feet thickness. These septaria break up readily in the weather; the outer portions consist of a blue argillaceous limestone of an earthy texture, traversed by cracks filled, or partly filled, with dogtooth calcite; in the centre is a heart almost entirely made up of calcite. This limestone was sent to Hull to be tried for hydraulic cement but proved worthless.[†]

Lime and Cement.

Lime for building and agricultural purposes is made from the limestones belonging to the Corallian, Portland, and Purbeck Beds.

The Corallian beds are thus employed in various localities—the lime from the coral-rocks being in some places of a stronger nature than that made from the Coralline Oolite.

Occasional bands of limestone in the Oxford and Ampthill Clays, as at Husborne Crawley, (where the rock is known as "hurlock,") and at Gamlingay, are burnt for lime.

Portland Beds on Portland, at Portisham and Upway, and in the Isle of Purbeck are burnt for lime. In the Vale of Wardour the chalky lime-stones are burnt at Chilmark and Wockley; so also are some of the Portland Beds in Oxfordshire and Buckinghamshire.

The cap-beds belonging to the Purbeck Beds of Portland, and some of the soft earthy Purbeck limestones at Wockley and Swindon are burnt for lime.

The Coralline Oolite of Westbury is used as a flux in the adjoining iron-furnaces.[‡]

Portland or Hydraulic cement has been made from the cement-stones in the Kimeridge clay of the Isle of Purbeck. Mr. B. Green remarks that "The only seam which has now any value is the 'East Ledge,' 2 ft. 10 ins. thick."[§] I am informed by Mr. Strahan that cement-stones were lately worked by levels driven into the cliff, close by where the Kimeridge "coal" was worked in a similar manner. Formerly the cement-stones of Broad Bench were employed for making cement, being sent over to Medina, in the Isle of Wight.

Hydraulic Lime, miscalled "Blue Lias Lime," is made from the Lower Purbeck Beds at Upway.

The limestones of the Purbeck Beds in Sussex, known as the Sussex limestone, were at one time largely worked for making lime both for building and agricultural purposes. An account of the beds and the method of working them has been published by Mr. C. Gould. The Blue limestone was considered to make by far the best lime for building-

---

* Trans. Geol. Soc., ser. 2, vol. iv. p. 256.

† Geol. Lincoln, p. 81.

‡ An analysis of the limestone has been published by G. C. Greenwell, Proc. S. Wales Inst. Civ. Eng., vol. i. p. 311.

§ Kimmeridge Shale, its Origin, &c., 1886, p. 21. Analyses of Septaria vary very much, see for instance that of Septaria from Oxford Clay of Bletchley by A. Timmins. Geol. Mag., 1889, p. 357.

purposes, but the " Greys " were by many considered equal to the " Blues " for agriculture.

In Archer Wood shafts were sunk to various depths up to 110 feet, and galleries were then driven in certain directions. In Limekiln Wood, and over a portion of Archer Wood, the stone was obtained by means of Bell Pits. A circular shaft 4 feet in diameter was sunk for 20 feet. The quarrymen then commenced to " bell " it out, increasing the diameter with the depth, so that eventually the base of a pit 50 or 55 feet deep was as much as 20 feet across. No timbering was necessary unless horizontal galleries were driven from the base of the pit. The lime made was more or less hydraulic in character, and some beds of the stone, especially the " Bastard Blues," were considered suitable for making hydraulic cement.*

Gypsum.

The Lower Purbeck Beds of Durlston Bay, near Swanage, have yielded a good deal of gypsum, but the material is not now worked, and during recent years only fallen masses have, from time to time, been collected upon the shore. Formerly the base of the cliff was cut away to work the gypsum, which was said to occur in masses from 2 to 20 feet in diameter, containing from 1 to 20 tons.†

Gypsum or white alabaster was discovered in the Purbeck Beds at Netherfield, near Battle, by the stream between Snep's Wood and Great Wood, during the Sub-Wealden Boring.

After passing through 119 feet of strata, a bed of gypsum 6 feet thick was penetrated, then 5 feet of shales, then 4 feet of gypsum, then 16 feet of shales, and afterwards a third seam of gypsum 7 feet thick was proved. In 1876, the amount raised was 234 tons, valued at 70l.

The gypsum is used in the preparation of Plaster of Paris, and in Keene's and Parian cements. The waste stuff has been sold at 20s. per ton for use as manure in parts of Kent and Surrey. Columns about one foot in length were fashioned for a font at Bexhill.

The following are the particulars of the production of Gypsum at the works of the Sub-Wealden Gypsum Company, Mountfield, near Battle:—

—	Tons.	Value at the Mine.
		£
1882	8,750	9,479
1883	8,720	9,450
1884	7,178	7,537
1885	5,767	5,478
1886	5,357	1,500
1887	4,944	3,615
1888	5,101	3,732
1889	3,166	2,170
1890	3,947	2,763
1891	4,109	2,876
1892	5,310	2,969
1893	6,162	3,083
1894	7,626	3,677

* Topley, Geology of the Weald, pp. 384–387.
† See also Englefield's Isle of Wight, p. xix.

Alum.

Alum was prepared from the Kimeridge Clay of Kimeridge in the 16th century, when Alum-works were erected.* It occurs as an efflorescence on the surface of some of the shales in this formation, and also in the Purbeck Beds.

Coprolites.

"Coprolites" or phosphatic concretions and phosphatized fossils are found in small quantities in the Kimeridge Clay, more especially at the base, and at the base of the Portland Stone. Coprolites have been observed in the Corallian Oolite at Osmington.†

Mr. Jukes-Browne thought that the phosphatic band at the base of the Spilsby Sandstone "might in some places be worth working for economic purposes" Analyses by Mr. Grant-Wilson and Mr. M. Staniland showed about 46 per cent. of phosphate of lime.‡

Brick Earth and Tile Clay.

Clays used for brick and tile-making and for drain-pipes are obtained from the Oxford Clay, the Corallian Beds, the Kimeridge Clay, and the Portland Beds. In all these formations the character of the clay is subject to much variation. Some examples of bricks and brick-clays are now exhibited in the Museum of Practical Geology.

The Oxford Clay is perhaps the most extensively dug over the area, although some of the smaller brickyards are being abandoned through the competition of larger works, where elaborate machinery is employed in the manufacture of the bricks.

The milder earths in the Kellaways Series or those passage-beds that occur at the junction of the Oxford Clay and Lower Corallian Beds, or at the junction of the Kimeridge Clay and Lower Portland Beds, are best adapted for brickmaking, for the sandy clays yield bricks that better withstand the weather than those made from stiffer clay.

The bricks, &c. made from the *Oxford Clay* are usually red, but in many cases they are mottled pale red and yellowish-white: the colour being dependent partly on the calcareous character of the clay; but mainly on their being burnt in a reducing atmosphere, which as Mr. A. B. Dick informs me, leads to the formation of the silicate of protoxide of iron. In some places chimney-pots, flower-pots, and ornamental bricks are made.

There are important brickyards at Weymouth, Melbury Osmund, Trowbridge, near Seend, Chippenham,§ Oxford, Fenny Stratford, Bedford, St. Neot's, Peterborough (Fletton, &c.), Ramsey, &c.

In Oxfordshire and Buckinghamshire, the men usually dig the clay "after the turn of the days," that is after Christmas.

Good brickearth is obtained from the *Kellaways Beds* near Bedford. As remarked by Mr. Cameron, the loam or "lam earth," as it is called, is mixed with the stiffer basement clay (Kellaways Clay), thus lessening the liability in the material to contract and crack in drying, as happens when the "strong" basement-clay alone is used.||

Clays belonging to the Kellaways Beds are worked at Oundle, Southwick, Benefield, Dogsthorpe, Uffington, Kate's Bridge near Thurlby, and Warmington; and Prof. Judd remarks that bricks made from these

---

* See Mansell-Pleydell, Proc. Dorset Nat. Hist. Club, vol. xv., p. 174.
† Buckland, Trans. Geol. Soc., ser. 2, vol. iii. p. 232.
‡ Geol. East Lincolnshire, p. 134.
§ An analysis of clay from the brickworks, at Canal Bank, Chippenham, is given by G. Maw, Quart. Journ. Geol. Soc., vol. xxiv. p. 357.
|| Geol. Mag., 1892, p. 67.

sandy beds are much superior in quality to those manufactured from the Oxford Clay, especially in respect to the amount of heat which they will bear.*

At Over Field, north of Long Stanton, clay has been dug for "clay-lumps" or sun-dried bricks, used for barns and hovels.

Bricks have been made from *Corallian Beds*, from the Sandsfoot Clays at Weymouth, and from loamy clay at Highworth; also from the Ampthill Clay at Boxworth, in Cambridgeshire, and at Gamlingay, where red and white bricks are made at the Belle Vue Steam brickworks.

At Boxworth the clay is of variable character, for that from the higher part of the pit was not found so good for bricks, as they swell after being shaped, and burn in uneven forms. Good 6-inch and other drain-pipes are made. I was informed that the brick-makers used to be paid 3s. 6d. per thousand bricks, 30 or 40 years ago; now they want 4s. 6d., and do not do their work so well.

The *Kimeridge Clay* is largely used for brick-making, and also for the manufacture of ornamental tiles, flower-pots, &c. Most of these bricks, tiles, and pipes are red, but in some places white bricks are made by the addition of chalk.

There are brickyards at Gillingham, Swindon, Headington, Brill, Ely, Downham Market, near Market Rasen, Horncastle, West Ashby, and many other places.†

At Roslyn Hole, Ely, the Clay has been used for mending the Fenland dykes, and repairing the banks of the River Cam.

At Brill where the lower Portland Beds pass down into the Kimeridge Clay the higher portion (equivalent to the Hartwell Clay) furnishes a "mild earth" well adapted for brick-making, while below the clay is stiffer and suitable for making tiles and drain-pipes.

From the Purbeck Beds of Poundsford, in Sussex, an unctuous blue clay "has been used with some success as a pigment."‡

IRON ORES.

Oolitic iron-ore, known commercially as "brown hæmatite," occurs in the Corallian Rocks at Abbotsbury, and Westbury, and it has for many years been worked at the latter place. A band of oolitic iron-ore has been proved to occur near the junction of Kimeridgian and Corallian Beds in the Dover Boring. (See pp. 150, 343.) At Claxby, in Lincolnshire, there is also an oolitic iron-ore, which occurs on the border-line between the Jurassic and Neocomian Strata, and this has at times been worked.

In the formation of these ironstones the evidence all tends to show that the ore was introduced subsequently to the accumulation of the strata, in the form of carbonate of iron, which has replaced the oolite grains where present, as well as certain organic remains and other calcareous portions of the strata. Along the outcrop the ironstone has been weathered and rusted into the form of brown hæmatite, while deeper beneath the surface, and especially where protected by coverings of clay, the iron-ore remains in the form of carbonate, and thus retains its original bluish-grey colour.

Messrs. Blake and Hudleston describe the Abbotsbury ironstone as containing particles of rounded and subangular quartz-grit and coffee-coloured granules about the size of a pin's head. The granule may perhaps be called an oolitic grain of the hydrous oxide of iron; "but

---

* Geol. Rutland, &c., pp. 232, 234, 235.

† An analysis of Kimeridge Clay from Westbury, Wiltshire, was given by J. D. Kendall, Trans. N. of Eng. Inst. of Mining Engineers, vol. xxxv. p. 156.

‡ Topley, Geol. Weald, p. 38.

even the outer skin is not pure oxide; and after treatment with acid there remains a residue presenting a sphere-on-sphere structure, hollow within, and composed of a uniform subcrystalline material, the greater part being silica." An analysis given by them showed that the granules contain about 73·57 per cent. of ferric oxide.[*]

The siliceous structures in the oolite grains are similar to those noticed by Mr. A. B. Dick in the Cleveland iron-ore; by Prof. Judd, · in the Northampton iron-ore; by Mr. Teall in those of Frodingham and Westbury; and by Mr. Staniland in that of Claxby.[†]

Near Sturminster Newton I found traces of iron-ore on the same horizon as that of Abbotsbury. (See p. 102.)

The beds at the Westbury Iron-works are described by Messrs. Blake and Hudleston as comprising "an oolitic ironstone more or less mixed up with black argillaceous ore. It is almost free from grit, and exists partly as ferrous carbonate, and partly as hydrated peroxide. . . . In the oolitic portions of the stone, oxidation has progressed to a greater extent than in the black earthy portions, which are almost in the condition of clay ironstone. The chief difference between this and the granules of Abbotsbury consists in the greater abundance of phosphoric acid."[‡]

Under the microscope the ironstone, as noted by Mr. Teall, shows reddish-brown oolitic grains with concentric structure in a dark yellowish-green matrix. The matrix contains a large number of colourless crystalline and detached grains of calcite of fairly uniform size, and much smaller than the oolitic grains. The nuclei of the larger oolite grains are in many instances broken fragments of other grains. The residue, after treatment with hydrochloric acid, consists of siliceous oolitic balls (soluble in a solution of caustic potash), and of minute angular fragments of oligoclase and quartz.[§]

The Westbury ore was first worked about the year 1856, and the annual quantity of ore raised, has varied from 5,000 tons (1856) to 105,929 tons (1873). Of later years the Mineral Statistics prepared by H.M. Inspectors of Mines, and published by the Home Department, give the following particulars up to 1885, after which, excepting in 1892, the returns have not been separately stated :—

Years.	No. of Furnaces.	In Blast.	Tons.	Average per-centage of Iron.	Value of the Ore at the Mine or Open Work.
1882	3	2	98,176	40·00	19,835
1883	3	2	92,114	40·00	18,445
1884	3	1	62,234	40·00	12,446
1885	3	1	67,489	40·00	13,498
1892	2	1	70,866	30·00	33,661

In 1890 and 1891 there were but two Blast Furnaces, and one only in blast, and the average per-centage of iron was then, as now, stated to be 30. This is no doubt due to the fact that most of the richer brown ore has been quarried.

* Quart. Journ. Geol. Soc., vol. xxxiii. pp. 274, 275.

† See also Memoir on the Lias of England and Wales, p. 302; Memoir on the Lower Oolitic Rocks of England, p. 496.

‡ Quart. Journ. Geol. Soc., vol. xxxiii. p. 285; see also A. C. Cruttwell, Geology of Frome, p. 18.

§ See Memoir on the Lower Oolitic Rocks of England, Plate II., Fig. 12, p. 28.

Analyses of Iron-ore.

—	I.	II.	III.	IV.
Peroxide of iron - - -	43·97	59·93	1·32	52·86
Protoxide of iron - -	—	—	47·25	—
Alumina - - - -	—	3·22	5·46	6·45*
Lime - - - -	Trace	1·84	0·74	2·00
Magnesia - - -	1·40	0·84	0·36	1·57
Chloride of sodium - -	Trace	—	—	—
Sulphuric acid - - -	—	—	0·10	Trace.
Carbonic acid - - -	—	1·45	30·01	2·94
Phosphoric acid - - -	—	—	Trace	0·24
Insoluble siliceous matter -	42·60	18·99	14·72	17·26
Water - - - -	11·88	13·70	—	16·16
	99·85	99·97	99·96	99·48
Metallic Iron - - -	30·78	41·95	37·67	37·00

I.—Analysis of Abbotsbury Ironstone, made (before 1860) by Prof. G. D. Liveing. See Damon's Geology of Weymouth, 1884, p. 48.

II.—Analysis of Brown Ironstone of Westbury, given by J. D. Kendall, Iron Ores of Great Britain and Ireland, 1893, p. 247. Another analysis recorded by G. C. Greenwell (Proc. S. Wales Inst. C. E., Vol. I., p. 311) gives 65 per cent. of peroxide of iron, and 45 of metallic iron.

III.—Analysis of Green Ironstone of Westbury, given by J. D. Kendall. The blue ironstone is said to yield from 49 to 55 per cent. of carbonate of protoxide of iron, and from 38 to 43 per cent. of metallic iron. Vanadium was noticed in an analysis by E. Riley, Journ. Chem. Soc., Vol. II., 1864, p. 21.

IV.—Analysis of the Claxby iron-ore, given by J. D. Kendall, op. cit., p. 250.

In the Claxby iron-ore, as described by Mr. M. Staniland, the grains appear "almost perfectly spherical, polished, and about the size of millet seed. They are made up of alternating concentric layers of clear silica and opaque hydrated sesquioxide of iron." An analysis of the grains gave the following results:—

Silica - - - - - -	5·34
Sesquioxide of iron - - - -	76·76
Moisture - - - - - -	16·17
	98·27

The per-centage of metallic iron was reckoned to be 53·73 per cent.[†]

More iron-ore is now obtained from the Jurassic rocks of this country than from other formations: and as already pointed out beds have been profitably worked in the Lower Lias, Middle Lias, Northampton Beds, and Corallian Beds.

In all these cases the iron-ore is local, equivalent beds being simply clays, sands, or limestones. The most wide-spread horizon is that of the Middle Lias which from Fawler in Oxfordshire to the Cleveland Hills has at intervals yielded a valuable iron-ore, and a bed of identical character was found by myself in the island of Raasay, in the Inner Hebrides.

* Including manganese.
[†] Strahan, Geol. Lincoln, p. 95. See also Analysis of the Hundleby iron-ore, by Grant-Wilson, Jukes-Browne, Geol. East Lincolnshire, p 133.

In all cases the iron-ore appears to be due to the local concentration of ferruginous matter and to the fact that this has replaced calcareous beds—limestones more or less oolitic and fossiliferous.

Gradations occur between the beds that are of value as an iron-ore and those which contain too small a per-centage of metal to be of service. Iron-shot limestones, beds of more or less earthy limestone with grains of iron-ore, are of frequent occurrence, and particular layers have been noted in the Lower Lias of Radstock, in the Marlstone of many areas, in the Cephalopoda-bed of the Cotteswold Hills, in the Inferior Oolite of Dorset, and in the Upper Coral-rag of Ringstead Bay. These limestones owe their iron-shot appearance mainly to the replacement of calcareous oolite grains by iron-ore, but sometimes to the similar replacement of tiny fragments of organisms.

Here and there we find that the iron-ore occurs in nodular bands as at the top of the Lower Lias in Gloucestershire, in the ironstone junction-bed at the base of the Upper Estuarine Series and occasionally in the Great Oolite Clay. In this form it appears to be due to local segregation, and some of the ochreous nodules suggest that the ferruginous matter gathered round nuclei of clay, such as might have originated as clay-pebbles.

The iron-shot limestones (Inferior Oolite) of Dorset pass in places into a bed having 35 per cent. of peroxide of iron.

In the Lincolnshire Limestone we find that the upper strata that lie beneath the Upper Estuarine Beds, are tinged various shades of red, as in the Crash Bed. This, however, may be usually attributed to staining from the ironstone junction-bed above.

The stratigraphical features of the main beds of ironstone are somewhat varied.*

The Frodingham iron-ore in the Lower Lias of Lincolnshire, lies between masses of shale and clay. The Northampton iron-ore rests on a mass of Upper Lias clay, and is covered usually by sands with occasional clay bands.

On the other hand the Marlstone iron-ore is covered by a mass of Upper Lias clay, and usually rests on semi-porous beds of sandy shale or even of sand; the very ferruginous oolite in Dorset rests on sands and is overlaid by buff earthy and oolitic limestones; the Corallian iron-ore of Westbury is overlaid by a mass of Kimeridge Clay, and rests on a thin semi-porous band of sand and laminated clay, beneath which are pale oolites and marls; the Claxby iron-ore rests on sands and sandstone (Spilsby Sandstone) and is overlaid by the Tealby Clay.

The organic remains in the various beds of iron-ore, though found for the most part as moulds and casts, are sometimes replaced with carbonate of iron, with the exception of shells like *Pecten* and *Ostrea* which retain their calcareous substance, and are found in layers in the Lias and Corallian iron ores. In the

---

* See a'so Ken'all, Iron Ores of Great Britain, 1893, pp. 252, 312.

Cleveland ore, as pointed out by Dr. Sorby, some of the shell-structures have been replaced by carbonate of iron.

That the replacement has been gradual and in some cases partial, is shown by the fact that both shell structures and oolite grains sometimes contain inner portions of carbonate of lime. Analogous partial replacement has been found in the chert with oolite grains (p. 234).

That the iron-ore was introduced subsequently to the deposition of the strata is amply proved by the state of preservation of the organic remains.*

Mr. Hudleston in drawing attention to the abundance in some iron-ores of large Monomyaria, such as *Ostrea, Lima, Pecten,* &c., remarked " that where this feature occurs there is often a large charge of phosphorus in the ore, due in part at least to the decomposition of the animal matter of these molluscs and associated organisms."† Referring to the Northampton iron-ores, he suggests that they have been formed by the replacement of lime-stone. The rock was originally a sandy oolitic limestone, and in the first place carbonate of iron replaced the carbonate of lime— the strata being saturated for a considerable period with waters holding bicarbonate of iron. Afterwards through the action of surface-waters the carbonate of the protoxide of iron has, along the outcrop of the beds, been converted into the ordinary hydrated peroxide. He considers that the Northampton ore was derived from the overlying Estuarine Sands, which contain much lignite, and that the action of organic acids, due to the decomposition of woody matter, would serve to leach out the iron from these overlying beds.‡

Mr. Kendall believes that the direct source of the iron was in many cases the clay contiguous to the deposits.

Neither of these views seems wholly satisfactory, but we may infer that the iron-ore was derived from ferruginous waters that percolated underground for long distances when the strata were buried deep beneath newer accumulations.

Lignite and Bituminous Shales.

Fruitless trials for coal, based sometimes on the occurrence of dark shales, lignite, or bituminous shale, were made in old times in several localities in the Oxford and Kimeridge Clays. (See also p. 22.)

Such trial-borings have been made in the Oxford Clay near Brewham§ (see p. 22) ; at Malmesbury (300 feet deep)‖; and also near Melksham.

---

* With reference to the original sources of iron-ore see Hudleston, Proc. Geol. Assoc., vol. xi. p. 108 ; and Review by J. J. H. Teall, of J. H. L. Vogt's, Formation of Iron-ore Deposits, Geol. Mag., 1892, p. 82.

† Proc. Geol. Assoc., vol. xiii. p. 138.

‡ Proc. Geol. Assoc., vol. iv. p. 123, and vol. xi. p. 123 ; see also Judd, Geol. Rutland, &c , pp 95, 131.

§ J. Phillips in Mem. William Smith, p. 66 ; Conybeare and Phillips, Geol. England and Wales, p. 195.

‖ J. Buckman, Geologist, vol. i. p. 184; Quart. Journ. Geol. Soc. vol. xiv. p. 125.

where the trial resulted in the finding of a mineral water and the establishment of the Melksham Spa. (See pp. 27, 340.)

Again, at Eynsham Heath, near Oxford, the presence of bituminous shale in the Oxford Clay led to a fruitless search for coal.*

At Abbots Ripton, near Huntingdon, a fruitless trial-boring was made ; and again at Ampthill, as I was informed by Mr. Cameron, a boring was carried though the Lower Greensand, and Ampthill Clay, into the Oxford Clay, to a total depth of about 400 feet.

A fruitless boring was made many years ago in the Kimeridge Clay, in ground called Richmond near Twiford and Whitings, to the west of Melbury Abbas, and to the south of Shaftesbury in Dorsetshire. At the depth of 40 feet "black grit-stone," and a large vertebra were found, and at a depth of 80 feet clay-stone (septarium).†

Other borings in search of coal have been made in Bagley Wood near Oxford, at Brill,‡ and again at Old Bolingbroke (600 feet), at Woodhall (1,020 feet), and Donnington-upon-Bain (309 feet) in Lincolnshire.

In the case of Woodhall the boring resulted in the finding of a saline spring and the establishment of the Woodhall Iodine Spa.§

Mr. R. N. Mantell noted the occurrence in the Oxford Clay at Studley, south of Trowbridge, of bituminous unctuous shales, remarking that "This clay emits a brilliant gas when burnt, and I have frequently used it to illuminate my room."‖

Thin beds of bituminous shale were proved in borings in the Oxford Clay at Swindon and also at Bletchley.

Lignite, like Kimeridge Coal, was stated by Fitton to occur near Steeple Ashton ;¶ and strata of a bituminous nature (said to have been used for coal) were discovered in sinking a shaft at the brickyard on the north side of the Bath road, near the foot of Devizes Hill.**

Mr. Jukes-Browne has noticed the occurrence of black bituminous shales at Old Bolingbroke, which led to the fruitless search for coal before mentioned ; he also states that at East Keal black shale that "would burn almost like coal" was found about 20 feet down in sinking wells. This "is evidently the same bed as that found near Bolingbroke."††

At the brick-kiln near South Willingham, according to Mr. A. Strahan,‡‡ "Below the layer of septaria there occur bands of hard inflammable oil shale, locally known as "dice." The bands are 4 to 6 inches thick, and are separated by blue clay. Fragments of dice readily blaze when dry, and leave a copious grey ash, giving off a most offensive smell while burning. It is said that the clay from this pit, when made up into bricks and ignited in the kiln, to a certain extent supports combustion, giving off a poisonous vapour that is highly prejudicial to vegetation. About Willingham water is got in some of the shallow wells from the beds of dice. Some of the water is ferruginous and smells offensively."

Similar beds of dice or inflammable shale were proved in the deep boring at Donnington. Mr. Strahan says these bands of shale probably run through the whole district, but they are very rarely seen from want of good sections.

Coaly matter has also been found in a well-boring at Driby in Lincolnshire ; and in the Acre House Mine, near Claxby, a thin bed of highly inflammable shale was met with below the Spilsby Sandstone.§§

---

* C. S. Read, Journ. R. Agric. Soc. vol. xv. p. 197.
† J. Woodward, Nat. Hist. Foss. Eng., Tome II. pp. 52 and 99.
‡ Phillips, Mem. William Smith, pp. 66, 67.
§ Strahan, Geol. Lincoln, pp. 81, 208.
‖ Quart. Journ. Geol. Soc., vol. vi. p. 313 ; Damon, Geol. Weymouth, 1884, p. 24.
¶ Trans. Geol. Soc., ser. 2, vol. iv. p. 269.
** Lonsdale, *Ibid.*, vol. iii. p. 265.
†† Geol. East Lincolnshire, p. 10.
‡‡ Geol. Lincoln, p. 81.
§§ Judd, Quart. Journ. Geol Soc., vol. xxvi. p. 331 ; and Ussher, Geol. N. Lincolnshire, p. 107.

Attempts have been made from time to time to found profitable works for the distillation of the Kimeridge Shale. Most of these have failed, although large quantities of paraffin, rough oils, candles, and mineral manure have been made and sold. Gas has also been made. About ten years ago the Kimeridge Oil and Carbon Company (Limited) established works at Sandford, near Wareham, and I was indebted to Mr. Charles Beaumont, the Manager, for particulars of the products obtained.

The shale is worked at Kimeridge Cliff east of the Bay, and the "best bed" or "black stone" is a two-foot seam of dense bituminous shale, which occurs some way down the cliff, overlaid by 4 feet 6 inches of clay and again by 16 feet of more or less bituminous shale, and soil. The beds are worked by tunnelling into the cliff. The Black-stone is the bed usually burnt by the villagers; it also yields carbon for filters. The common shale is considered better as a material for yielding products for decolorising and deodorising. (See Fig. 80, p. 159.)

The beds yield from 30 to 66 gallons of crude oil per ton, and about 10 cwt. of carbon. Other products are Naphtha, Lubricating Oils, Varnish, Asphalt, Pitch, Wax, Carbolic disinfectants, Animal charcoal, and Sulphate of Ammonia.

The shale is at times shipped from Kimeridge to be burnt in the potteries at Poole. The expense of carriage has no doubt seriously hindered the success of the various undertakings to utilize the beds.

Analyses of the Black-stone show from 55 to upwards of 60 per cent. of volatile matter.

The following shows the general composition :—

Volatile matter - - - - - -	61·00
Carbon - - - - - - -	13·15
Ash - - - - - - -	25·85
	100·0

The bulk of the shale on the other hand yields about 20 per cent. of volatile matter and 70 per cent. of ash.

It has been proved that from 9,000 to 10,000 cubic feet of purified gas can be obtained from one ton of the Black-stone. The residue, which amounts to about half a ton, is of no value as a fuel. The coal yields a reddish ash which contains peroxide of iron.

The following analyses have been published by Mr. Burton Green :—[*]

—	Black-stone.	Shale.
Naphtha - - - - - -	2·3	2·7
Heavy Oil (containing 1·3°/₀ of Paraffin) - -	36·7	9·5
Gas Water - - - - - - -	18·0	16·3
Carbon - - - - - - -	19·5	15·0
Mineral Matter - - - - - -	23·5	54·1
Pitch - - - - - - -	—	2·4
	100·0	100·0

[*] Kimeridge Shale, 1886, p. 8, Analysis by Prof. Keates; p. 17, Analysis from Dr. Ure's Dictionary (ed. 1867, p. 780); see also Mansel-Pleydell, Geol. Mag., 1878, p. 408; and Proc. Dorset Nat. Hist. Club, vol. xv., p. 178.

Y 2

"It is generally supposed that the hydrocarbons have been distilled from the putrefaction of animal matter."[*]

Dr. E. D. Burrowes, Vicar of Kimeridge, writing in 1879, says, "This shale, locally called Black-stone, is burnt by the villagers generally, and, whilst it throws out a great amount of heat, from its intensely clear and strong light it renders candles in a great measure unnecessary to them. It is worked by a company, from mines in the face of the sea-cliff, and its selling price is twelve shillings per ton; three tons being considered equivalent to one ton of ordinary coal; its ashes are used as a garden manure. Jet is often found in connexion with it."[†] As might be expected the shale emits an unpleasant odour while burning. The ashes spread over the meadows are considered a good manure.[‡]

Mr. Burton Green mentions that "Round the side of the hill 'Cuddle Head,' upon the line at which the Blackstone seam crops to the surface, are several excavations and heaps where adits have been driven for the purpose of winning this fuel; and upon the top of the hill also are several shafts, apparently connected with these adits." This would be to the north and north-west of Hen Cliff, where traces of old workings are to be seen. Here no doubt the Black-stone came to the surface, and one would infer that its outcrop might be traced from Hen Cliff, to the valley south-west of Smedmore, westwards about half-way between the village of Kimeridge and Kimeridge Bay, and thence to the coast on the west of the Bay, where it has been observed by Mr. Strahan. The strata are, however, so much faulted that actual trial-borings would be necessary to prove its exact position. That bituminous beds are persistent for a great distance, may be inferred from their presence near Portisham; but whether any particular band could be traced so far is, of course, doubtful.

The following statistics of the working of oil-shale in Dorsetshire have been published:—

Years.	Tons.	Average Price per Ton at the Works.	Value at the Works.
1883 - -	1,000	6s.	£300
1885 - -	200	£1	£200
1886 - -	250	£1	£250
1890 - -	150		
1891 - -	300		
1892 - -	70		

The so-called "coal-money" is found in various places near the surface at Smedmore near Kimeridge.

It is now regarded as of Romano-British age, and to be the useless discs of larger pieces which had been turned on the lathe into ornamental

---

[*] Hudleston, Proc. Geol. Assoc., vii. p. 385; see also A. Smee, Proc. Geol. Soc., vol. ii. p. 673.

[†] Guide to Swanage, 1879, p. 39. See also Robinson, A Royal Warren, or Picturesque Rambles in the Isle of Purbeck, pp. 54, 55.

[‡] Dr. J. F. Berger. Trans. Geol. Soc., vol. i. p. 264.

objects; such as centres of armlets, bracelets, or rings, and bases of vases or bowls.* The discs are about 2 inches in diameter, with usually one square perforation, and sometimes two, three, or four round holes.

In his excavations at Rushmore, among Romano-British remains, General Pitt-Rivers discovered a decorated tablet of Kimeridge Shale, which appeared to be of the kind used for writing upon with the stylus, by means of a coating of wax spread over it. Spindle whorls and bangles of Kimeridge Shale were also obtained.†

In September 1826 spontaneous combustion took place in the Kimeridge clay near the east extremity of Ringstead Bay, at Holworth Cliff, adjacent to the promontory of White Nore. This combustion continued until 1829, although the extent of the surface of the clay, which was burnt, did not exceed 50 feet square.

Buckland and De la Beche state that " within this space are many small fumaroles that exhale bituminous and sulphureous vapours, and some of which are lined with a thin sublimation of sulphur; much of the shale near the central parts has undergone a perfect fusion, and is converted to a cellular slag. In the parts adjacent to this ignited portion of the cliff where the effect of the fire has been less intense, the shale is simply baked and reduced to the condition of red tiles, like that on the shore near Portland Ferry."‡ The occurrence of the burnt shale at Portland Ferry indicates that there a similar combustion formerly took place.

It has been stated that at the northern end of Portland Isle there " is a stratum of black imflammable slate. It is found on the outside of the cliffs, and dug horizontally. The upper stratum is 14 feet of natural black earth; the next is a greyish soft pavingstone, six inches thick; under this is black earth; and at the bottom of the cliff, clay.* * * It is used to heat ovens, piled up as turf, gives a clear light, and yields a strong bituminous smell, and is in all respects iike the Kimeridge coal." A somewhat similar statement was made by J. Woodward in 1729; he noted that the " Coal-stone " was about a foot in thickness, that it " holds and endures the fire much longer than coal," and contains Ammonites (*Cornu Ammonis*) and other shells.§

The Purbeck shales near Poundsford, Sussex, are occasionally bituminous. Mr. R. Hallett, the owner of Swife's Farm, had several tons of shale operated upon, and this yielded tar, pitch, grease, oil, naphtha and paraffin. Mr. Topley called my attention to the fact that the gypsum becomes bituminous along a line of fault at the Gypsum-works at Netherfield.‖

Miscellaneous Minerals.

Mention has been made of the occurrence of Pyrites, Selenite, Glauconite, &c., in various formations. There are but few other minerals to which special attention need be drawn. (See pp. 49, 167, 325.)

---

* See Evans, Ancient Bronze Implements, &c. 1881 ; and Figures on Plate 20 of Damon's Supplement to Geology of Weymouth, &c.; Mansel-Pleydell, Proc. Dorset Nat. Hist. Club, vol. xiii. p. 178.

† Excavations in Bokerly and Wansdyke, Dorset and Wilts, 1892.

‡ Trans. Geol. Soc., ser. 2., vol. iv. p. 23.

§ Hutchins. History and Antiquities of Dorset, Ed. 3, vol. ii., 1863, p. 819; J. Woodward, Nat. Hist. Foss. England, Tome II. p. 17 ; also T. Webster, in Englefield's Isle of Wight, &c., p. 196.

‖ Topley, Geology of the Weald, p. 38.

CHAPTER XVII.

SPRINGS, WATER SUPPLY, AND DEEP BORINGS.

On the tracts of Oxford Clay and Kimeridge Clay there are few important towns and villages, excepting where the strata are overlaid by superficial deposits of gravel which have furnished supplies of water from shallow wells.

Thus on the Oxford Clay, Melcomb Regis (the northern part of Weymouth), is on marine sand and shingle: and Melksham, Oxford, Fenny Stratford, St. Neot's, Huntingdon, St. Ives, Peterborough, and Brigg are situated on Alluvial or Glacial gravels and sands. The same is the case with regard to Gillingham, Abingdon, Sutton near Ely, Market Rasen, and other large villages and towns situated on the Kimeridge Clay.

Oxford Clay and Kellaways Rock and underlying strata.

Water is found in many places in the Kellaways Rock at the base of the Oxford Clay, though it may be difficult to say if in deep wells it may be entirely derived from that rock, as some supply may also be yielded from the Cornbrash, &c. beneath. As a rule, however, there is about 10 feet of clay at the base of the Kellaways sands and rock-beds, separating them from the Cornbrash. The Kellaways Beds are very variable in character, there being alternations of sands, sometimes with doggers, and loams and clays, so that they cannot be depended upon for any large supply of water.

Some particulars of borings in the Vale of Blackmore, made for the Blackmore Vale Dairy Company, have been communicated to me by the Secretary of that Company, and by Messrs. Rawlence and Squarey, of the Digby Estate Office, Sherborne. At the Fox Inn, south of Holwell, in Dorset, a well was sunk through " marl " (Oxford Clay) 178 feet, and " sand rock " (probably Kellaways Rock) 8 feet, and a full supply of water was obtained.

By the Green Man, west of Kingstag, in Lidlinch parish, a boring was carried through similar marl 145 feet, " much harder rock " 25 feet, to close sand, when water of a saline character, rose to within 38 feet of the surface. After pumping it was found that the saline character of the water was maintained.

The occurrence of saline water is interesting, for similar water was found in a well at Woodbridge Farm, Stock Gaylard.

Water was found at a depth of about 150 feet at Grange Farm, north-east of Pulham; but a boring at East Pulham further to the north-east was carried to a depth of 228 feet without finding water. There a sandy rock full of shells (probably Kellaways Rock), was found at a depth of about 200 feet, and beneath blue marly clay, the same as the upper strata. No doubt faults affect the water-levels in this Dorset area.

Saline water has been met with in the Great Oolite series close to Templecombe railway-station, in a bore-hole carried to a depth of 172 feet. The water contained 74 grains of saline matter per gallon; chiefly sulphate of soda.

The well-waters of Wiltshire are notably saline* (see p. 27).

---

* See H. B. Woodward, Quart. Journ. Geol. Soc., vol. xlii. p. 299; and Memoir on the Lower Oolitic Rocks of England, p. 514.

From information kindly supplied by Prof. Edward Kinch, I learn that slightly saline waters were met with in the Kellaways Rock at the Manor House, South Cerney, where at a depth of 110 feet the yield was about 10,000 gallons per day. In this case analysis showed that out of 39 grains per gallon, no less than 29 were carbonate of soda.

In a boring at Cricklade carried to a depth of 105 feet, the water contained 194 grains per gallon, including 107 of chloride of sodium and 62 of carbonate of soda.

Again at Somerford, near Malmesbury, a boring carried to a depth of 90 feet through Kellaways Rock, tapped water containing 139 grains per gallon, including 102 of chloride of sodium and about 29 of carbonate of soda.

The occurrence of saline waters in the Kellaways Beds at Bletchley, and in Bedfordshire has been mentioned (see pp. 48, 341).

Again at Boston, in Lincolnshire, saline water was found in sinking a well in 1783, the strata, as interpreted by Mr. Cameron, being Oxford Clay 470 feet, and sand (Kellaway's Beds) 8 feet.[*]

The Inferior Oolite below Bourn has proved one of the most prolific sources of water in this country. Attention has previously been drawn to this locality,[†] but further information has since been obtained. The water is doubtless derived from the rainfall on the large area of Lincolnshire Limestone exposed to the north-north-west.

Springs yielding a little over 5,000,000 gallons a day, at a pressure of 10lb. to the square inch, were tapped at depths of 100 to 134 feet from the surface, by means of an artesian tube-well 13 ins. in diameter. The water is conveyed to Spalding, 10 miles distant, by gravitation through a line of pipes. This is said to be the largest overflowing spring in existence. Chalybeate water was found at 65 ft. 10 in. from the surface. This was safely excluded by the driving of the 13 in. pipes. The main springs were tapped at depths of 78 ft. 6 in., 100 feet, 120, and 134 feet.

The following is the account of this recent artesian well at Bourn, made for the supply of Spalding by Messrs. C. Isler & Co. :—[‡]

		THICKNESS. Ft. In.	DEPTH. Ft. In.
Made Ground	-	2 0	
Alluvium. {	Clay	1 6	3 6
	Gravel	1	4 6
Kellaways {	Clay	2	6 6
Beds. {	Loamy Clay	1	7 6
Cornbrash. {	Rock and Shells	2	9 6
	Limestone	6	16 0
	Hard blue clay	4	20 0
Great	Mottled clay	10	30 0
Oolite	Blue and green shaly clay	1	31 0
Clay.	Hard blue rock	2	33 0
	Dark blue soft rock with shells	1	34 0
	Hard blue clay	2	36 0

---

* Geol. Mag., 1892, p. 66. See also Phil. Trans., vol. xvi. p. 183.
† Memoir on the Lower Oolitic Rocks of England, p. 505.
‡ Engineering, vol. lvi. Nov. 24, 1893, p. 649 ; and Engineer, Jan. 12, 1894, p. 23.

		THICKNESS. Ft. In.	DEPTH. Ft. In.
Great Oolite Limestone.	Limestone - - -	7 0	43 0
	Ditto light colour - -	4 0	47 0
	Ditto green, very hard -	1 0	48 0
Upper Estuarine Series.	Dark green clay - - -	7 0	55 0
	Hard blue rock - - -	1 0	56 0
	Dark and light green clay -	9 0	65 0
	Hard rock, with water - -	0 10	65 10
	Light green sandy clay - -	9 8	75 6
	Black clay and peat [lignite] -	0 6	76 0
Lincolnshire Limestone.	Grey porous rock - -	1 6	77 6
	Hard oolite limestone - -	56 6	134 0

A bore-hole at Tongue End Farm, about 5 miles E.S.E. of Bourn, and nearly 5 miles S.W. of Spalding, proved the following strata, the record of which was communicated by Mr. E. Easton through Mr. M. W. B. Ffolkes :—

		Ft. In.
Alluvium and Oxford Clay.	Blue clay - - - - -	50 0
	Hard brown clay - - - -	46 0
Kellaways Beds.	Hard sandstone - - - -	10 0
	Blue clay - - - - -	10 0
Cornbrash.	Hard stone - - - - -	7 6
Great Oolite Clay	Clay - - - - - -	17 0
Great Oolite Limestone.	Hard stone - - - - -	13 6
Upper Estuarine Series.	Clay - - - - - -	11 0
	Stone - - - - - -	1 0
	Hard brown clay - - - -	7 0
	Clay, stones, and bed of shells - -	10 0
Lincolnshire Limestone.	Hard stone like " granite " - -	15 0
		198 0

Water rose 62 feet above the surface, and a good supply was yielded. From the same authorities I have received the record of a boring at the Great Eastern railway-station at March : this was carried to a depth of 284 feet—through 11 feet of made ground, and 274 feet of blue clay with 9 bands of stone at depths between 65 and 175 feet. One of these stone-bands, reached at a depth of 68 feet, was 6 feet thick. This may be a Corallian rock-bed. The boring was commenced in Kimeridge Clay and thus doubtless was carried through the Ampthill Beds well into the Oxford Clay.

A boring at Littleworth, near Market Deeping, reached a hard rock-bed (probably Cornbrash) at a depth of 226 feet; and water-bearing strata (Lincolnshire Limestone) were proved at a depth of 285 feet. The water rose to the surface.

In Cambridgeshire limited supplies of water are occasionally obtained from the thin rock-bands that occur in the Corallian Clays, in the upper part of the Oxford Clay, or at the base of the Kimeridge Clay. Wells and borings have been made S.S.W. of Elsworth (150 feet), Rampton (11 feet), Lolworth (135 feet), Over (215 feet), and Stretham Fen (141 feet).[*]

On the other hand a well sunk to a depth of 300 feet in Oxford Clay, etc., at Ramsey, failed to find water : it was evidently not carried deep enough to reach the base of the Great Oolite Series. (See p. 60.)

---

* Penning and Jukes-Browne, Geol. Cambridge, pp. 162, 163, 165, 168 ; Seeley, Ann. Nat. Hist., ser. 3, vol. x. p. 100 ; and Whitaker, Geol. parts of Cambridgeshire and Suffolk, p. 113.

Again in the Kimeridge clay at Downham Market, a boring made by Messrs. C. Isler & Co., proved the following beds:—

		FEET.
Lower Greensand		29
Kimeridge Clay.	Rock -	2
	Blue clay	72
	Rock -	6
	Blue clay	99
	Blue clay and stone -	6
	Blue clay	2
		216

The water-level was 35 feet down, and the yield, which apparently came from the stone-beds, was 360 gallons per hour.[*] It is possible that beds of Corallian age were reached.

A well at Langworth, north-east of Lincoln, sunk through Boulder Clay and Oxford Clay to a depth of 90 feet, found water in the Kellaways Beds, and it rose nearly to the surface.[†]

A well-boring made by Messrs. Le Grand and Sutcliff, at the Market Place, Winslow, proved the following strata, according to Mr. W. H. Dalton:—

	FEET.
Drift } Oxford Clay }	238
Kellaways Beds	32
Great Oolite Series	9

Water was obtained from the Kellaways Beds, and stood at 140 feet from the surface.

Corallian Beds.

The Corallian rocks are not noted for yielding any very large supplies of water, comparable with those obtained in places from the Lower Oolites. The rock-beds are parted by clays that perhaps interfere with the free circulation of water. Moreover the beds where much jointed, sometimes have the fissures infilled with clay that has been washed in from overlying beds.

In the Weymouth district water has been obtained from wells sunk into the Corallian Beds, but the main supply of the town is derived from a Chalk spring that issues from the base of Moignes Down, above Sutton Pointz.[‡]

A well-boring for a dairy, south of Boywood Farm, north of Mappowder, was carried through 168 feet of marl, and 8 feet of rock: the strata said to be of similar character to those by the Fox Inn and Green Man (p. 332). Water was plentiful. Judging from the Geological Survey Map the strata passed through would be Kimeridge Clay and Corallian.

A well made by Messrs. Le Grand and Sutcliff at Ansty, north of Melcomb Bingham, found a good supply of water at a depth of 103 feet, the water rising to within two feet of the surface. It seems probable that Corallian Beds were touched at the base of the boring, and that the beds penetrated were Cretaceous 21 feet, and Kimeridge Clay 82 feet. Particulars were communicated to me by Mr. Whitaker.

A well sunk and bored at Gillingham through (Kimeridge) clay and rock 60 feet, and "clay, hard sand and rock" (partly Corallian) 26 feet, yielded about 20 gallons per minute.[§]

---

[*] Whitaker, Geol. S.W. Norfolk, &c. p. 157.
[†] De Rance, Proc. Geol. and Polyt. Soc., vol. xii. p. 49.
[‡] Damon, Geol. Weymouth, 1884, p. 133.
[§] De Rance, Rep. Brit. Assoc. for 1879, p. 160.

Reference has elsewhere been made to a well at Westbury, in Wiltshire (p. 111).

Springs from the Corallian rocks at Loxwell or Lokeswell south-east of Chippenham, were found to yield from 65 to 70 thousand gallons of water a day.

At the Beaufort Brewery, Wootton Bassett, a little north of the railway-station a well-boring was carried to a depth of 114 feet through Kimeridge Clay into Corallian Rocks. (See p. 116.) Water rose to within 8 or 9 feet of the surface, the yield being equal to 540 gallons per day. The water was found to contain 85 grains per gallon of saline matter, including 43 of sulphate of soda, 19 of carbonate of lime, 13 of carbonate of soda, and 9 of chloride of sodium. This water after long pumping became less saline. A plentiful supply of good water has since been obtained, in the same neighbourhood, the yield being 360 gallons an hour.

In 1816 a well was commenced by the Wilts and Berks Canal Company near Toot Hill between Swindon and Wootton Bassett to supply water for the canal. The strata passed through were as follows :—

			FEET.
Kimeridge Clay.	{ Clay with septaria ; a small chaly-	{ Sunk	- 138
	beate spring at depth of 42 feet	{ Bored about	- 102
Corallian Rocks.	{ Coral Rag - - -	}	
	{ Pisolite - - -	} ...	- 21
	{ Sand and Sandstone - -	}	

| | 261 |

William Smith, who was subsequently consulted, remarked that "here the depth was, of course, expected to be great, from the known depth of several deep wells in the neighbourhood, all of which produced water which ascended to their tops; and the deepest and nearest to this experiment having done so and continued to overflow ever since it was sunk, afforded data for such a proceeding. Besides the water found at Mr. King's of Mannington Farm, I find that water has been obtained at another farm of his, and at Costar and Whitefield, along the course of the same clay-ridge which extends to Wootton Bassett : and that at three of these wells, like that of Mr. Edwards's well (at Even Swindon) the water is of a mineral quality. All of them I am informed, have a copious supply of water, and stand full to the surface, or nearly so."

The water at the above-mentioned well was found at first to be considerable and it brought up a great quantity of sand, but in time the amount of water became very limited and the well was abandoned. It is very interesting to learn that at this early date (about 1817) Smith advised headings to be driven. He said "The first sudden rise of water into the well through such a small aperture, and the subsequent rush of water through the sand above referred to, shows that the water comes freely to this hole ; but to obtain a supply sufficient for the regular work of a 50 horse-power engine will require headings driven in the level course of the stratum which produces the water, unless the natural working of the water through the sand, with an enlarged aperture, should of itself make a sufficient aperture. Headings may be driven in this loose sand, under such a rock roof, to any extent and at a moderate price. Such subterraneous cavities or headings, if made capacious enough to hold from 10 to 20 locks full, would serve as regulators between the flowing of the spring and the inequalities of consumption."[*]

The account of this well is interesting in connexion with the sinking at Swindon made by the Great Western Railway Company, and I am sorry to have been unacquainted with the old record when giving some account of the newer well.[†]

A well at Arkell's Brewery, by the cross roads about a mile N.W. of Stratton St. Margaret's, was sunk 53 feet in Oxford Clay, obtaining but

---

[*] J. Phillips, Memoirs of William Smith, pp. 83–86.
[†] Quart. Journ. Geol. Soc., vol. xlii. p. 287 ; and Memoir on the Lower Oolitic Rocks of England, p. 514.

a poor supply of water. Another well only 12 feet distant to the N.E., was sunk 16 feet into Corallian Rocks and obtained a good supply. This information was communicated by Mr. W. H. Dalton who inferred the presence of a fault running in a northerly direction from the hamlet of Upper Stratton towards the valley west of Kingsdown; and having a downthrow on the east.

At Faringdon a well at the Eagle Brewery, bored 113 feet through the Corallian Beds to the Oxford Clay, yielded a supply (from a 3-inch bore) of 12,700 gallons per day of 11 hours. The water rose to within 13 feet of the surface, but was reduced by pumping to 30 feet. Another well, bored to a depth of 114 feet 6 inches, yielded about 70 gallons per minute.[*]

Kimeridge Clay.

Reference has been made to a boring in Kimeridge Clay that yielded a supply of water from stone-beds. As a rule no yield could be expected from the formation, and even where the Corallian stone-beds occur in the Fenland area between the Oxford Clay and Kimeridge Clay, we could not anticipate any store of water (See pp. 328, 335.)

Mr. Jukes-Browne informs me that at Ascott, near Wing, south-west of Leighton Buzzard, a shaft was sunk 44 feet through Drift and Gault into Kimeridge Clay. A moderate supply of somewhat saline water, was found at a depth of 42 feet, when a hard band of clayey rock had been penetrated. A boring was then carried to a depth of 145 feet in stiff clay, when the well was abandoned.

At Puttenham three miles north-west of Tring a boring was made to a depth of 225 feet as follows:—[†]

		FEET.
Gault - -	Clay, with rock at bottom -	about 150
Lower Greensand or Portland Beds? -	Sand (without water)	75
Kimeridge Clay -	Dark Clay with water-stones [probably septaria]	

A deeper boring was made some years ago at the Asylum at Stone, near Aylesbury, and the following record has been preserved at the Geological Survey Office:—

		FEET.
	Soil - - - - -	
Portland Beds.	Shelly oolitic limestone - - - -	
	Yellow sand - - - -	25
	Shelly oolitic limestone - - - -	
	Yellow sand - - - -	
	Blue limestone - - - -	
	Blue clay - - - - -	
	Limestone - - - -	
	Dark sand with water - - -	
	Blue clay with fossils - - -	145
	Limestone with Ammonites - - -	
	Limestone - - - -	
	Blue clay - - - - -	45
	Limestone - - - -	
	Blue clay - - - -	
	Limestone - - - -	40
	Blue clay - - - -	
	Limestone - - - -	

---

[*] See De Rance, Rep. Brit. Assoc. for 1878, p. 414, and 1879, p. 159.
[†] Whitaker, Trans. Hertfordshire Nat. Hist. Soc., vol. vi. p. 60.

FEET.

Blue clay	.	.	.	.	}	
Limestone	.	.	.	.		
Limestone	.	.	.	.	} 75	
Blue clay	.	.	.	.		
Limestone	.	.	.	.	}	
Blue clay	.	.	.	.	} 200	
Sandy clay	.	.	.	.		
Oolitic coralline limestone		.	.	} 10		
Blue clay -	.	.	.	.		
Oolitic rock, various hardness		.	.	}		
Sand and pebbles	.	.	.	} 30		
Oolitic rock	.	.	.	.		
Clay -	.	.	.	.	}	

Oolitic rock containing fossils, at bottom of boring.

570

A shaft was sunk to the depth of 145 feet, and the rest bored.

In reference to this boring at the "Aylesbury Asylum," John Phillips remarked that it "showed zones of *Gryphæa dilatata* and other characteristic fossils;"[*] but the meagre record, and the want of information concerning the thicknesses of the beds of limestone (probably in many cases septaria), render it difficult to classify the strata.

The main mass appears to have been Kimeridge and Oxford Clays, and another record published by Mr. C. S. Read[†] suggests the following grouping of the strata :—

FEET.

[Portland Beds.]	Limestone and yellow sand	-	- 20
[Kimeridge, Corallian, and Oxford Clays.]	Clay with a few narrow bands of limestone - - - -		- 500
	Great Oolite - .-	-	- 30

The depth here assigned to the boring is 20 feet less than that given above. The well failed to produce an adequate supply of water.

Portland Beds.

The Portland Beds are well calculated to store up supplies of water, for the stone-beds are usually traversed by well-marked joints and fissures, and in many places there is a considerable thickness of sand in the lower beds and locally in the upper beds. The limited outcrop of the strata however forbids their being regarded as of more than very local value for the supply of water. (See p. 196.)

Portland Island possesses no surface streams. The rain-waters penetrate generally down to the Portland Sands and are supported by clayey beds in these strata, or by the Kimeridge Clay. Springs issue from the cliffs, and two more prominent sources have been utilized at Fortune's Well and Southwell; these in old times have supplied all the wants of the inhabitants. More recently a deep well has been made and this has proved successful. (See p. 202.)

Dry valleys occur in the Isle of Purbeck, as at Seacombe, where the rain-waters are conducted underground through the rubbly and porous Portland Beds; but at the sea-margin, springs issue over steps and platforms of the cherty Portland rocks, which are eroded in a very irregular way owing to the bosses of chert.

---

[*] Geol. Oxford, &c., p. 297.
[†] Journ. R. Agric. Soc., vol. xvi. p. 280.

It seems likely that the Portland Beds of Thame are continuous with those of Great Milton; but no evidence from well-borings has, so far as I know, proved the point. The subject is not without its importance to those who would seek a supply of water from beneath the Gault along the borders of the Chiltern Hills, in the neighbourhood of Tetsworth; but in any case the gathering ground would be somewhat limited.

In old times the hardness of the water at Thame was not praised, for Plot remarked, "there is never a Well in the whole Town whose water will wash, or (which is worse) brew."[*]

At Whitchurch in Buckinghamshire copious springs are thrown out from the base of the Portland Beds. There is one beneath the moat of the Castle.

Purbeck Beds.

In Dorsetshire springs are thrown out here and there by the Lower Purbeck Beds, water being held in the Swanage stone-beds. Some of the old quarries are reservoirs of water.

Locally, as in the Vale of Wardour, supplies of water may be obtained from the Purbeck stone-beds. (See p. 272.)

In East Lincolnshire the Spilsby Sandstone is a good water-bearing formation, and a number of villages are situated on it. As remarked by Mr. Jukes-Browne "Strong springs gush out at many points along the line of its junction with the Kimeridge Clay, especially in the dales on the western side of the Steeping Valley." These springs are occasionally ferruginous.[†]

At Skegness a well to which reference has been made (p. 291), was sunk, through the Spilsby Sandstone, to a depth of 425 feet. The yield of water was 8 gallons a minute, and it rose above the surface.

A good supply of water from the Spilsby Sandstone, was also yielded at Willoughby, near Alford. The boring was carried to a depth of 245 feet, and the yield of water was 100,000 gallons per day, the water overflowing.[‡]

Reservoirs and Ornamental Waters.

There are Ornamental Waters on the Oxford Clay in the parks of Longleat, Bowood, Wotton Underwood, Middle Claydon, and Southill west of Biggleswade; and on the Kimeridge Clay at Encombe. There is a Reservoir to the south-east of Swindon that is situated partly on the Lower Portland Beds.

The large tracts of Oxford and Kimeridge Clay furnish many sites suitable for excavated reservoirs, but only in a few cases have sites been observed that would be adapted for embanked reservoirs.[§]

SPRINGS.

Of the freshwater springs there are no very copious and noteworthy examples that can be compared with those which issue from the Lower Oolites.

Holy Wells, in the district under consideration, are found here and there, and from their permanent character they have in old times been dedicated to various Saints.[||] There is a Holywell parish in Oxford, while the village of Holywell, east of St. Ives, derives its name from a well in the churchyard: the water occurs in a bed of gravel that rests on the Oxford Clay. The Wishing Well of Upway, near Weymouth, derives its water from

* Natural History of Oxfordshire, 1677, p. 36.
† Jukes-Browne, Geol. East Lincolnshire, pp. 18, 135.
‡ Ibid., p. 169 : and Quart. Journ. Soc., vol. xlix. p. 469.
§ See Report Royal Commission on Water Supply, 1869, p. xxx; Evidence of Bailey Denton, pp. 94, &c., also Report Royal Commission on Metropolitan Water Supply, 1893; Evidence of T. Hawksley and others, pp. 255–263, 320–349.
|| See R. C Hope, in the "Antiquary," vols. xxi. to xxv.

the Chalk range on the north. The Lady Well at West Keal, and the Holy Well at Somersby in East Lincolnshire, are derived from the Spilsby Sandstone.

"*Petrifying Springs*" have been noted in a few places in the area occupied by the Middle and Upper Oolites; at St. Clement's, and at Carfax (pump) Oxford; at Headington, near Oxford (Corallian Beds); Osbournby, south of Sleaford (probably the same as Aswarby Spa); at Blashenwell, near Corfe Castle, and at Poundsford (Pounceford) near Battle, from the Purbeck Limestone.

A number of springs have obtained a certain notoriety as Spas, but few if any of these retain the fame accorded to them in old days, though whatever virtue they possessed may still be with them.

Chalybeate Springs.

The following may be noted with reference to the formations from which they issue :—

Oxford Clay.

Melksham, at Drinkwater House (well said to yield 800 gallons per day).
Clapham, Bedford.
Somersham Spa, Huntingdonshire (from hard bed in the Oxford Clay):
Oundle (base of Kellaways Rock). Bourn, Blind Well, in Bourn Wood.
Kingerby Spa, north-west of Market Rasen (at base of Drift Sand on Oxford Clay).

Corallian Beds.

Rodwell, Weymouth.
Heywood, north of Westbury, Wiltshire.
Wootton Bassett Spa, by Whitehill farm, north-west of town (said to contain traces of Iodine).
Abingdon.
Headington Wick, near Headington, Oxford.

Portland Beds.

Dorton Spa, Dorton Park, near Brill.

Saline Springs.

Saline waters have proved to be more abundant than desirable in the Kellaways Beds, Oxford Clay, and Corallian Beds. The following include all the localities that are known to me.

Lidlinch, Dorset (well-boring nearly 200 feet deep into Kellaways Rock, by Green Man).
Stock Gaylard, near King Stag (well at Woodbridge Farm).
Templecombe (well-boring 172 feet deep into Fullonian? ; 31 sulphate of soda, 12 chloride of sodium, 12 carbonate of lime, &c. Total 74 grains per gallon).
Horwood Wells, south-east of Wincanton.
West Ashton, near Steeple Ashton, Wiltshire (saline chalybeate; chloride of sodium, &c.).
Holt, near Melksham (saline chalybeate ; 146 grains per gallon).
Somerford, near Malmesbury (139 grains per gallon ; chloride of sodium, 102, carbonate of soda, 29, &c.).
Purton Spa, Salt's Hole, west of Purton Stoke, between Purton and Cricklade* (yield 120 gallons per day ; 350 grains per gallon, chiefly sulphates of soda, lime, and magnesia).

---

* See Analyses, Quart. Journ. Chem. Soc., vol. xiv. p. 43; and Quart. Journ. Geol. Soc., vol. xlii. p. 299.

Braydon Manor Farm, near Purton,* (445 grains per gallon, same salts as above).

Cricklade (boring 105 feet; 194 grains per gallon, chiefly chloride of sodium).

Highworth.

Wootton Bassett (well at town, 85 grains sulphate of soda, &c.).

Swindon (well at Great Western Railway Co.'s works, from Corallian Rocks, 144 grains per gallon, chloride of sodium (86), carbonate of soda (49), &c., more saline waters were obtained deeper from the Forest Marble).

Rodbourn Lane, near Rodbourn Cheney, Swindon (112 grains per gallon; 45 carbonate of soda, 56 chloride of sodium).

South Cerney, Manor House (boring 124 feet to Kellaways Beds; 39 grains per gallon, chiefly carbonate of soda. Supply at depth of 110 feet about 10,000 gallons per day).

Cumner, west of Oxford (well).

Bletchley (boring at depths of 390 and 410 feet in Kellaways Beds; 340 to 392 grains per gallon of chloride of sodium, sulphate of soda, &c. (See p. 48.)

Stony Stratford. (See p. 49.)

Kempston, Wootton, Stanton, and Lower Shelton near Marston Moreteyne (saline waters, 180 to 250 grains per gallon, in well-waters from Kellaways Beds).†

At Eaton Socon, near St. Neots (brewery); and at St. Neots (well at paper-mill 300 feet deep; chloride of sodium and sulphate of soda).

Soham Fen (boring 120 feet to rock, water overflowed but was brackish).

Stainfield, north-west of Bourn, Lincolnshire (264 grains per gallon).

Cawthorpe, north of Bourn (saline chalybeate spring, 160 grains per gallon).

Aswarby Spa, between Sleaford and Folkingham (480 grains per gallon).

Walcott, north-west of Billinghay, Lincolnshire (256 grains per gallon).

Waters impregnated with mineral matter have been noticed at Christian Malford, north-east of Chippenham, Hail Weston (Spa), near St. Neots, Huntingdon, St. Ives, and at Haverholme, near Anwick, north-east of Sleaford (medicinal spring), in the Oxford Clay area; and at North Willingham, near Market Rasen (medicinal spring) in the Kimeridge Clay area.

"*Sulphur Springs*" giving off sulphuretted hydrogen, due to the decomposition of pyrites, are met with here and there in the Oxford Clay. Some of these springs are slightly saline, as at Radipole and Nottington Spa near Weymouth (34 grains per gallon).‡

Water from a well at Abingdon, sunk about 60 feet through Kimeridge Clay into Corallian beds, was slightly inpregnated with sulphuretted hydrogen and iron-salt.§

A spring at Catley Abbey, east of Digby, and north-west of Sleaford, is advertised as the "Only natural British Seltzer Water." The water is obtained from a depth of about 80 feet from beds below the Oxford Clay.

Springs having a temperature of about 66° to 74° were noticed at Chatteris by Mr. F. W. Harmer. Their origin, which is very doubtful, has been discussed elsewhere.||

Reference has previously been made to the mineral waters in the Jurassic Rocks, and to the possible sources of supply of the saline ingredients.¶ In many cases we have to look to deep-seated

* A. Voelcker, Journ. R. Agric. Soc., ser. 2, vol. ii. p. 393.
† Jukes-Browne, Geol. Mag., 1889, p. 360; and Midland Nat., vol. xiv. p. 205.
‡ Buckland and De la Beche, Trans. Geol. Soc., ser. 2, vol. iv. p. 49, Explan. of Plate.
§ Rev. J. C. Clutterbuck, Journ. R. Agric. Soc., ser. 2, vol. i. p. 281.
|| Skertchly, Geol. Fenland, p. 243; and Whitaker, Geol. S.W. Norfolk; &c., p. 152.
¶ Memoir on the Lias of England and Wales, pp. 320-326; Memoir on the Lower Oolitic Rocks of England, pp. 511-516. See also R. Warrington, Journ. Chem. Soc., vol. li. p. 500.

DOVER BORING.	THICKNESS.		DEPTH.	
	FT.	IN.	FT.	IN.
Dark sandy clay - -	6	0	775	0
Sandstone - - -	2	0	777	0
Black marl - - -	2	0	779	0
Sandstone - - -	1	0	780	0
Marl with *Ostrea gregaria* -	5	0	785	0
Shelly sandstone with *Modiola imbricata, Ostrea gregaria, Rhynchonella spathica* - - -	1	0	786	0
Brown sandy marl - -	6	0	792	0
Calcareous sandstone -	3	0	795	0
Shale - - -	43	0	838	0
Marly limestone - -	1	0	839	0
Shale - - -	7	0	846	0
Shelly limestone - -	2	0	848	0
Dark grey marl - -	34	0	882	0
Marly limestone - -	1	0	883	0
Brown sandy clay, with *Gryphæa dilatata* - -	22	0	905	0
Calcareous sand - -	2	0	907	0
Sandstone - - -	3	0	910	0
Brown sandy clay, with *Belemnites, Rhynchonella socialis* - - -	1	0	911	0
Brown sandstone, with small quartz pebbles, ganoid Fish-scales, *Belemnites* (broken and possibly derived) - - -	4	0	915	0
Calcareous sand - -	3	0	918	0
Hard grey limestone -	3	0	921	0
Calcareous sand, with quartz grains, iron-pyrites, and glauconite - -	1	0	922	0
Brown sandstone with lignite and shells - -	4	0	926	0
Calcareous sand - -	3	0	929	0
Sandstone - - -	2	0	931	0
Clay with vegetable impressions and bivalve shells	4	0	935	0
Calcareous sand, with *Gryphæa* - - -	3	0	938	0
Dark grey marl - -	2	0	940	0
Bluish-grey clay - -	12	0	952	0
Brown sand - -	1	0	953	0
Hard grey marl - -	4	0	957	0
Shelly limestone - -	2	0	959	0
Grey marl - - -	4	0	963	0
Siliceous limestone - -	3	0	966	0
Sandstone - - -	1	0	967	0
Marl - - -	1	0	968	0
Shelly limestone - -	1	0	969	0
Dark marl - - -	5	0	974	0
Earthy limestone - -	1	0	975	0
Marl - - -	2	0	977	0
Sandstone - - -	1	0	978	0
Oolitic limestone with *Terebratula globata? Ostrea Sowerbyi; Gryphæa* and *Belemnites* - - -	8	0	986	0

Oxfordian Beds. 188 feet.

DOVER BORING.		THICKNESS.	DEPTH.
		FT. IN.	FT. IN.
	Sandstone - - -	1 0	987 0
	White shelly oolitic lime- stone; *Gryphæa, Belemnites*	9 0	996 0
	Marl - - -	2 0	998 0
	White shelly oolitic lime- stone; *Ostrea Sowerbyi, Rhynchonella varians* -	15 0	1,013 0
	Calcareous sand - -	5 0	1,018 0
	Hard grey limestone, with *Chemnitzia, Gryphæa, Rhynchonella concinna, Serpula, Clypeus Ploti* -	14 0	1,032 0
Bathonian. 156 feet.	Sand - - -	1 0	1,033 0
	Hard limestone - -	3 0	1,036 0
	Calcareous sand - -	1 0	1,037 0
	Dark grey limestone -	10 0	1,047 0
	Calcareous sand - -	1 0	1,048 0
	Dark grey marl - -	5 0	1,053 0
	Calcareous grit - -	3 0	1,056 0
	Bedded sandstone - -	3 0	1,059 0
	Calcareous grit - -	4 0	1,063 0
	Dark grey sand - -	13 0	1,076 0
	Hard grey clay - -	7 0	1,083 0
	Hard dark grey limestone -	2 0	1,085 0
	Dark grey marl - -	1 0	1,086 0
	Grey argillaceous limestone	3 0	1,089 0
	Marl - - - -	1 0	1,090 0
	Hard grey limestone -	1 0	1,091 0
	Marl - - - -	1 0	1,092 0
	Hard grey limestone -	1 0	1,093 0
	Marl - - - -	1 0	1,094 0
	Hard grey limestone -	2 0	1,096 0
	Dark grey clay - -	5 0	1,101 0
	Earthy limestone - -	1 0	1,102 0
	Dark grey clay - -	6 0	1,108 0
	Limestone - - -	1 0	1,109 0
	Marl - - - -	1 0	1,110 0
	Limestone - - -	1 0	1,111 0
	Marl - - - -	1 0	1,112 0
	Sandstone - -	1 0	1,113 0
Coal-measures (shales, sandstones, and coal-seams, dip 2°) - - - - -		1,068 6	2,181 6

Sub-Wealden Boring.

The Sub-Wealden exploration was commenced in Sussex to commemorate the visit of the British Association to Brighton in 1872. It owes its origin to Mr. H. Willett of Brighton, and its main object was to reach if possible the Palæozoic rocks, which it was hoped would be found within about 1700 feet of the surface. The great development of Oolitic rocks was however anticipated by S. V. Wood, jun., and also by Mr. Bauerman and Mr. Kinahan.

Mr. Topley, who examined the cores, says that one noteworthy fact was the very gradual transition from one kind of rock to another. " Shale passed gradually into sandy shale and shaly sandstone, sometimes becoming a true sandstone. On the other hand shale gradually became calcareous, passing into cement stone and sometimes into a tolerably good limestone. The chief cases of fairly abrupt change were at the occurrence of some highly fossiliferous bands, when it would seem that forms of life, chiefly oysters, suddenly migrated to the area and as suddenly departed."[*]

Two borings were made, the first, begun by Mr. Bosworth and continued by the Diamond Rock Boring Company, was carried to a depth of 1,030 feet; the second was carried out by the same Company and completed in 1875 to a depth of 1,905 feet. The two borings were but a few yards apart and started practically from the same level.

Various accounts of the boring have from time to time been published. The actual depth bored has been somewhat differently stated at 1,903 and 1,905 feet; but we may take the latter as the correct figure, as it is so stated by Mr. Henry Willett.[†]

The interpretations put upon the record have also differed very much, not only those given by different authorities, but by the same individual. After having the benefit of considering all these views, and having seen the cores preserved in the Brighton Museum I venture to make the following grouping of the strata :—

SUB-WEALDEN BORING.		THICKNESS.	DEPTH.	FOSSILS.
		FEET.	FEET.	
Purbeck Beds. 177 feet.	Clay and shale with bands of limestone or cement-stone -	123	—	
	Shaly limestone with veins of gypsum - - -	4	127	
	Gypsum and gypseous marls, &c. - - - -	42	169	
	Fragmentary beds of shale and chert - - - -	8	177	
Portland Beds. 115 feet.	Sandy shale with chert - -	23	200	
	Soft whitish sandstone - -	52	252	
	Darker sandstone - -	5	257	
	Sandy shale - - -	35	292	Am. biplex.
	Shales, sometimes calcareous and sandy, with septaria and cement-stones - -	668	960	Am. Callisto. Am. mutabilis.
	Bituminous shales - -	40	1,000	Am. longispinus.
	Clays and calcareous sandy beds	4	1,004	
	Sandstone - - -	41	1,045	Am. orthocera.
	Sandy shale - - -	19	1,064	
	Sandy shale with nodules of limestone - - -	28	1,092	Am. biplex.

* Science Conferences, South Kensington Museum, 1876 (1877), p. 380.
† The Record of the Sub-Wealden Exploration, 1878. See also Dixon's Geology of Sussex, Ed. 2, pp. xxiii., 6, 153; J. F. Blake, Proc. Geol. Assoc., vol. vii. p. 358; Dawkins, Nature, July, 31, 1890; and De Rance, Rep. Brit. Assoc., for 1880, p. 18.

SUB-WEALDEN BORING.	THICKNESS.	DEPTH.	FOSSILS.
	FEET.	FEET.	
Kimeridgian Beds. 1,278 feet.			
Shaly and quartzose sandstone, &c. - - - .	45	1,137	*Exogyra virgula.*
Shaly limestone - - -	27	1,164	
Limestone - - -	4	1,168	
Shaly limestone - - -	14	1,182	
Sandy and calcareous shale -	94	1,276	
Soft gritty limestone - -	29	1,305	
Calcareous shale - -	20	1,325	
Calcareous grit and bands of limestone - - -	45	1,370	
Limestone and shale - -	27	1,397	*Am. lallierianus, Gervillia aviculoides.*
Shale - - - -	23	1,420	
Limestone with *Ostrea* - -	10	1,430	*O. deltoidea.*
Calcareous shale - -	16	1,446	
Shale - - - -	80	1,526	
Shale with *Pentacrinus,* &c. -	20	1,546	
Shale with calcareous beds -	19	1,565	*Am. mutabilis, A. Callisto.*
Corallian Beds. 241 feet.			
Calcareous sandstone with quartz pebbles, and limestone -	95	1,660	*Pinna lanceolata, Trichites, Rhynchonella pinguis.*
Calcareous shale and cementstone - - - -	40	1,700	
Dark shale - - -	69	1,769	
Oolitic limestone - -	17	1,786	*Trigonia.*
Sandy shale - - -	20	1,806	
Oxfordian Beds. 99 feet.			
Dark shale - - -	59	1,865	*Am. Chamusseti.*
Limestone, some oolitic - -	15	1,880	*Am. Chamusseti.*
Dark shale - - -	25	1,905	

SUMMARY OF SOME OF THE DEEP BORINGS IN THE JURASSIC ROCKS.

	Mickleton, Gloucestershire.	Burford, Oxfordshire.	Witney, Oxfordshire.	Wytham, near Oxford.	St. Clements, Oxford.	Stone, near Aylesbury.	Richmond, Surrey.	Meux's Brewery, London.	Streatham, Surrey.	Chatham.	Dover.	Sub-Wealden Boring, near Battle.
Superficial Deposits	—	—	—	15	—	—	10	22	10	42	—	—
Tertiary	—	—	—	—	—	—	242	134	236	—	—	—
Cretaceous	—	—	—	—	—	—	897	844	775½	901	505	—
Purbeck Beds	—	—	—	—	—	20	—	—	—	—	—	177
Portland Beds	—	—	—	—	—	—	—	—	—	—	32	116
Kimeridgian	—	—	—	—	—	500	—	—	—	—	73	1,273
Corallian	—	—	—	—	—	—	—	—	—	—	159	241
Oxfordian	—	—	—	255	265	30	—	—	—	20	188	99
Cornbrash	—	—	10	19	135	—	—	—	—	—	—	—
Forest Marble	—	—	33	24½	—	—	87½	64	38½	—	156	—
Great Oolite	—	62½	141½	96	—	—	—	—	—	—	—	—
Inferior Oolite Series	—	27	30	38½	—	—	—	—	—	—	—	—
Upper Lias	—	82	55½	14½	—	—	—	—	—	—	—	—
Middle Lias	250	98		170½	—	—	—	—	—	—	—	—
Lower Lias	961	447½	—	—	—	—	—	—	—	—	—	—
New Red Series	101	486	—	—	—	—	—	—	—	—	—	—
Coal Measures	—	207	—	—	—	—	207½	80	138	—	1,068	—
Devonian	—	—	—	—	—	—	—	—	—	—	—	—
Feet	1,348	1,410	270	633	400	550	1,444	1,144	1,258	963	2,181	1,905

APPENDIX.

APPENDIX I.

CATALOGUE OF FOSSILS

FROM THE

MIDDLE AND UPPER OOLITIC ROCKS OF ENGLAND.

Explanatory remarks on the preparation of the lists of Fossils were made in the Memoir on the Lias; and it need only be repeated that the names of species, whose occurrence or identification is doubtful, are either omitted, or they are inserted with a query.

The species in the Museum of Practical Geology are recorded; and I have received much help in noting them from Mr. H. A. Allen. Many specimens obtained by Mr. J. Rhodes and myself, during the progress of the work, have been named by Messrs. G. Sharman and E. T. Newton. To these colleagues, as well as to Mr. Allen, I am indebted for sundry corrections in the list.

Great assistance in the matter of nomenclature has been received from the British Museum Catalogues of Fossil Vertebrata, by Mr. R. Lydekker, and Mr. A. Smith Woodward; from the Catalogue of British Fossil Vertebrata, by Messrs. A. S. Woodward and C. D. Sherborn; and from the Catalogue of British Jurassic Gasteropoda, by Messrs. W. H. Hudleston and E. Wilson.

Species are recorded from the volumes of the Palæontographical Society, and from the later Memoirs of the Geological Survey, as well as from certain papers included in the bibliographical list, and to which special reference has previously been made.

With regard to local museums and private collections of fossils, it may be stated that, in addition to those recorded in the Memoirs on the Lias and Lower Oolitic Rocks, there are at Swanage, Corfe Castle, Salisbury, Ely, and Wisbech, museums which contain fossils from the Upper Jurassic Rocks.

Some of the specimens collected by the Rev. O. Fisher from the Purbeck Beds, and by the late Dr. H. Porter from the Oxford Clay, are in the Woodwardian Museum at Cambridge. Many Saurian remains obtained by Mr. Alfred N. Leeds (of Eyebury), are in the British Museum, and some are in the Oxford Museum. The Purbeck Mammals collected by the late S. H. Beckles are in the British Museum. The collection of Mr. William Cunnington is preserved partly in the Devizes Museum, partly in the British Museum. That of the late Dr. John Lee is at Hartwell; that of Mr. Marshall Fisher is at Ely; that of the Rev. W. R. Andrews is at Eastbourne; and that of Mr. J. C. Mansel-Pleydell at Dorchester.

The species of *Ammonites* (as before) are indexed under this generic name ; and a list of the sub-genera is given for convenience of reference.

Many fossils of the Middle and Upper Oolitic formations are illustrated in Damon's Supplement to the Geology of Weymouth and the Isle of Portland, of which a 3rd edition was published in 1888 ; and in Phillip's Geology of Oxford and the Valley of the Thames, 1871.

Some of the Purbeck Mollusca are figured by P. de Loriol, *Mem. Soc. Phys. et Hist. Nat. Genève*, vol. xviii. 1865, p. 80.

REFERENCES TO LOCALITIES.

D. Dorsetshire.	N. Northamptonshire.
S. Somersetshire.	H. Huntingdonshire.
W. Wiltshire.	C. Cambridgeshire.
G. Gloucestershire.	Nf. Norfolk.
O. Oxfordshire.	L. Lincolnshire.
Be. Berkshire.	Y. Yorkshire.
Bu. Buckinghamshire.	Sx. Sussex.
Bd. Bedfordshire.	K. Kent (Borings).

The grouping of the sub-divisions is arranged as follows, in ascending order :—

1. Zone of *Ammonites calloviensis*; Kellaways Beds.

2. Zone of *A. ornatus.*
3. Zone of *A. cordatus.* } Oxford Clay.

4. Zone of *A. perarmatus.*
5. Zone of *A. plicatilis.* } Corallian.

6. Zone of *A. alternans.*
7. Zone of *A biplex.* } Kimeridge Clay.

8. Zone of *A. gigas.*
9. Zone of *A. giganteus.* } Portland Beds.

10. Lower
11. Middle } Purbeck Beds.
12. Upper

× Occurs in the Lower Oolitic Rocks, or in the Wealden Beds, according to the column.

It should be mentioned that a full list of the Yorkshire Jurassic fossils has been published by Mr. Fox-Strangways, *Mem. Jurassic Rocks*, vol. ii. The species recorded in the present list from Yorkshire are those which also occur elsewhere in England.

Species.	Lower Oolite.	Am. calloviensis, Kellaways Rock.	Am. ornatus.	Am. cordatus.	Am. perarmatus.	Am. plicatilis.	Lower Beds.	Upper Beds.	Lower Beds.	Upper Beds.	Lower Beds.	Middle Beds.	Upper Beds.	Wealden Beds.	Localities.	Remarks.
			Oxford Clay.		Corallian.		Kimeridge Clay.		Portland Beds.		Purbeck Beds.					
	×	1	2	3	4	5	6	7	8	9	10	11	12	×		
MAMMALIA.																
Marsupialia.																
Achyrodon nanus, Owen	-	-	-	-	-	-	-	-	-	-	-	11	-	-	D.	
—— pusillus, Owen	-	-	-	-	-	-	-	-	-	-	-	11	-	-	D.	
Amblotherium ? dubius, Owen.	-	-	-	-	-	-	-	-	-	-	-	11	-	-	D.	
—— mustelula, Owen	-	-	-	-	-	-	-	-	-	-	-	11	-	-	D.	
—— soricinum, Owen	-	-	-	-	-	-	-	-	-	-	-	11	-	-	D.	
—— talpoides, Owen	-	-	-	-	-	-	-	-	-	-	-	11	-	-	D.	
Bolodon crassidens, Owen	-	-	-	-	-	-	-	-	-	-	-	11	-	-	D.	
Kurtodon pusillus, Osborn	-	-	-	-	-	-	-	-	-	-	-	11	-	-	D.	
Leptocladus. See Peramus.																
Peralestes longirostris, Owen	-	-	-	-	-	-	-	-	-	-	-	11	-	-	D.	
Peramus dubius, Owen	-	-	-	-	-	-	-	-	-	-	-	11	-	-	D.	
—— minor, Owen	-	-	-	-	-	-	-	-	-	-	-	11	-	-	D -	*Spalacotherium minus,* Owen.
—— tenuirostris, Owen	-	-	-	-	-	-	-	-	-	-	-	11	-	-	D.	
Peraspalax. See Amblotherium.																
Phascolestes. See Amblotherium and Peralestes.																
Plagiaulax Becklesi, Falc.	-	-	-	-	-	-	-	-	-	-	-	11	-	-	D.	
—— Falconeri, Owen	-	-	-	-	-	-	-	-	-	-	-	11	-	-	D.	
—— medius, Owen	-	-	-	-	-	-	-	-	-	-	-	11	-	-	D.	
—— minor, Falc.	-	-	-	-	-	-	-	-	-	-	-	11	-	-	D.	
Spalacotherium tricuspidens, Owen.	-	-	-	-	-	-	-	-	-	-	-	11	-	-	D.	
Stylodon pusillus, Owen	-	-	-	-	-	-	-	-	-	-	-	11	-	-	D.	
—— robustus, Owen	-	-	-	-	-	-	-	-	-	-	-	11	-	-	D.	
Triacanthodon serrula, Owen	-	-	-	-	-	-	-	-	-	-	-	11	-	-	D.	
Triconodon ferox, Owen	-	-	-	-	-	-	-	-	-	-	-	11	-	-	D.	
—— major, Owen	-	-	-	-	-	-	-	-	-	-	-	11	-	-	D.	
—— mordax, Owen	-	-	-	-	-	-	-	-	-	-	-	11	-	-	D.	
—— occisor, Owen	-	-	-	-	-	-	-	-	-	-	-	11	-	-	D.	
Cetacea.																
Palæocetus Sedgwicki, Seeley	-	-	-	-	-	-	-	-	-	-	-	-	-	-	C	{ *Palæobalæna,* Ex Boulder Clay. ? Kim. Cl. or Oxf. Clay. }
REPTILIA.																
Ornithosauria.																
Doratorhynchus validum, Owen.	-	-	-	-	-	-	-	-	-	-	-	11	-	-	D -	{ *Pterodactylus macrurus,* Seel. *Ornithocheirus?* *validus,* Owen. }
Pterodactylus Manseli, Owen	-	-	-	-	-	-	-	7	-	-	-	-	-	-	D.	

SPECIES.	Lower Oolites.	Am. callovienais. (Kellaways Rock.)	Am. ornatus. (Oxford Clay.)	Am. cordatus.	Am. perarmatus. (Corallian.)	Am. plicatilis.	Lower Beds. (Kimeridge Clay.)	Upper Beds.	Lower Beds. (Portland Beds.)	Upper Beds.	Lower Beds. (Purbeck Beds.)	Middle Beds.	Upper Beds.	Wealden Beds.	LOCALITIES.	REMARKS.
	×	1	2	3	4	5	6	7	8	9	10	11	12	×		
REPTILIA—*cont.*																
Ornithosauria—cont.																
Pterodactylus Pleydelli, Owen.	-	-	-	-	-	-	-	7	-	-	-	-	-	-	D.	
Rhamphorhynchus Jessoni, Lyd.	-	-	-	3	-	-	-	-	-	-	-	-	-	-	H.	
Dinosauria.																
Bothriospondylus suffossus, Owen.	-	-	-	-	-	-	-	7	-	-	-	-	-	-	W.	
Camptosaurus Leedsi, Lyd. -	-	-	2	-	-	-	-	-	-	-	-	-	-	-	N.	
—— Prestwichi, Hulke -	-	-	-	-	-	-	-	7	-	-	-	-	-	-	Be.	
Cetiosaurus longus, Owen -	-	-	-	-	-	-	-	-	-	9	-	-	-	-	O. -	Kim. Cl.
—— sp. - - - -	-	-	2	-	-	-	-	-	-	-	-	-	-	-	W.	
—— See also Pelorosaurus.																
Cryptodraco eumerus, Seeley	-	-	2	-	-	-	-	-	-	-	-	-	-	-	H. -	*Cryptosaurus.*
Cumnoria. See Camptosaurus.																
Echinodon Becklesi, Owen -	-	-	-	-	-	-	-	-	-	-	-	11	-	-	D.	
Gigantosaurus megalonyx, Seeley.	-	-	-	-	-	-	-	7	-	-	-	-	-	-	C.	
Iguanodon Mantelli, Meyer -	-	-	-	-	-	-	-	-	-	-	-	11	-	×	D.	
—— See also Camptosaurus.																
Ischyrosaurus. See Pelorosaurus.																
Megalosaurus Bucklandi, von Meyer.	×	-	2	-	-	5	-	7	-	-	-	-	-	-	D. W. O. Y.	
—— insignis, Desl. - -	-	-	-	-	-	-	-	-	-	-	-	-	-	-	W.	Kim. Cl.
—— sp. - - -	-	-	-	-	-	-	-	-	-	-	-	-	-	×	D. -	Purbeck.
Nuthetes destructor, Owen -	-	-	-	-	-	-	-	-	-	-	-	11	-	-	D.	
Omosaurus armatus. Owen -	-	-	-	-	-	-	-	7	-	-	-	-	-	-	W.	
—— durobrivensis, Hulke -	-	-	2	-	-	-	-	-	-	-	-	-	-	-	N.	
—— hastiger, Owen - -	-	-	-	-	-	-	-	-	-	-	-	-	-	-	W. -	Kim. Cl.
Ornithopsis. See Pelorosaurus.																
Pelorosaurus humerocristatus, Hulke.	-	-	-	-	-	-	-	-	-	9	-	-	-	-	D. Bu. C. -	Kim. Cl.
—— Leedsi, Hulke -	-	-	2	-	-	-	-	-	-	-	-	-	-	-	N.	
—— Manseli, Hulke -	-	-	-	-	-	-	-	-	-	-	-	-	-	-	D. C. -	Kim. Cl.
Sarcolestes Leedsi, Lyd. -	-	-	2	-	-	-	-	-	-	-	-	-	-	-	N.	
Stegosaurus. See Omosaurus.																
Crocodilia.																
Brachydectes. See Oweniasuchus.																
Dacosaurus - - -	-	-	3	-	-	-	-	-	-	-	-	-	-	-	O. Bd. -	Ampthill Cl.

SPECIES.	Lower Oolites.	Am. calloviensis. Kellaways Rock.	Am. ornatus. (Oxford Clay)	Am. cordatus. (Oxford Clay)	Am. perarmatus. (Corallian)	Am. lips iilis. (Corallian)	Lower Beds. (Kimeridge Clay)	Upper Beds. (Kimeridge Clay)	Lower Beds. (Portland Beds)	Upper Beds. (Portland Beds)	Lower Beds. (Purbeck Beds)	Middle Beds. (Purbeck Beds)	Upper Beds. (Purbeck Beds)	Wealden Beds.	LOCALITIES.	REMARKS.
	x	1	2	3	4	5	6	7	8	9	10	11	12	x		
REPTILIA—*cont.*																
Crocodila—cont.																
Geosaurus maximus, Plien.	-	-	-	-	-	-	-	7	-	-	-	-	-	-	D. O. C.	Dacosaurus lisso-cephalus Seeley. Steneosaurus Manseli, Hulke.
Goniopholis crassidens, Owen	-	-	-	-	-	-	-	-	-	-	-	11	12	-	D W.	
— sima, Owen	-	-	-	-	-	-	-	-	-	-	-	11	-	-	D.	
— tenuidens, Owen	-	-	-	-	-	-	-	-	-	-	-	11	-	-	D.	
— sp.	-	-	-	-	-	-	-	-	-	9	-	-	-	-	W.	
Machimosaurus mosæ, Liénard.	-	-	-	-	-	-	-	-	-	-	-	-	-	-	D.	Kim. Cl.
Metriorhynchus gracile, Phil.	-	-	-	-	-	-	-	-	-	9	-	-	-	-	O.	
— Moreli, Deal.	-	-	-	-	-	-	-	-	-	-	-	-	-	-	W.	Ox. Cl.
— palpebrosum, Phil.	-	-	-	-	-	-	-	7	-	-	-	-	-	-	D. O. Bu. N.	Ox. Cl. Steneosaurus. M. rostro-minor, Geoff.
— superciliosum, de Blainv.	-	-	-	-	-	-	-	-	-	-	-	-	-	-	O.	Ox. Cl. St. dasy-cephalus? Seeley.
Nannosuchus gracilidens, Owen.	-	-	-	-	-	-	-	-	-	-	-	11	-	-	D.	
Oweniasuchus major, Owen	-	-	-	-	-	-	-	-	-	-	-	11	-	-	D.	
— minor, Owen	-	-	-	-	-	-	-	-	-	-	-	11	-	-	D.	
Petrosuchus lævidens, Owen.	-	-	-	-	-	-	-	-	-	-	-	11	-	-	D.	
Plesiosuchus. See Geosaurus.																
Steneosaurus megarhinus, Hulke.	-	-	-	-	-	-	-	7	-	-	-	-	-	-	D.	
— ? purbeckensis, Mansel-Pleydell.	-	-	-	-	-	-	-	-	-	-	-	-	-	-	D.	Purbeck.
— robustus, Owen	x	-	-	-	-	5	-	-	-	-	-	-	-	-	O.	S. oxoniensis, De la Beche. S. Boutilieri, Desl.
— See also Metriorhynchus.																
Suchodus durobrivensis, Lyd.	-	-	2	-	-	-	-	-	-	-	-	-	-	-	N.	
Teleosaurus asthenodeirus, Owen.	-	-	-	-	-	-	-	-	-	-	-	-	-	-	O.	Kim. Cl.
— sp.	-	-	?	-	-	-	-	-	-	9	-	-	-	-	D. Bu.	Purbeck? Cor.
— See also Steneosaurus.																
Theriosuchus pusillus, Owen	-	-	-	-	-	-	-	-	-	-	-	11	-	-	D. W.	
Lacertilia?																
Macellodus Brodiei, Owen	-	-	-	-	-	-	-	-	-	-	10	11	-	-	D. W.	Saurillus obtusus, Owen.
Homœosauria.																
Homœosaurus	-	-	-	-	-	-	-	-	-	-	-	-	-	-	D.	Purbeck.
Ichthyopterygia (Ichthyosauria).																
Ichthyosaurus æqualis, Phil.	-	-	-	-	-	-	-	-	-	-	-	-	-	-	O.	Kim. Cl.
—? dilatatus, Phil.	-	-	-	3	-	-	-	7	-	-	-	-	-	-	D. W. O.	Kim. Cl.

SPECIES.	Lower Oolites.	Kellaways Rock. Am. calloviensis.	Oxford Clay. Am. ornatus.	Oxford Clay. Am. cordatus.	Corallian. Am. perarmatus.	Corallian. Am. pheatilis.	Kimeridge Clay. Lower Beds.	Kimeridge Clay. Upper Beds.	Portland Beds. Lower Beds.	Portland Beds. Upper Beds.	Purbeck Beds. Lower Beds.	Purbeck Beds. Middle Beds.	Purbeck Beds. Upper Beds.	Wealden Beds.	LOCALITIES.	REMARKS.
	×	1	2	3	4	5	6	7	8	9	10	11	12	×		
REPTILIA—cont.																
*Ichthyopterygia (Ichthyosauria)—*cont.																
Ichthyosaurus entheciodon, Hulke.	-	-	?	-	-	-	-	7	-	-	-	-	-	-	D. W. N.	
—— megalodeirus, Seeley	-	-	2	-	-	-	-	-	-	-	-	-	-	-	N.	
—— ovalis, Phil.	-	-	-	-	-	-	-	7	-	-	-	-	-	-	D. W. O.	
—— thyreospondylus, Owen	-	-	-	3	-	-	-	7	-	9	-	-	-	-	D. W. O. Bu. C.	
—— trigonus, Owen	-	-	-	-	-	-	-	7	-	-	-	-	-	-	D. W. O. C.	Ox. Cl.
Ophthalmosaurus icenicus, Seeley.	-	-	2	-	-	-	-	7	-	-	-	-	-	-	W. O. H.	
—— Pleydelli, Lyd.	-	-	-	-	-	-	-	7	-	-	-	-	-	-	D.	
Sauropterygia (Plesiosauria).																
Cimoliosaurus brachisto-spondylus, Hulke.	-	-	-	-	-	-	-	7	-	-	-	-	-	-	D.	
—— brevior, Lyd.	×	-	-	-	-	-	-	7	-	-	-	-	-	-	D.	
—— durobrivensis, Lyd.	-	-	2	-	-	-	-	-	-	-	-	-	-	-	H.	
—— eurymerus, Phil.	-	-	2	-	-	-	-	-	-	-	-	-	-	-	W. Bd. H.	
—— oxoniensis, Phil.	-	-	-	3	-	-	-	-	-	-	-	-	-	-	O.	
—— plicatus, Phil.	-	-	2	3	-	-	-	-	-	-	-	-	i	-	W. O. Bu. H. C.	*Murænosaurus Leedsi,* Seal. *Pl. hexagonoli* Phil. *Pl. infraplanu* Phil.
—— portlandicus, Owen.	-	-	-	-	-	-	-	-	9	10	-	-	-	-	D. W. Bu.	*Plesiosaurus car natus,* Phil. *P. Phillips* Sauv. *P. winspitensi* Seeley.
—— Richardsoni, Lyd.	-	-	2	-	-	-	-	-	-	-	-	-	-	-	W. H. C.	
—— trochanterius, Owen	-	-	-	3	-	-	-	7	-	-	-	-	-	-	D. to C.	*Ples. validus,* Phil. *P. affinis,* Owen. *P. Manseli,* Hulke. *Colymbosaurus megadeirus,* Seeley. *Murænosaurus Manseli,* Sauvage.
—— truncatus, Owen	-	?	-	-	-	-	-	7	-	9	-	-	-	-	D. W. O.	*Plesiosaurus dæd comus,* Owen.
Colymbosaurus. See Cimoliosaurus.																
Cryptoclidus oxoniensis, Phil.	-	-	2	-	-	-	-	-	-	-	-	-	-	-	H.	
Murænosaurus. See Cimoliosaurus.																
Peloneustes æqualis, Phil.	-	-	-	-	-	-	-	7	-	-	-	-	-	-	W. C.	*Plesiosaurus ste deirus,* Seeley.
—— Evansi, Seeley	-	-	2	-	-	-	-	-	-	-	-	-	-	-	H.	
—— philarchus, Seeley	-	-	2	-	-	-	-	-	-	-	-	-	-	-	W. Bd. N. C.	

Species.	Lower Oolite. (×)	Am. calloviensis. Kellaways Rock. (1)	Am. ornatus. Oxford Clay. (2)	Am. cordatus. Oxford Clay. (3)	Am. perarmatus. Corallian. (4)	Am. plicatilis. Corallian. (5)	Lower Beds. Kimeridge Clay. (6)	Upper Beds. Kimeridge Clay. (7)	Lower Beds. Portland Beds. (8)	Upper Beds. Portland Beds. (9)	Lower Beds. Purbeck Beds. (10)	Middle Beds. Purbeck Beds. (11)	Upper Beds. Purbeck Beds. (12)	Wealden Beds. (×)	Localities.	Remarks.
REPTILIA—cont.																
Sauropterygia (*Plesiosauria*)—cont.																
Plesiosaurus? ellipsospondylus, Phil.	-	-	-	-	-	-	-	-	-	-	-	-	-	-	O.	Kim. Cl.
—— See also Cimoliosaurus, Peloneustes, Pliosaurus, and Thaumatosaurus.																
Pliosaurus brachydeirus, Owen.	-	-	-	-	-	-	-	7	-	9	-	-	-	-	D.W.O.C.L.	
—— brachyspondylus (Owen), Phil.	-	-	-	-	-	-	-	7	-	-	-	-	-	-	D. C.	
—— ferox, Sauvage	-	-	2	-	-	-	-	-	-	-	-	-	-	-	D. W. N. C.	{Kim. Cl.? P. pachydeirus? Seeley.
—— gamma, Owen	-	-	-	3	-	-	-	7	-	-	-	-	-	-	D. O	
—— Grossouvrei, Sauvage	-	-	-	-	-	5	-	-	-	-	-	-	-	-	W. O. Y.	
—— macromerus, Phil.	-	-	-	-	-	-	-	7	-	-	-	-	-	-	D. to N.	P. grandis, Owen.
—— nitidus, Phil.	-	-	-	-	-	-	-	7	-	-	-	-	-	-	O. C.	
—— simplex, Phil.	-	-	-	-	-	-	-	7	-	-	-	-	-	-	O.	Ox. Clay; Cor.
—— See also Cimoliosaurus and Peloneustes.																
Thaumatosaurus carinatus, Cuv.	-	-	-	-	-	-	-	7	-	-	-	-	-	-	D. W. O.	
—— See also Peloneustes.																
Chelonia.																
Chelone obovata, Owen	-	-	-	-	-	-	-	-	-	-	-	11	-	-	D.	
—— See also Stegochelys.																
Enaliochelys. See Thalassemys.																
Eurysternum. See Thalassemys.																
Hylæochelys emarginata, Owen	-	-	-	-	-	-	-	-	-	-	-	11	-	-	D.	
—— latiscutata, Owen	-	-	-	-	-	-	-	-	-	-	-	11	-	-	D. W.	
Pelobatochelys Blakei, Seeley	-	-	-	-	-	-	-	7	-	-	-	-	-	-	D.	
Platemys. See Hylæochelys and Pleurosternum.																
Platychelys? anglica, Lyd.	-	-	-	-	-	-	-	-	-	-	-	11	-	-	D.	
Plesiochelys	-	-	-	-	-	-	-	7	-	-	-	11	-	-	D.	
—— See also Hylæochelys.																
Pleurosternum Bullocki, Owen.	-	-	-	-	-	-	-	-	-	-	-	11	-	-	D.	
—— concinnum, Owen	-	-	-	-	-	-	-	-	-	-	-	11	-	-	D.	
—— ovatum, Owen	-	-	-	-	-	-	-	-	-	-	-	11	-	-	D.	
—— Oweni, Seeley	-	-	-	-	-	-	-	-	-	-	-	11	-	-	D.	
—— portlandicum, Lyd.	-	-	-	-	-	-	-	-	-	9	-	-	-	-	D.	
Stegochelys planiceps, Owen	-	-	-	-	-	-	-	-	-	9	-	-	-	-	D.	Chelone.
Thalassemys Hugi, Rütim.	-	-	-	-	-	-	-	7	-	-	-	-	-	-	W. C.	

Species.	Lower Oolites.	Am. calloviensis. (Kellaways Rock.)	Am. ornatus. (Oxford Clay.)	Am. cordatus.	Am. perarmatus. (Corallian.)	Am. plicatilis.	Lower Beds. (Kimeridge Clay.)	Upper Beds.	Lower Beds. (Portland Beds.)	Upper Beds.	Lower Beds. (Purbeck Beds.)	Middle Beds.	Upper Beds.	Wealden Beds.	Localities.	Remarks.
	×	1	2	3	4	5	6	7	8	9	10	11	12	×		
REPTILIA—*cont.*																
Chelonia—cont.																
Thalassemys Rütimeyeri, Lyd.	–	–	–	–	–	–	–	–	–	–	–	11	–	–	D.	
Tretosternum punctatum, Owen.	–	–	–	–	–	–	–	–	–	–	–	11	–	–	D.	*T. Bakewelli*, Mant.
Tropidemys? Langi, Rutim.	–	–	–	–	–	–	–	7	–	–	–	–	–	–	D.	
PISCES.																
Thrissops	–	–	–	–	–	–	–	7	–	9	–	?	–	–	D.	?Teleostean.
Ganoidei.																
Aspidorhynchus euodus, Eg.	–	–	2	–	–	–	–	–	–	–	–	–	–	–	W.	
—— Fisheri, Eg.	–	–	–	–	–	–	–	–	–	–	–	–	–	–	D.	Purb.
—— sp.	–	–	–	–	–	–	–	7	–	–	–	–	–	–	D. Bu.	Purb.
Athrodon intermedius, A. S. Woodw.	–	–	–	–	–	–	–	–	–	–	10	–	–	–	Bu.	
Caturus angustus, Ag.	–	–	–	–	–	–	–	–	9	–	–	–	–	–	O.	
—— purbeckensis, A. S. Woodw.	–	–	–	–	–	–	–	–	–	–	–	11	–	–	D.	
—— tenuidens, A. S. Woodw.	–	–	–	–	–	–	–	–	–	–	–	–	–	–	D.	Purb.
Ceramurus macrocephalus, Eg.	–	–	–	–	–	–	–	–	–	–	–	–	–	–	W.	Purb.
Coccoderma substriolatum, (Huxley) O. Reis.	–	–	–	–	–	–	–	–	–	–	–	–	–	–	C.	Kim. Cl.
Coccolepis Andrewsi, A. S. Woodw.	–	–	–	–	–	–	–	–	–	–	10	–	–	–	W.	
Coelodus Mantelli, Ag.	–	–	–	–	–	–	–	–	–	–	–	–	–	×	D.	Purb.
Ditaxiodus impar, Owen	–	–	–	–	–	–	–	–	–	–	–	–	–	–	W. O. C.	Kim. Cl.
Eurycormus Egertoni, A. S. Woodw.	–	–	2	–	–	–	–	–	–	–	–	–	–	–	N.	
—— grandis, A. S. Woodw.	–	–	–	–	–	–	–	–	–	–	–	–	–	–	C.	Kim. Cl.
Gyrodus coccoderma, Eg.	–	–	–	–	–	–	–	7	–	–	–	–	–	–	D.	
—— Cuvieri, Ag.	–	–	–	–	–	–	–	7	–	–	–	–	–	–	D. C. Y.	Cor.
—— ornatissimus, Blake	–	–	–	–	–	–	–	7	–	–	–	–	–	–	D.	
—— sp.	–	–	–	3	–	–	–	–	–	–	–	–	–	–	O.	
Histionotus angularis, Eg.	–	–	–	–	–	–	–	–	–	–	–	11	–	–	D.	
Hypsocormus Leedsi, A. S. Woodw.	–	–	2	–	–	–	–	–	–	–	–	–	–	–	N.	
—— tenuirostris, A. S. W.	–	–	2	–	–	–	–	–	–	–	–	–	–	–	N.	
—— sp.	–	–	–	–	–	–	–	–	–	–	–	–	–	–	D.	Kim. Cl.
Leedsia problematica, A. S. W.	–	–	2	–	–	–	–	–	–	–	–	–	–	–	N.	*Leedsichthys.*
Lepidotus latifrons, A. S. Woodw.	–	–	2	–	–	–	–	–	–	–	–	–	–	–	N.	
—— macrocheirus, Eg.	–	–	2	3	–	–	–	–	–	–	–	–	–	–	W. O.?	
—— ?Manseli, Eg.	–	–	–	–	–	–	–	7	–	–	–	–	–	–	D.	
—— Mantelli Ag.	–	–	–	–	–	–	–	–	–	–	–	11	–	×	D. Sx.	Purb. *Tetragonolepis mastodonteus*, Ag.

Species.	Lower Oolite	Am. calloviensis (Kellaways Rock)	Am. ornatus (Oxford Clay)	Am. cordatus	Am. perarmatus (Corallian)	Am. plicatilis	Lower Beds (Kimeridge Clay)	Upper Beds	Lower Beds (Portland Beds)	Upper Beds	Lower Beds (Purbeck Beds)	Middle Beds	Upper Beds	Wealden Beds	Localities.	Remarks.
	×	1	2	3	4	5	6	7	8	9	10	11	12	×		
Pisces—*cont.*																
Ganoidei—cont.																
Lepidotus maximus, Wagner	-	-	-	-	-	-	6	7	-	9	-	-	-	-	D. O.	*Sphærodus gigas,* Ag.
—— minor, Ag.	×	-	-	-	-	-	-	-	-	9	-	11	-	-	D. W. Bu.	
—— palliatus, Ag.	-	-	-	-	-	-	-	7	-	-	-	-	-	-	D.	
Leptolepis Brodiei, Eg.	-	-	-	-	-	-	-	-	-	-	10	-	-	-	D. W.	
—— costalis, Eg.	-	-	2	-	-	-	-	-	-	-	-	-	-	-	W.	
—— macrophthalmus, Eg.	-	-	2	-	-	-	-	-	-	-	-	-	-	-	W.	
—— nanus, Eg.	-	-	-	-	-	-	-	-	-	-	-	-	-	-	W	Purb.
Macropoma. See Coccoderma.																
Macrosemius Andrewsi, A. S. Woodw.	-	-	-	-	-	-	-	-	-	-	-	11	-	-	W.	
Megalurus Austeni, Eg.	-	-	-	-	-	-	-	-	-	-	-	-	-	-	D	Purb.
—— Damoni, Eg.	-	-	-	-	-	-	-	-	-	-	-	-	-	-	D.	Purb.
Mesodon Bucklandi, Ag.	×	-	-	-	-	-	-	-	-	-	-	-	-	-	O	Cor. { *Pycnodus Buck- landi, Damon* (non Ag.). }
—— Damoni, A. S. Woodw.	-	-	-	-	-	-	-	-	-	9	-	-	-	-	D	Purb.
—— Daviesi, A. S. Woodw.	-	-	-	-	-	-	-	-	-	-	-	11	-	-	D	Purb.
—— macropterus, Ag. var. parvus, A. S. Woodw.	-	-	-	-	-	-	-	-	-	-	-	11	-	-	W.	
Microdon pagoda, Blake	-	-	-	-	-	-	-	7	-	9	-	-	-	-	D.	
—— quincuncialis, Blake	-	-	-	-	-	-	-	7	-	-	-	-	-	-	D.	
—— radiatus, Ag.	×	-	-	-	-	-	-	-	-	-	-	11	-	-	D. W.	
Œonoscopus	-	-	-	-	-	-	-	-	-	-	-	-	-	-	D.	Purb.
Oligopleurus vectensis, A. S. Woodw.	-	-	-	-	-	-	-	-	-	-	-	11	-	×	D.	
Ophiopsis breviceps, Eg.	-	-	-	-	-	-	-	-	-	-	10	-	-	-	D. W.	
—— dorsalis, Ag.	-	-	-	-	-	-	-	-	-	-	-	-	-	-	D	Purb.
—— penicillatus, Ag.	-	-	-	-	-	-	-	-	-	-	10	-	-	-	D. W.	
Oxygonius tenuis, Ag.	-	-	-	-	-	-	-	-	-	?	-	-	-	-	W.	
Pachycormus	-	-	2	-	-	-	-	-	-	-	-	-	-	-	N. C. L	Kim. Cl.
Pholidophorus brevis, Davies	-	-	-	-	-	-	-	-	-	-	-	-	12	-	D.	
—— granulatus, Eg.	-	-	-	-	-	-	-	-	-	-	-	-	-	-	D	Purb.
—— ornatus, Ag.	-	-	-	-	-	-	-	-	-	-	-	-	-	-	D. W	Purb.
—— purbeckensis, Davies	-	-	-	-	-	-	-	-	-	-	10	-	-	-	D.	
Pleuropholis attenuata, Eg.	-	-	-	-	-	-	-	-	-	-	10	11	-	-	D. W.	
—— crassicauda, Eg.	-	-	-	-	-	-	-	-	-	-	-	11	-	-	D.	
—— longicauda, Eg.	-	-	-	-	-	-	-	-	-	-	10	11	-	-	D. W.	
—— serrata, Eg.	-	-	-	-	-	-	-	-	-	-	10	-	-	-	Bu.	

SPECIES.	Lower Oolites.	Am. calloviensis. Kellaways Rock.	Am. ornatus.	Am. cordatus.	Am. perarmatus.	Am. plicatilis.	Lower Beds.	Upper Beds.	Lower Beds.	Upper Beds.	Lower Beds.	Middle Beds.	Upper Beds.	Wealden Beds.	LOCALITIES.	REMARKS.
	x	1	2	3	4	5	6	7	8	9	10	11	12	x		
PISCES—*cont.*																
Ganoidei—cont.																
Pycnodus. See Coelodus, Mesodon, and Microdon.																
Semionotus. See Lepidotus.																
Sphærodus. See Lepidotus.																
Strobilodus suchoides, Owen	-	-	-	-	-	-	-	-	-	-	-	-	-	-	N. - -	Kim. Cl.
—— See also Caturus.																
Tetragonolepis. See Lepidotus.																
Thlattodus. See Strobilodus.																
Dipnoi.																
Ceratodus . .	x	-	-	-	-	-	-	-	-	-	-	-	-	-	D. - -	Kim. Cl.
Selachii.																
Acrodus . . .	x	-	-	-	-	-	-	-	-	-	-	-	-	-	.	Kim. Cl.
Asteracanthus acutus, Ag. .	x	-	-	-	-	-	-	-	-	-	-	-	-	-	D. - -	Kim. Cl.
—— granulosus, Eg. .	.	-	-	-	-	-	-	-	-	-	-	11	-	x	D.	
—— lepidus, Dollf. .	.	-	-	-	-	-	-	7	-	-	-	-	-	-	D.	
—— ornatissimus, Ag. .	x	-	2	-	-	5	6	7	-	-	-	-	-	-	D. to N. -	And var. *Rettenensis, A. 1 Woodw.*
—— semiverrucosus, Eg. .	-	-	-	-	-	-	-	-	-	-	-	-	-	-	D. - -	Purbeck.
—— verrucosus, Eg. .	x	-	-	-	-	-	-	-	-	-	-	-	-	-	D. W. -	Ox. Cl. includes *A. Stutchburyi, Ag*
estracion falcifer ?, Wagner	-	-	-	-	-	-	-	-	-	-	-	-	-	-	D. - -	Kim. Cl.
Hybodus acutus, Ag. . .	.	-	-	-	-	-	6	-	-	-	-	-	-	-	D. W. O.	
—— dorsalis, Ag. . .	.	-	-	-	-	-	-	-	-	-	-	-	-	-	D. - -	Purb.
—— Fisheri, Newton .	.	-	-	-	-	-	-	-	-	-	-	-	-	-	C. - -	Kim. Cl.
—— grossiconus, Ag. .	.	-	2	-	4	-	-	-	-	-	-	-	-	-	D. O. H. -	Kim. Cl., Purb.
—— leptodus, Ag. . .	.	-	-	-	-	-	-	-	-	-	-	-	-	-	O. - -	Kim. Cl.
—— obtusus, Ag. . .	.	-	-	3	-	-	6	-	-	-	-	-	-	-	D. W. O. H. Y.	Cor.
—— strictus, Ag. .	.	-	-	-	-	-	-	-	-	9	-	11	-	-	D. W.	
—— See also Notidanus.																
Notidanus Daviesi, A. S. Woodw.	-	-	-	3	-	-	-	-	-	-	-	-	-	-	O. - -	*Hybodus polyprion* Phil. (non Ag.).
—— Muensteri, Ag. . .	-	-	2	-	-	-	-	-	-	-	-	-	-	-	H.	
Ptychacanthus. See Astera-canthus.																
Rhinobatis . . .	-	-	-	-	-	-	-	-	-	-	-	-	-	-	O. - - .	Kim. Cl.
Sphenonchus. See Hybodus.																
Strophodus. See Astera-canthus.																

SPECIES.	Lower Oolites. (×)	Am. calloviensis. Kellaways Rock. (1)	Am. ornatus. Oxford Clay. (2)	Am. cordatus. Oxford Clay. (3)	Am. perarmatus. Corallian. (4)	Am. plicatilis. Corallian. (5)	Kimeridge Clay. Lower Beds. (6)	Kimeridge Clay. Upper Beds. (7)	Portland Beds. Lower Beds. (8)	Portland Beds. Upper Beds. (9)	Purbeck Beds. Lower Beds. (10)	Purbeck Beds. Middle Beds. (11)	Purbeck Beds. Upper Beds. (12)	Wealden Beds. (×)	LOCALITIES.	REMARKS.
PISCES—cont.	×	1	2	3	4	5	6	7	8	9	10	11	12	×		
Chimæroidei.																
Brachymylus altidens, A. S. Woodw.	—	—	2	—	—	—	—	—	—	—	—	—			N.	
— minor, A. S. Woodw.	—	—	—	—	—	—	—	—	—	—	—	—			D.	Kim. Cl.
Chimæra. See Ischyodus.																
Elasmodectes secans, A. S. Woodw.	—	—	—	—	—	—	—	—	—	—	—	—			D.	Kim. Cl.
Ischyodus Beaumonti, Eg.	—	—	—	—	—	—	—	7	—	—	—	—			D.	
— Egertoni, Buckl.	—	—	—	3	—	—	6	7	—	—	—	—			D. O. H.	
— Townsendi, Buckl.	—	—	—	—	—	—	—	—	—	9	—	—			D. W. O.	
— sp.	—	—	—	—	—	—	—	—	—	—	—	—		×		Ampthill Clay.
Pachymylus Leedsi, A. S. Woodw.	—	—	2	—	—	—	—	—	—	—	—	—			N.	
MOLLUSCA.																
Cephalopoda.																
Acanthoteuthis speciosa, Münst.	—	—	—	—	—	—	—	—	—	—	—	—		×		Kim. Cl.
— See also Belemnoteuthis.																

INDEX TO SUB-GENERIC NAMES OF AMMONITES.

Amaltheus.	Craspedites.	Ochetoceras.	Patoceras.	Proplanulites.
Aspidoceras.	Harpoceras.	Œkotraustes.	Peltoceras.	Quenstedticeras.
Cadoceras.	Hoplites.	Olcostephanus (Holcostephanus).	Perisphinctes.	Reineckeia.
Cardioceras.	Macrocephalites.	Oppelia.	Polyptychites.	Stephanoceras.
Cosmoceras.				

SPECIES	×	1	2	3	4	5	6	7	8	9	10	11	12	×	LOCALITIES	REMARKS
Ammonites Achilles, d'Orb.	—	—	—	3	4	5	6	—	—	—	—	—	—		D. to Y.	*Perisph.*
— alternans, von Buch.	—	—	—	—	—	—	6	7	—	—	—	—	—		D. C. L. Y.	*Card. Amalth.*
— anceps, Rein.	—	—	—	2	3	—	—	—	—	—	—	—	—		W. Bu.	*Reineckeia.*
— annularis, Rein.	—	—	—	—	—	—	—	—	—	—	—	—	—		D.	Ox. Cl.
— arduennensis, d'Orb.	—	—	1	—	3	—	—	—	—	—	—	—	—		D. N. Y.	*Peltoceras.*
— athleta, Phil.	—	—	1	2	3	—	—	—	—	—	—	—	—		D. to Y.	*Pelt.*
— babeanus, d'Orb.	—	—	—	—	3	—	—	—	—	—	—	—	—		H.	
— Bakeriæ, Sow.	—	×	1	2	3	—	—	—	—	—	—	—	—		W. H. N. Y.	*Perisph.*
— — var. fluctuosus, Pratt	—	—	2	—	—	—	—	—	—	—	—	—	—		W. N.	
— Berryeri, Lesueur	—	—	—	—	5	6	—	—	—	—	—	—	—		W. L. Y.	
— biplex, Sow.	—	—	—	—	—	6	7	8	9	—	—	—	—		D. to Y. Sx.	*Perisph.*
— Bleicheri? de Lor.	—	—	—	—	—	—	—	9	—	—	—	—	—		W.	
— Boisdini, de Lor.	—	—	—	—	—	—	—	?	9	—	—	—	—		D. W.	
— boloniensis, de Lor.	—	—	—	—	—	—	—	—	9	—	—	—	—		D. W. O. B.	*Perisphinctes bononiensis.*
— *Brighti,* Pratt. See A. hectious.																

E 82428.

A A

Species.	Lower Oolites	Am. callaviensis (Kellaways Rock)	Am. ornatus (Oxford Clay)	Am. cordatus (Oxford Clay)	Am. perarmatus (Corallian)	Am. plicatilis (Corallian)	Lower Beds (Kimeridge Clay)	Upper Beds (Kimeridge Clay)	Lower Beds (Portland Beds)	Upper Beds (Portland Beds)	Lower Beds (Purbeck Beds)	Middle Beds (Purbeck Beds)	Upper Beds (Purbeck Beds)	Wealden Beds	Localities.	Remarks.
	×	1	2	3	4	5	6	7	8	9	10	11	12	×		
MOLLUSCA—cont.																
Cephalopoda—cont.																
Ammonites Callisto, d'Orb.	-	-	-	-	-	5	6	7	-	-	-	-	-	-	D. C. Sx.	Hoplites.
— calloviensis, Sow.	-	1	2	-	-	-	-	-	-	-	-	-	-	-	W. G. O. Bu. N. Y.	Cosmoceras: Patoceras.
— canaliculatus, Munst.	-	-	2	-	4	5	6	-	-	-	-	-	-	-	O. Bu. C. Y.	Harpoceras: Ochet.
— catena, Sow. See A. perarmatus.																
— cawtonensis, Blake and Hudl.	-	-	-	-	-	5	-	-	-	-	-	-	-	-	O. Bd. Y.	
— Chamusseti, d'Orb.	-	1	2	3	-	-	-	-	-	-	-	-	-	-	D. W. G. Y. Sx.	Cosm. Card.
— chauvinianus, d'Orb.	-	1	-	-	-	-	-	-	-	-	-	-	-	-	D. Y.	
— Comptoni, Pratt	-	-	2	-	-	-	-	-	-	-	-	-	-	-	W. N. Y.	
— Constanti, d'Orb.	-	-	2	-	-	-	-	-	-	-	-	-	-	-	N.	Peltoc.
— convolutus, Quenst.	-	-	-	-	4	-	-	-	-	-	-	-	-	-	C.	
— — ornatus, Quenst.	-	1	2	-	-	-	-	-	-	-	-	-	-	-	N. Y.	
— cordatus, Sow.	-	-	2	3	4	5	6	-	-	-	-	-	-	-	D. to Y.	Amaltheus, Card.
— var excavatus, Sow.	-	1	2	3	-	-	6	-	-	-	-	-	-	-	D. to Y.	
— coronatus, Brug.	-	-	2	-	-	-	-	-	-	-	-	-	-	-	H.	Steph.
— crenatus, Brug.	-	1	2	3	-	-	-	-	-	-	-	-	-	-	D. to Y.	Ækot. Oppelia.
— cymodoce, d'Orb.	-	-	-	-	-	5	6	-	-	-	-	-	-	-	D. L.	
— decipiens, Sow.	-	-	-	-	-	5	6	-	-	-	-	-	-	-	D. W. Be. L. N. Y.	{ Kim. Cl. non Lytoceras decipiens, Buckm.
— dentatus, Ziet. See A. crenatus.																
— Duncani, Sow.	-	-	2	3	-	-	-	-	-	-	-	-	-	-	D. to H.	Cosm.
— var spinosus, Sow.	-	-	2	-	-	-	-	-	-	-	-	-	-	-	D.	
— Elizabethæ, Pratt.	-	-	2	-	-	-	-	-	-	-	-	-	-	-	W. Bu. N. Y.	? var. of A. Jason.
— eudoxus, d'Orb.	-	-	-	-	-	-	6	7	-	-	-	-	-	-	D. C. Y. Sx.?	Perisph. Rein. Hoplites.
— Eugeni, Rasp.	-	-	2	3	-	-	-	-	-	-	-	-	-	-	H. Y.	Pelt.
— eumelus, d'Orb.	-	-	-	-	-	-	6	-	-	-	-	-	-	-	O. Y.	
— eupalus, d'Orb.	-	-	-	-	-	?	-	7	-	-	-	-	-	-	D. O.	
— excavatus, Sow. See A. cordatus.																
— flexicostatus, Phil.	-	1	-	-	-	-	-	-	-	-	-	-	-	-	D. Y.	
— flexuosus, Munst. See A. oculatus.																
— fluctuosus, Pratt. See A. Bakeriæ.																
— gemmatus, Phil. See A. Duncani.																
— giganteus, Sow.	-	-	-	-	-	-	-	-	-	9	-	-	-	-	D. W. B.	
— gigas, Ziet.	-	-	-	-	-	-	-	-	-	9	-	-	-	-	Bu. Y.	Steph.
— goliathus, d'Orb.	-	-	2	3	4	5	-	-	-	-	-	-	-	-	D. to Y.	Card.
— gowerianus, Sow.	-	1	2	-	-	-	-	-	-	-	-	-	-	-	D. W. G. O. Y.	Steph. Cosm.

Species	Lower Oolites	Am. calloviensis	Am. ornatus	Am. coriatus	Am. perarmatus	Am. plicatilis	Kim. Lower Beds	Kim. Upper Beds	Portl. Lower Beds	Portl. Upper Beds	Purb. Lower Beds	Purb. Middle Beds	Purb. Upper Beds	Wealden Beds	Localities	Remarks
	x	1	2	3	4	5	6	7	8	9	10	11	12	x		
MOLLUSCA—*cont.*																
Cephalopoda—cont.																
Ammonites gravesianus, d'Orb.	–	–	–	–	–	–	–	–	–	–	–	–	–	–	L. Y.	Speeton Beds.
—— *Gulielmi*, Sow. See A. Jason.																
—— hecticus, Rein.	–	1	2	3	–	–	–	–	–	–	–	–	–	–	D. to Y.	Harp.
—— —— var lunula, Rein.	–	1	2	–	–	–	–	–	–	–	–	–	–	–	W. Y.	
—— Hector, d'Orb.	–	–	–	–	–	5	6	–	–	–	–	–	–	–	D. O. L.	
—— Henrici, d'Orb.	–	–	–	–	4	–	–	–	–	–	–	–	–	–	C. H.	
—— *Herveyi*, Sow. See A. macrocephalus.																
—— Jason, Rein.	–	1	2	3	–	–	–	–	–	–	–	–	–	–	D. to Y.	Cosm.
—— —— var. Gulielmi, Sow.	–	1	2	–	–	–	–	–	–	–	–	–	–	–	D. W. Bu. Y.	Cosm.
—— Kapffi, Oppel	–	–	–	–	–	–	7	–	–	–	–	–	–	–	D. L. Y.	
—— Koenigi, Sow.	–	1	2	–	–	–	–	–	–	–	–	–	–	–	D. W. G. Y.	Cosm. Propl. Perisph.
—— lallierianus, d'Orb	–	–	–	–	–	6	–	–	–	–	–	–	–	–	Sx.	
—— Lamberti, Sow.	–	–	2	3	–	–	–	–	–	–	–	–	–	–	D. to Y. K.	Amal. Card. Quenst.
—— *Leachi*, Sow. See A. Lamberti.																
—— *lenticularis*, Phil. See A. Chamusseti.																
—— longispinus, Sow.	–	–	–	–	–	6	7	–	–	–	–	–	–	–	D. C. Y. Sx.	Aspid.
—— *Lonsdalei*, Pratt. See A. hecticus.																
—— lophotus, Ziet.	–	–	–	3	–	–	–	–	–	–	–	–	–	–	H.	
—— macrocephalus, Schloth.	x	1	2	–	–	–	–	–	–	–	–	–	–	–	D. to Y.	Macroc.
—— Mariæ, d'Orb.	–	1	2	3	–	–	–	–	–	–	–	–	–	–	D. W. G. O. H. Y.	Card.
—— modiolaris, Lhwyd	–	1	2	–	–	–	–	–	–	–	–	–	–	–	D. W. G. Y.	Cosm. Cadoc.
—— multiplicatus, Roem.	–	–	–	–	–	–	–	–	–	–	–	–	–	–	L.	Spilsby Sandstone.
—— mutabilis, Sow.	–	–	–	–	–	5	6	7	–	–	–	–	–	–	D. to Y. Sx.	Perisph.
—— oculatus, Phil.	–	–	–	3	–	–	–	–	–	–	–	–	–	–	O. H. Y.	
—— ornatus, Schloth.	–	–	2	–	–	–	–	–	–	–	–	–	–	–	W. H. C. Y.	Cosm.
—— orthocera, d'Orb.	–	–	–	–	–	6	–	–	–	–	–	–	–	–	Sx.	Aspid.
—— pectinatus, Phil.	–	–	–	–	–	–	–	–	8	9	–	–	–	–	W. O. Bu.	Kim. Cl.
—— perarmatus, Sow.	–	–	2	3	4	5	–	–	–	–	–	–	–	–	D. to Y.	Aspid.
—— placenta, Leck.	–	1	–	–	–	–	–	–	–	–	–	–	–	–	D. Y.	
—— plicatilis, Sow.	–	–	2	3	4	5	6	–	–	–	–	–	–	–	D. to Y. K.	Perisph.
—— plicomphalus, Sow.	–	–	–	–	–	–	7	–	–	–	–	–	–	–	L. Y.	Spilsby Sandstone.
—— polyplocus, Rein.	–	–	–	–	–	–	?	–	–	–	–	–	–	–	O.	Perisph.
—— pseudocordatus, Blake and Hudl.	–	–	–	–	–	5	–	–	–	–	–	–	–	–	W.	
—— pseudogigas, Blake	–	–	–	–	–	–	–	–	–	9	–	–	–	–	D. O. Bu.	

Species.	Lower Oolites	Kellaways Rock	Oxford Clay		Corallian		Kimeridge Clay		Portland Beds		Purbeck Beds			Wealden Beds	Localities.	Remarks.
		Am. calloviensis.	Am. ornatus.	Am. cordatus.	Am. perarmatus.	Am. plicatilis.	Lower Beds.	Upper Beds.	Lower Beds.	Upper Beds.	Lower Beds.	Middle Beds.	Upper Beds.			
	×	1	2	3	4	5	6	7	8	9	10	11	12	×		
MOLLUSCA—cont.																
Cephalopoda—cont.																
Ammonites pseudomutabilis, De Lor.						5	6								D.	Perisph. Rein. Hoplites.
—— radisensis, d'Orb.			2												H.	
—— Reginaldi, Morris			2												D. W.	
—— rotundus, Sow.							6		8						Bu. C. L. Y.	Spilsby Sandstone.
—— rupellensis, d'Orb.				3											H.	
—— Sedgwicki, Pratt			2												W.	
—— serratus, Sow. See A. cordatus var excavatus.																
—— spinosissimus, Damon								7							D.	
—— spinosus. See A. ornatus.																
—— stenomphalus, Pavl.															L.	Spilsby Sandstone.
—— stephanoides, Damon. See A. subundorae.																
—— striolaris, Quenst.															D.	Ox. Cl.
—— Stutchburyi, Pratt			2												W.	
—— subditus, Trautsch.															L.	SpilsbySdst. Perisp Oleos. Craspedite Hoplites.
—— subundorae, Pavlow								7							D.	Perisp Hoplites.
—— sublaevis, Sow. See A. modiolaris.																
—— superstes, Phil.					?										W.	
—— sutherlandiae, Sow.					3	4									W. O. Y.	Amalth.
—— tatricus, Pusch.			2												N.	
—— Thurmanni, Cont.								7							D.	
—— trifidus, Sow.			2	3		5	6	7		9					D. O. C. H. Y.	
—— triplex, Sow. See A. trifidus.																
—— triplicatus, Sow.							6	7		9					D. W. Bc. L. Y.	Olcostephanus.
—— varicostatus, Buckl.					4	5		7							W. Be. O.	Ox. Cl.
—— vertebralis, Sow.			2	3	4										D. to Y.	Card.
—— var. cawtonensis, Blake and Hudl.						5									C. Y.	
—— vertumnus, Leck. See A. Mariae.																
—— Williamsoni, Phil.					3	4	5								D. O. Y.	
Ancyloceras calloviense, Morris.															W.	
Aptychus biplex, Sow.								7							L.	Trigonellites.
—— lamellosus, Park.								7							W.	Ox. Cl.
—— latus, Park.							6	7							D. O. Bu. C.	

Species.	Lower Oolites ×	Am. callovensis (Kellaways Rock) 1	Am. ornatus (Oxford Clay) 2	Am. cordatus 3	Am. perarmatus (Corallian) 4	Am. plicatilis 5	Kimeridge Clay Lower Beds 6	Upper Beds 7	Portland Beds Lower Beds 8	Upper Beds 9	Purbeck Beds Lower Beds 10	Middle Beds 11	Upper Beds 12	Wenlden Beds ×	Localities.	Remarks.
MOLLUSCA—*cont.*																
Cephalopoda—cont.																
Aptychus Münsteri?	-	-	-	-	-	-	-	-	-	-	-	-	-	-	O.	Kim. Cl.
—— politus, Phil.	-	-	-	-	-	-	-	-	-	-	-	-	-	-	W.	Ox. Cl.
—— sp.	-	-	-	-	-	-	-	-	-	-	-	-	-	-	Sx.	Kim. Cl.
Belemnites abbreviatus, Mill.	×	-	2	3	4	5	6	7	-	-	-	-	-	-	D. to Y.	
—— —— var. excentricus, Blainv.	×	-	-	3	4	5	6	7	-	-	-	-	-	-	O.	
—— —— var. oxyrhynchus, Phil.	-	-	-	-	-	5	-	-	-	-	-	-	-	-	W. O. Y.	
—— aripistillum, Lhwyd	×	-	-	3	-	-	-	-	-	-	-	-	-	-	O.	
—— Blainvillei, Voltz.	×	-	-	-	-	-	6	7	-	-	-	-	-	-	D.	
—— *explanatus,* Phil. See B. nitidus.																
—— explanatoides, Pavl.	-	-	-	-	-	-	-	-	-	-	-	-	-	-	L. Y.	Kim. Cl.
—— *fusiformis,* Park. See B. aripistillum.																
—— *gracilis,* Phil. See B. hastatus.																
—— hastatus, Blainv.	-	1	2	3	4	-	-	-	-	-	-	-	-	-	D. to Y.	? Kim. Cl.
—— —— var. bulbosus, Phil.	-	-	2	-	-	-	-	-	-	-	-	-	-	-	N. H.	
—— lateralis, Phil.	-	-	-	-	-	-	-	-	-	-	-	-	-	-	L. Y.	Spilsby Sandstone.
—— nitidus, Dollf.	-	-	-	-	-	5	6	7	8	-	-	-	-	-	D. to Y.	(Non B. *nitidus* Phil.)
—— obeliscus, Phil.	-	-	2	-	-	-	-	-	-	-	-	-	-	-	H. Y.	
—— Oweni, Pratt	-	1	2	3	-	5	-	-	8	-	-	-	-	-	D. to Y.	Hartwell Clay.
—— —— var. puzosianus, d'Orb.	-	-	-	-	-	-	-	-	-	-	-	-	-	-		See p. 17.
—— —— var. tornatilis, Phil.	-	1	-	-	-	-	-	-	-	-	-	-	-	-	W. Y.	
—— —— var. verrucosus, Phil.	-	-	2	-	-	-	-	-	-	-	-	-	-	-	H.	
—— russiensis, d'Orb.	-	-	-	-	-	-	-	-	-	-	-	-	-	-	L.	Spilsby Sandstone.
—— Souichi, d'Orb.	-	-	-	-	-	-	-	7	8	-	-	-	-	-	D. Bu.	
—— strigosus, Phil.	-	-	-	3	-	-	-	-	-	-	-	-	-	-	O.	
—— sulcatus, Mill.	-	1	2	3	-	-	-	-	-	-	-	-	-	-	D. W. O. N.	
—— troslayanus? d'Orb.	-	-	-	-	-	-	-	-	-	-	-	-	-	-	D. Y.	Kim. Clay.
Belemnoteuthis antiquus, Pearce.	-	1	2	-	-	-	-	-	-	-	-	-	-	-	W.	*Acanthoteuthis.*
—— sp.	-	-	-	3	-	-	-	7	-	-	-	-	-	-	H. L.	
Coccoteuthis latipennis, Owen.	-	-	-	-	-	-	-	7	-	-	-	-	-	-	D.	
—— sp.	-	-	-	2	-	-	-	-	-	-	-	-	-	-	W.	Ox. Cl.
Geoteuthis	-	-	-	2	-	-	-	-	-	-	-	-	-	-	W.	
Hibolites. See Belemnites.																
Nautilus calloviensis, Oppel.	-	1	-	3	4	-	-	-	-	-	-	-	-	-	W. Bu. H. Bk. Y.	

SPECIES.	Lower Oolites	Kellaways Rock Am. calloviensis	Oxford Clay Am. ornatus	Oxford Clay Am. cordatus	Corallian Am. perarmatus	Corallian Am. plicatilis	Kimeridge Clay Lower Beds	Kimeridge Clay Upper Beds	Portland Beds Lower Beds	Portland Beds Upper Beds	Purbeck Beds Lower Beds	Purbeck Beds Middle Beds	Purbeck Beds Upper Beds	Wealden Beds	LOCALITIES.	REMARKS.
	×	1	2	3	4	5	6	7	8	9	10	11	12	×		
MOLLUSCA—*cont.*																
Cephalopoda—cont.																
Nautilus hexagonus, Sow. -	×	1	2	–	4	–	6	–	–	–	–	–	–	–	D. to Y.	
—— perinflatus, Foord and Crick.	×	–	–	3	4	–	–	–	–	–	–	–	–	–	C.	
—— portlandicus, Foord and Crick.	–	–	–	–	–	–	–	–	–	9	–	–	–	–	D.	*Hercoglossa.*
Teudopsis Brodiei, Carr. -	–	–	–	–	–	–	–	–	–	–	–	–	–	–	D.	Purbeck.
Trachyteuthis. See Coccoteuthis.																
Trigonellites. See Aptychus.																
Gasteropoda.																
Actæon retusus, Phil. -	–	1	2	–	–	–	–	–	–	–	–	–	–	–	W. O. Y.	Cor.
—— secalinus, Buv. -	–	–	–	–	–	–	6	–	–	–	–	–	–	–	L.	
—— sp. -	–	–	–	–	–	–	–	–	–	–	–	–	–	–	D.	Purb.
Actæonina miliola, d'Orb. -	–	–	–	–	–	5	–	–	–	–	–	–	–	–	D.	
—— signum, Hudl. -	–	–	–	–	–	–	–	–	–	9	–	–	–	–	W.	
Alaria Arsinoe, d'Orb. -	–	–	2	–	–	–	–	–	–	–	–	–	–	–	W.	
—— *Beaugrandi,* de Lor. See Aporrhais.																
—— bispinosa, Phil. -	–	1	2	–	4	–	–	–	–	–	–	–	–	–	W. to Y.	
—— cingulata, Koch and Dunk.	–	–	–	–	–	–	–	7	–	–	–	–	–	–	D.	
—— composita, Sow. -	–	–	–	–	–	–	–	–	–	–	–	–	–	–	D.	Cor. & Oxf.
—— ? deshayesea, Buv.	–	–	–	–	–	5	–	–	–	–	–	–	–	–	D. Y.	
—— ? mosensis, Buv. -	–	–	–	–	–	–	6	–	–	–	–	–	–	–	D. L. Y.	
—— nodifera, Koch and Dunk.	–	–	–	–	–	–	–	–	–	–	–	–	–	–	D.	Kim. Cl.
—— rasenensis, Blake	–	–	–	–	–	–	6	–	–	–	–	–	–	–	D. L.	
—— seminuda, Heb. and Deal.	–	–	–	–	–	–	5	–	–	–	–	–	–	–	D. O.	Ox. Cl. ?
—— *tenuistria,* Buv. See Aporrhais.																
—— *Thurmanni,* Contej. See Aporrhais.																
—— trifida, Phil. -	–	1	2	3	4	–	6	–	–	–	–	–	–	–	D. to Y.	
—— See also Aporrhais -																
Amberleya princeps, Roem. -	–	–	–	–	–	5	–	–	–	–	–	–	–	–	C. Y.	
Aporrhais Beaugrandi, de Lor. & Pellat.	–	–	–	–	–	–	–	–	–	9	–	–	–	–	D. Bu.	
—— (cf.) tenuistria, Buv. -	–	–	–	–	–	5	–	–	–	–	–	–	–	–	W. Y.	
—— Thurmanni, Contej. -	–	–	–	–	–	–	–	–	8	–	–	–	–	–	W.	
—— sp. -	–	–	–	–	–	–	–	–	–	–	–	–	–	–	D. Sx.	Kim. Cl.
—— Bythinia? -	–	–	–	–	–	–	–	–	–	–	–	–	–	–	W.	Purb.

SPECIES.	Lower Oolites.	Kellaways Rock.	Oxford Clay.		Corallian.		Kimeridge Clay.		Portland Beds.		Purbeck Beds.			Wealden Beds.	LOCALITIES.	REMARKS.
		Am. callovienais.	Am. ornatus.	Am. cordatus.	Am. perarmatus.	Am. plicatilis.	Lower Beds.	Upper Beds.	Lower Beds.	Upper Beds.	Lower Beds.	Middle Beds.	Upper Beds.			
	×	1	2	3	4	5	6	7	8	9	10	11	12	×		
MOLLUSCA—cont.																
Gasteropoda—cont.																
Bourguetia Buvignieri, d'Orb.	-	-	-	-	-	5	-	-	-	-	-	-	-	-	D. to Y.	
—— striata, Sow.	-	-	-	-	-	5	6	-	-	-	-	-	-	-	D. to Y.	
Buccinum? angulatum, Sow.	-	-	-	-	-	-	-	-	?	9	-	-	-	-	D. W. O.	Figured by Damon as a synonym of Pterocera oceani, Brongn.
—— naticoides, Sow. See Natica elegans. Bulla. See Cylindrites.																
Ceritella costata, Whit.	-	-	-	-	-	-	-	-	-	-	-	-	-	-	O.	Cor.
Cerithium abbreviatum, Leck.	-	-	-	-	-	-	-	-	-	-	-	-	-	-	Be. Y.	Cor. Oxf.
—— bifurcatum, Blake	-	-	-	-	-	-	-	-	8	-	-	-	-	-	W.	
—— Boisdini, de Lor & Pellat.	-	-	-	-	-	-	-	-	-	9	-	-	-	-	D.	
—— bouchardianum, de Lor. & Pellat.	-	-	-	-	-	-	-	-	-	9	-	-	-	-	W.	
—— Carabœufi, de Lor.	-	-	-	-	-	-	-	-	-	9	-	-	-	-	D.	
—— concavum, Sow.	-	-	-	-	-	-	-	-	-	9	-	-	-	-	W.	
—— crebrum, Blake	-	-	-	-	-	-	6	-	?	-	-	-	-	-	Bu. ? L.	
—— Culleni, Leck.	-	1	-	-	-	-	-	-	-	-	-	-	-	-	W. G. Y.	
—— Damonis, Lyc.	-	1	2	3	-	-	-	-	-	-	-	-	-	-	D. W. N. H.	Cor.
—— forticostatum, Blake	-	-	-	-	-	-	6	-	-	-	-	-	-	-	D. L.
—— Hudlestoni, Blake	-	-	-	-	-	-	-	-	-	9	-	-	-	-	Bu.	
—— humbertinum, Buv.	-	-	-	-	-	5	-	-	-	-	-	-	-	-	W. Y.	
—— inornatum, Buv.	-	-	-	-	-	5	-	-	-	-	-	-	-	-	D. W. Y.	
—— Lamberti, de Lor. & Cott.	-	-	-	-	-	-	-	-	8	-	-	-	-	-	W.	
—— limæforme, Roem.	-	-	-	-	-	5	-	-	-	-	-	-	-	-	D. W. Y.	Kim. Cl.
—— Manseli, de Lor.	-	-	-	-	-	-	-	-	-	-	-	-	-	-	D.	Purbeck.
—— multiplicatum, Blake	-	-	-	-	-	-	6	-	8	-	-	-	-	-	W. Bu. Y.	
—— muricatum, Sow.	-	1	2	-	4	5	-	-	-	-	-	-	-	-	D. to Y.	
—— Pellati, de Lor.	-	-	-	-	-	5	-	-	-	-	-	-	-	-	D.	
—— portlandicum, Sow.	-	-	-	-	-	-	-	-	-	9	-	-	-	-	D. W. O. Bu.	
—— septemplicatum, Roem.	-	-	-	-	-	5	-	-	-	9	-	-	-	-	D.	
—— trinodule, Buv.	-	-	-	-	-	-	-	-	-	9	-	-	-	-	D.	
Chemnitzia Beaugrandi, de Lor.	-	-	-	-	-	-	-	-	8	9	-	-	-	-	D. Bu.	
—— ferruginea, Blake & Hudl.	-	-	-	-	-	5	-	-	-	-	-	-	-	-	D.	
—— heddingtonensis, Sow.	-	-	-	-	-	5	6	-	-	-	-	-	-	-	D to Y.	Pseudomelania.
—— ? inflata, Phil.	-	-	-	-	-	-	6	-	-	-	-	-	-	-	O.	
—— limbata, Contej.	-	-	-	-	-	5	-	-	-	-	-	-	-	-	D.	

SPECIES.	Lower Oolites.	Kellaways Rock. / Am. calloviensis.	Oxford Clay. / Am. ornatus.	Am. cordatus.	Corallian. / Am. perarmatus.	Am. plicatilis.	Kimeridge Clay. / Lower Beds.	Upper Beds.	Portland Beds. / Lower Beds.	Upper Beds.	Purbeck Beds. / Lower Beds.	Middle Beds.	Upper Beds.	Wealden Beds.	LOCALITIES.	REMARKS.
	×	1	2	3	4	5	6	7	8	9	10	11	12	×		
MOLLUSCA—*cont.*																
Gasteropoda—cont.																
Chemnitzia. See also Cerithium and Pseudomelania.																
Cloughtonia percincta, Hudl.										9					W.	
Cylindrites elongatus, Phil.						5									D. O. Y.	
—— Luidi, Whit.					4										Be. O.	
—— See also Actæon																
Delphinula funiculata, Phil.	?					5									D. W. Y.	
—— globata, Buv.									8						W.	
—— nassoides, Buv.							6								D.	
Emarginula Goldfussi, Roem.						5									C.	
Fissurella corallensis, Buv.						5									C.	
Hydrobia. See Rissoa.																
Limnæa, sp.											10	11	12		D. W.	
Littorina ? lævissima, Whit.															O.	Cor.
—— Meriani, Goldf.					4	5									C. H.	
—— muricata, Sow.		1			4	5		7		9					D. W. Be. O. C. Y.	
—— —— var. pulcherrima, Doll.						5	6								D. W. Y.	
—— paucisulcata, Phil.										9					D. O.	Kim. Cl.
Melania. See Bourguetia, Melanopsis, & Pseudomelania.																
Melanopsis attenuata, Sow.												11		×	D. Sx.	
—— harpæformis, Koch & Dunk.												11			D. Sx.	
—— Popei, Sow.												11		×	W. Sx.	
—— rugosa, Dunk.												11			D. Sx.	
—— tricarinata, Sow.												11			D. Sx.	
—— sp.															W.	Ox. Cl.
—— See also Chemnitzia, Phasianella and Pseudomelania.																
Murex? haccanensis, Phil.		1													W. Y.	Cor.
Natica arguta, Phil.						5									Be. O. Y.	
—— calypso, d'Orb. var. tenuis, Hudl.					4										C. H. Y.	
—— ceres, de Lor.										9					Bu. O.	
—— clio, d'Orb.						5									D. W. O.	
—— clymenia, d'Orb.						5									Y.	

Species	Lower Oolites (×)	Am. calloviensis / Kellaways Rock (1)	Am. ornatus / Oxford Clay (2)	Am. cordatus / Oxford Clay (3)	Am. perarmatus / Corallian (4)	Am. plicatilis / Corallian (5)	Kimeridge Clay Lower Beds (6)	Kimeridge Clay Upper Beds (7)	Portland Beds Lower Beds (8)	Portland Beds Upper Beds (9)	Purbeck Beds Lower Beds (10)	Purbeck Beds Middle Beds (11)	Purbeck Beds Upper Beds (12)	Wealden Beds (×)	Localities	Remarks
MOLLUSCA—cont.	×	1	2	3	4	5	6	7	8	9	10	11	12	×		
Gasteropoda—cont.																
Natica clytia, d'Orb.						5									D. to Y.	
— corallina (Damon), Lyc.						5									D. W. Be.	
— dejanira, d'Orb.						5									D.	
— elegans, Sow.									8	9					D.W.O.Bu.	
— eudora, d'Orb.						5	6								D.	
— felina, Blake and Hudl.					4										W.	
— incisa, Blake									8	9					D. W. Bu.	
— marchamensis, Bl. and Hudl.					4	5									Be.	
— microscopica, Contej.							6								L.	
— punctulata, Blake							6								L.	
— punctura, Bean	×	1													W. Y.	
— turbiniformis, Roem.										9					Bu. O.	
Neridomus transversus, Seeb.										9					W.	
Nerinæa bernardiana, d'Orb.															D.	Kim. Cl., Cor.
— Desvoidyi, d'Orb.						5									D.	
— fasciata, Voltz and Bronn.															D. ? Y.	Cor.
— Goodhalli, Sow.						5									D. O. Y.	Kim. Cl.
— Roemeri, Phillippi						5									D. W. Y.	
Nerita brevispiralis, Phil.															O.	Cor.
— See also Neritoma and Neritopsis.																
Neritina															D.	Purb.
Neritoma sinuosa, Sow.										9					D. W. Bu.	Nerita angulata, Sow. N. corallensis, Buv.
Neritopsis decussata, Münst.						5									W. C. Y.	
— delphinula, d'Orb.							6		8						Bu. L.	
— Guerrei, Heb. and Deal.						5									O. C. Y.	
Orthostoma. See Actæonina.																
Paludina carinifera, Sow.												11	12	×	D. W. Sx.	
— elongata, Sow.											10		12	×	D.W.O.Bu.	
— sussexiensis, J. Sow.													12	×	D.	
Patella portlandica, Damon										9					D.	
— sp.															D. W.	Ox. Cl.
Phasianella. See Bourguetia.																
Physa Bristovii, Forbes												11			D.	

SPECIES.	Lower Oolites. (×)	Am. callovensis (Kellaways Rock) (1)	Am. ornatus (Oxford Clay) (2)	Am. cordatus (3)	Am. perarmatus (Corallian) Lower Beds (4)	Am. plicatilis Upper Beds (5)	Kimeridge Clay Lower Beds (6)	Upper Beds (7)	Portland Beds Lower Beds (8)	Upper Beds (9)	Purbeck Beds Lower Beds (10)	Middle Beds (11)	Upper Beds (12)	Wealden Beds (×)	LOCALITIES.	REMARKS.
MOLLUSCA—cont.																
Gasteropoda—cont.																
Physa wealdiana, Coq. -		–	–	–	–	–	–	–	–	–	–	–	–	–	D. - -	Purb.
Planorbis Fisheri, Forbes -		–	–	–	–	–	–	–	–	–	10	11	–	–	D.W.O.Bu.	
Pleurotomaria Agassizi, Münst.		–	–	–	–	5	–	–	–	–	–	–	–	–	D.Y.	
—— depressa, Phil. -		1	2	–	–	–	–	–	–	–	–	–	–	–	W.G.O.Y.	Kim. Cl.
—— granulata, Lyc. -		–	–	–	4	–	–	–	–	–	–	–	–	–	C.H.	
—— Münsteri, Roem. -		–	–	–	4	5	–	–	–	–	–	–	–	–	D.W.C.H.Y.	
—— peles, d'Orb, -		–	–	–	–	–	–	–	–	–	–	–	–	–	D. - -	Cor.
—— reticulata, Sow. -		–	2	–	–	5	6	–	8	–	–	–	–	–	D. to Y. Sx.	
—— Rozeti, de Lor. -		–	–	–	–	–	–	–	–	9	–	–	–	–	D.	
—— rugata, Benett -		–	–	–	–	–	–	–	8	9	–	–	–	–	D.W. Bu. O.	
—— See also Trochotoma and Turbo.																
Pseudomelania abbreviata, Roem.		–	–	–	4	5	–	–	–	–	–	–	–	–	D. Be.	
—— cæcilia, d'Orb. -		–	–	–	–	–	6	–	–	–	–	–	–	–	D.	
—— collisa, de Lor. -		–	–	–	–	–	–	–	–	9	–	–	–	–	Bu.	
—— corallina, d'Orb. -		–	–	–	–	5	–	–	–	–	–	–	–	–	W.Y.	
—— decussata, Hudl. -		–	–	–	–	–	–	–	–	9	–	–	–	–	W.	
—— delia, d'Orb. -		–	–	–	–	5	–	–	–	–	–	–	–	–	D.	
—— gigantea, Leym. -		–	–	–	–	5	–	–	–	–	–	–	–	–	W. - -	Kim. Cl.
—— multispirata, Etall. -		–	–	–	–	–	–	–	–	9	–	–	–	–	D.	
—— naticoides, Hudl. -		–	–	–	–	–	–	–	–	9	–	–	–	–	W.	
—— paludinæformis, de Lor. -		–	–	–	–	–	–	–	–	9	–	–	–	–	D.	
—— pseudolimbata, Bl. and Hudl.		–	–	–	–	5	–	–	–	–	–	–	–	–	D.	
—— cf. rupellensis, d'Orb. -		–	–	–	–	5	–	–	–	–	–	–	–	–	D.	
—— teres, Hudl.		–	–	–	–	–	–	–	–	9	–	–	–	–	W.	
—— See also Bourguetia and Cloughtonia.																
Purpuroidea portlandica, Hudl.		–	–	–	–	–	–	–	–	9	–	–	–	–	Bu. - -	? external form of *Buccinum angulatum* (Hudl.)
Rissoa acuticarina, Blake -		–	–	–	–	–	–	–	–	9	–	–	–	–	Bu.	
—— mosensis, Buv. -		–	–	–	–	–	6	–	–	–	–	–	–	–	L.	
—— sp. -		–	–	–	–	–	–	–	–	–	10	11	–	–	D.W. -	*Hydrobia.*
Rostellaria. See *Alaria.*																
Spinigera -		–	–	–	–	–	–	–	–	–	–	–	–	–	W. - -	Ox. Cl.?
Tornatella Le Blanchi, Pict. and Lor.		–	–	–	–	–	–	–	–	9	–	–	–	–	D.	

Species.	Lower Oolite (×)	Am. calloviensis, Kellaways Rock (1)	Am. ornatus, Oxford Clay (2)	Am. cordatus (3)	Am. perarmatus, Corallian (4)	Am. plicatilis (5)	Kimeridge Clay, Lower Beds (6)	Kimeridge Clay, Upper Beds (7)	Portland Beds, Lower Beds (8)	Portland Beds, Upper Beds (9)	Purbeck Beds, Lower Beds (10)	Purbeck Beds, Middl. Beds (11)	Purbeck Beds, Upper Beds (12)	Wealden Beds (×)	Localities.	Remarks.
MOLLUSCA—cont.	×	1	2	3	4	5	6	7	8	9	10	11	12	×		
Gasteropoda—cont.																
Tornatella ? *purbeckensis*, Forbes MS. See Melanopsis Popei. —— See also Actæon and Melania.																
Tornatina oppeliana, de Lor.	-	-	-	-	-	-	-	-	-	9	-	-	-	-	D,	
Trochotoma tornata, Phil.	×	-	-	-	-	5	6	-	-	-	-	-	-	-	D. W. C. Y.	
Trochus dædalus, d'Orb.	-	-	-	-	-	5	-	-	-	-	-	-	-	-	W.	
—— excavatus, Blake	-	-	-	-	-	-	6	-	-	-	-	-	-	-	W. L.	
—— permedius, de Lor.	-	-	-	-	-	-	-	-	-	9	-	-	-	-	W.	
—— *tornatilis*, Phil. See Trochotoma tornata.																
—— retrorsus, Blake	-	-	-	-	-	-	-	7	-	-	-	-	-	-	L.	
—— sp.	-	1	-	-	-	-	-	-	-	-	-	-	-	-	W.	Purb.
—— See also Pleurotomaria.																
Turbo apertus, Blake	-	-	-	-	-	-	-	-	-	9	-	-	-	-	D. W.	
—— Archiaci, d'Orb.	-	1	-	-	-	-	-	-	-	-	-	-	-	-	W.	
—— exiguus, Roem.	-	-	-	-	5	-	-	-	-	-	-	-	-	-	D.	
—— Foucardi, Cott.	-	-	-	-	-	-	7	8	-	-	-	-	-	-	W. Sx.	
—— Julii, Etall.	-	-	-	-	5	-	-	-	-	-	-	-	-	-	D.	
—— See also Amberleya, Delphinula, and Littorina.																
Turritella jurassica, Quenst.	-	-	-	-	5	-	-	-	-	-	-	-	-	-	D.	
—— minuta, Koch and Dunk.	-	-	-	-	-	-	6	-	-	9	-	-	-	-	D. W. Sx.	Purb.
—— sp.	-	-	-	-	-	-	-	-	-	-	-	-	-	-	H.	Ox. Cl.
Valvata helicoides, Forbes	-	-	-	-	-	-	-	-	-	-	10	11	-	-	D. W.	
Scaphopoda.																
Dentalium cinctum, Goldf.	-	-	-	-	-	5	-	-	-	-	-	-	-	-	D.	
—— Quenstedti, Blake	-	-	-	-	-	-	6	-	-	-	-	-	-	-	W. C. L.	
—— sp.	-	1	-	-	-	-	-	-	-	-	-	-	-	-	D. G. W.	Kim. Cl., Purb.
Lamellibranchiata (Pelecypoda).																
Alectryonia. See Ostrea.																
Amphidesma. See Myacites and Pleuromya.																
Anatina minuta, Blake	-	-	-	-	-	-	6	-	-	-	-	-	-	-	L.	
—— parvula, Bean	-	-	-	-	-	-	6	-	-	-	-	-	-	-	L. Y.	
—— siliqua, Ag.	×	1	-	-	-	?	6	-	-	-	-	-	-	-	W. O. Y.	

SPECIES.	Lower Oolites.	Kellaways Rock. (Am. callovianensis.)	Oxford Clay. (Am. ornatus.)	Corallian. (Am. cordatus.)	Corallian. (Am. perarmatus.)	Corallian. (Am. plicatilis.)	Kimeridge Clay. Lower Beds.	Kimeridge Clay. Upper Beds.	Portland Beds. Lower Beds.	Portland Beds. Upper Beds.	Purbeck Beds. Lower Beds.	Purbeck Beds. Middle Beds.	Purbeck Beds. Upper Beds.	Wealden Beds.	LOCALITIES.	REMARKS.
	×	1	2	3	4	5	6	7	8	9	10	11	12	×		
MOLLUSCA—cont.																
Lamellibranchiata (Pelecypoda)—cont.																
Anatina cf. striata, Ag.	-	-	-	-	-	-	-	-	-	-	-	-	-	-	Sx. - -	Cor.
—— undulata, Sow. See A. siliqua.																
Anisocardia pulchella, de Lor.	-	-	-	-	-	-	-	-	9	-	-	-	-		W. Bu.	
Anodonta	-	-	-	-	-	-	-	-	-	-	-	-	-	-	D. - -	Purb.
Anomia Dollfusi, Blake	-	-	-	-	-	-	6	-	8	-	-	-	-		D. Bu. L.	
—— radiata, Phil.	-	-	-	-	4	-	-	-	-	-	-	-	-		D. O. Y. -	Discina.
—— suprajurensis, Buv.	-	-	-	-	-	5	-	-	-	-	-	-	-		W. C.	
—— sp.	-	-	-	-	-	-	-	-	-	-	-	-	-		D. - -	Ox. Cl., Purb.
Arca æmula, Phil.	×	1	-	-	4	5	6	-	8	-	-	-	-		D. to Y.	
—— anomala, Blake and Hudl.	-	-	-	-	-	5	-	-	-	-	-	-	-		C.	
—— Beaugrandi, de Lor.	-	-	-	-	-	-	-	-	8	9	-	-	-		W. Bu.	Kim. Cl.
—— catalaunica, de Lor.	-	-	-	-	-	-	-	-	-	-	-	-	-		D. W. -	
—— contracta, Phil.	-	-	-	-	-	5	6	-	-	-	-	-	-		C.	
—— Langi, Thurm.	-	-	-	-	-	-	-	7	-	-	-	-	-		D.	
—— longipunctata, Blake	-	-	-	-	-	5	6	7	8	-	-	-	-		D. to L.	
—— cf. menandellensis, de Lor.	-	-	-	-	-	-	-	-	-	9	-	-	-		Bu.	
—— minuscula, Cont.	-	-	-	-	-	-	6	-	-	-	-	-	-		D. C. L.	
—— mosensis, Buv. See A. æmula.																
—— pectinata, Phil.	-	-	-	-	-	5	-	-	-	-	-	-	-		C.	
—— quadrisulcata, Sow.	-	-	-	-	-	5	-	-	-	-	-	-	-		C. Y.	
—— Quenstedti, Lyc.	-	1	2	-	-	-	6	-	-	-	-	-	-		D. W.	
—— reticulata, Blake	-	-	-	-	-	-	6	7	8	-	-	-	-		C. L.	
—— rhomboidalis, Contej.	-	-	-	-	-	5	6	-	-	-	-	-	-		Bu. Bd. C. L.	
—— rustica, Contej.	-	-	-	-	-	-	-	-	-	-	-	-	-		W.	
—— sublata, d'Orb.	-	-	-	-	-	5	-	-	-	-	-	-	-		D.	
—— subtetragona, Mor.	-	1	2	-	4	5	-	-	-	-	-	-	-		W. O. Bd. H.	
—— terebrans, Buv.	-	-	-	-	4	-	-	-	-	-	-	-	-		C.	
—— velledæ, de Lor.	-	-	-	-	-	-	-	-	8	-	-	-	-		D. W.	
—— sp.	-	-	-	-	-	-	-	-	-	-	-	-	-		W. - -	Purb.
—— See also Cucullæa.																
Arcomya	-	-	-	-	-	5	-	-	-	-	-	-	-		D.	

Species	Lower Oolites	Am. calloviensis (Kellaways Rock)	Am. ornatus (Oxford Clay)	Am. cordatus (Oxford Clay)	Am. perarmatus (Corallian)	Am. plicatilis (Corallian)	Lower Beds (Kimeridge Clay)	Upper Beds (Kimeridge Clay)	Lower Beds (Portland Beds)	Upper Beds (Portland Beds)	Lower Beds (Purbeck Beds)	Middle Beds (Purbeck Beds)	Upper Beds (Purbeck Beds)	Wealden Beds	Localities	Remarks
	x	1	2	3	4	5	6	7	8	9	10	11	12	x		
MOLLUSCA—*cont* *Lamellibranchiata* (*Pelecypoda*)—cont.																
Astarte aliena, Phil.	x	-	-	-	-	5	-	-	-	-	-	-	-	-	D. Y.	
—— autissiodorensis, Cott.	-	-	-	-	-	-	-	-	8	-	-	-	-	-	8x.	
—— aytonensis, Lyc.	x	-	-	-	-	5	-	-	-	-	-	-	-	-	C. Y.	
—— carinata, Phil.	-	1	2	-	-	-	-	-	-	-	-	-	-	-	D. W. Bu. Y.	
—— *curvirostris*, Roem. See A. extensa.																
—— depressa, Goldf.	x	-	-	-	4	-	-	-	-	-	-	-	-	-	W. L. Y.	Kim. Cl.
—— duboisiana, d'Orb.	-	-	-	-	-	-	-	-	-	-	-	-	-	-	D. Y.	Cor.
—— extensa, Phil.	-	-	-	-	-	5	-	-	-	-	-	-	-	-	D. O.	
—— hartwellensis, Sow.	-	-	-	-	-	-	6	7	8	-	-	-	-	-	O. Bu. 8x.	Hartwell Cl.
—— lineata, Sow.	-	-	-	-	-	-	6	7	-	-	-	-	-	-	D. O. L. Y.	
—— *lurida*, Phil. See A. ungulata.																
—— micheaudiana, d'Orb.	-	-	-	-	-	-	6	-	-	-	-	-	-	-	D. W. L.	
—— minima, Phil.	x	1	-	-	-	-	-	-	-	-	-	-	-	-	W. Y.	
—— modiolaris, Lam.	-	-	-	-	-	-	-	-	-	-	-	-	-	-	D.	Cor.
—— mysis, d'Orb.	-	-	-	-	-	-	-	7	8	-	-	-	-	-	D. Bu.	
—— ovata, W. Smith	-	-	2	3	4	5	6	7	-	9	-	-	-	-	D. to Y. Sx.	
—— pesolina, Cont.	-	-	-	-	-	-	6	-	-	-	-	-	-	-	L.	
—— polymorpha, Cont.	-	-	-	-	-	5	-	-	8	-	-	-	-	-	D. W.	
—— regularis, Cont.	-	-	-	-	-	-	6	-	-	-	-	-	-	-	D.	
—— rhomboidalis, Phil.	x	-	-	-	-	-	-	-	-	-	-	-	-	-	W. Be. Y.	Kim. Cl., Cor.
—— robusta, Lyc.	-	1	2	-	4	-	-	-	-	-	-	-	-	-	W. G. C. H. Y.	
—— *rugosa*, Sow. See Cytherea.																
—— Sæmanni, de Lor	-	-	-	-	-	-	-	-	8	-	-	-	-	-	W. Bu.	
—— sequana, Contej.	-	-	-	-	-	-	-	7	-	-	-	-	-	-	8x.	
—— subdepressa, Blake and Hudl.	-	-	-	-	-	5	-	-	-	-	-	-	-	-	W.	
—— supracorallina, d'Orb.	-	-	-	-	-	5	6	-	-	-	-	-	-	-	D. to L. 8x.	
—— Thomsoni, Damon	-	-	-	-	-	-	-	-	-	-	-	-	-	-	D.	Cor.
—— ungulata, Lyc.	x	1	-	-	-	-	-	-	-	-	-	-	-	-	G. C.	Cor.
—— zonata, Roem.	-	-	-	-	-	-	-	-	-	-	-	-	-	-	D.	Ox. Cl.
—— sp.	-	-	-	-	-	-	-	-	-	-	10	-	-	-	Bu.	
—— See also Cytherea.																
Aucella Pallasi, Keys	-	-	-	-	-	-	-	-	-	-	-	-	-	-	L.	Kim. Cl.
Avicula ædilignensis, Blake	-	-	-	-	-	5	-	-	-	-	-	-	-	-	D. L. Y. •	Kim. Cl.

SPECIES.	Lower Oolites.	Kellaw sysRock. Am. calloviensis.	Oxford Clay. Am. ornatus.	Oxford Clay. Am. cordatus.	Corallian. Am. perarmatus.	Corallian. Am. plicatilis.	Kimeridge Clay. Lower Beds.	Kimeridge Clay. Upper Beds.	Portland Beds. Lower Beds.	Portland Beds. Upper Beds.	Purbeck Beds. Lower Beds.	Purbeck Beds. Middle Beds.	Purbeck Beds. Upper Beds.	Wealden Beds.	LOCALITIES.	REMARKS.
	×	1	2	3	4	5	6	7	8	9	10	11	12	×		
MOLLUSCA—*cont.*																
Lamellibranchiata (Pelecypoda)—cont.																
Avicula braamburiensis, Sow.	×	1	–	–	4	–	–	–	–	–	–	–	–	–	W. C. Y.	
—credneriana, de Lor.	–	–	–	–	–	–	–	–	–	9	–	–	–	–	W.	
—dorsetensis, Blake	–	–	–	–	–	6	–	–	–	–	–	–	–	–	D. W. L.	
—echinata, Sow.	×	–	–	–	–	6	–	–	–	–	–	–	–	–	C.	
—expansa, Phil.	–	1	2	–	4	5	–	–	–	–	–	–	–	–	W. to Y.	
—? Gesneri, Thurm.	–	–	–	–	–	–	–	–	–	9	–	–	–	–	W.	
—inæquivalvis, Sow.	×	1	2	3	4	–	–	7	–	–	–	–	–	–	D. to Y. Sx.	
—lævis, Blake & Hudl.	–	–	–	–	4	5	–	–	–	–	–	–	–	–	O. Y.	
—*modiolaris*, Roem. See A. Gesneri.																
—Münsteri, Goldf.	×	1	–	–	–	–	–	–	–	–	–	–	–	–	W. G.	
—nummulina, Blake	–	–	–	–	–	–	6	–	–	–	–	–	–	–	L.	
—octavia, d'Orb.	–	–	–	–	–	–	–	–	8	–	–	–	–	–	D.	
—ovalis, Phil.	–	1	2	–	4	5	–	–	–	–	–	–	–	–	D. to Y.	
—— var. obliqua, Blake & Hudl.	–	–	–	–	–	5	–	–	–	–	–	–	–	–	O. Y.	
—pteropernoides, Blake & Hudl.	–	–	–	–	4	5	–	–	–	–	–	–	–	–	D. to Y.	
—pterosphena, Seeley	–	×	–	–	–	–	–	–	–	–	–	–	–	–	C. -	Cor.
—Struckmanni, de Lor.	–	–	–	–	–	5	–	–	–	–	–	–	–	–	D. Y.	
—vellicata, Blake	–	–	–	–	–	–	–	7	–	–	–	–	–	–	L.	
Cardita boloniensis, de Lor.	–	–	–	–	–	–	–	–	–	9	–	–	–	–	D.	
—ovalis, Quenst.	–	–	–	–	–	5	–	–	–	–	–	–	–	–	C.	
Cardium calcareum, Blake	–	–	–	–	–	–	–	–	8	9	–	–	–	–	D. W. Bu.	
—cognatum, Phil.	×	1	–	–	–	–	–	–	–	–	–	–	–	–	W. Bu. Y.	
—Crawfordi, Leck.	–	1	2	3	4	–	–	–	–	–	–	–	–	–	D. to. Y.	
—cyreniforme, Buv.	–	–	–	–	–	5	–	–	–	–	–	–	–	–	D. Y.	
—? decussatum, Sow.	–	–	–	–	–	–	–	–	–	–	–	–	–	–	W. - -	Purb.
—delibatum, de Lor.	–	–	–	–	–	5	–	–	–	–	–	–	–	–	D. W. O. Y.	
—dissimile, Sow.	–	–	–	–	–	–	–	–	8	9	–	–	–	–	D. to Bu.	
—isocardioides, Blake & Hudl.	–	–	–	–	–	–	–	–	–	–	–	–	–	–	O. Y.	Cor. *Cypricardia isocardina*, Buv.
—morinicum, de Lor.	–	–	–	–	–	–	–	7	8	–	10	–	–	–	W. Bu. Sx.	? var. of C. striatulum.
—orthogonale, Buv.	–	–	–	–	–	–	–	7	–	–	–	–	–	–	D.	
—Pellati, de Lor.	–	–	–	–	–	–	–	–	–	9	–	–	–	–	W. O. Ba.	
—pesolinum, Contej.	–	–	–	–	–	–	6	7	–	–	–	–	–	–	Sx.	
—pseudoaxinum, Thurm.	–	–	–	–	–	–	6	–	–	–	–	–	–	–	D.	

Species.	Lower Oolites.	Kellaways Rock. Am. calloviensis.	Oxford Clay. Am. ornatus.	Am. cordatus.	Corallian. Am. perarmatus.	Am. plicatilis.	Kimeridge Clay. Lower Beds.	Upper Beds.	Portland Beds. Lower Beds.	Upper Beds.	Purbeck Beds. Lower Beds.	Middle Beds.	Upper Beds.	Wealden Beds.	Localities.	Remarks.
	×	1	2	3	4	5	6	7	8	9	10	11	12	×		
MOLLUSCA.—cont.																
Lamellibranchiata (Pelecypoda)—cont.																
Cardium purbeckense, de Lor.	-	-	-	-	-	-	-	-	-	-	-	-	-	-	D. W.	Purb.
—— striatulum, Sow.	-	-	-	-	-	5	6	7	8	-	-	-	-	-	D. to Y. 8x.	
—— —— var. lepida, Sauv. & Rig.	-	-	-	-	-	-	-	-	-	-	-	-	-	-	Sx.	Kim. Cl.
—— subhillanum, Leym.	-	-	-	-	-	-	-	-	-	-	-	-	-	-	L.	Spilsby Sandstone.
—— sp.	-	-	-	-	-	-	-	-	-	-	10	-	-	-	D. W. 8x.	
Ceromya excentrica, Ag.	×	-	-	-	-	5	-	-	-	-	-	-	-	-	D. Y.	Kim. Cl.
—— inflata, Volts. See C. orbicularis.																
—— orbicularis, Roem.	-	-	-	-	-	5	6	-	-	-	-	-	-	-	D. L.	
Corbicella lævis, Sow.	-	-	-	-	-	5	-	-	-	-	-	-	-	-	O. Y.	C. depressa, Desh.
—— morœana, Buv.	-	-	-	-	-	-	-	-	-	9	-	-	-	-	W.	
—— ovalis, Phil.	×	1	-	-	-	-	-	-	-	-	-	-	-	-	W. Y.	
—— portlandica, Damon	-	-	-	-	-	-	-	-	-	9	-	-	-	-	D.	
Corbis concinna, Damon	-	-	-	-	-	-	6	7	-	-	-	-	-	-	D.	
Corbula alata, J. Sow.	-	-	-	-	-	-	-	-	-	-	10	11	-	×	D. W. 8x.	
—— dammariensis, Buv.	-	-	-	-	-	-	-	-	8	9	-	-	-	-	W.	
—— Deshayesea, Buv.	-	-	-	-	-	5	6	-	8	-	-	-	-	-	D. Bu. Bd. C. L.	
—— fallax, Cont.	-	-	-	-	-	-	6	-	-	-	-	-	-	-	L.	
—— forbesiana, de Lor.	-	-	-	-	-	-	-	-	-	-	-	-	-	-	D.	Purb.
—— inflexa? Roem.	-	-	-	-	-	-	-	-	-	-	-	-	-	-	D.	Purb. Nucula.
—— Macneilli, Mor.	-	1	-	-	-	-	-	-	-	-	-	-	-	-	W.	
—— saltans, Blake	-	-	-	-	-	-	-	-	8	9	-	-	-	-	D. W. B.	
Corimya. See Thracia.																
Cucullæa clathrata, Leck.	-	-	-	-	4	-	-	-	-	-	-	-	-	-	C. Y.	
—— concinna, Phil.	×	1	2	3	-	-	-	-	-	-	-	-	-	-	D. to Y.	
—— contracta, Phil.	-	-	-	-	4	5	-	-	-	-	-	-	-	-	D. to Y.	
—— corallina, Damon	-	-	-	-	4	5	-	-	-	-	-	-	-	-	D. Y.	C. oblonga, Phil.
—— donningtonensis, Keeping.	-	-	-	-	-	-	-	-	-	-	-	-	-	-	L.	Spilsby Sandstone.
—— elongata, Sow.	×	-	-	-	4	5	-	-	-	-	-	-	-	-	W. C.	
—— errans, Keeping	-	-	-	-	-	-	-	-	-	-	-	-	-	-	L.	Spilsby Sandstone.
—— oblonga, Sow.	×	-	-	-	4	-	-	-	-	-	-	-	-	-	C.	
—— superba, Cont.	-	-	-	-	-	5	-	-	-	-	-	-	-	-	D.	
—— sp.	-	-	-	-	-	-	-	-	-	-	-	-	-	-	Bu.	Portl.
Cyclas. See Cyrena.																

SPECIES.	Lower Oolites.	Am. calloviensis. / Kellaways Rock.	Am. ornatus. / Oxford Clay.	Am. cordatus.	Am. perarmatus. / Corallian.	Am. plicatilis.	Lower Beds. / Kimeridge Clay.	Upper Beds.	Lower Beds. / Portland Beds.	Upper Beds.	Lower Beds. / Purbeck Beds.	Middle Beds.	Up er Beds.	Wealden Beds.	LOCALITIES.	REMARKS.
	×	1	2	3	4	5	6	7	8	9	10	11	12	*		
MOLLUSCA—*cont*																
Lamellibranchiata (Pelecypoda)—cont.																
socardia striata, d'Orb,	-	-	-	-	-	5	-	-	-	-	-	-	-	-	D. L. -	Kim. Cl.
—— See also Ceromya.																
Isodonta Deshayesea, Buv. See Sowerbya.																
Leda lachryma, Sow.	×	1	-	3	-	-	-	-	-	-	-	-	-	-	H. Y.	
—— Phillipsi, Mor.	-	1	2	-	-	-	-	-	-	-	-	-	-	-	W. Bu.	
—— venusta, Sauv. and Rig.	-	-	-	-	-	-	7	-	-	-	-	-	-	-	L. -	*L. lineata.* Blake.
—— sp.	-	-	-	-	-	-	-	-	-	9	10	-	-	-	D. W.	
Lima ædilignensis, Blake	-	-	-	-	-	-	-	-	-	-	-	-	-	-	L. -	Kim. Cl.
—— bifurcata, Blake	-	-	-	-	-	-	-	-	9	-	-	-	-	-	Bu.	
—— boloniensis, de Lor.	-	-	-	-	-	-	-	-	8	9	-	-	-	-	D. W. Bu.	
—— cf. dammariensis, Buv.	-	-	-	-	-	-	-	-	-	-	-	-	-	-	Sx.	Kim. Cl.
—— densepunctata, Roem.	-	-	-	-	-	5	-	-	-	-	-	-	-	-	W. Y.	
—— duplicata, Sow.	×	1	-	-	4	5	-	-	8	-	-	-	-	-	W. Bu. C. Y.	
—— elliptica, Whit.	-	-	-	-	4	5	-	-	-	-	-	-	-	-	D. to Y.	
—— gibbosa, Sow.	×	-	-	-	-	5	-	-	-	-	-	-	-	-	W. C. Y.	
—— læviuscula, Sow.	×	-	-	-	4	5	-	-	-	-	-	-	-	-	W. to Y.	
—— ornata, Buv.	-	-	-	-	-	-	-	-	-	9	-	-	-	-	W. Bu.	
—— pectiniformis, Schloth.	×	1	-	-	4	5	6	7	-	-	-	-	-	-	D. to Y.	
—— rigida, Sow.	×	-	-	3	4	5	-	-	-	-	-	-	-	-	D. to Y.	
—— rigidula, Phil.	×	-	2	-	-	-	-	-	-	-	-	-	-	-	N.	
—— rudis, Sow.	×	1	-	-	-	5	-	-	-	-	-	-	-	-	W. O. C. Y.	
—— rustica, Sow.	-	-	-	-	-	-	-	7	8	9	-	-	-	-	D. to Y. -	Cor.
—— subantiquata, Roem.	-	-	-	-	-	5	-	-	-	-	-	-	-	-	D. W. Y.	
—— tombeckiana, d'Orb.	-	-	-	-	-	-	-	-	-	-	-	-	-	-	L. -	Spilsby Sandstone.
—— virgulina, Cont.	-	-	-	-	-	-	6	-	-	-	-	-	-	-	D.	
Lithodomus inclusus, Phil.	×	-	-	-	-	5	-	-	-	-	-	-	-	-	W. to Y. K.	*Modiola.*
—— Porteri, Lyc.	×	1	-	-	-	-	-	-	-	-	-	-	-	-	W.	
—— portlandicus, (Damon) Lyc.	-	-	-	-	-	-	-	-	-	9	-	-	-	-	D.	
Lopha. See Ostrea.																
Lucina aliena, Phil.	-	-	-	-	-	5	-	-	-	-	-	-	-	-	D. W. Bd. C. Y.	Cor.
—— Beani, Lyc.	-	1	-	-	4	5	-	-	-	-	-	-	-	-	W. C. Y. -	Cor.
—— bellona, d'Orb.	×	1	-	-	-	-	-	-	-	-	-	-	-	-	W. Y. -	Cor.
—— circumcisa, Zit. and G.	-	-	-	-	-	4	-	-	-	-	-	-	-	-	W.	

SPECIES.	Lower Oolite.	Am. calloviensis. (Kellaways Rock)	Am. ornatus. (Oxford Clay)	Am. cordatus. (Corallian)	Am. perarmatus.	Am. plicatilis.	Lower Beds. (Kimeridge Clay)	Upper Beds.	Lower Beds. (Portland Beds)	Upper Beds.	Lower Beds. (Purbeck Beds)	Middle Beds.	Upper Beds.	Wealden Beds.	LOCALITIES.	REMARKS.
	×	1	2	3	4	5	6	7	8	9	10	11	12	×		
MOLLUSCA—*cont.*																
Lamellibranchiata (Pelecypoda)—cont.																
Lucina crassa, Sow.	×	–	–	–	–	–	–	–	–	–	–	–	–	–	L.	Spilsby Sandstone.
—— despecta, Phil.	×	1	–	–	–	–	–	–	–	–	–	–	–	–	W. G.	
—— *elgaudiæ*, Thurm. See L. sub-striata.																
—— fragosa, de Lor	–	–	–	–	–	–	–	–	8	–	–	–	–	–	W.	
—— *globosa*, Buv. See L. Beani.																
—— lirata, Phil.	–	1	–	–	–	–	–	–	–	–	–	–	–	–	W. O. L. Y.	Cor. Spilsby Sandstone.
—— minuscula, Blake	–	–	–	–	–	–	–	7	–	–	–	–	–	–	D. to Y. Sx.	
—— moreana, Buv.	–	–	–	–	–	5	–	–	–	–	–	–	–	–	D. to Y.	
—— plebeia, Contej.	–	–	–	–	–	–	6	–	–	9	–	–	–	–	D. W.	
—— portlandica, Sow.	–	–	–	–	–	–	–	–	–	9	–	–	–	–	D. W. Bu. Y.	
—— rotundata, Roem.	×	1	–	–	–	–	–	–	–	–	–	–	–	–	D. W.	
—— rugosa, Roem.	–	–	–	–	–	–	6	–	–	–	–	–	–	–	W.	
—— substriata, Roem.	–	–	–	–	–	5	6	–	–	–	–	–	–	–	D. Y.	
—— See also Astarte.																
Macrodon	–	–	–	–	–	–	–	–	–	–	–	–	–	–	Sx.	Kim. Cl.
Mactra tenuissima, Cont.	–	–	–	–	–	5	–	–	–	–	–	–	–	–	D.	
Mactromya rugosa, Roem. See Lucina.																
Modiola *autissiodorensis*, Cott. See Mytilus.																
—— bipartita, Sow.	×	1	2	3	4	5	–	7	–	9	–	–	–	–	D. to Y.	
—— boloniensis, de Lor.	–	–	–	–	–	–	–	–	–	9	–	–	–	–	D. Bu.	
—— cancellata, Roem.	–	–	–	–	4	5	–	–	–	–	–	–	–	–	W. Be. H. Y.	
—— cuneata, Sow.	×	1	–	–	–	–	–	7	–	–	–	–	–	–	D. W. O. Y.	Cor.
—— imbricata, Sow.	×	1	–	–	–	–	–	–	–	–	–	–	–	–	D.W.O.Y. K.	Kim. Cl., Cor
—— Lonsdalei, Mor. & Lyc.	×	1	–	–	–	–	–	–	–	–	–	–	–	–	W. Y.	Cor.
—— Lycetti, Morris	×	–	–	–	–	5	–	–	–	–	–	–	–	–	O. Y.	
—— pallida, Sow.	–	–	–	–	–	–	6	–	–	9	–	–	–	–	D. W. Bu.	
—— pulchra, Phil.	–	1	–	–	–	–	–	–	–	–	–	–	–	–	W. Y.	
—— (cf.) rauraciensis, Greppin.	–	–	–	–	–	5	–	–	–	–	–	–	–	–	C.	
—— semiplicata, Buv.	–	–	–	–	–	–	6	–	–	–	–	–	–	–	W. Bu.	
—— subæquiplicata, Goldf.	–	–	–	–	4	5	–	–	–	–	–	–	–	–	D. W. C. Y.	
—— unguiculata, Phil.	–	–	–	–	–	–	–	–	8	9	–	–	–	–	W. Bu.	
—— varians, Roem.	–	–	–	–	–	5	–	–	–	–	–	–	–	–	D.	
—— sp.	–	–	–	–	–	–	–	–	–	–	10	11	–	–	W. O. Bu.	

BB 2

SPECIES.	Lower Oolites.	Am. callloviensis. (Kellaways Rock)	Am. ornatus. (Oxford Clay)	Am. cordatus. (Oxford Clay)	Am. perarmatus. (Corallian)	Am. plicatilis. (Corallian)	Lower Beds. (Kimeridge Clay)	Upper Beds. (Kimeridge Clay)	Lower Beds. (Portland Beds)	Upper Beds. (Portland Beds)	Lower Beds. (Purbeck Beds)	Middle Beds. (Purbeck Beds)	Upper Beds. (Purbeck Beds)	Wealden Beds.	LOCALITIES.	REMARKS.
	×	1	2	3	4	5	6	7	8	9	10	11	12	×		
MOLLUSCA.—*cont.*																
Lamellibranchiata. (*Pelecypoda*)—cont.																
Modiola. See also Lithodomus and Mytilus.																
Myacites æquatus, Phil.	×	1	-	-	-	-	-	-	-	-	-	-	-	-	W. Y.	
—— calceiformis, Phil.	×	1	-	-	-	-	-	-	-	-	-	-	-	-	W. O. Y.	Cor.
—— decurtatus, Phil.	×	1	-	-	4	5	-	-	-	-	-	-	-	-	D. W. Bd. C, Y.	
—— Jurassi, Brongn.	×	-	-	-	4	5	-	-	-	-	-	-	-	-	D. C. H. Y.	Portl.
—— —— var portlandica, Damon.	-	-	-	-	-	-	-	-	-	9	-	-	-	-	D.	
—— oblatus, Sow.	-	-	-	-	4	5	6	-	-	-	-	-	-	-	D. C. Y.	
—— recurvus, Phil.	×	1	2	3	4	5	6	7	-	-	-	-	-	-	D. to Y.	
—— securiformis, Phil.	×	1	-	-	-	5	-	-	-	-	-	-	-	-	D. W. Y.	
Myoconcha portlandica, Blake	-	-	-	-	-	-	-	-	8	9	-	-	-	-	W. Bu.	
—— Sæmanni, Dollf.	-	-	-	-	-	5	-	-	8	-	-	-	-	-	W. O. Bu. C.	
—— texta, Buv.	-	-	-	-	-	5	-	-	8	-	-	-	-	-	Bu. C. Y.	*Mytilus.*
Mytilus autissiodorensis, Cott.	-	-	-	-	-	-	-	7	8	9	-	-	-	-	D.W. Bu. Sx.	
—— boloniensis, de Lor.	-	-	-	-	-	-	-	-	8	9	-	-	-	-	D. W. Bu.	
—— jurensis, Mérian	-	-	-	-	-	5	-	-	8	9					D.W. Bu.C. Y.	
—— longævus, Contej.	-	-	-	-	-	-	-	-	8	-	-	-	-	-	W.	
—— Lyelli, J. Sow.	-	-	-	-	-	-	-	-	-	-	10	11	-	×	Bu. Sx.	
—— morinicus, de Lor.	-	-	-	-	-	-	-	-	-	9	-	-	-	-	D.	
—— pectinatus, Sow.	×	-	-	-	-	5	-	7	8	-	-	-	-	-	D. to Y. Sx.	
—— *subpectinatus,* d'Orb. See M. pectinatus.																
—— ungulatus, Y. & B.	-	-	-	-	-	5	-	-	-	-	-	-	-	-	W. C.	
—— varians, Roem.	-	-	-	-	-	5	-	-	-	-	-	-	-	-	D.	
—— See also Modiola and Myoconcha.																
Neæra portlandica, de Lor.	-	-	-	-	-	-	-	-	-	9	-	-	-	-	W. Bu.	
Nucula elliptica, Phil.	-	-	2	3	-	-	-	-	-	-	-	-	-	-	O. Bu. H. Y.	
—— Menkei, Roem.	×	-	-	-	5	6	7	8	-	-	-	-	-	-	D. Bu. Bd. L.	
—— nuda, Bean	-	-	-	3	-	5	-	-	-	-	-	-	-	-	D. to Y.	
—— obliquata, Blake	-	-	-	-	-	6	-	-	9	-	-	-	-	-	W. Bu. L.	
—— ornata, Quenst.	-	1	2	3	-	-	-	-	-	-	-	-	-	-	D. W. Bu. H.	Cor.
—— *Phillipsi,* Morris. See Leda.																
—— turgida, Bean	-	-	-	3	-	-	-	-	-	-	-	-	-	-	H. Y.	
Opis angulosa, d'Orb.	-	-	-	-	4	-	-	-	-	-	-	-	-	-	C.	
—— arduennensis, d'Orb.	-	-	-	-	-	5	-	-	-	-	-	-	-	-	C.	

Species.	Lower Oolites.	Kellaways Rock. Am. calloviensis.	Oxford Clay. Am. ornatus.	Am. cordatus.	Corallian. Am. perarmatus.	Am. plicatilis.	Kimeridge Clay. Lower Beds.	Upper Beds.	Portland Beds. Lower Beds.	Upper Beds.	Purbeck Beds. Lower Beds.	Middle Beds.	Upper Beds.	Wealden Beds.	Localities.	Remarks.
	×	1	2	3	4	5	6	7	8	9	10	11	12	×		
MOLLUSCA—*cont.*																
Lamellibranchiata (Pelecypoda)—cont.																
Opis corallina, Damon.	–	–	–	–	–	5	–	–	–	–	–	–	–	–	D. O. C. Y.	Kim. Cl.
—— lunulata, Roem.	×	–	–	–	–	5	–	–	–	–	–	–	–	–	C. Y.	
—— (cf.) paradoxa, Buv.	–	–	–	–	–	5	–	–	–	–	–	–	–	–	C.	
—— Phillipsi, Mor.	–	–	–	–	4	5	–	–	–	–	–	–	–	–	D. to Y.	
—— suprajurensis, Cont.	–	–	–	–	–	5	–	–	–	–	–	–	–	–	D.	
—— virdunensis, Buv.	–	–	–	–	–	5	–	–	–	–	–	–	–	–	C.	
—— See also Corbula.																
Ostrea bononiæ, de Lor.	–	–	–	–	–	–	–	–	8	–	–	–	–	–	W. Bu.	
—— bullata, Sow.	–	–	–	–	–	5	–	–	–	–	–	–	–	–	N. L.	
—— deltoidea, Sow.	–	–	–	–	–	5	6	7	–	–	–	–	–	–	D. to Y. Sx.	
—— discoidea, Seeley	–	–	–	–	4	5	–	–	–	–	–	–	–	–	Bd. H. C.	
—— distorta, Sow.	–	–	–	–	–	–	–	–	–	–	–	11	–	×	D. W. Bu. Sx.	
—— duriuscula, Phil.	–	–	–	–	–	5	–	7	8	–	–	–	–	–	D. O. Bu.	
—— expansa, Sow.	–	–	–	–	–	–	–	?	8	9	–	–	–	–	D. to Bu. Sx.?	
——*falcata*, Sow. See O. expansa.																
—— fiabelloides, Lam.	×	1	–	–	4	–	–	–	–	–	–	–	–	–	D. to Y.	Kim. Cl.
—— fiabellosa, Damon	–	–	–	–	–	–	–	–	–	9	–	–	–	–	D.	
—— gibbosa, Les.	–	–	–	–	–	–	–	7	–	–	–	–	–	–	L. Y. Sx.	
—— gregaria, Sow.	×	1	2	3	4	5	6	–	–	–	–	–	–	–	D. to Y. K.	
—— læviuscula, Sow.	–	–	–	–	–	–	6	7	8	9	–	–	–	–	D. W. Bu.	
—— *Marshi*, Sow. See O. fiabelloides.																
——monsbeliardensis, Cont.	–	–	–	–	–	–	6	–	–	–	–	–	–	–	L.	
—— moreana, Buv.	–	1	–	–	–	–	–	–	–	–	–	–	–	–	D. W. Y.	Cor.
—— multiformis, Koch and Dunk.	–	–	–	–	–	–	–	–	8	9	–	–	–	–	D. W.	
—— *palmetta*, Sow. See O. gregaria.																
—— Roemeri, Quenst.	–	–	–	–	–	–	–	–	–	–	–	–	–	–	D.	Cor.
—— *sandalina*, Goldf. See Exogyra nana.																
—— solitaria, Sow.	–	–	–	–	4	5	6	7	8	9	–	–	–	–	D. to Y. Sx.	
—— spinigera, Phil.	–	–	–	–	–	–	–	–	–	–	–	–	–	–	O.	Cor.
—— Thurmanni, Etall.	–	–	–	–	–	–	–	7	–	–	–	–	–	–	Sx.	
—— unciformis, Buv.	–	–	–	–	–	–	6	–	–	–	–	–	–	–	Sx.	
—— undosa, Phil.	–	1	–	–	–	–	–	–	–	–	–	–	–	–	W. Y.	
—— sp.	–	–	–	–	–	–	–	–	–	–	10	–	–	–	O. Bu. Sx.	

Species.	Lower Oolites.	Am. calloviensis. (Kellaways Rock.)	Am. ornatus. (Oxford Clay.)	Am. cordatus.	Am. perarmatus. (Corallian.)	Am. plicatilis.	Lower Beds. (Kimeridge Clay.)	Upper Beds.	Lower Beds. (Portland Beds.)	Upper Beds.	Lower Beds. (Purbeck Beds.)	Middle Beds.	Upper Beds.	Wealden Beds.	Localities.	Remarks.
	×	1	2	3	4	5	6	7	8	9	10	11	12	*		
MOLLUSCA—*cont.*																
Lamellibranchiata (Pelecypoda)—cont.																
Ostrea. See also Exogyra and Gryphæa.																
Panopæa. See Gresslya.																
Pecten anisopleurus, Buv.	×	-	-	-	-	-	-	-	-	-	-	-	-	-	D. . .	Cor.
—— arcuatus, Sow.	×	-	-	-	-	-	6	7	-	-	-	-	-	-	D. L. Sx. -	? Cor.
—— articulatus, Schloth.	×	-	-	-	4	5	-	-	-	-	-	-	-	-	D. to Y. K.	
—— *cancellatus*, Bean. See P. intertextus.																
—— cinctus, Sow.	-	-	-	-	-	-	-	-	-	-	-	-	-	-	L. . .	Spilsby Sandstone.
—— demissus, Phil.	-	1	2	-	-	5	6	-	-	-	-	-	-	-	D. to Y.	
—— *distriatus*, Laym. See P. suprajurensis.																
—— fibrosus, Sow.	×	1	2	-	4	5	-	-	-	-	-	-	-	-	D. to Y.	Kim. Cl.
—— Grenieri, Contej.	-	-	-	-	-	-	6	-	-	-	-	-	-	-	C. L.	
—— inæquicostatus, Phil.	×	-	-	-	-	5	-	-	-	-	-	-	-	-	W. O. Y.	
—— intertextus, Roem.	×	-	-	-	-	5	-	-	-	-	-	-	-	-	D. W. Y.	
—— kimmeridgensis, Cott.	-	-	-	-	-	-	-	-	-	-	-	-	-	-	Sx. . .	Kim. Cl.
—— lamellosus, Sow.	-	-	-	-	-	-	-	-	8	9	-	-	-	-	D. to Bu.	
—— lens, Sow.	×	1	-	-	4	5	6	7	-	-	-	-	-	-	D. to Y. Sx.	
—— —— var. Morini. de Lor.	-	-	-	-	-	-	-	7	-	-	-	-	-	-	D. W. Bu. Y.	
—— midas. d'Orb.	-	-	-	-	-	5	6	-	-	-	-	-	-	-	D. W. Y.	
—— minerva, d'Orb.	-	-	-	-	-	-	-	-	-	-	-	-	-	-	D. . .	Cor.
—— *Morini*, de Lor. See P. lens. var.																
—— nitescens, Phil.	-	-	-	-	-	-	-	7	8	-	-	-	-	-	O. Bu. Be.	
—— orbicularis, Sow.	-	-	-	-	-	-	-	-	-	-	-	-	-	-	L. . .	Spilsby Sandstone.
—— qualicosta, Etall.	-	-	-	-	4	5	-	-	-	-	-	-	-	-	D. W. Be. O. Y.	
—— Quenstedti, Blake	-	-	-	-	-	5	-	-	-	-	-	-	-	-	D.	
—— similis, Sow.	-	-	-	-	-	-	-	-	-	-	-	-	-	-	O. . .	Cor.
—— solidus, Roem.	-	-	-	-	-	-	-	-	8	9	-	-	-	-	D. to Bu.	
—— strictus, Münst.	-	-	-	-	-	5	-	-	-	-	-	-	-	-	D. W.	
—— subtextorius, Münst.	-	-	-	-	4	5	-	-	-	-	-	-	-	-	W. C. Y.	
—— subtextularis, Münst.	-	-	-	-	-	5	-	-	-	-	-	-	-	-	D.	
—— suprajurensis, Buv.	-	-	-	-	-	5	6	7	8	9	-	-	-	-	D. W. Bu.	
—— Thurmanni, Cont.	-	-	-	-	-	5	6	-	8	-	-	-	-	-	Bu. Bd. L.	
—— vagans, Sow.	×	1	2	-	4	-	-	-	-	-	-	-	-	-	D. to Y. K.	

Species.	Lower Oolites	Kellaways Rock — Am. calloviensis	Oxford Clay — Am. ornatus	Am. cordatus	Corallian — Am. perarmatus	Am. plicatilis	Kimeridge Clay — Lower Beds	Upper Beds	Portland Beds — Lower Beds	Upper Beds	Purbeck Beds — Lower Beds	Middle Beds	Upper Beds	Wealden Beds	Localities.	Remarks.
	×	1	2	3	4	5	6	7	8	9	10	11	12	×		
MOLLUSCA—cont.																
Lamellibranchiata (*Pelecypoda*)—cont.																
Pecten *vimineus*, Sow. See P. articulatus.																
—— virdunensis, Buv.						5									D.	Near to P. lens.
Perna? Bayani, de Lor.									8						W.	
Perna Bouchardi, Oppel.									8	9					D. W. Bu.	
—— Flambarti, Dollf.							6		8						Bu. C.	
—— mytiloides, Lam.	×	1			4	5	6	7	8	9					D. to Y.	
—— octavia, d'Orb. See Avicula.																
—— quadrata, Sow.	×				4	5									D. W. Y.	
—— subplana, Etall.						5									C.	
—— See also Avicula.																
Pholadidea abbreviata, Blake							6								L.	
—— compressa, Sow.							6	7							D. Sx.	Ox. Clay.
Pholadomya acuticosta, Sow.	×	1					6								W. N. L. Y.	
—— æqualis, Sow.					4	5	6	7							D. to Y.	
—— angustata, Sow.															D.	Ox. Cl.
—— concentrica, Roem.					4										C. L.	
—— concinna, Ag.					4										D. W.	
—— decemcostata, Roem.						5									D. W. C. Y.	
—— decorata, Hartm.		1													W.	
—— deltoidea, Sow.	×	1	2												W. G.	
—— hemicardia, Roem.						5									D. W.	
—— inæqualis, Phil.															O.	Portl.
—— læviuscula, Ag.							6								W.	
—— Murchisoni, Sow.		1													W.	
—— obsoleta, Phil.		1													G. O.	Cor.
—— ovalis, Sow.	×														C.	Cor.
—— paucicosta, Roem.		1				5	6								D. W. Y.	?=P. paucicosta, Ag.
—— pelagica, Ag.							6								D.	
—— Phillipsi, Morris	×	1		3			6								D. G. H.	
—— Protei, Ag.									8						Bu.	
—— rustica, Phil. See P. tumida.																
—— tricostata, Seeley															C.	Cor.
—— tumida, Ag.								7	8	9					W. O. Bu.	

Species	Lower Oolites	Kellaways Rock — Am. calloviensis	Oxford Clay — Am. ornatus	Oxford Clay — Am. cordatus	Corallian — Am. perarmatus	Corallian — Am. plicatilis	Kimeridge Clay — Lower Beds	Kimeridge Clay — Upper Beds	Portland Beds — Lower Beds	Portland Beds — Upper Beds	Purbeck Beds — Lower Beds	Purbeck Beds — Middle Beds	Purbeck Beds — Upper Beds	Wealden Beds	Localities	Remarks
	×	1	2	3	4	5	6	7	8	9	10	11	12	×		
MOLLUSCA—cont. *Lamellibranchiata (Pelecypoda)*—cont.																
Pholadomya. See also Homomya.																
Pholas. See Gastrochæna and Pholadidea.																
Pinna ampla, Sow.	×					5									D.	
—— granulata, Sow.						5	6	7	8						D. Be. Bu. Y.	
—— lanceolata, Sow.	×				4	5	6	7							D. to Y. 8x.	Ox. Cl.
—— mitis, Phil.		1	2	3											W. to Y.	Cor.
—— pesolina, Cont.						5									D.	
—— suprajurensis, d'Orb.									8						W. O. Bu.	
—— sp.												11				Purb.
Pisidium															D.	Purb.
Placunopsis Lycetti, de Lor									8						W. Bu.	
—— similis, Whit.															O.	Cor.
Plagiostoma. See Lima.																
Plectomya rugosa, Roem.									8						D. W.	
Pleuromya donacina, Ag.						5	6								D.	
—— tellina, Ag.						5	6		8	9					D. W. Bu. Y.	
—— Voltzi, Ag.						5	6		8	9					D. to Y.	
Plicatula Boisdini, de Lor.										9					W. Bu.	
—— echinoides, Blake										9					D. Bu.	
—— fistulosa, Mor. & Lyc.	×				4	5									D. W. C. Y.	
—— semiarmata, Etall.						5									D.	
—— tubifera, Lam.															D.	Cor.
—— weymouthiana, Damon															D.	Cor.
Protocardium isocardioides, Blake and Hudl.						5									D. Y (?).	
Psammobia tellinoides, Sow.											10	11			O. 8x.	
Pteroperna pygmæa, K. and D.	×														O. Y	Cor.
Quenstedtia lævigata, Phil.	×					5									D. to Y.	
—— —— var. gibbosa, Hudl.						5									C.	
—— sp.										9					Bu.	
Sowerbya Deshayesea, Buv.						5									D. O. Y.	
—— Dukei, Lyc. (Damon)										9					D. W.	
—— longior, Blake									8	9					W. Bu.	
—— triangularis, Phil.	×				4	5									D. W. O. Y.	

Species.	Lower Oolites.	Kellaways Rock (Am. callovienus) 1	Oxford Clay (Am. ornatus) 2	Am. cordatus 3	Corallian (Am. perarmatus) 4	Am. plicatilis 5	Kimeridge Clay Lower Beds 6	Upper Beds 7	Portland Beds Lower Beds 8	Upper Beds 9	Purbeck Beds Lower Beds 10	Middle Beds 11	Upper Beds 12	Wealden Beds.	Localities.	Remarks.
MOLLUSCA—*cont.*	×	1	2	3	4	5	6	7	8	9	10	11	12	×		
Lamellibranchiata (Pelecypoda)—cont.																
Sphæra ·															D	Portl.
Tancredia brevis, Mor. and Lyc.															O. Y	Cor.
— curtansata, Phil.	×					5									D. W. O. Y.	
— disputata, Blake and Hudl.						5									D.	
— planata, Mor. and Lyc.	×					5									D.	
— sp.															D	Kim. Cl.
Tellina. See Thracia.																
Thracia depressa, Sow.		1	2	3	4	5	6	7	8						D. to Y. 8x.
— hartwellensis, Sow.								7	8						Bu. 8x.
— incerta, Thurm.	×							7							O. Y. 8x. ?	Cor.
— *Studeri,* Ag. See T. incerta.																
— *suprajurensis,* Desh. See T. incerta.																
— tenera, Ag.									8	9					D. W. Bu.
Trichites Ploti, Lhwyd	×					5									W. Be. O. Y. 8x.
Trigonia alina, Contej.								7	8	9					D. O.
— Carrei, Mun.-Chal.								?	9						D. W. Bu.
— clavellata, Sow.					4	5									D. to Y. K.	Including *T. Bronni.* Ag.
— corallina, d'Orb.					4										D. W. Y.
— costata, Sow.	×		2			5									D. to C.
— cymba, Cont.								7	8						D.
— damoniana, de Lor.										9					D. W. O. Bu.	? var. of *T. gibbosa.*
— densinoda, Eth.												11			W.
— elongata, Sow.				3	4	6									D. B. H. C. Y.
— gibbosa, Sow.									8	9		11			D. to Bu.
— Hudlestoni, Lyc.				3	4	5									D. O. C. Y.
— incurva, Benett									8	9					D. W. Bu.
— irregularis, Seeb.		1		3					8						D. W.
— juddiana, Lyc.							6								L. 8x.
— Keepingi, Lyc.															L	Spilsby Sandstone.
— Manseli, Lyc.										9					D. W. Bu.
— marginata, Damon							6								D.
— Medlycotti, Damon						5									D.
— Meriani, Ag.						5	6								D. to Y.

Species.	Lower Oolites.	Am. callovienais.	Kellaways Rock. Am. ornatus.	Oxford Clay. Am. cordatus.	Am. perarmatus.	Corallian. Am. plicatilis.	Kimeridge Clay. Lower Beds.	Upper Beds.	Portland Beds. Lower Beds.	Upper Beds.	Purbeck Beds. Lower Beds.	Middle Beds.	Upper Beds.	Wealden Beds.	Localities.	Remarks.
	×	1	2	3	4	5	6	7	8	9	10	11	12	×		
MOLLUSCA—cont.																
Lamellibranchiata (*Pelecypoda*)—cont.																
Trigonia Micheloti, de Lor.	-	-	-	-	-	-	-	-	?	9	-	-	-	-	W. Bu.	
—— monilifera, Ag.	-	-	-	-	-	5	6	-	-	-	-	-	-	-	D. W.	Ox. Cl.
—— muricata, Goldf.	-	-	-	-	-	?	-	-	8	9	-	-	-	-	D. W. Bu.	
—— paucicosta, Lyc.	-	1	-	-	-	5	-	-	-	-	-	-	-	-	W. Bd. Y.	
—— Pellati, Mun.-Chal.	-	-	-	-	-	-	-	7	8	9	-	-	-	-	D. to C. Sx.	
—— perlata, Ag.	-	-	-	-	4	5	-	-	-	-	-	-	-	-	D. W. Be. C. Y.	Ox. Cl.
—— radiata, Benett	-	-	-	-	-	-	-	-	8	-	-	-	-	-	W.	
—— robinaldina, d'Orb.	-	-	-	-	-	-	-	-	-	-	-	-	-	-	L.	Spilsby Sandstone.
—— rupellensis, d'Orb.	-	1	-	-	-	-	-	-	-	-	-	-	-	-	W. Y.	
—— swindonensis, Blake	-	-	-	-	-	-	-	-	8	9	-	-	-	-	W. O.	
—— tealbiensis, Lyc.	-	-	-	-	-	-	-	-	-	-	-	-	-	-	L.	Spilsby Sandstone.
—— tenuitexta, Lyc.	-	-	-	-	-	-	-	-	-	9	-	-	-	-	D. W. Bu.	
—— triquetra, Seeb.	-	-	-	-	-	-	-	-	-	-	-	-	-	-	O. Y.	Cor.
—— truncata, Ag.	-	-	-	-	-	-	-	-	-	-	-	-	-	-	Sx.	Cor.
—— variegata, Credn.	-	-	-	-	-	-	-	-	8	-	-	-	-	-	W.	
—— Voltzi, Ag.	-	-	-	-	-	-	-	7	8	9	-	-	-	-	D. to L. Sx.	
—— Woodwardi, Lyc.	-	-	-	-	-	-	-	7	-	-	-	-	-	-	D. W.	
Unicardium circulare, d'Orb.	-	-	-	-	-	-	-	-	-	9	-	-	-	-	D. Bu.	
—— depressum, Phil.	×	-	-	-	4	5	-	-	-	-	-	-	-	-	W. C.	
—— excentricum, d'Orb.	-	-	-	-	-	-	-	7	-	-	-	-	-	-	Sx.	
—— gibbosum, Mor. & Lyc.	×	1	-	-	-	-	-	-	-	-	-	-	-	-	W. C. Y.	Cor.
—— plenum, Blake & Hudl.	-	-	-	-	-	5	-	-	-	-	-	-	-	-	W. Y.	
—— sulcatum, Leck.	-	1	-	-	-	5	6	-	8	-	-	-	-	-	D. W. G. Y.	
Unio compressus, Sow.	-	-	-	-	-	-	-	-	-	-	-	-	12	×	D.	
—— Valdensis, Mantell	-	-	-	-	-	-	-	-	-	-	-	-	12	×	D.	
Venus. See Cyprina.																
BRACHIOPODA.																
Discina elevata, Blake	-	-	-	-	-	-	6	-	-	-	-	-	-	-	L.	
—— humphriesiana, Dav.	-	-	-	-	-	5	-	7	8	-	-	-	-	-	D. to Y. Sx.	
—— latissima, Dav.	-	-	-	-	-	-	6	7	-	-	-	-	-	-	D. O. to Y. Sx.	Ox. Cl.
Dictyothyris. See Rhynchonella.																
Lingula Brodiei, Dav.	-	-	-	-	-	5	-	-	-	-	-	-	-	-	O.	
—— Cranœ, Dav.	-	-	2	-	-	-	-	-	-	-	-	-	-	-	W. Bu.	

Species	Lower Oolites (×)	Kellaways Rock — Am. calloviensis (1)	Oxford Clay — Am. ornatus (2)	Oxford Clay — Am. cordatus (3)	Corallian — Am. perarmatus (4)	Corallian — Am. plicatilis (5)	Kimeridge Clay — Lower Beds (6)	Kimeridge Clay — Up er Beds (7)	Portland Beds — Lower Beds (8)	Portland Beds — Up er Beds (9)	Purbeck Beds — Lower Beds (10)	Purbeck Beds — Middle Beds (11)	Purbeck Beds — Up er Beds (12)	Wealden Beds (×)	Localities	Remarks
BRACHIOPODA—cont.																
Lingula ovalis, Sow.	–	–	–	–	–	5	6	7	8	–	–	–	–	–	D. O. to Y. 8x.	
Orbicula. See Discina.																
Rhynchonella corallina, Leym.	–	–	–	–	–	5	–	–	–	–	–	–	–	–	D. W.	
— inconstans, Sow.	–	–	–	–	–	5	6	–	–	–	–	–	–	–	D. to Y.	
— lacunosa, Schloth.	–	–	–	–	–	–	–	–	–	–	–	–	–	–	D. Y.	Cor.
— lævirostris, McCoy	–	–	2	–	–	–	–	–	–	–	–	–	–	–	H. N.	
— obsoleta, Sow.	×	–	–	–	–	–	–	–	–	–	–	–	–	–	D.	Cor.
— pinguis, Roem.	–	–	–	–	–	5	–	–	–	–	–	–	–	–	D. 8x.	
— — var. pectunculoides, Etall.	–	–	–	–	–	5	–	–	–	–	–	–	–	–	D.	
— portlandica, Blake	–	–	–	–	–	–	–	–	8	–	–	–	–	–	D.	
— subvariabilis, Dav.	–	–	–	–	–	–	6	7	8	–	–	–	–	–	W. Bu. Y. 8x.	
— varians, Schloth.	×	1	2	3	–	–	–	–	–	–	–	–	–	–	D. to Y.	Cor.
— — var. Smithi, Walk.	–	–	4	–	–	–	–	–	–	–	–	–	–	–	O.	Ox. Cl.
— — var. socialis, Phil.	–	1	2	3	–	–	–	–	–	–	–	–	–	–	W. Y. K.	
— — var. spathica, Lam.	–	1	–	3	–	–	–	–	–	–	–	–	–	–	W. O. H. K.	
— — var. Thurmanni, Voltz.	–	–	–	–	4	5	–	–	–	–	–	–	–	–	D. W. Y.	Ox. Cl.
Terebratula bisuffarcinata, Ziet.	–	–	–	–	–	–	7	–	–	–	–	–	–	–	D.	
— insignis, Schübl.	–	–	–	–	4	5	–	–	–	–	–	–	–	–	D. to Y.	Ox. Cl.
— — var. maltonensis, Oppel.	–	–	–	–	–	5	–	–	–	–	–	–	–	–	C. Y.	
— intermedia, Sow.	×	–	–	–	–	–	–	–	–	–	–	–	–	–	D.	Ox. Cl.
— oxoniensis, Walk.	–	–	–	3	–	–	–	–	–	–	–	–	–	–	H.	
— subsella, Leym.	–	–	–	–	–	5	–	–	–	–	–	–	–	–	D. W.	
Thecideum ornatum, Moore	–	–	–	–	–	5	–	–	–	–	–	–	–	–	W.	
— pygmæum, Moore	–	–	–	–	–	–	–	–	–	–	–	–	–	–	W.	Cor.
— triangulare, d'Orb.	–	–	–	–	–	–	–	–	–	–	–	–	–	–	W. Y.	Cor.
Waldheimia bernardiana, d'Orb.	–	–	2	–	–	–	–	–	–	–	–	–	–	–	C.	
— boloniensis, Sauv. & Rig.	–	–	–	–	–	–	–	8	9	–	–	–	–	–	D. O. Bu. Y.	Ox. Cl.? : Cor.
— bucculenta, Sow.	–	–	–	–	4	–	–	–	–	–	–	–	–	–	O. C. H. Y.	
— dorsetensis, Dav.	–	–	–	–	–	5	–	–	–	–	–	–	–	–	D.	
— Gesneri, Etall.	–	–	–	–	–	–	6	–	–	–	–	–	–	–	O. L. Y.	Cor.
— Hudlestoni, Walk.	–	–	–	–	4	–	–	–	–	–	–	–	–	–	C. Y.	
— impressa, von Buch.	–	–	–	3	–	–	–	–	–	–	–	–	–	–	O. N H. Y.	
— lampas, Sow.	–	–	–	–	–	5	–	–	–	–	–	–	–	–	D.	

Species.	Lower Oolites.	Am. calloviensis, Kellaways Rock.	Oxford Clay.	Am. ornatus.	Am. cordatus.	Am. perarmatus, Corallian.	Am. plicatilis.	Kimeridge Clay Lower Beds.	Upper Beds.	Purbeck Beds Lower Beds.	Lower Beds.	Middle Beds.	Purbeck Beds Upper Beds.	Wealden Beds.	Localities.	Remarks.
	×	1	2	3	4	5	6	7	8	9	10	11	12	×		
BRACHIOPODA—*cont.*																
Waldheimia margarita, Oppel.						5									W. Be. Y.	
— obovata, Sow.	×	1		3											W. to Y.	
— ornithocephala, Sow.	×	1													D. W. Y.	? Cor.
— — var. calloviensis, Dav.		1													W.	W. Kelloway
— umbonella, Lam.		1													W. Y.	
Zellania globata, Moore															W.	Cor.
POLYZOA.																
Diastopora diluviana, M. Edw.	×														O.	Ox. Cl.
Lepralia															Bu.	Kim. Cl.
Parendea bullata, Etall.						5									W.	
CRUSTACEA.																
Macroura.																
Callianassa isochela, H. Woodw.							6								Sx.	
Eryma Babeaui, Etallon				3		5									D. H.	
— Georgei, Carter				3											H.	
— Mandelslohi, Meyer				3											H.	
— ? pulchella, Carter				3											H.	
— ventrosa, Meyer				3											H.	
— Villersi, Moriére				3											H.	
— sp.										9					Bu.	
Eryon sublevis, Carter				3											H.	
Gastrosacus Wetzleri. Meyer						5									C.	
Glyphea ferruginea, Blake and Hudl.						5									D.	
— hispida, Carter				3											H.	
— regleyana, Meyer				3											H.	
— rostrata, Phil.	×		2												O. H. Y.	Cor.
— scabrosa, Phil.				3											H. Y.	
— sp.										9					Bu.	
Magila dissimilis, Carter				3											H.	
— levimana, Carter				3											H.	
— Pichleri, Oppel.				3											H.	

Species	Lower Oolites (x)	Am. calloviensis (1)	Kellaways Rock (2)	Am. ornatus / Oxford Clay (3)	Am. cordatus (4)	Corallian (5)	Kimeridge Clay Lower Beds (6)	Upper Beds (7)	Portland Beds Lower Beds (8)	Upper Beds (9)	Purbeck Beds Lower Beds (10)	Middle Beds (11)	Upper Beds (12)	Wealden Beds (x)	Localities	Remarks
CRUSTACEA—*cont.*																
Macroura—cont.																
Mecocheirus Pearcei, McCoy			2												W. Y.	
—— Peytoni, H. Woodw.							6								Sx.	
—— socialis, Meyer				3											H. Y.	
Pseudastacus ? socialis, Carter.				3											H.	
Pseudoglyphea							6								Sx.	
Brachyura.																
Goniocheirus cristatus, Carter.				3	4										H. Y.	
Prosopon marginatum, Meyer						5									C.	
Isopoda.																
Archæoniscus Brodiei, M. Edw.											10	11			D. W.	
Cirripedia.																
Pollicipes concinnus, Morris		1	2												W. Y. Sx.	Kim. Cl.
—— Hausmanni, Koch and Dunk.							6								C. Y.	
—— planulatus, Morris			2												W.	
Scalpellum reticulatum, Blake.							6								W.	
Ostracoda.																
Bairdia			2												K.	
Candona ansata, Jones										9	10				D. W. Bu.	
—— bononiensis, Jones										9	10				D. W. Bu.	
Cypridea Dunkeri, Jones											10	11	12	x	D. W. Sx.	
—— granulosa, Sow.												11			D. W. Sx.	
—— —— var. fasciculata, Forbes.												11			D. W. Sx.	
—— —— var. paucigranulata, Jones.												11			D.	
—— punctata, Forbes											10	11	12		D. W. Sx.	
—— —— var. gibbosa, Forbes												?	12		D.	
—— —— var. posticalis, Jones												11	12		D.	
—— spinigera, Sow.			2											x	W.	Also Oligocene.
—— tuberculata, Sow.												11	12	x	D.	
—— —— var. adjuncta, Jones													12		D.	
—— valdensis, Fitton												11	12	x	D. W. Sx.	
—— ventrosa, Jones												11	12		D.	

SPECIES.	×	Kellaways Rock. Am. calloviensis.	Oxford Clay. Am. ornatus.	Am. cordat us.	Coralline. Am. perarmatus.	Am. plicatilis.	Kimeridge Clay. Lower Beds.	Upper Beds.	Portland Beds. Lower Beds.	Up er Beds.	Purbeck Beds. Lower Beds.	Middle Beds.	Upper Beds.	Wealden Beds. ×	LOCALITIES.	REMARKS.
	×	1	2	3	4	5	6	7	8	9	10	11	12	×		
CRUSTACEA—*cont.*																
Ostracoda—cont.																
Cypridea, ventrosa var. globosa, Jones.												11			D.	
Cyprione Bristovi, Jones												11	12	×	D. W.	
Cypris purbeckensis, Forbes										9	10				D. W. Bu.	
—— See also Metacypris.																
Cythere retirugata, Jones										9	10				D. W. Bu.	
—— var. rugulata, Jones										9	10				W. Bu.	
—— var. textilis, Jones											10				Bu.	
—— transiens, Jones										9	10				W.	
Cythereis lonsdaleana, Jones										9					D.	
Cytherella							6								D. L.	
Cytheridea							6								D. C. L.	
Darwinula leguminella, Forbes.												11	12	×	D. W. Sx.	
Metacypris Forbesi, Jones												11			D.	*Cypris. striato-punctata* Forbes.
—— var. verrucosa, Jones												11			D.	
Phyllopoda.																
Estheria Andrewsi, Jones												11			W.	
—— elliptica, Dunk, var. subquadrata, Sow.											?	11		×	W. Sx.	
INSECTA.																
Coleoptera.																
Agrilus cyllabacus, Westw.											10	11			D.	Agrilium.
—— cyllarus, Westw.											10	11			D.	
—— strombus, Westw.											10				D.	
—— stomphax, Westw.											10	11			D.	
—— See also Buprestium.																
Ancylocheira teleas, Westw.											10	11			D.	
Buprestium bolbus, Westw.											10	11			D.	
—— dardanus, Westw.											10	11			D.	
—— gorgus, Westw.											10	11			D.	
—— stygnus, Westw.											10	11			D.	
—— teleas, Westw. See Ancylocheira.																
—— valgus, Westw											10	11			D.	

Species.	Lower Oolite (×)	Am. calloviensis (1)	Am. ornatus / Kellaways Rock (2)	Am. cordatus / Oxford Clay (3)	Am. perarmatus / Corallian (4)	Am. plicatilis (5)	Lower Beds / Kimeridge Clay (6)	Upper Beds (7)	Lower Beds / Portland Beds (8)	Upper Beds (9)	Lower Beds / Purbeck Beds (10)	Middle Beds (11)	Upper Beds (12)	Wealden: Beds (×)	Localities.	Remarks.
INSECTA—cont.																
Coleoptera—cont.																
Buprestium Woodleyi, West.	×	-	-	-	-	-	-	-	-	-	10	11	-	-	D.	
Camptodontus angliæ, Gieb.		-	-	-	-	-	-	-	-	-	10	-	-		W.	
Carabidium dejeanianum, Westw.		-	-	-	-	-	-	-	-	-	10	11	-	-	D.	
—— elongatum, Brod.		-	-	-	-	-	-	-	-	-	10	-	-		W.	
—— Westwoodi, Gieb.		-	-	-	-	-	-	-	-	-	10	11	-	-	D.	
Cerylon striatum, Brod.		-	-	-	-	-	-	-	-	-	10	-	-		W -	Purb.
Chrysomela dubia, Gieb.		-	-	-	-	-	-	-	-	-	10	11	-	-	D.	
—— Dunkeri, Gieb.		-	-	-	-	-	-	-	-	-	10	11	-	-	D.	
—— ignota, Gieb.		-	-	-	-	-	-	-	-	-	10	11	-	-	D.	
Coccinella Neptuni, Gieb.		-	-	-	-	-	-	-	-	-	10	11	-	-	D.	
Colymbetes. See Hydrophilus.																
Curculionites syrichthus, Westw.		-	-	-	-	-	-	-	-	-	10	11	-	-	D.	
—— tuberculatus, Gieb.		-	-	-	-	-	-	-	-	-	10	11	-	-	D.	
—— Westwoodi, Gieb.		-	-	-	-	-	-	-	-	-	10	11	-	-	D.	
Crypticus Ungeri, Gieb.		-	-	-	-	-	-	-	-	-	10	11	-	-	D.	
Ctenicerium blissus, Westw.		-	-	-	-	-	-	-	-	-	10	11	-	-	D.	
—— hylastes, Westw.		-	-	-	-	-	-	-	-	-	10	11	-	-	D.	
Cymindis antiqua, Gieb.		-	-	-	-	-	-	-	-	-	10	11	-	-	D. W.	
—— Beyrichi, Gieb.		-	-	-	-	-	-	-	-	-	10	11	-	-	D.	
Cyphon vetustus, Gieb.		-	-	-	-	-	-	-	-	-	10	-	-		W.	
Diaperidium mithrax, Westw.		-	-	-	-	-	-	-	-	-	10	11	-	-	D. W.	
Dytiscus		-	-	-	-	-	-	-	-	-	-	11	-	-	D.	
Elaterium barypus, Westw.		-	-	-	-	-	-	-	-	-	10	11	-	-	D.	
—— Morrisi, Gieb,		-	-	-	-	-	-	-	-	-	10	11	-	-	D.	
—— Oweni, Gieb.		-	-	-	-	-	-	-	-	-	10	11	-	-	D.	
—— pronæus, Westw.		-	-	-	-	-	-	-	-	-	10	11	-	-	D.	
—— purbeckensis, Gieb.		-	-	-	-	-	-	-	-	-	10	-	-		W.	
—— triopas, Westw.		-	-	-	-	-	-	-	-	-	10	11	-	-	D.	
—— Werneri, Gieb.		-	-	-	-	-	-	-	-	-	10	-	-		W. Bu.	
Elmis Brodiei, Gieb.		-	-	-	-	-	-	-	-	-	10	-	-		W.	
Harpalidium anactus, Westw.		-	-	-	-	-	-	-	-	-	10	11	-	-	D -	Harpalus.
—— Burmeisteri, Gieb.		-	-	-	-	-	-	-	-	-	10	11	-	-	D.	
—— Ewaldi, Gieb.		-	-	-	-	-	-	-	-	-	10	11	-	-	D.	

SPECIES.	Lower Oolites.	Kellaways Rock. Am. calloviensis.	Oxford Clay. Am. ornatus.	Am. cordatus.	Corallian. Am. perarmatus.	Am. plicatilis.	Kimeridge Clay. Lower Beds.	Upper Beds.	Portland Beds. Lower Beds.	Upper Beds.	Purbeck Beds. Lower Beds.	Middle Beds.	Upper Beds.	Wealden Beds.	LOCALITIES.	REMARKS.
	x	1	2	3	4	5	6	7	8	9	10	11	12	x		
INSECTA—*cont.*																
Coleoptera—cont.																
Harpalidium Knorri, Gieb.	-	-	-	-	-	-	-	-	-	-	10	11	-	-	D.	
— nothrus, Westw.	-	-	-	-	-	-	-	-	-	-	10	11	-	-	D.	
Helophorus antiquus, Gieb.	-	-	-	-	-	-	-	-	-	-	10	-	-	-	W.	
— Brodiei, Gieb.	-	-	-	-	-	-	-	-	-	-	10	-	-	-	W.	
Helopium agabus, Westw.	-	-	-	-	-	-	-	-	-	-	10	11	-	-	D.	Helopidium.
— neoridas, Westw.	-	-	-	-	-	-	-	-	-	-	10	-	-	-	D.	
Hydrobius purbeckensis, Gieb.	-	-	-	-	-	-	-	-	-	-	10	-	-	-	W.	
Hydrophilus Brodiei, Gieb.	-	-	-	-	-	-	-	-	-	-	10	-	-	-	W.	
— Westwoodi, Gieb.	-	-	-	-	-	-	-	-	-	-	10	-	-	-	W.	
Hydrophorus Neptuni, Gieb.	-	-	-	-	-	-	-	-	-	-	10	11	-	-	D.	
Hypera antiqua, Gieb	-	-	-	-	-	-	-	-	-	-	10	-	-	-	W.	
Lamia Schroeteri, Gieb.	-	-	-	-	-	-	-	-	-	-	10	11	-	-	D.	
Limnius	-	-	-	-	-	-	-	-	-	-	10	-	-	-	W.	
Prionus antiquus, Gieb.	-	-	-	-	-	-	-	-	-	-	10	11	-	-	D.	
Telephorium abgarus, Westw.	-	-	-	-	-	-	-	-	-	-	10	11	-	-	D.	
Tenebrio rugosa-striatus, Gieb.	-	-	-	-	-	-	-	-	-	-	10	-	-	-	W.	
Tentyridium peleus, Westw,	-	-	-	-	-	-	-	-	-	-	10	11	-	-	D.	
(Elytra of beetles)	-	-	-	-	-	-	-	7	-	-	-	-	-	-	D.	
Hymenoptera.																
Formicium Brodiei, Westw.	-	-	-	-	-	-	-	-	-	-	10	11	-	-	D.	Giant ant.
Myrmicium Heeri, Westw.	-	-	-	-	-	-	-	-	-	-	10	11	-	-	D.	Myrmica.
Ponera Brodiei, Westw. See Formicium.																
Neuroptera.																
Abia duplicata, Gieb.	-	-	-	-	-	-	-	-	-	-	10	11	-	-	D.	
— See also Pteroblattina (Orthoptera).																
Æschna perampla, Brodie	-	-	-	-	-	-	-	-	-	-	10	-	-	-	W.	
— sp.	-	-	-	-	-	-	-	-	-	-	10	-	-	-	D.	
Æschnidium bubas, Westw. See Libellulium antiquum.																
Agrionidium aetna, Westw.	-	-	-	-	-	-	-	-	-	-	10	11	-	-	D.	Agrion.
Chimarrha pytho, Westw. See Phryganidium.																
Corydalis. See Orthopblebia.																
Cyllonium boisduvalianum, Westw.	-	-	-	-	-	-	-	-	-	-	10	11	-	-	D.	

Species.	Lower Oolite.	Am. callloviensis, Kellaways Rock.	Am. ornatus.	Am. cordatus.	Am. perarmatus.	Am. plicatilis.	Lower Beds.	Upper Beds.	Lower Beds.	Upper Beds.	Lower Beds.	Middle Beds.	Upper Beds.	Wealden Beds.	Localities.	Remarks.
	×	1	2	3	4	5	6	7	8	9	10	11	12	×		
Insecta—cont.																
*Neuroptera—*cont.																
Cyllonium hewitsonianum, Westw.	-	-	-	-	-	-	-	-	-	-	10	11	-	-	D.	
Ela brephos, Westw.	-	-	-	-	-	-	-	-	-	-	10	11	-	-	D.	
Elcana. See under *Orthoptera.*																
—— See also Panorpidium.																
Estemoa. See Æschnidium.																
Flata Haidingeri, Gieb.	-	-	-	-	-	-	-	-	-	-	10	-	-	-	W.	
Gomphus petrificatus, Hagen	-	-	-	-	-	-	-	-	-	-	10	-	-	-	W.	
Hagla ignota, Westw.	-	-	-	-	-	-	-	-	-	-	10	11	-	-	D.	
Leptoceridæ	-	-	-	-	-	-	-	-	-	-	10	-	-	-	W.	
Libellulium agrias, Westw.	-	-	-	-	-	-	-	-	-	-	10	11	-	-	D.	Libellula.
—— antiquum, Brod.	-	-	-	-	-	-	-	-	-	-	10	11	-	-	D. W.	
—— jurassica, Gieb.	-	-	-	-	-	-	-	-	-	-	10	-	-	-	W.	
—— Kaupi, Westw.	-	-	-	-	-	-	-	-	-	-	10	11	-	-	D.	
—— *petrificata,* Hagen. See Gomphus.																
Lindenia. See Libellula.																
Meloe Hoernesi, Gieb. See Termes.																
Orthophlebia bifurcata, Gieb.	-	-	-	-	-	-	-	-	-	-	10	-	-	-	W.	
—— minuta, Gieb.	-	-	-	-	-	-	-	-	-	-	10	-	-	-	W.	
Panorpidium gracile, Gieb.	-	-	-	-	-	-	-	-	-	-	10	-	-	-	W.	Panorpa.
—— *tesselatum,* Westw. See Elcana.																
Philonthus Kneri, Gieb. See Termes grandævus.																
Phryganidium pytho, Westw.	-	-	-	-	-	-	-	-	-	-	10	11	-	-	D.	
Prognatha crassa, Gieb. See Termes grandævus.																
Raphidium. See Ela brephos.																
Sialina. See Ela brephos.																
Termes grandævus. Brod.	-	-	-	-	-	-	-	-	-	-	10	-	-	-	W.	
Termitidium ignotum. Westw. See Hagla.																
Trichoptera. See Phryganidium.																
Zalmona Brodiei, Gieb.	-	-	-	-	-	-	-	-	-	-	10	-	-	-	W.	Purb.
Lepidoptera ?	-	-	-	-	-	-	-	-	-	-	10	-	-	-	D.	"Butterfly."
Cyllonium. See under Neuroptera.																
Diptera.																
Adonia Fittoni, Brod.	-	-	-	-	-	-	-	-	-	-	10	-	-	-	W.	

E 82428.

Species.	Lower Oolites.	Am. callovienus, Kellaways Rock.	Oxford Clay.		Corallian.		Kimeridge Clay.		Portland Beds.		Purbeck Beds.			Wealden Beds.	Localities.	Remarks.
			Am. ornatus.	Am. cordatus.	Am. perarmatus.	Am. plicatilis.	Lower Beds.	Upper Beds.	Lower Beds.	Upper Beds.	Lower Beds.	Middle Beds.	Upper Beds.			
	×	1	2	3	4	5	6	7	8	9	10	11	12	×		
INSECTA—*cont.*																
Diptera—cont.																
Asuba Brodiei, Gieb.	–	–	–	–	–	–	–	–	–	–	10	–	–	–	W.	
Bria prisca, Brod. See Rhyphus.																
Campylomyza. See Cecidomium.																
Cecidomium grandævum, Westw.	–	–	–	–	–	–	–	–	–	–	10	11	–	–	D.	
Chenesia. See Macropeza																
Chironomus arrogans, Gieb.	–	–	–	–	–	–	–	–	–	–	10	–	–	–	W.	
—— extinctus, Brod.	–	–	–	–	–	–	–	–	–	–	10	–	–	–	W.	
Corethrium pertinax, Westw.	–	–	–	–	–	–	–	–	–	–	10	11	–	–	D.	
Culex fossilis, Brod.	–	–	–	–	–	–	–	–	–	–	10	–	–	–	W.	
Dara. See Culex.																
Hasmona leo, Gieb.	–	–	–	–	–	–	–	–	–	–	10	–	–	–	W.	*Empidæ*
Macrocera rustica, Brod. See Sama.																
Macropeza prisca, Gieb.	–	–	–	–	–	–	–	–	–	–	10	–	–	–	W.	
Platyura Pittoni, Brod. See Adonia.																
Remalia sphinx, Gieb.	–	–	–	–	–	–	–	–	–	–	10	–	–	×	W.	
Rhyphus priscus, Brod.	–	–	–	–	–	–	–	–	–	–	10	–	–	–	W.	
Sama rustica, Bird.	–	–	–	–	–	–	–	–	–	–	10	–	–	–	W.	
Sciophila ossa, Brod. See Thimna.																
Simulidium humidum, Brod.	–	–	–	–	–	–	–	–	–	–	10	–	–	–	W.	Simulium.
—— priscum, Westw.	–	–	–	–	–	–	–	–	–	–	10	11	–	–	D.	
Tanypus dubius, Brod.	–	–	–	–	–	–	–	–	–	–	10	–	–	–	W.	
Thimna defossa, Brod.	–	–	–	–	–	–	–	–	–	–	10	–	–	–	W.	
Thiras Westwoodi, Gieb.	–	–	–	–	–	–	–	–	–	–	10	11	–	–	D.	
Tipula	–	–	–	–	–	–	–	–	–	–	10	–	–	–	D.	
Rhynchota (*Hemiptera, Homoptera*).																
Aphis dubia, Gieb.	–	–	–	–	–	–	–	–	–	–	10	–	–	×	W.	
—— plana, Brod.	–	–	–	–	–	–	–	–	–	–	10	–	–	–	W.	
—— valdensis, Brod.	–	–	–	–	–	–	–	–	–	–	10	–	–	–	W.	
Asira Egertoni, Brod.	–	–	–	–	–	–	–	–	–	–	10	–	–	–	W.	Asiraca.
—— *Kenngotti*, Gieb. See A. Egertoni.																
Cercopidium Hahni, Westw.	–	–	–	–	–	–	–	–	–	–	10	11	–	–	D.	
—— lanceolata, Heer	–	–	–	–	–	–	–	–	–	–	10	–	–	–	W.	

Species.	Lower Oolites.	Am. calloviensis. / Kellaways Rock.	Am. ornatus.	Am. cordatus.	Am. perarmatus.	Am. plicatilis.	Lower Beds.	Upper Beds.	Lower Beds.	Upper Beds.	Lower Beds.	Middle Beds.	Upper Beds.	Wealden Beds.	Localities.	Remarks.
	×	1	2	3	4	5	6	7	8	9	10	11	12	×		
Insecta—*cont.*																
Rhynchota (Hemiptera, Homoptera)—cont.																
Cercopidium mimas, Westw.	-	-	-	-	-	-	-	-	-	-	10	11	-	-	D.	
—— Schaefferi, Westw.	-	-	-	-	-	-	-	-	-	-	10	11	-	-	D.	
—— signoreti, Westw.	-	-	-	-	-	-	-	-	-	-	10	11	-	-	D.	
—— telesphorus, Westw.	-	-	-	-	-	-	-	-	-	-	10	11	-	-	D.	
—— trigonale, Gieb.	-	-	-	-	-	-	-	-	-	-	10	11	-	-	D.	
Cercopsis. See Cercopidium.																
Cicadellium dipsas, Westw.	-	-	-	-	-	-	-	-	-	-	10	11	-	-	D.	Cicada.
—— psocus, Westw.	-	-	-	-	-	-	-	-	-	-	10	11	-	-	D.	
—— punctatum, Brod.	-	-	-	-	-	-	-	-	-	-	10	-	-	-	W.	
Cimicidium Dallasi, Westw. See Lygæites.																
Cixius maculatus, Brod.	-	-	-	-	-	-	-	-	-	-	10	-	-	-	W.	
Delphax pulcher, Brod.	-	-	-	-	-	-	-	-	-	-	10	-	-	-	W.	
Hydrometra	-	-	-	-	-	-	-	-	-	-	10	-	-	-	W.	
Kleidocerys	-	-	-	-	-	-	-	-	-	-	10	-	-	-	W.	
Lygæites Dallasi, Westw.	-	-	-	-	-	-	-	-	-	-	10	11	-	-	D.	
—— furcatus, Gieb.	-	-	-	-	-	-	-	-	-	-	10	11	-	-	D.	
—— priscus, Gieb.	-	-	-	-	-	-	-	-	-	-	10	11	-	-	D.	
Molobius	-	-	-	-	-	-	-	-	-	-	10	-	-	-	D.	
Nepidium stolones, Westw.	-	-	-	-	-	-	-	-	-	-	10	11	-	-	D.	
Neurocoris	-	-	-	-	-	-	-	-	-	-	10	11	-	-	D.	
Pachymerus. See Kleidocerys.																
Ricania fulgens, Brod.	-	-	-	-	-	-	-	-	-	-	10	-	-	-	W.	
Velia	-	-	-	-	-	-	-	-	-	-	10	-	-	-	W.	
Orthoptera.																
Abia sipylus, Westw. See Pterinoblattina.																
Acheta Sedgwicki, Brodie	-	-	-	-	-	-	-	-	-	-	10	-	-	-	W.	
Aporoblattina anceps, Westw.	-	-	-	-	-	-	-	-	-	-	10	11	-	-	D.	Blattina.
—— eatoni, Scud.	-	-	-	-	-	-	-	-	-	-	-	-	-	-	Locality unknown.	Purb., England.
—— exigua, Scud.	-	-	-	-	-	-	-	-	-	-	10	-	-	-	D.	
—— Kollari, Gieb.	-	-	-	-	-	-	-	-	-	-	10	-	-	×	W.	Blatta.
—— Maclachlani, Scud.	-	-	-	-	-	-	-	-	-	-	-	-	-	-	Locality unknown.	Purb., England.
—— recta, Gieb.	-	-	-	-	-	-	-	-	-	-	10	11	-	×	D. W.	

SPECIES.	Lower Oolites.	Am. calloviensis. (Kellaways Rock)	Am. ornatus. (Oxford Clay)	Am. cordatus.	Am. perarmatus. (Corallian)	Am. plicatilis.	Lower Beds. (Kimeridge Clay)	Upper Beds.	Lower Beds. (Portland Beds)	Middle Beds.	Lower Beds. (Purbeck Beds)	Middle Beds.	Upper Beds.	Wealden Beds.	LOCALITIES.	REMARKS.
	×	1	2	3	4	5	6	7	8	9	10	11	12	×		
INSECTA—*cont.*																
Orthoptera—cont.																
Aporoblattina Westwoodi, Scud.	–	–	–	–	–	–	–	–	–	–	–	–	–	–	D.	Purb.
Bittacus dubius Brod.	–	–	–	–	–	–	–	–	–	10	–	–	–	–	W.	Locustariæ.
Blattidium acholous, Westw.	–	–	–	–	–	–	–	–	–	10	11	–	–	–	D.	Blatta.
—— nogaus, Westw.	–	–	–	–	–	–	–	–	–	10	11	–	–	–	D.	
—— pinna, Gieb.	–	–	–	–	–	–	–	–	–	10	–	–	–	–	W.	
—— symyrus, Westw.	–	–	–	–	–	–	–	–	–	10	11	–	–	–	D.*	
—— See also Rithma.																
Blattina. See Aporoblattina, Diechoblattina, Mesoblattina.																
Clathrotermes. See Bittacus.																
Corydalis. See Blattidium.																
Ctenoblattina arcta, Scud.	–	–	–	–	–	–	–	–	–	–	–	–	–	–	Loc. unknown.	Purb. England.
—— See also Blattidium.																
Diechoblattina Ungeri, Gieb.	–	–	–	–	–	–	–	–	–	10	–	–	–	–	D. W.	
—— Wallacei, Scud.	–	–	–	–	–	–	–	–	–	–	–	–	–	–	Loc. unknown.	Purb. England.
Dipluroblattina Bailyi, Scud.	–	–	–	–	–	–	–	–	–	–	–	–	–	–	Loc. unknown.	Purb. England.
Elcana tessellata, Westw.	–	–	–	–	–	–	–	–	–	10	11	–	–	–	D.	{ Locustariæ. *E. Beyrichi*, Gi
—— See also Bittacus.																
Elisama Bucktoni, Scud.	–	–	–	–	–	–	–	–	–	–	–	–	–	–		Purb.
—— Kirkbyi, Scud.	–	–	–	–	–	–	–	–	–	–	–	–	–	–		Mesozoic, Engla
—— Kueri, Gieb.	–	–	–	–	–	–	–	–	–	10	–	–	×		W.	
—— minor, Gieb.	–	–	–	–	–	–	–	–	–	10	–	–	×		W.	
—— molossus, Westw.	–	–	–	–	–	–	–	–	–	10	11	–	–	–	D.	
—— See also Blattidium.																
Gryllidium Oweni, Westw.	–	–	–	–	–	–	–	–	–	10	11	–	–	–	D.	Gryllus.
—— See also Acheta.																
Mesoblattina antiqua, Gieb.	–	–	–	–	–	–	–	–	–	10	11	–	–	–	D.	
—— Brodiei, Scud.	–	–	–	–	–	–	–	–	–	–	–	–	–	–		Purb.
—— Bucklandi, Scud	–	–	–	–	–	–	–	–	–	–	–	–	–	–		Purb.? England.
—— elongata, Gieb.	–	–	–	–	–	–	–	–	–	10	11	–	–	–	D.	
—— Higginsi, Scud.	–	–	–	–	–	–	–	–	–	–	–	–	–	–	Loc. unknown.	Purb. England.
—— Hopei, Scud.	–	–	–	–	–	–	–	–	–	–	–	–	–	–	Loc. unknown.	Purb. England.
—— Mantelli, Scud.	–	–	–	–	–	–	–	–	–	–	–	–	–	–	Loc. unknown.	Purb. England.

SPECIES.	Lower Oolite.	Am. callovienis. (Kellaways Rock)	Am. ornatus.	Am. cordatus. (Oxford Clay)	Am. perarmatus. (Corallian)	Am. plicatilis.	Lower Beds. (Kimeridge Clay)	Upper Beds.	Lower Beds. (Portland Beds)	Upper Beds.	Lower Beds. (Purbeck Beds)	Middle Beds.	Upper Beds.	Wealden Beds.	LOCALITIES.	REMARKS.
	×	1	2	3	4	5	6	7	8	9	10	11	12	×		
INSECTA—*cont.*																
Orthoptera—cont.																
Mesoblattina Murchisoni, Gieb.	-	-	-	-	-	-	-	-	-	-	10	11	-	-	D.	
—— Murrayi, Scud.	-	-	-	-	-	-	-	-	-	-	-	-	-	-	Loc. unknown.	Purb. England.
—— Peachi, Scud.	-	-	-	-	-	-	-	-	-	-	-	-	-	-	Loc. unknown.	Purb. England.
—— Swintoni, Scud	-	-	-	-	-	-	-	-	-	-	-	-	-	-	Loc. unknown.	Purb. England.
Nannoblattina Prestwichi, Scud.	-	-	-	-	-	-	-	-	-	-	-	-	-	-	-	Purb.
—— similis, Gieb.	-	-	-	-	-	-	-	-	-	-	10	-	-	-	W.	
—— Woodwardi, Scud.	-	-	-	-	-	-	-	-	-	-	-	-	-	-	W.	Purb.
—— See also Blattidium.																
Nethania. See Blattidium and Elisama.																
Panorpidium. See Elcana.																
Pterinoblattina Binneyi, Scud.	-	-	-	-	-	-	-	-	-	-	10	-	-	-	D.	
—— penna, Scud.	-	-	-	-	-	-	-	-	-	-	-	-	-	-	Loc. unknown.	Purb. England.
—— pluma, Gieb.	-	-	-	-	-	-	-	-	-	-	10	11	-	-	D.	
—— sipylus, Westw.	-	-	-	-	-	-	-	-	-	-	10	11	-	-	D.	
Rithma Daltoni, Scud.	-	-	-	-	-	-	-	-	-	-	-	-	-	-	Loc. known.	Purb. England.
—— disjuncta, Scud.	-	-	-	-	-	-	-	-	-	-	-	-	-	-	W.	Purb.
—— Gossi, Scud.	-	-	-	-	-	-	-	-	-	-	-	-	-	-	Loc. unknown.	Purb. England.
—— minima, Scud.	-	-	-	-	-	-	-	-	-	-	-	11	-	-	D.	
—— Morrisi, Gieb.	-	-	-	-	-	-	-	-	-	-	10	11	-	-	D.	
—— purbeckensis, Gieb.	-	-	-	-	-	-	-	-	-	-	10	11	-	-	D.	
—— ramificata, Gieb.	-	-	-	-	-	-	-	-	-	-	10	11	-	-	D.	
—— Stricklandi, Brod.	-	-	-	-	-	-	-	-	-	-	10	-	-	-	W.	
—— Westwoodi, Gieb.	-	-	-	-	-	-	-	-	-	-	10	11	-	-	D.	
—— See also Blattidium and Mesoblattina.																
Stalium. See Abia and Pterinoblattina.																
ANNELIDA.																	
Serpula coacervata, Blum.	-	-	-	-	-	-	-	-	9	-	11	-	-	D.	
—— gordialis, Schloth.	-	-	-	-	5	-	-	9	-	-	-	-	D. Bu.	
—— intestinalis, Phil.	×	-	-	-	-	5	6	7	-	-	-	-	-	D. Bd. C. L. Y.	
—— Royeri, de Lor.	-	-	-	-	-	5	-	-	-	-	-	-	-	D.	
—— runcinata, Sow.	-	-	-	-	-	5	6	7	-	-	-	-	-	D. O.	
—— squamosa, Phil.	×	-	-	-	-	5	-	-	-	-	-	-	-	O. Y.	Cor.

Species.	Lower Oolites.	Kellaways Rock — Am. calloviensis	Oxford Clay — Am. ornatus	Oxford Clay — Am. cordatus	Corallian — Am. perarmatus	Corallian — Am. plicatilis	Kimeridge Clay — Lower Beds	Kimeridge Clay — 7	Upper Beds	Portland Beds	Purbeck Beds — Lower Beds	Middle Beds	Upper Beds	Wealden Beds.	Localities.	Remarks
	×	1	2	3	4	5	6	7	8	9	10	11	12	×		
Annelida—cont.																
Serpula tetragona, Sow.	×	1	-	-	-	5	6	7	-	-	-	-	-	-	D. to Y.	
—— tricarinata, Sow.	×	-	-	3	4	5	6	7	-	-	-	-	-	-	D. to Y.	
—— triserrata, J. Sow.	-	-	-	-	-	-	-	-	-	-	-	-	-	-	D.	Portl.
—— variabilis. Sow.	-	-	-	-	-	5	-	-	-	-	-	-	-	-	D.C.	Kim. Cl.
—— vertebralis, Sow.	×	1	2	3	-	-	-	-	-	-	-	-	-	-	D. to H.	
—— See also Vermilia.																
Vermicularia contorta, Blake	-	-	-	-	-	-	-	-	-	-	-	-	-	-	O.	Kim. Cl.
—— ovata, Sow.	-	-	-	-	-	-	-	-	-	-	-	-	-	-	O.	Cor.
Vermilia quinquangularis, Goldf.	×	-	-	-	-	-	-	-	-	9	-	-	-	-	O. Bu.	
—— sulcata, Sow.	×	-	-	3	4	5	6	7	-	-	-	-	-	-	D.W.O.H.C.	
Echinodermata. *Crinoidea.*																
Antedon calloviensis, P. H. Carp.	-	1	-	-	-	-	-	-	-	-	-	-	-	-	W.	*Actinometra.*
Apiocrinus polycyphus, Desor	-	-	-	-	-	5	-	-	-	-	-	-	-	-	C.	
Extracrinus	-	-	-	-	-	-	-	-	-	-	-	-	-	-	O	Ox. Cl.
Millericrinus echinatus, Phil.	-	-	-	-	4	5	-	-	-	-	-	-	-	-	D. O. C. Y.	
Pentacrinus Fisheri, Baily	-	-	2	3	-	-	-	-	-	-	-	-	-	-	D. O. K.	
—— sigmaringensis, Quenst.	-	-	-	-	-	-	-	-	-	-	-	-	-	-	Sx.	Kim. Cl.? or Cor. =P. Fisheri?
—— sp.	-	1	-	-	-	-	-	7	-	-	-	-	-	-	W. C. H.	Cor. Purb.
Echinoidea.																
Acrosalenia decorata, Haime	-	-	-	-	-	5	-	-	-	-	-	-	-	-	D. W. Y.	
—— hemicidaroides, Wr.	×	-	-	-	-	5	-	-	-	-	-	-	-	-	W.	
—— Koenigi, Desmoulins	-	-	-	-	-	-	-	-	8	-	-	-	-	-	W. O.	
—— spinosa, Ag.	×	-	-	-	-	5	-	-	-	-	-	-	-	-	D. O.	Ox. Cl.
Cidaris boloniensis, Wr.	-	-	-	-	?	-	-	7	-	-	-	-	-	-	D. Sx.	
—— florigemma, Phil.	-	-	-	-	4	5	6	7	-	-	-	-	-	-	D. to Y. K.	
—— horrida, Wr.	-	-	-	-	-	-	6	-	-	-	-	-	-	-	W.	
—— insperata, Phil.	-	-	-	-	-	-	-	-	-	-	-	-	-	-	O.	Ox. Cl.
—— Smithi, Wr.	-	-	-	4	5	-	-	-	-	-	-	-	-	-	D. to Y.	
—— spinosa, Ag.	-	-	-	-	-	-	-	7	8	-	-	-	-	-	W. Bu.	
—— sp.	-	1	-	-	-	-	-	-	-	-	-	-	-	-	W.	
Collyrites bicordata, Leake	-	-	-	-	4	5	-	-	-	-	-	-	-	-	C. H.	
Clypeus subulatus, Y. & B.	-	-	-	-	-	-	-	-	-	-	-	-	-	-	O. Y.	Cor.

Species.	Lower Oolites	Am. calloviensis, Kellaways Rock	Oxford Clay: Am. ornatus	Am. cordatus	Corallian: Am. perarmatus	Am. plicatilis	Kimmeridge Clay: Lower Beds	Upper Beds	Portland Beds: Lower Beds	Upper Beds	Purbeck Beds: Lower Beds	Middle Beds	Upper Beds	Wealden Beds	Localities.	Remarks.
	x	1	2	3	4	5	6	7	8	9	10	11	12	x		
ECHINODERMATA—cont.																
Echinoidea—cont.																
Diadema See Pseudodiadema																
Dysaster. See Collyrites.																
Echinobrissus Brodiei, Wr.	-	-	-	-	-	-	-	-	-	9	-	-	-	-	W. Bu.	
—— dimidiatus, Phil.	-	-	-	-	-	5	-	-	-	-	-	-	-	-	D. O. Y.	
—— orbicularis, Phil.	x	-	-	-	-	-	-	-	-	-	-	-	-	-	W. · ·	Cor.
—— scutatus, Lam.	-	-	-	-	4	5	-	-	-	-	-	-	-	-	D. to Y.	
Echinus. See Stomechinus.																
Glypticus hieroglyphicus, Goldf.	-	-	-	-	-	-	-	-	-	-	-	-	-	-	W. Y. ·	Cor.
Hemicidaris brillensis, Wr.	-	-	-	-	-	-	-	-	-	9	-	-	-	-	Bu.	
—— Davidsoni, Wr.	-	-	-	-	-	-	-	8	-	-	-	-	-	-	D.?	
—— intermedia, Flem.	-	-	-	-	-	5	-	-	-	-	-	-	-	-	D. to Y.	
—— purbeckensis, Forbes	-	-	-	-	-	-	-	-	-	-	-	11	-	-	D. W.	
Hemipedina Cunningtoni, Wr.	-	-	-	-	-	-	-	8	-	-	-	-	-	-	Bu.	
—— marchamensis, Wr.	-	-	-	-	4	-	-	-	-	-	-	-	-	-	O. Be.	
—— Morrisi, Wr.	-	-	-	-	-	-	-	8	-	-	-	-	-	-	Bu.	
—— tuberculosa, Wr.	-	-	-	-	-	-	-	-	-	-	-	-	-	-	W. · ·	Cor.
—— See also Pelanechinus.																
Holectypus depressus, Leske	x	-	-	-	4	5	-	-	-	-	-	-	-	-	C. H.	
Hyboclypus gibberulus, Ag.	-	-	-	-	-	3	-	-	-	-	-	-	-	-	C.	
—— stellatus, Desor.	-	-	-	-	-	-	-	-	-	-	-	-	-	-	W. ·	Cor.
Nucleolites. See Echinobrissus.																
Pelanechinus corallinus, Wr.	-	-	-	-	-	5	-	-	-	-	-	-	-	-	W. Y.	
Pseudodiadema hemisphærica, Ag.	-	-	-	-	-	5	-	-	-	-	-	-	-	-	W. Y.- ·	*Diadema pseudodiadema*, Ag.
—— mamillanum, Roem.	-	-	-	-	-	5	-	-	-	-	-	-	-	-	D. W.- ·	*Diadema Davidsoni*, Wr.
—— radiatum, Wr.	-	-	-	-	-	5	-	-	-	-	-	-	-	-	D. W.	
—— versipora, Phil.	-	-	-	-	4	5	-	-	-	-	-	-	-	-	D. to Y.	
Pygaster umbrella, Ag.	-	-	-	-	4	5	-	-	-	-	-	-	-	-	D. to Y.	
Pygurus Blumenbachi, Koch. and Dunk.	-	-	-	-	-	5	-	-	-	-	-	-	-	-	D. O.	
—— costatus, Wr.	-	-	-	-	4	5	-	-	-	-	-	-	-	-	W. O.	
—— pentagonalis, Phil.	-	-	-	-	-	5	-	-	-	-	-	-	-	-	O. Be. Y. ·	Cor.
Rhabdocidaris maxima, Münst.	x	-	-	-	-	6	-	-	-	-	-	-	-	-	C.	
Stomechinus gyratus, Ag.	-	-	-	-	-	5	-	-	-	-	-	-	-	-	W. O.	
—— nudus, Wr.	-	-	-	-	-	-	-	-	-	-	-	-	-	-	W. ä	?Cor.

Species.	Kellaways Rock. Am. calloviensis.	Oxford Clay. Am. ornatus.	Am. cordatus.	Corallian. Am. perarmatus.	Am. plicatilis.	Kimeridge Clay. Lower Beds.	Upper Beds.	Portland Beds. Lower Beds.	Upper Beds.	Purbeck Beds. Lower Beds.	Middle Beds.	Upper Beds.		Localities.	Remarks.
	×	1	2	3	4	5	6	7	8	9	10	11	12	×	
ECHINODERMATA—cont.																
Ophiuroidea.																
Amphiura Pratti, Forbes -	–	–	2	–	–	–	–	–	–	–	–	–	–	–	W.	
Ophioderma weymouthiensia, Damon.	–	–	–	–	–	–	–	–	–	–	–	–	–	–	D.	Ox. Cl.
Ophiurella nereida, Wr. -	–	–	–	–	–	5	–	–	–	–	–	–	–	–	D.	
Asteroidea.																
Astropecten rectus, McCoy -	–	–	–	–	–	–	–	–	–	–	–	–	–	–	W. O. Y.	Cor.
ACTINOZOA.																
Anabacia complanata, Defr. -	×	–	–	3	–	5	–	–	–	–	–	–	–	–	W. O.	*A. orbulites,*
Astrocœnia major, Tomes -	–	–	–	–	–	5	–	–	–	–	–	–	–	–	W. O. Be.	
Calamophyllia Stokesi, E. & H.	–	–	–	–	–	5	–	–	–	–	–	–	–	–	W.	
Caryophyllia. See Clado-phyllia and Thecosmilia.																
Cladophyllia Conybearei, E. and H.	–	–	–	–	–	5	–	–	–	–	–	–	–	–	W. O. Y.	
Comoseris irradians, E. and H.	–	–	–	–	–	5	–	–	–	–	–	–	–	–	D. W. Y.	
Crateroseris fungiformis, Tomes.	–	–	–	–	–	5	–	–	–	–	–	–	–	–	W.	
Dendrophyllia. See Gonio-cora.																
Dimorpharæa -	–	–	–	–	–	5	–	–	–	–	–	–	–	–	C.	
Goniocora socialis, Roem.	–	–	–	–	–	5	–	–	–	–	–	–	–	–	W. Y.	
Isastræa explanata, Goldf.	–	–	–	–	–	5	–	–	–	–	–	–	–	–	D. to Y.	
—— Greenoughi, E. and H. -	–	–	–	–	–	5	6	7	–	–	–	–	–	–	D. Be.	
—— heliantholdes, Goldf.	–	–	–	–	–	5	–	–	–	–	–	–	–	–	W.	—
—— oblonga, Flem. -	–	–	–	–	–	–	–	–	–	9	–	–	–	–	D. W.	
Latimæandrarea corallina, E. de From.	–	–	–	–	–	5	–	–	–	–	–	–	–	–	W. Be.	
Microphyllia -	–	–	–	–	–	5	–	–	–	–	–	–	–	–		
Microsolena Gresslyi, Etall. -	–	–	–	–	–	5	–	–	–	–	–	–	–	–		
Montlivaltia dispar, Phil. -	–	–	–	–	–	5	–	–	–	–	–	–	–	–	W. O. C. Y.	
Oroseris. See Latimæan-drarea.																
Protoseris Waltoni, E. and H.	–	–	–	–	–	5	–	–	–	–	–	–	–	–	D	
Rhabdophyllia Phillipsi, E. and H.	–	–	–	–	–	5	–	–	–	–	–	–	–	–	W. to Y.	*R. Edwardsi,*
Siderastræa. See Thamnas-træa.																
Stylina De la Bechei, E. and H.	–	–	–	–	–	5	–	–	–	–	–	–	–	–	W,	
—— tubulifera, Phil. -	–	–	–	–	–	5	–	–	–	–	–	–	–	–	W. O. Y.	
Thamnastræa arachnoides, Park.	–	–	–	–	–	5	–	–	–	–	–	–	–	–	D. W. C. Y.	*formis,*

SPECIES.	Lower Oolite	Am. calloviensis (Kellaways Rock)	Am. ornatus (Oxford Clay)	Am. cordatus (Oxford Clay)	Am. perarmatus (Corallian)	Am. plicatilis (Corallian)	Lower Beds (Kimeridge Clay)	Upper Beds (Kimeridge Clay)	Lower Beds (Portland Beds)	Upper Beds (Portland Beds)	Lower Beds (Purbeck Beds)	Middle Beds (Purbeck Beds)	Upper Beds (Purbeck Beds)	Wealden Beds	LOCALITIES.	REMARKS.
	×	1	2	3	4	5	6	7	8	9	10	11	12	×		
ACTINOZOA—*cont.*																
Thamnastræa concinna, Goldf.	-	-	-	-	-	5	-	-	-	-	-	-	-		D. W. C. Y.	
—— Lyelli, M. Edw.	-	-	-	-	-	5	-	-	-	-	-	-	-		W.	
—— micrastron, Phil.	-	-	-	-	-	5	-	-	-	-	-	-	-		O.	
Thecosmilia annularis, Flem.	-	-	-	-	-	5	6	7	-	-	-	-	-		D. to Y.	
Trochocyathus	-	-	-	-	-	5	-	-	-	-	-	-	-		D.	
SPONGIDA.																
Geodites	-	-	-	-	-	-	-	-	-	9	-	-	-		D.	
Holcospongia floriceps, Phil.	-	-	-	-	4	5	-	-	-	-	-	-	-		W.	*Stellispongia corallina,* de From (in part).
Pachastrella antiqua, Moore	×	-	-	-	-	-	-	-	-	9	-	-	-		D.	*Grantia.*
Rhaxella perforata, Hinds	-	-	-	-	4	5	-	-	-	-	-	-	-		D. W. Y.	*Rhaxella Sorbyana,* Blake.
"Scyphia"	-	-	-	-	-	5	-	-	-	-	-	-	-		C.	
Spongilla purbeckensis, Young.	-	-	-	-	-	-	-	-	-	-	-	11	-		D.	
FORAMINIFERA.																
Bolivina punctata, d'Orb.	-	-	-	-	-	-	-	-	-	-	-	-	-			Kim. Cl.
Cristellaria acutauricularis, Fich. & Moll.	-	-	-	-	-	-	-	-	-	-	-	-	-		Sx.	Kim. Cl.
—— crepidula, F. & M.	×	-	2	-	-	-	-	-	-	-	-	-	-		W. K.	Kim. Cl.
—— decorata, Reuss	-	-	-	-	-	-	-	-	-	-	-	-	-		Sx.	Kim. Cl.
—— italica, Defr.	-	-	-	-	-	-	-	-	-	-	-	-	-		Sx.	Kim. Cl.
—— lævigata, d'Orb.	-	-	-	-	-	-	6	7	-	-	-	-	-		D. C. L.	
—— navicula, d'Orb.	-	-	-	-	-	-	-	7	-	-	-	-	-		D.	
—— nummulitica, Gümb.	-	-	-	-	-	-	-	-	8	-	-	-	-		Bu.	
—— rotulata, Lam.	×	-	2	-	-	-	-	-	8	9	-	-	-		D. W. Bu. K.	Kim. Cl.
—— sp.	-	-	-	3	-	-	-	-	-	-	-	-	-		D. O.	
Dentalina communis, d'Orb.	-	-	-	-	-	5	6	-	-	-	-	-	-		D. W.	
—— jurensis, Terq.	-	-	-	-	-	-	6	-	-	-	-	-	-		L.	
—— pauperata, d'Orb.	-	-	-	-	-	-	6	-	-	-	-	-	-		D.	
—— sp.	-	-	-	3	-	-	-	-	-	-	-	-	-		O.	
Fabularia	-	-	-	-	-	-	-	-	-	-	-	-	-		W.	Purb.
Flabellina rugosa, d'Orb.	-	-	-	-	-	5	6	-	-	-	-	-	-		D.	
—— sp.	-	-	-	3	-	-	-	-	-	-	-	-	-		O.	
Frondicularia nodosaria, Terq.	-	-	-	-	-	-	-	-	-	-	-	-	-			Kim. Cl.
—— sp.	-	-	-	3	-	-	-	-	-	-	-	-	-		O	

Species	Lower Oolite (×)	Kellaways Rock — Am. calloviensis (1)	Oxford Clay — Am. ornatus (2)	Oxford Clay — Am. cordatus (3)	Kimeridge Clay — Lower Beds (4)	Kimeridge Clay — Upper Beds (5)	Portland Beds — Lower Beds (6)	Portland Beds — Upper Beds (7)	Purbeck Beds — Lower Beds	Purbeck Beds — Middle Beds (11)	Purbeck Beds — Upper Beds (12)	Wealden Beds (×)	Localities	Remarks
FORAMINIFERA—*cont.*														
Glandulina tenuis, Born.	-	-	-	-	-	-	6	-	-	-	-	-	D.	
Globigerina	-	-	-	-	-	-	-	-	-	-	-	-	W.	Kim. Cl.
Haplophragmium canariense, d'Orb.	-	-	-	-	-	-	-	-	-	-	-	-	W.	Kim. Cl.
Lagena apiculata, Reuss	-	-	-	-	-	-	6	-	-	-	-	-	D. L.	
— clavata, d'Orb.	-	-	-	-	-	-	6	-	-	-	-	-	D.	
— globosa, Brown	-	-	-	-	-	-	-	-	-	-	-	-	D.	
Lingulina carinata, d'Orb.	-	-	-	-	-	-	-	-	-	-	-	-	W.	Kim. Cl.
— sp.	-	-	-	3	-	-	-	-	-	-	-	-	D. O.	Kim. Cl.
Lituola globigeriniformis, Parker and Jones.	-	-	-	-	-	-	6	-	-	-	-	-	O.	
— nautiloidea, Lam.	-	-	-	-	-	-	6	7	-	-	-	-	D. L. Bu.	
— sp	-	-	-	3	-	-	-	-	-	-	-	-	D. O.	
Marginulina dispar, Reuss	-	-	-	-	-	-	6	-	-	-	-	-	O. Sx.	
— ensis, Reuss	-	-	-	-	-	-	-	-	-	-	-	-	Sx.	Kim. Cl.
— gracilis, Corn.	-	-	-	-	-	-	6	-	-	-	-	-	D. C.	
— inæquistriata, Terq.	-	-	-	-	-	-	6	-	-	-	-	-	L.	
— lata, Corn.	-	-	-	-	-	-	6	7	-	-	-	-	D. C. L.	
— raphanus, d'Orb.	×	-	-	-	-	-	6	7	-	-	-	-	D.	
— sp.	-	-	-	3	-	-	-	-	-	-	-	-	D. O.	
Nodosaria mutabilis, Terq.	-	-	-	-	-	-	6	-	-	-	-	-	D.	
— cf. nitidula, Gümbel	-	-	-	-	-	-	-	-	-	-	-	-	W.	Kim. Cl.
— radicula, Linn.	-	-	-	-	-	-	6	-	-	-	-	-	D.	
— cf. raphanistrum, Linn.	-	-	-	-	-	-	-	-	-	-	-	-	Bu.	
— sp.	-	-	-	-	-	-	-	-	-	-	-	-	W.	Purb.
Orthocerina hæringenæ, Gümb.	-	-	-	-	-	-	6	-	-	-	-	-	D.	
— sp.	-	-	-	3	-	-	-	-	-	-	-	-	O.	
Placospilina. See Lituola.														
Planularia Bronni, Roem	-	-	-	-	-	-	6	-	-	-	-	-	D.	
— reticulata, Corn.	×	-	-	-	-	-	6	-	-	-	-	-	D. W. L.	
— strigillata, Reuss	-	-	-	-	-	5	-	-	-	-	-	-	C.	
— sp.	-	-	-	3	-	-	-	-	-	-	-	-	O.	
Planulina ornata, Roem.	-	-	-	-	-	-	6	-	-	-	-	-	D.	
Plecanium. See Textilaria.														
Polymorphina fusiformis, Roem.	-	-	-	-	-	5	6	-	-	-	-	-	D.	

Species	Lower Oolites. (×)	Kellaways Rock. Am. calloviensis. (1)	Oxford Clay. Am. ornatus. (2)	Am. cordatus. (3)	Corallian. Am. perarmatus. (4)	Am. plicatilis. (5)	Kimeridge Clay. Lower Beds. (6)	Upper Beds. (7)	Portland Beds. Lower Beds. (8)	Upper Beds. (9)	Purbeck Beds. Lower Beds. (10)	Middle Beds. (11)	Upper Beds. (12)	Wealden Beds. (×)	Localities.	Remarks.
FORAMINIFERA—cont.																
Pulvinulina caracolla, Roem.	-	-	-	-	-	-	-	-	-	-	-	-	-	-	D. Y.	Ox. Cl., Kim. Cl.
—— pulchella, d'Orb.	-	-	-	-	-	-	6	-	-	-	-	-	-	-	D. L.	
Quinqueloculina bermentstoriensis, Kübler.	-	-	-	-	-	-	6	-	-	-	-	-	-	-	D.	
Robulina Münsteri, Roem.	-	-	-	-	-	5	6	-	-	-	-	-	-	-	D. L.	
Textilaria agglutinans, d'Orb.	-	-	-	-	-	-	6	7	-	-	-	-	-	-	D.	
—— gibbosa, d'Orb.	-	-	-	-	-	-	6	-	-	-	-	-	-	-	L.	
Trochammina (cf.) gordialis J. & P.	-	-	-	-	-	-	-	-	-	9	-	-	-	-	D. W. -	Kim. Cl.
—— incerta, d'Orb.	×	-	-	-	-	-	-	-	-	9	-	-	-	-	D. W. -	Kim. Cl.
—— cf. oligogyra, Hantken	-	-	-	-	-	-	-	-	-	-	-	-	-	-	W. -	Kim. Cl.
Vaginulina badenensis, d'Orb.	-	-	-	-	-	-	6	-	-	-	-	-	-	-	D. L.	
—— cristellaroides, Reuss	-	-	-	-	-	-	-	-	-	-	-	-	-	-	W. -	Kim. Cl.
—— harpa, Roem.	×	-	-	3	-	-	-	7	8	-	-	-	-	-	D. W. O. C. Bu.	
—— orthonota, Reuss	-	-	-	-	-	-	-	-	-	-	-	-	-	-	W. -	Kim. Cl.
—— striata, d'Orb.	-	-	-	-	-	5	6	-	-	-	-	-	-	-	D. C.	
—— tricarinata, d'Orb.	-	-	-	-	-	-	-	-	-	-	-	-	-	-	Sr. -	Kim. Cl.
Webbina irregularis, d'Orb.	-	-	-	-	4	5	6	-	-	-	-	-	-	-	D. Bd.	
PLANTÆ.																
ANGIOSPERMÆ.	-	-	-	-	-	-	-	-	-	-	-	-	-	-	W. -	Purb.
GYMNOSPERMÆ.																
Coniferæ.																
Araucarites -	-	-	-	-	-	-	-	-	-	9	10	-	-	-	D. -	Ox. Cl.
Brachyphyllum mammillare, Brongn.	-	-	2	-	-	-	-	-	-	-	-	-	-	-	W. O.	
Carpolithes plenus, Phil.	-	-	-	-	-	-	-	-	-	-	-	-	-	-	Be. Y. -	Cor.
Cedroxylon -	-	-	-	-	-	-	-	-	-	-	-	-	-	-	D. -	Purb.
Dammarites. See Pinites.																
Palæocyparis -	-	-	-	-	-	-	-	-	-	-	10	-	-	-	W.	
Pinites dejectus, Carr.	-	-	-	-	-	-	6	-	-	-	-	-	-	-	D. -	P. depressus, Carr.
—— Fittoni, Ung.	-	-	-	-	-	-	-	-	-	-	-	-	-	-	D. -	Purb.
Thuyites -	-	-	-	-	-	-	-	-	-	-	-	-	-	-	D. W. -	Purb.
Cycadeæ.																
Bennettites portlandicus, Carr.	-	-	-	-	-	-	-	-	-	-	10	-	-	-	D.	
Cycadeoidea. See Mantellia.																

SPECIES.	Lower Oolites	Am. callloviensis (Kellaways Rock)	Am. ornatus (Oxford Clay)	Am. cordatus	Am. perarmatus (Corallian)	Am. plicatilis	Lower Beds (Kimeridge Clay)	Upper Beds	Lower Beds (Portland Beds)	Upper Beds	Lower Beds	Middle Beds	Upper Beds	Wealden Beds	LOCALITIES.	REMARKS.
	×	1	2	3	4	5	6	7	8	9	10	11	12	×		
GYMNOSPERMÆ—*cont.*																
Cycadeæ—cont.																
Cycadeostrobus sphæricus, Carr.	-	-	-	-	-	-	-	-	-	-	-	-	-	-	W.	Ox. Cl.
Fittonia	-	-	-	-	-	-	-	-	-	-	-	-	-	-		Purb.
Mantellia intermedia, Carr.	-	-	-	-	-	-	-	-	-	10	-	-	-		D.	
—— microphylla, Buckl.	-	-	-	-	-	-	-	-	-	10	-	-	-		D. W.	
—— nidiformis, Brongn.	-	-	-	-	-	-	-	-	-	10	-	-	-		D. W.	{ Cycadeoidea megalophylla, Buckl. Purb.
Williamsonia	-	-	-	-	-	-	-	-	-	-	-	-	-	-		Purb.
Zamites. See Mantellia.																
CRYPTOGAMÆ.																
Filices.																
Alethopteris elegans, Dunk. See Matonidium.																
Lonchopteris. See Weichselia.																
Matonidium Göpperti, Ett.	-	-	-	-	-	-	-	-	-	-	-	-	-	×	Sx.	Purb.
Onychiopsis Mantelli, Brongn.	-	-	-	-	-	-	-	-	-	-	-	-	-	×	Sx.	Purb.
Sphenopteris. See Onychiopsis .																
Weichselia Mantelli, Brongn.	-	-	-	-	-	-	-	-	-	-	-	-	-	×	D.?	Purb.
Equisetaceæ.																
Clathraria. See Equisetites.																
Equisetites Lyelli, Mant.	-	-	-	-	-	-	-	-	-	-	-	11	-	×	W. Sx.	
Characeæ.																
Chara Jaccardi, Heer.	-	-	-	-	-	-	-	-	-	-	-	-	-	-	D.	{ Purb. C. purbeckensis, Forbes.
—— sp.	-	-	-	-	-	-	-	-	-	-	-	11	-	-	D. Sx.	
Algæ.																
Caulerpa Carruthersi, G. Murray.	-	-	-	-	-	-	-	-	-	-	-	-	-	-	D.	Kim. Clay.

APPENDIX II.

LIST OF WORKS

JURASSIC ROCKS OF ENGLAND AND WALES,

COMPILED BY

H. B. WOODWARD and C. FOX-STRANGWAYS.

In the two lists now given it has been the object to record all important works dealing with the stratigraphy and palæontology of the Jurassic Rocks of England and Wales. In the first list the publications of the Geological Survey are enumerated; in the second list the principal papers and other works on the subject are given in chronological order. The titles of those papers which have been found to contain no original information are omitted. Moreover it has been decided not to insert references to works that deal only with economic matters, such as limes and cements, and also those on agriculture, mineral waters, &c. In the text references are given to works that have afforded special information.

Papers that treat particularly of the Jurassic Rocks of Scotland and Ireland will be recorded in the concluding volume of the Memoir on the Jurassic Rocks of Britain, a work which will deal with those areas.

LIST OF AUTHORS.

(The figures refer to the dates of publications, those in brackets indicating Geological Survey Works enumerated in List No. I.)

A.

ABEL, F. A., 1863.
ADDY, J., 1883.
AGASSIZ, LOUIS, 1833–44.
ALLISON, T., 1869.
AMMON, L. VON, 1877.
ANDREWS, C. W., 1895.
——, REV. W. R., 1881, 1891, 1894.
ANNING, MARY, 1839.
ANON., 1826, 1829, 1837, 1862–66, 1868-71, 1873–76, 1879–81.
ANSTED, PROF. D. T., 1866.
ANSTIE, JOHN, (1873–75).
ARGALL, W., 1874.
ARMITAGE, J., 1871.
AUSTEN, REV. J. H., 1851.
AUSTIN, FORT-MAJOR T., 1867.
AVELINE, W. T., (1844–45, 1857–61, 1874–75).

B.

BADCOCK, P., 1862.
BAILY, W. H., 1860, 1863.
BAKER, H., 1754.
——, J. G., 1863.
——, T. B. LLOYD, 1856.

BARR, T. M. 1874.
BARRETT, C. 1878.
BARROIS, DR. C. 1894.
BARROW, GEORGE, (1878–86, 1888–89, 1891) 1877, 1880.
BARRY, C. 1832.
BATE, C. SPENCE, 1884.
BATES, E. F., 1886.
BATHER, F. A., 1886.
BAUERMAN, H., (1858–59, 1862–63, 1865).
BAXTER, R. C., 1881.
BEAN, WILLIAM., 1836, 1839.
BEDFORD, W., 1839, 1843.
BEESLEY, THOMAS., 1873–73, 1877, 1883.
BELL, J. L., 1872.
BENNETT, F. J., (1891).
BESWICK, J., 1858, 1861.
BINNEY, E. W., 1859.
BIRD, C., 1881.
——, J., 1818, 1822.
BLACKWELL, J. K., 1856.
BLAINVILLE, H. D. de., 1838.
BLAKE, C. CARTER., 1863.
——, PROF. J. F., 1872–73, 1875–81, 1885, 1887–88, 1891–93.
——, J. H., (1873), 1872.

BLANFORD, W. T., 1885.
BOGG, EDWARD, 1816.
BONNEY, PROF. T. G., 1875, 1877.
BOTT, A., 1870.
BOULANGER, G. A., 1887, 1891.
BOULGER, PROF. G. S., 1876.
BOWERBANK, DR. J. S., 1840, 1841, 1848.
BOYD, C. 1807, 1810.
BRADY, H. B., 1864–65.
BRANNON, P., 1872.
BRAVENDER, JOHN, 1868.
BRIGHT, DR. R., 1817.
BRISTOW, H. W., (1845, 1850–52, 1855–59, 1865–67, 1870–71, 1873, 1875), 1867, 1869.
BROCKBANK, W., 1866.
BRODERIP, W. J., 1828, 1835, 1839.
BRODIE, REV. P. B., 1839, 1842, 1843, 1845, 1847, 1849–51, 1853–54, 1857–58, 1861, 1865–69, 1870, 1874–75, 1888–90, 1893–94.
——, W. R., 1876.
BRONGNIART, AD., 1825, 1828, 1850.
BROWETT, A., 1889.
BROWN, A., 1894.
——, CAPT. T., 1843, 1849
——, T. C., 1873.
——, W., 1861.
BROWNE, M., 1889–91, 1893.
BRUCE, A. C., 1868–69.
BUCKLAND, REV. PROF. W., 1823–24, 1828–30, 1835–36, 1838–40.
BUCKMAN, PROF. J., 1842–43, 1845, 1849–50, 1853–55, 1857–58, 1860, 1865–68, 1875, 1877–80.
——, S. S., 1878, 1880–81, 1886–95.
BUNBURY, SIR C. J. F., 1851.
BURR, FREDERICK, 1838.
BURROWS, H. W., 1893.
BUTLER A. G., 1873–74.
BUTLIN, W. H., 1883.

C.

CAMERON, A. C. G., (1880, 1883–86, 1888, 1890), 1888–91.
CARPENTER, P. H., 1877, 1880, 1882, 1887.
——, DR. W. B., 1845, 1848.
CARR, W. D., 1883–84.
CARRUTHERS, WILLIAM, 1866–71, 1875–76.
CARTE, DR. A., 1863.
CARTER, H. J., 1878.
——, JAMES, 1886.
CASLEY, G., 1880.
CHADWICK, S., 1886.
CHAPMAN, W., 1759.
CHARLESWORTH, EDWARD, 1837, 1839, 1845–47.
CHURCH, PROF. A. H., 1864, 1872.
CLAPHAM, R. C., 1864.
CLARKE, J. F. M., 1891.
——, REV. W. B., 1838.
CLEMINSHAW, E., 1868–69.
CLUTTERBUCK, REV. J. C., 1861, 1863.
COBBOLD, E. S., 1880.

COCKBURN, W., 1870.
COLE, REV. E. M., 1886, 1890–91.
COLEMAN, REV. W. H., 1846.
COLES, HENRY, 1853.
COLWAL, D., 1679.
CONYBEARE, REV. J. J., 1822.
——, REV. W. D., 1821–24, 1829, 1833, 1840.
COOPER, B., (1848).
COPE, PROF. E. D., 1884, 1887.
CRICK, G. C., 1890–91.
——, W. D., 1883, 1887, 1889, 1891–92, 1894–95.
CROSS, REV. J. E., 1875.
CROSSE, H., 1875.
CROWDER, W., 1856–57.
CUMBERLAND, G., 1821, 1824, 1829.
CUMMING, L., 1893.
CUNNINGTON, WILLIAM, 1847, 1859, 1869, 1872.
CUVIER, BARON G., 1821–24.

D.

DA COSTA, E. M., 1757–58.
DAGLISH, T., 1875.
DAKYNS, J. R., (1864, 1874, 1884, 1886).
DALTON, W. H., (1886, 1888), 1887.
DAMON, ROBERT, 1860.
D'ARCHIAC, E. J. A., 1838.
DAVIDSON, DR. THOMAS, 1847, 1849–51, 1877.
DAVIES, WILLIAM, 1864, 1867, 1871–72, 1876, 1887.
DAVIS, J. W., 1884, 1887.
DAWKINS, PROF. W. BOYD, (1867), 1890, 1892, 1894.
DAY, E. C. H., 1863–65.
DE LA BECHE, SIR H. T., (1838–39, 1844–46, 1848, 1855), 1822–23, 1826, 1830, 1835, 1839, 1845.
DELESSE, M., 1851.
DENNIS, REV. J. B. P., 1857.
DE RANCE, C. E., 1876–95.
DESNOYERS, J., 1825.
DICK, A. B., (1856), 1856.
DIXON, F., 1878.
DOVE, G., 1876.
DREW, F., (1864, 1875).
DUNCAN, PROF. P. M., 1867, 1869–70, 1877, 1884–86.
DUNN, J., 1831.

E.

EGERTON, SIR P. DE M. G., (1852, 1855, 1858, 1872), 1835–37, 1839, 1843, 1845, 1847, 1850, 1854, 1858, 1868–69, 1871–73, 1876.
ENGLEFIELD, SIR HENRY C., 1781, 1816.
ENNISKILLEN, EARL OF, 1869.
ESKRIGGE, R. A., 1864.
ETHERIDGE, ROBERT, (1865, 1873), 1860, 1871–72, 1875, 1877, 1879, 1881–82, 1885, 1887, 1889, 1893.
EUNSON, H. J., 1883–84, 1886.
——, J., 1882.
EVANS, SIR JOHN, 1879.

F.

FALCONER, DR. HUGH, 1857, 1862, 1868.
FAREY, JOHN, 1810, 1815, 1819.
FISCHER, P., 1879.
FISHER, REV. OSMOND, (1857), 1850, 1856, 1861.
FITTON, DR. W. H. 1818, 1824, 1827-28, 1832-33, 1835-36.
FLEWKER, J 1831.
FOORD, A. H. 1890-91, 1895.
FORBES, PROF. EDWARD, (1848-50, 1852, 1856), 1844, 1850.
FOSTER, DR. C. LE NEVE, (1867, 1875), 1881.
FRAAS, DR. OSCAR, 1851.
FRECHEVILLE, R. J., 1881.

G.

GARDNER, J. STARKIE, 1886-87
GAUDRY, DR. ALBERT, 1853.
GAVEY, G. E. 1853.
GEIKIE, SIR A., (1858-59, 1862), 1861.
GIEBEL, C. G., 1856.
GILL, W. H 885.
GODWIN-AUSTEN, R. A. C., 1850, 1856, 1873.
GOODRICH, E.S., 1894.
GOSS, HERBERT, 1878.
GOULD, C., (1864, 1875).
GOWEN, R., 1843.
GRAHAM, W. B., 1877.
GRATEAU, E., 1863.
GRAY, DR. J. E. 1828.
——, W., 1862.
GREAVES, J. 1832
GREEN, PROF. A. H., (1863-65, 1886), 1894.
——, BURTON, 1886.
GREENWELL, G. C., 1859, 1894.
GREGORY, DR. J. W., 1894.
GREGSON, C. S., 1863.
GROOM, T. T., 1887
GROOM-NAPIER. C. O., 1868.
GROVES, T. B., 1887
GUISE, SIR W. V., 1860, 1862-65, 1867-69, 1872-75, 1877-85, 1887.
GÜNTHER, DR. A., (1872).
GUNN, REV. JOHN, 1864.

H.

HAEUSLER, DR. R. 1887.
HAGEN, H. A., 1849-50, 1862.
HAIME, JULES, 1851, 1854.
HARKER, ALFRED, 1885, 1890.
——, PROF. ALLEN, 1884-85, 1888, 1890-91.
HARLAN, DR. R., 1823.
HARRIS, G. F., 886, 1895.
——, W. H., 1879.
HARRISON, W. J., 1877 1882.
HASTINGS, DR. C., 1834.
HATCHETT, C., 1798.
HAWKSLEY, T., 1875.

HAWKESWORTH, E., 1891.
HAWKINS, THOMAS, 1834, 1840.
HEER, REV. DR. O.,1864-65.
HERMAN W. D., 1871.
HILL, J., 1748.
HINDE, DR. G. J., 1883-84, 1888-90, 1893.
HINDERWELL, T., 1798.
HODGES, L., 1886.
HOLDSWORTH, JOSEPH, 1833-34.
HOLL, DR. H. B., 1863.
HOLLOWAY, W. H., (1879, 1881, 1887).
HOLMES, T. V., (1888), 1881, 1889,
HOME, SIR EVERARD, 1814, 1816, 1818-20.
HOMERSHAM, C., 1884-85.
HOPEWELL, E. W., 1875.
HORNER, LEONARD, 1816.
HORTON, W. S., 1860-61, 1864.
HOWELL, H. H., (1845, 1852, 1854-56, 1858-60, 1863-64, 1870, 1880, 1883, 1884).
HOWES, PROF. G. B., 1891.
HOWSE, R. 1875.
HUDLESTON, W. H., 1874, 1876-82, 1885-89, 1892-95.
HUGHES, PROF. T. McK., 1879, 1884.
HULKE, J W 1869-72, 1874, 1878-80, 883, 1887, 1889, 1892.
HULL, PROF. E., (1852, 1855-61, 1868, 1867, 1869-70), 1857, 1860, 1862, 1865-66, 1877. 1882.
HUNT, ROBERT, (1857, 1859-60, 1867), 1861, 1868.
HUNTER, W. P, 1835-36.
HUNTON, LOUIS, 1837.
HUTTON, W. 1831.
HUXLEY, RIGHT HON. T. H., (1864-66, 1872), 1858-59, 1868-69.
HYATT, PROF. A. 1874-76, 1883.
HYETT, W. H., 1867.

I.

IBBETSON, CAPT. L. L. B., 1847-48.

J.

JAEKEL, O., 1891.
JAMES, TREVOR E., (1844-45).
JARDINE, SIR WILLIAM, 1858.
JECKS, C., 1869.
JELLY, REV. H., 1833, 1839.
JOHNSTONE, SIR J., 1843.
JONES, JOHN, (of Gloucester)., 1858, 1863-64.
——, J., 1860.
——, PROF. T. RUPERT, 1853, 1862-63, 1871, 1875, 1878, 1882, 1884-86, 1888, 1890-91, 1894.
JUDD, PROF. J. W., (1870-73, 1875, 1886), 1868, 1870-71, 1884-85.
JUKES, J. B., (1852), 1838, 842.
JUKES-BROWNE, A. J. (1881-82, 1884-88, 1890, 1891, 1893), 1884, 1889, 1891, 1893-94.

K.

KEEPING, H., 1882.
——, W., 1878.
KENDALL, J. D., 1886, 1893.
KENNEDY, L., 1867.
KENT, A. U., 1879.
KINGDON, J., 1825.
KIRBY, W. F., 1890.
KNOX, R., 1855.
KÖNIG, CHARLES, 1825.

L.

LAMPLUGH, G. W., 1888–92.
LECKENBY, JOHN, 1855, 1859, 1864, 1873.
LEE, J. E., 1881.
LEIGHTON, THOMAS, 1891, 1894.
LEWIS, REV. J., 1728.
LHWYD, EDWARD, 1693, 1698, 1699.
LIMBIRD, J., 1787.
LINDLEY, DR. JOHN, 1831.
LISTER, DR. MARTIN, 1671, 1674, 1675, 1678, 1684, 1693. See also 1819.
LOBLEY, PROF. J. LOGAN, 1870–71, 1874.
LONGE, F. D., 1881.
LONSDALE, WILLIAM, 1827, 1829, 1832–33.
LORIEUX, E., 1892.
LOWE, W. B., 1872–73.
LUCAS, JOSEPH, 1891.
——, REV. S., 1862.
——, W., 1845.
LUCY, W. C., 1869, 1882–86, 1888–89, 1891.
LYCETT, DR. JOHN, 1848–51, 1853, 1855, 1857, 1862, 1864, 1872, 1881.
LYDEKKER, R., 1885, 1887–91, 1893.

M.

MACALISTER, J. H., 1861.
McCOY, PROF. F., 1848–49, 1858.
MACDAKIN, CAPT., 1877.
MACKIE, S. J., 1858, 1863.
McMURTRIE, JAMES, (1875), 1883.
MANSEL-PLEYDELL, J. C., 1873, 1877–79, 1881, 1885, 1888–90, 1892, 1894.
MANTELL, DR. G. A., 1833, 1846–48, 1850, 1852.
——, R. N., 1848, 1850.
MARCOU, JULES, 1857, 1859.
MARLEY, J., 1857, 1870.
MARR, J. E., 1894.
MARRAT, W., 1840.
MARRIOTT, J., 1884.
MASON, J. WOOD, 1869.
MAW, GEORGE, 1868.
MAY, C., 1852.
MEADE, RICHARD, 1876, 1882.
MERRYWEATHER, DR. GEORGE, 1853.
MICHELL, REV. JOHN, 1760.
MILLER, J. S., 1821, 1826.
MILLS, H. M., 1891.
MILNE-EDWARDS, PROF. H., 1848, 1851.

MITCHELL, DR. JAMES, 1834, 1837.
——, W. S., 1871.
MOBERLY, REV. DR., 1849.
MONCKTON, H. W., 1894.
MOORE, CHARLES, 1852–53, 1855, 1857, 1860–61, 1863–65, 1867, 1870, 1873, 1875, 1877–82.
——, C. A., 1884–85.
MORGAN, PROF. C. LLOYD, 1887.
MORIÈRE, J., 1880.
MORRIS, PROF. JOHN, 1843, 1845–51, 1853, 1856, 1868–69, 1873, 1875, 1878.
MORTON, G. H., 1864.
——, REV. JOHN, 1712.
MURCHISON, SIR R. I., 1831–32, 1834–35, 1839.
MURRAY, GEORGE, 1892.
——, DR. P., 1828, 1855.

N.

NATHORST, A. G., 1880.
NEWTON, E. T., 1878, 1881, 1886, 1888, 1891.
NICHOL, W., 1834–35.
NIKITIN, S., 1889.
NORWOOD, REV. T. W., 1859.

O.

ODLING, DR. W., 1875.
OGILBY, WILLIAM, 1839.
OLDHAM, R. D., 1875.
——, T. B., 1879–80.
OPPEL, DR. A., 1856.
ORLEBAR, A. B., 1850–51.
OSBORN, H. F., 1888.
OWEN, EDWARD, 1754.
——, SIR RICHARD, 1838–44, 1846, 1854–55, 1857, 1859–62, 1866, 1871, 1878–79, 1883–84.

P.

PARKER, JAMES, 1874, 1884.
——, W. K., 1871, 1875.
PARKIN, C., 1882.
PARKINSON, JAMES, 1804.
PARSONS, DR. H. F., 1879.
PATTINSON, J., 1864, 1870.
PAUL, J. D., 1883–84.
PAVLOW, A., 1889, 1892.
PEARCE, J. CHANING, 1833, 1842, 1846–47.
PENNING, W. H., (1878, 1881–82, 1886, 1888).
PEYTON, J. E. H., 1873.
PHILLIPS, PROF. JOHN, (1845), 1828–29, 1831, 1844, 1853–55, 1857–58, 1860, 1863, 1865–66, 1870–71, 1873.
——, JOHN ARTHUR, 1881.
——, RICHARD, 1824.
——, WILLIAM, 1818, 1822.
PILBROW, J., 1884.
PLANT, JAMES, 1874, 1877, 1884.
PLATNAUER, H. M., 1886, 1888, 1891, 1894.
PLATT, J., 1759, 1765.

PLAYNE, G. F., 1869.
PLOT, DR. R, 1677.
POLWHELE, T. R., (1859, 1863).
PORTER, DR. H., 1861, 1863.
PRATT, C., 1861.
——, S. P., 1841.
PRESTWICH. PROF. J., 1876, 1878–80.
PREVOST, CONSTANT, 1825, 1830, 1839.
PRIME, REV. A. DE LA, 1700.
PRIOR, DR. C. E., 1888.

Q.

QUILTER, H. E., 1881, 1883–84, 1886, 1889–90.

R.

RAMSAY, SIR A. C., (1845–46, 1852, 1855, 1858–59, 1862, 1867), 1864, 1871.
RANSOME, T., (1848).
READE, T. MELLARD, 1876.
REEKS, TRENHAM, (1855, 1871).
REID, CLEMENT, (1883, 1885, 1890–91).
Report, 1874, 1893.
RICHARDS, J T., 1884.
RICHARDSON, T., 1864.
RILEY, E., 1864.
ROBERTS, GEORGE, 1834, 1840.
——, G. E., 1860, 1862.
——, R. W. B., 1891.
——, THOMAS, 1887, 1889, 1892.
ROBINSON, C. E., 1882.
——, H., 1868.
ROSE, C. B., 1835, 1855, 1864.
RUDLER, F. W., (1867, 1871, 1876, 1893, 1895).
RUTLEY, FRANK, 1893.

S.

SÆMANN, B , 1854.
SALTER, J. W., 1865.
SANDERS, WILLIAM, (1844–45), 1842, 1847, 1868.
SAUNDERS, JAMES, 1890.
SAUVAGE, DR. H.E., 1878, 1880–81 1883.
SCHWARZ, E. H. L., 1894.
SCOTT, W. L., 1865.
SCROPE, G. P., 1831, 1858.
SCUDDER, S. H., 1874–75, 1886, 1891.
SEDGWICK, REV. PROF. A., 1822, 1824, 1826, 1846, 1861, 1869.
SEELEY, PROF. H. G., 1861–62, 1865, 1869–71, 1874–77, 1879–80, 1888–89, 1891, 1893.
SELWYN, A. R. C., (1855, 1857).
SEWARD, A. C., 1895.
SHARP, SAMUEL, 1869–71, 1873, 1881.
SHARPE, DANIEL, 1830.
SHERBORN, C. D., 1886, 1888, 1890–92.
SIMPSON, MARTIN, 1843, 1855, 1859, 1865–66.
SKERTCHLY, S. B. J., (1877, 1882–83, 1886, 1891, 1893).
SLADEN, W. P., 1880.
SLATTER, T. J., 1882.
SMEE, ALFRED, 1838.

e 82423.

SMITH, C. H., 1839–40.
—— T. McD., 1869.
—— WILLIAM, 1815–17, 1819, 1832, 1839.
SMITHE, REV. F., 1862, 1867, 1877, 1883, 1891.
SMYTH, SIR W. W., (1856).
SOLLAS, PROF. W. J., 1880–83, 1885.
SOLLY, H. S., 1890.
SORBY, DR. H. C., 1851–52, 1554, 1857–58, 1879–80.
SOWERBY, JAMES 1812–23.
——, J. DE CARLE, 1825–46.
SPILLER, J., (1856.)
STEPHENS, J., 1762.
STEVENSON, J. C., 1864.
STODDART, W. W., 1865, 1867–68, 1871–72, 1877, 1879.
STOKES, C., 1829.
STRACHEY, JOHN, 1727.
STRAHAN, AUBREY (1884, 1886, 1888), 1895.
STRANGWAYS, C. FOX-, (1874, 1876, 1878–87, 1890–91), 1888, 1894.
STRICKLAND, H. E., 1834–35, 1839–40, 1842, 1844–46, 1849–50.
STRUCKMANN, C., 1881.
STUART, M. G., 1889.
STUKELY, DR. W., 1719.
STUTCHBURY, S., 1839, 1842, 1846.

T.

TATE, PROF. RALPH, 1867, 1869–72, 1875–76.
TAUNTON, J. H., 1868, 1872, 1877, 1887.
TAWNEY, E. B., 1866, 1873–75, 1878.
TAYLOR, H., 1855.
——, W., 1866.
THIESSING, J. B., 1891.
THOMPSON, BEEBY, 1880, 1884, 1887, 1889–92, 1894–95.
THOMSON, J. G., 1844.
THORNE, J., 1869.
THORP, REV. W., 1841, 1856.
TOMES, R. F. 1864, 1878–79, 1882–86, 1888–89, 1893.
TOPLEY, WILLIAM, (1867, 1875), 1868, 1873–74, 1876, 1882.
TOWNSEND, REV. JOSEPH, 1813.
TRAQUAIR, DR. R. H., 1887.
TRAVIS, DR. W., 1798.
TRENCH, RICHARD, (1859–60).
TRIBOLET, M. DE, 1876.
TRIGER, —, 1855.
TRIMMER, JOSHUA, 1853.
TUTE, REV. J. S., 1871.

U.

USSHER, W. A. E., (1873, 1875, 1886–88, 1890).

V.

VALENCIENNES, PROF. A., 1838.
VEITCH, W. Y., 1886.
VERNON, REV. W., 1826, 1829.
VINE, G. R., 1880–81, 1883–84, 1887.

D D

W.

WAAGEN, DR. W., 1865.
WALCOTT, JOHN, 1779.
WALFORD, E. A., 1878, 1882–83, 1885, 1887, 1889, 1894–95.
WALKER, J. F., 1869–70, 1875–78, 1888, 1890, 1892–93.
WALLIS, A. M., 1891.
WALTER, R., 1854.
WALTON, JOHN, 1847.
WARD, L. F., 1884.
WARRINGTON, R., 1866.
WATSON, J., 1861.
WATTS, J. K., 1853.
WEAVER, THOMAS, 1824.
WEBSTER, THOMAS, 1826.
WESTON, C. H., 1848–49, 1852.
WESTWOOD, PROF. J. O., 1841, 1854.
WETHERED, EDWARD B., 1886, 1888–91, 1895.
WHIDBORNE, REV. G. F., 1883.
WHITAKER, WILLIAM, (1859, 1861, 1863, 1865, 1869, 1875–89, 1891, 1893), 1878, 1886, 1889–90, 1893.
WHITEAVES, J. F., 1861
WICKES, W. H., 1893.
WILLETT, E. W., 1881.
——, H., 1878, 1889.
WILLIAMS, D. H., (1845, 1848).
WILLIAMSON, PROF. W. C., 1836–37, 1841, 1849, 1855, 1870, 1883.
WILLS, H., 1871.
WILSON, EDWARD, 1879, 1885–87, 1889–93.
——, DR. G., 1861.
——, REV. J. M., 1868–69, 1872, 1875.
WILTON, REV. C. P., 1826, 1830.
WINCH, N. J., 1821–22.

WINKLER, DR. T. C., 1873, 1876.
WINTER, R., 1810.
WINWOOD, REV. H. H., 1872, 1874–76, 1878–79, 1888, 1891–93, 1895.
WITCHELL, EDWIN, 1862, 1867, 1868, 1875, 1880, 1882, 1886–88.
WITHAM, H., 1831–33.
WOOD, REV. H. H., 1877.
——, N., 1859.
——, S. V., JUN., 1862–63.
WOODALL, J., 1857.
WOODHOUSE, REV. T., 1878.
WOODS, H., 1822, 1829.
——, HENRY, 1891.
WOODWARD, A. SMITH, 1885–95.
——, C. J., 1881.
——, DR. HENRY, 1863, 1865–66, 1868–70, 1874, 1876–78, 1881, 1885, 1888–90, 1892–93.
——, HORACE B., (1873–76, 1882, 1886, 1891), 1870–72, 1874, 1885–89, 1891–93.
——, DR. JOHN, 1728–29.
——, SAMUEL, 1833.
——, S. P., (1856), 1848, 1860.
WOOLER, —, 1759.
WORTHINGTON, J. K., 1881–82.
WORTHY, G. S., 1861.
WRIGHT, DR. THOMAS, 1851, 1854–58, 1860–62, 1864–65, 1867, 1870–71, 1875–76, 1878, 1880–82, 1887.

Y.

YATES, J., 1849.
YOUNG, REV. G, 1817–19, 1821–22, 1824–25.
——, J. T., 1878.

I.—PUBLICATIONS OF THE GEOLOGICAL SURVEY, AND OF THE MUSEUM OF PRACTICAL GEOLOGY.

Maps.—Scale, 1 inch to a mile.

Sheet 5.—[Purbeck Beds, at Battle, near Hastings, by F. DREW, and C. GOULD.] 1864.

Sheet 13.—[Cornbrash to Purbeck : Faringdon, Bampton, Abingdon, Oxford, and Thame, by E. HULL.] 1860.

Sheet 14.—[Great Oolite to Purbeck Beds : Trowbridge, Westbury, and Melksham, by W. T. AVELINE and H. W. BRISTOW.] 1857.

Sheet 15.—[Corallian to Purbeck Beds : Tisbury, Vale of Wardour, by H. W. BRISTOW.] 1856.

Sheet 16.—[Kimeridge Clay to Purbeck Beds : Swanage, St. Aldhelms Head, Kimeridge, and Mewps Bay, by H. W. BRISTOW.] 1855.

Sheet 17.—[Lower Lias to Purbeck Beds : Charmouth, Bridport, Burton Bradstock, Weymouth, Portland, and Lulworth Cove, by H. W. BRISTOW.] 1850.

Sheet 18.—[Lower Lias to Kimeridge Clay : Langport, Somerton, Wincanton, Stalbridge, South Petherton, Yeovil, Sherborne, Sturminster Newton, Crewkerne and Beaminster, by H. W. BRISTOW.] 1850.

———.—New edition. [Lias, &c, by H. W. BRISTOW, H. B. WOODWARD, and W. A. E. USSHER.] 1875.

Sheet 19.—[Lower Lias to Kimeridge Clay : Bath, Dundry, Bradford-on-Avon, Frome, Bruton, Shepton Mallet, Glastonbury, Castle Cary, by H. T. DE LA BECHE, J. PHILLIPS, D. H. WILLIAMS, A. C. RAMSAY, H. W. BRISTOW, W. T. AVELINE, and T. E. JAMES.] 1845.

———.—New edition. [By H. W. BRISTOW, H. B. WOODWARD, W. A. E. USSHER and J. H. BLAKE]. 1873.

Sheet 20.—[Lower Lias, &c. : Watchet, Puriton, Brent Knoll, Uphill, Lavernock, Aberthaw, by H. T. DE LA BECHE.] Before 1839.

Sheet 21.—[Lower Lias, &c. : Ilminster, Axminster, by H. T. DE LA BECHE.] Before 1839.

Sheet 22.—[Lower Lias : Axmouth to Lyme Regis, by H. T. DE LA BECHE.] Before 1839.

Sheet 34.—[Lower Lias to Purbeck Beds : Calne, Chippenham, Corsham, Swindon, Highworth, Cricklade, Malmesbury, Tetbury, Cirencester, Fairford, Stroud, Michinhampton, by W. T. AVELINE and E. HULL.] 1857.

Sheet 35.—[Lower Lias to Forest Marble : Marshfield, Chipping Sodbury, Wotton Underedge, Dursley, Frocester, Purton Passage, Patchway, and Gold Cliff, by D. H. WILLIAMS, JOHN PHILLIPS, A. C. RAMSAY, H. W. BRISTOW, and WILLIAM SANDERS.] 1845.

———.—New edition. [By H. W. BRISTOW.] 1865.

Sheet 36.—[Lower Lias : Lliswerry near Newport, Penarth, Cowbridge, Bridgend, Sutton, and Southerndown, by H. T. DE LA BECHE,, and others.] Before 1845.
———.—New edition. [By H. W. BRISTOW, and H. B. WOODWARD.] 1873.
Sheet 43 N.E.—[Lower Lias : near Hasfield and Berrow, by H. H. HOWELL.] 1855.

Sheet 43 S.E.—[Lower Lias to Inferior Oolite : Gloucester, Haresfield, and Westbury-on-Severn, by A. C. RAMSAY and H. H. HOWELL.] 1845 and 1855.

Sheet 44.—[Lower Lias to Cornbrash : Cheltenham, Tewkesbury, Winchcomb, Evesham, Chipping Campden, Shipston-on-Stour, Moreton-in-the-Marsh, Stow-on-the-Wold, and Burford, by E. HULL and H. H. HOWELL.] 1856.

Sheet 45 N.W.—[Lower Lias to Forest Marble : Chipping Norton, Deddington, and Banbury, by H. BAUERMAN and T. R. POLWHELE.] 1859.

———.—New edition. [Revision (colour only) of Northampton Sand, by J. W. JUDD.] 1871.

Sheet 45 N.E.—[Lower Lias to Oxford Clay : Brackley and Buckingham, by T. R. POLWHELE and A. H. GREEN.] 1863.

———.—New edition. [Revision of Northampton Sand, by J. W. JUDD.] 1871.

Sheet 45 S.W.—[Lower Lias to Corallian : Witney, Stonesfield, and Woodstock, by E. HULL.] 1859.

Sheet 45 S.E.—[Great Oolite to Purbeck Beds : Bicester, Brill, and Quainton, by E. HULL, H. BAUERMAN, W. WHITAKER, and T. R. POLWHELE.]   1863.

Sheet 46 S.W.—[Oxford Clay to Purbeck Beds : Aylesbury and Whitchurch, by E. HULL, H. BAUERMAN, W. WHITAKER, and A. H. GREEN.]   1865.

Sheet 46 N.E.—[Oxford Clay : near Ampthill, by W. Whitaker.]   1869.

Sheet 46 N.W.—[Upper Lias to Oxford Clay : Winslow, Stony Stratford, Fenny Stratford, Newport Pagnell, by A. H. GREEN and J. R. DAKYNS.]   1864.

Sheet 51 N.E.—[Kimeridge Clay : West of Mildenhall Road Station, by S. B. J. SKERTCHLY.]   1883.

Sheet 51 N.W.—[Oxford Clay to Kimeridge Clay : St. Ives, Ely, and Upware by H. B. WOODWARD, W. H. PENNING, S. B. J. SKERTCHLY, and A. J. JUKES-BROWNE.]   1882.

Sheet 51 S.W.—[Oxford and Kimeridge Clays : Elsworth and Boxworth, by W. H. PENNING and A. J. JUKES-BROWNE.]   1881.

Sheet 52 N.E.—[Great Oolite to Corallian : Kimbolton and Huntingdon, by H. H. HOWELL.]   1864.

Sheet 52 N.W.—[Upper Lias Clay to Oxford Clay : Wellingborough, Kettering, Rothwell, Thrapston, and Higham Ferrers, by H. H. HOWELL.]   1864.

———.—New edition. [Additional Lines and Revisions, by J. W. JUDD.]   1870

Sheet 52 S.W.—[Upper Lias to Oxford Clay : Northampton, Olney, and Harrold, by H. H. HOWELL.]   1863.

———.—New edition. [Revisions at Northampton, by J. W. JUDD.]   1871.

Sheet 52 S.E.—[Great Oolite to Kimeridge Clay : Bedford and St. Neots, by H. H. HOWELL.]   1864.

Sheet 53 N.E.—[Lower Lias to Great Oolite : Braunston, Kilsby, East and West Haddon, by W. T. AVELINE and H. H. HOWELL.]   1859.

———.—New edition. [Revision of Oolites, by J. W. JUDD.]   1870.

Sheet 53 S.E.—[Lower Lias to Forest Marble : Towcester, Daventry, by W. T. AVELINE and R. TRENCH.]   1859.

———.—New edition. [Revision of Northampton Sand, by J. W. JUDD.]   1870.

Sheet 53 S.W.—[Lower and Middle Lias and Northampton Sand : Kineton, Hornton, Edge Hill, Chipping Warden, Fenny Compton and Southam, by H. H. HOWELL.]   1856.

———.—New edition. [Revision of Northampton Sand, by J. W. JUDD.]   1870.

Sheet 53 N.W.—[Lower Lias : Stockton, Church Lawford and Rugby, by H. H. HOWELL.]   1855.

Sheet 54 N.W.—[Lower Lias : near Hanbury, by A. C. RAMSAY, J. B. JUKES, H. H. HOWELL, and E. HULL.]   1852.

Sheet 54 N.E.—[Lower Lias : Near Wotton Wawen and Knowle, by H. H. HOWELL.]   1855.

Sheet 54 S.W.—[Lower Lias : Pershore, Crowle and Himbleton, by H. H. HOWELL.]   1854.

Sheet 54 S.E.—[Lower and Middle Lias : Cleeve Prior, Temple Grafton, Wilmcote, Stratford-upon-Avon, &c., by H. H. HOWELL.]   1854.

Sheet 63 N.E.—[Lower and Middle Lias : near Leicester, Barrow-on-Soar, and Loseby, by H. H. HOWELL.]   1855.

———.—New edition. [Additional lines (Middle Lias), by J. W. JUDD.]   1873.

Sheet 63 S.W.—[Lower Lias : near Willy, by H. H. HOWELL.]   1855.

Sheet 63 S.E.—[Lower, Middle, and Upper Lias : Lutterworth, Market Harborough and Wigston Magna, by W. T. AVELINE and H. H. HOWELL.]   1859.

Sheet 64.—[Lower Lias to Oxford Clay : Melton Mowbray, Oakham, Uppingham, Stamford, Peterborough, Oundle, Rockingham, Whittlesea and Ramsey, by J. W. JUDD.]   1872.

Sheet 65.—[Oxford and Kimeridge Clays : March, Downham Market, King's Lynn, by W. WHITAKER and S. B. J. SKERTCHLY.]   1886.

Sheet 70.—[Lower Lias to Kimeridge Clay : Vale of Belvoir, Grantham, Corby, Sleaford, Ancaster, by J. W. JUDD, W. H. HOLLOWAY, W. H. PENNING, W. H. DALTON, and A. J. JUKES-BROWNE.]   1886.

Sheet 71 S.E.—[Lower and Middle Lias : Wimeswold, Old Dalby and Owthorpe, by E. HULL.]   1855.

———.—New edition.   [By W. T. AVELINE.]   1879.

Sheet 73 N.E.—[Lower Lias : Adderley, near Audlem, by A. R. SELWYN.] 1857.

Sheet 73 S.W.—[Lower and Middle Lias : Prees, by A. R. SELWYN and E. HULL.] 1855.

Sheet 73 NW.—[Lower Lias : Burley Dam and Ightfield, by A. R. SELWYN and E. HULL.] 1855.

Sheet 73 S.E.—[Lower Lias : near Moreton Say, by A. R. SELWYN and E. HULL.] 1855.

Sheet 83.—[Lower Lias to Kimeridge Clay : Lincoln, Gainsborough, Horncastle and Market Rasen, by W. H. PENNING, W. H. DALTON, A. C. G. CAMERON, W. A. E. USSHER, A. J. JUKES-BROWNE, and A. STRAHAN.] 1886.

Sheet 84.—[Kimeridge Clay : Spilsby, by A. J. JUKES-BROWNE and A. STRAHAN.] 1884.

Sheet 86.—Lower Lias to Kimeridge Clay : Kirton-in-Lindsey, Glamford Briggs (Brigg), Whitton, Cave, and Brough-on-the-Humber, by C. FOX-STRANGWAYS and W. A. E. USSHER.] 1887.

Sheet 93 N.E. (N.S. 63).—[Lower Lias to Kimeridge Clay : Castle Howard, North Grimston, and Kirkby Underdale, by C. FOX-STRANGWAYS.] 1882.

Sheet 93 N.W. (N.S. 62).—[Lias : Easingwold, by W. T. AVELINE, J. R. DAKYNS and C. FOX-STRANGWAYS.] 1874.

Sheet 93, S.E. (N.S. 71).—[Lower Lias : Pocklington, by C. FOX-STRANGWAYS and A. C. G. CAMERON.] 1885.

Sheet 94 N.W. (N.S. 64).—[Lower Calc. Grit to Kimeridge Clay : North Grimston and Wharram, by J. R. DAKYNS and C. FOX-STRANGWAYS.] 1884.

Sheet 94 S.W. (N.S. 72).—[Lower Lias to Kimeridge Clay : Londesborough, Market Weighton, and Cave, by J. R. DAKYNS and C. FOX-STRANGWAYS.] 1884.

Sheet 95 N.W. (N.S. 44).—[Lower Lias to Upper Calc. Grit : Hackness, Coast between Whitby and Scarborough, by C. FOX-STRANGWAYS and G. BARROW.] 1881.

Sheet, 95 S.W. (N.S. 54).—Lower Oolites to Kimeridge Clay : Scarborough, Filey and Eastern portion of Vale of Pickering, by C. FOX-STRANGWAYS.] 1881.

Sheet 95 S.E. (N.S. 55).—[Kimeridge Clay and Portlandian Beds : Speeton, by C. FOX-STRANGWAYS.] 1881.

Sheet, 96 S.W. (N.S. 52).—[Lower Lias to Kimeridge Clay : Thirsk, Hambleton Hills, and Coxwold, by H. H. HOWELL, C. FOX-STRANGWAYS, A. C. G. CAMERON and G. BARROW.] 1884.

Sheet 96 N.E. (N.S. 43).—[Lower Lias to Middle Calc. Grit : Egton, Rosedale, and Helmsley Moors, by C. FOX-STRANGWAYS, C. REID, and G. BARROW.] 1883.

Sheet 96 N.W. (N.S. 42).—[Lower Lias to Lower Limestone: Stokesley, Northallerton, and Hambleton Hills, by H. H. HOWELL, C. FOX-STRANGWAYS, A. C. G. CAMERON and G. BARROW.] 1883.

Sheet 96 S.E. (N.S. 53).—[Lower Lias to Kimeridge Clay : Malton, Pickering, Helmsley, &c., by C. FOX-STRANGWAYS.] 1882.

Sheet 103 S.E. (N.S. 33).—[Lias and Lower Oolite : Eston, by H. H. HOWELL, A. C. G. CAMERON, and G. BARROW.] 1880.

———.—New edition. 1889.

Sheet 104 S.E. (N.S. 35).—[Lias and Lower Oolites : Whitby, by G. BARROW.] 1880.

Sheet 104 S.W. (N.S. 34.)—Lower Lias to Kellaways Rock : Guisborough, Coast from Redcar to Runswick Bay, by G. BARROW.] 1880.

Sheet 107 S.E.—[Lower Lias : West of Carlisle, by T. V. HOLMES.] 1888.

Maps.—Scale, 6 inches to a mile.

YORKSHIRE.

Sheet 7.—[Lower and Middle Lias : Coast-line adjoining Redcar, by G. BARROW.] 1881.

Sheet 8.—[Lias and Lower Oolite : Saltburn, by G. BARROW.] 1888.

Sheet 9.—[Lias and Lower Oolite : Coast-line adjoining Staithes, by G. BARROW.] 1878.

Sheet 17.—[Lias and Lower Oolite: Guisborough, by G. Barrow.] 1888.

Sheet 20.—[Lias and Lower Oolite : Coast line near Lythe, by G. BARROW.] 1879.

Sheet 32.—[Lias and Lower Oolite : Whitby, by G. BARROW.] 1880.

Sheet 33.—[Lias and Lower Oolite : Coast north of Robin Hood's Bay, by G. BARROW.] 1878.

Sheet 46.—[Lias to Kellaways Rock : District west of Robin Hood's Bay, by G. BARROW.] 1879.

Sheet 47.—[Lias and Lower Oolite : Robin Hood's Bay, by G. BARROW.] 1879.

Sheet 62.—[Lower Oolite to Lower Calc. Grit : Cloughton and Harwood-dale Moor, by C. FOX-STRANGWAYS.] 1879.

Sheet 77.—[Lower Oolite to Upper Calc. Grit : Hackness, by C. FOX-STRANGWAYS.] 1878.

Sheet 78.—[Upper Est. Series to Lower Limestone : Scarborough, by C. FOX-STRANGWAYS.] 1878.

Sheet 93.—[Lower Oolite to Kimeridge Clay : Brompton and Seamer, by C. FOX-STRANGWAYS.] 1879.

Sheet 94.—[Lower Oolite to Kimeridge Clay : Coast between Scarborough and Filey, by C. FOX-STRANGWAYS.] 1879.

MS. Copies of other Six-maps of Yorkshire, and of parts of the Midland Counties, Dorsetshire, &c. are deposited for reference in the Geological Survey Office, London.

Horizontal Sections—Scale, 6 inches to a mile.

Sheet 11.—Section near the Bendick Rock, near Barry Island, Glamorgan, to Allt Llwyd, Brecknock, by Sir H. T. DE LA BECHE and D. H. WILLIAMS. 1846.

Sheet 12.—From the Ebwy River, near Cefn Crib, Monmouth, across the Forest of Dean, to Garden Cliff, Gloucester, by D. H. WILLIAMS. [N.D.]

Sheet 13. No. 3.—Section through Keys End Hill and Berrow Hill, by JOHN PHILLIPS. [N.D.]

Sheet 14. No. 1.— Section from Pyrton Passage through Cam Long Down and Uley Bury to Kingscote Park, Gloucestershire, by H. W. BRISTOW.

——— No. 2.—From the Great Western Railroad near Saltford Station, over Round Hill, Charmy Down, &c., in the direction of the Box Valley, near Slaughterford, by A. C. RAMSAY.

——— No. 3.—Section from Dodington Park, by Wapley and Winterbourn, to the Severn Flats near Chittening Wharf, by H. W. BRISTOW and D. H. WILLIAMS. 1845.

———.—New edition. 1867.

Sheet 15. No. 1.—Section from Ridge Barn Hill near Castle Cary, Somerset-shire, to Jay Hill, near Bitton, Gloucestershire, by Sir H. T. DE LA BECHE, A. C. RAMSAY, and D. H. WILLIAMS. 1845.

———.—New edition (by H. W. BRISTOW). 1871. (Continuation to Wincanton, &c. in Sheet 22.)

Sheet 16.—Section from Mere, Wiltshire, to Vobster, Somerset, by H. T. DE LA BECHE and D. H. WILLIAMS. 1845.

———.—New edition (by H. W. BRISTOW). 1871.

Sheet 17.—Section from Glastonbury Tor across the Mendip Hills, and by Dundry Hill, Clifton, Bristol, and Blaize Castle, to the Severn Flats near Compton Green-field, by H. T. DE LA BECHE, TREVOR E. JAMES, W. TALBOT AVELINE, and WILLIAM SANDERS. 1844.

———.—New edition (by H. W. BRISTOW). 1871.

Sheet 19.—Section from the Hill east of Honiton Church, to the Tertiary Strata east of Dorchester, by H. W. BRISTOW. 1851.

———.—New edition. 1866.

Sheet 20.—Section from the Bill of Portland, Dorsetshire, to Ridge Barn Hill, near Castle Cary, Somersetshire, by H. W. BRISTOW. 1652. (Continuation in Sheet 22.)

Sheet 21.—Section from the River Brue, South of Glastonbury Tor, Somerset, to the Sea at Golden Cap, Dorsetshire, by H. W. BRISTOW. 1852.

———.—New edition. 1873.

Sheet 22.—No. 1. Section from Cadbury Castle to the north-east; No. 2. Section from the Coast, east of Lulworth Cove, Dorsetshire, to Cadbury Castle, Somersetshire, by H. W. BRISTOW. 1855.

Sheet 41.—Section from south-west to north-east across Lower Lias (Shavington Park), &c., by E. HULL. 1857. [With Explanation by E. HULL and A. H. GREEN. 1864.]

Sheet 46.—No. 2. Section from Nailstone Church, through Bagworth Colliery . . . to the Lias Limestone quarries near Barrow-on-Soar, by H. H. HOWELL. 1858. [With Explanations. 1859.]

Sheet 48.—From Lazy Hill . . . near Stanford Hall, to Wysall, by H. H. HOWELL. 1858. [With Explanation, 1859].

Sheet 51.—Section through the [Lower Lias,] New Red Sandstone, &c. and the Warwickshire Coal-field, by H. H. HOWELL. 1858. [With Explanation, 1859].

Sheet 56.—No. 1. Section from the Sea at Broad Bench, to Worgret Heath, near Wareham, Dorsetshire; No. 2. Section from the Sea at St. Alban's Head, to Russel Quay in Wareham Bay; No. 3. From the Sea at Tilly Whim, across Branksea Island, to Poole Harbour; No. 4. From the Sea, east of Lulworth Cove, to Bindon Hill; by H. W. BRISTOW. 1859.

Sheet 59.—From Marlborough Downs, Wiltshire, on the south, to the River Avon, Worcestershire, on the north; crossing the Cretaceous, Oolitic, and Liassic formations of Barbury Hill, Swindon Hill, the Cotteswold Hills, and the outlier of Bredon Cloud, by E. HULL. 1860. [With Explanation, 1861.]

Sheets 71 and 72.—No. 1. From Nettlebed Hill, Oxfordshire, on the south to the Burton Dassett Hills, Worcestershire, on the north; by Cuddesden, and Shotover Hill, Oxford, Islip, Kirtlington, Deddington, Crouch Hill, near Banbury, and Warmington: showing Tertiary, Cretaceous, Oolitic, and Liassic formations, including the Middle Lias Ironstone. No. 2. Drawn from Lambourn Downs, Berkshire, on the south, to Parsonage Farm near Kencott, Gloucestershire, on the north; through Woolstone, Longcott, and Great Coxwell, near Faringdon; showing Cretaceous and Oolitic formations. By E. HULL. 1867. [With Explanation, 1869.]

Sheet 78.—Section from the Chalk of the South Downs at Beachy Head, Sussex; to the Chalk of the north Downs near Boxley, north-east of Maidstone, by H. W. BRISTOW, C. LE NEVE FOSTER, W. TOPLEY, and W. BOYD DAWKINS. 1867. (Purbeck Beds near Battle.)

Sheet 81.—From White Hill, near Kingsclere, in Hampshire, to Pinsley Wood, near Handborough, in Oxfordshire, by H. W. BRISTOW and E. HULL. 1870.

Sheet 82.—From Handborough, in Oxfordshire, to Milverton near Warwick, by H. H. HOWELL and E. HULL. 1870.

Sheet 103.—Section across the northern part of the Somersetshire Coal-field, from Broadfield Down, by Chew Magna, Stanton Wick, Pensford, Marksbury, and Newton St. Loe, to Tiverton near Bath, by JOHN ANSTIE and H. B. WOODWARD. 1874.

Sheet 104.—Section across the Somersetshire Coal-field from Chewton Mendip by Farrington Gurney, Paulton, Camerton, and Dunkerton, to Combe Down, near Bath, by JOHN ANSTIE, J. McMURTRIE, and H. B. WOODWARD. 1875.

Sheet 105.—Section across the southern part of the Somersetshire Coal-field from the Mendip Hills, near Binegar, by Stratton-on-the-Fosse, Midsummer Norton, Radstock, Clandown, Braysdown, and Foxcote, to Norton St. Philip, by JOHN ANSTIE and H. B. WOODWARD. 1875.

Sheet 107.—Section from Portskewet, in Monmouthshire, across the River Severn at New Passage, by Stoke Giffard, in Gloucestershire, across the Bristol Coal-field through Westerleigh, to Wapley, by JOHN ANSTIE and H. B. WOODWARD. 1875.

Sheet 111.—Section across the Bristol Coal-field and the north part of the Somersetshire Coal-field, from Cromhall at the north, through Iron Acton to Mangotsfield, through Warmley to Willsbridge at the south of the Bristol Coal-field, thence to Keynsham and Compton Dando, in Somersetshire, by JOHN ANSTIE and H. B. WOODWARD. 1875.

Sheet 112.—Section across the Somersetshire Coal-field, from Houndstreet on the North, through Farmborough, Timsbury, Clandown, Radstock, Kilmersdon, and Mells, crossing the east end of the Mendip Hills at Whatley, to Long Knoll near Maiden Bradley, by JOHN ANSTIE and H. B. WOODWARD. 1875.

Sheet 121.—Section across the Chalk of Hertfordshire, from 3 miles eastward of Ware, by Baldock, across the Gault and Lower Greensand of Bedfordshire, to

Everton, and across the Oxford Clay to a point 4 miles N.W. of Kimbolton, by
W. H. Penning. 1878.

Sheet 122.—Section from the Three Shire Stone 4 miles N.W. of Kimbolton
(Hunts), through Thrapston, Rockingham Forest (Northamptonshire), Uppingham
(Rutland), and across Burrow Hill to the River Eye, west of Melton Mowbray
(Leicestershire), by William H. Holloway. 1879.

Sheet 124.—Section from Budden Wood, across Mount Sorrel, Burrow Hill
(Leicestershire) The Vale of Catmos, through Oakham and Ketton (Rutland), to
the Fenland at Peterborough (Northamptonshire), by W. H. Holloway. 1887.

Sheet 125.—Section north-north-westward from the Chalk (S. of Dunstable),
across the Gault, the Lower Greensand (S. and W. of Woburn) and the Oxford
Clay to just E. of Newport Pagnell, and thence northward across the Lower
Oolites, and the Lias, to beyond Mears Ashby, by W. Whitaker and W. H.
Holloway. 1881.

Sheet 130.—From the Lower Lias at Staithes, through the Lias and Oolites of
Whitby, Robin Hood's Bay, the Oolites of Hackness and Hutton Bushel; and
across the Vale of Pickering, to the Chalk of Willerby Wold, by C. Fox-
Strangways and G. Barrow. 1883. [With Explanation, 1891.]

Sheet 131.—From the Yorkshire coast at Redcar, across the Ironstone Measures
of Upleatham, and the Lower Oolites of Skelton, Stanghow and Danby, with the
outliers of Kellaways Rock at Freeborough Hill and Danby Beacon; thence across
the valley of the Esk, and the Lower Oolites of Egton, and Pickering Moors, to the
Middle Oolites of Levisham, Kingthorpe and Thornton Dale, and across the Vale of
Pickering, to the Chalk at Knapton, by C. Fox-Strangways, C. Reid, and
G. Barrow. 1883. [With Explanation, 1891.]

Sheet 132.—From the estuary of the Tees at Middlesborough, across the Ironstone
Measures and Lower Oolites of Eston, Roseberry Topping, and Easby Moor, the
Lias of Ingleby Greenhow, the Oolites of Helmsley and Rievaulx Moors, and the
Howardian Hills, to the Lias of Brandsby and Stillington, by C. Fox-Strangways
and G. Barrow, 1883. [With Explanation, 1891.]

Sheet 133 —From the Coal Measures and Magnesian Limestone of Ferryhill,
across the Trias north of Darlington and the valley of the Tees, to the Lias and
Lower Oolite of Whorlton and Osmotherley Moors; thence across the Middle
Oolites of Black Hambleton, the Lower Oolite and Lias of Boltby, Hood Hill,
Kilburn, Coxwold and Husthwaite, to the Lias and Trias of Easingwold, by H. H.
Howell, C. Fox-Strangways, and G. Barrow. 1884. [With Explanation,
1891.]

Sheet 134.—From the Cod Beck east of Northallerton, across the Lias and Lower
Oolite of Over Silton, the Middle Oolite of Black Hambleton, and the Lower Oolite
and Lias of Snilesworth Moor, Bilsdale, Bransdale, Farndale and Rosedale; thence
across Egton High Moor, the Murk Esk, Sleights Moor, and Sneaton, to the Coast at
Hawsker Bottoms south of Whitby, by C. Fox-Strangways, C. Reid and
G. Barrow. 1883. [With Explanation, 1891.]

Sheet 135.—From the Lower Lias of Upsall near Kirkby Knowle, across the
Lower Oolite of Boltby Moor, the Middle Oolite of the Hambleton Hills and the
outliers near Hawnby; thence along the edge of the Tabular escarpment by
Cropton, Saltersgate, Hackness and Seamer Moor, to the Lower Oolite of Gristhorpe
Bay, by C. Fox-Strangways and G. Barrow. 1884. [With Explanation,
1891.]

Sheet 136.—From the Lias below Whitestone Cliff, across the Oolites of the
Hambleton Hills, Helmsley, Kirkby Moorside, Pickering, Allerston, Brompton and
Ayton to the Coast at Scarborough; illustrating the general structure of the Middle
Oolites on the north side of the Vale of Pickering, by C. Fox-Strangways. 1884.
[With Explanation, 1891.]

Sheet 137.—From near Knaresborough, across the Trias of the Vale of York, the
Lias and Oolite of Crayke and the Howardian Hills, the Oolite of the Vale of
Pickering, Kirkby Moorside, Lastingham, Pickering and Fylingdale Moors, to the
Lias of Robin Hood's Bay, by C. Fox-Strangways and G. Barrow. 1884.
[With Explanation, 1891.]

Sheet 138.—Section A.—From Hovingham Spa, across Terrington and Stittenham
to Flaxton; Section B.—From Slingsby across Castle Howard Park to Barton-le-
Willows; Section C.—From Amotherby and Swinton across Hildenley, Huttons-
Ambo, and Westow to Leppington; Section D.—From Wintringham, along the edge
of the Wolds by Settrington, North Grimston, and Birdsall, to Water Dale and
Kirkby Underdale; by C. Fox-Strangways. 1884. [With Explanation, 1891.]

Sheet 139.—Along the western escarpment of the Wolds; illustrating the over-lap of the Cretaceous Rocks; from the Oolites of Malton, Langton, Burythorpe, Leavening, Acklam and Kirby Underdale, across the Lias and Keuper Marl of Bishop Wilton, Kildwick Percy, Londesborough and Market Weighton, to the Oolites of Sancton, Newbald, South Cave and Brough-on-the-Humber, by C. Fox-Strangways. 1884. [With Explanation, 1891.]

Sheet 140.—Section from Bishopstone near Hartwell, through the Oolites and Lias of Bucks and Northamptonshire (near Buckingham and Daventry), and through the Lias and New Red Marl of Warwickshire, from Rugby to near Wibtoft (5 miles S.S.E. of Hinckley), by Prof. A. H. Green, H. B. Woodward, and W. H. Penning. 1886. [With Explanation, by H. B. Woodward, 1891.]

Vertical Sections.

Sheet 22.—Comparative Sections of the Purbeck Strata of Dorset : 1. Durlston Bay; 2. Worbarrow Bay; 3. Mewps Bay; 4. Ridgway Hill, railway cutting; by H. W. Bristow and the Rev. O. Fisher. 1857.

Vertical Section of the Purbeck Strata of Lulworth Cove, Dorsetshire, by H. W. Bristow and W. Whitaker. 1859. (On Horizontal Sections, Sheet 56.)

Sheet 46.—Vertical Sections of the Lower Lias and Rhætic or Penarth Beds of Somerset and Gloucester-shires, (Puriton, Horfield, Uphill, Upper Knowle, Whit-church (Somerset), Aust Cliff, Westbury Cliff, Patchway, Saltford, Weston near Bath, Paulton, Wells (Somerset), Shepton Mallet, and Clutton), by H. W. Bristow, R. Etheridge, and H. B. Woodward. 1873.

Sheet 47.—Vertical Sections of the Lower Lias and Rhætic or Penarth Beds of Glamorgan, Somerset, and Gloucester-shires, (Penarth, Street (Somerset), Laver-nock, Penarth Roads, Curry Rivell, St. Audries' Slip, Combe Hill (Gloucestershire), and Watchet), by H. W. Bristow and R. Etheridge. 1873.

Sheet 48.—Somersetshire and Gloucestershire Coal-fields. (Braysdown Pit, Upper Writhlington Pit, Tyning Pit, Wellsway Pit—Lias and Inferior Oolite), by John Anstie. 1873.

Sheet 49.—Somersetshire and Gloucestershire Coal-fields. (Norton Hill Pit, and Farmborough Pit—Lias), by John Anstie. 1873.

Sheet 50.—Somersetshire and Gloucestershire Coal-fields. (New Mells Pit—Lias and Inferior Oolite), by John Anstie. 1873.

Sheet 51.—Somersetshire and Gloucestershire Coal-fields. General Vertical Section near Keynsham (Queen Charlton—Lias), by John Anstie. 1873.

Sheet 52.—Somersetshire and Gloucestershire Coal-fields. (General Vertical Section near Claudown—Lias and Inferior Oolite), by John Anstie. 1873.

Sheet 67.—Oolites of the Yorkshire Coast, from Filey to Cloughton, by C. Fox-Strangways. 1876.

Memoirs, Reports, Decades, &c.

1839.

De la Beche, H. T.—Report on the Geology of Cornwall, Devon, and West Somerset. 8vo. London.

1846.

De la Beche, Sir H. T.—On the Formation of the Rocks of South Wales and South-western England. Mem. Geol. Surv., vol. i., pp. 1–296.

Ramsay, [Sir] A. C.—On the Denudation of South Wales and the adjacent Counties of England. Mem. Geol. Surv., vol. i., pp. 297–335.

1848.

Forbes, Prof. E.—On the Asteriadæ found fossil in British Strata. Mem. Geol. Surv., vol. ii., pt. 2, pp. 457–482. 8vo. London.

Ransome, T., and B. Cooper.—On the Composition of some of the Limestones used for Building purposes . . Mem. Geol. Surv., vol. ii., part 2, pp. 685–702.

1849–72.

Forbes, E., and others.—Figures and Descriptions illustrative of British Organic Remains. 4to. and 8vo. London.

Decade 1. Echinodermata, by E. Forbes. 1849.
  „   3.   Do.   by E. Forbes. 1850.
  „   4.   Do.   by E. Forbes. 1852.

Decade 5. Echinodermata, by E. FORBES, S. P. WOODWARD, and J. W. SALTER
        1856.
  „   6. FISHES, by Sir P. de M. G. EGERTON. 1852.
  „   8. Do. by Sir P. de M. G. EGERTON. 1855.
  „   9. Do. by Sir P. de M. G. EGERTON. 1858.
  „ 12. Do. by T. H. HUXLEY. 1866.
  „ 13. Do. by A. GÜNTHER, Sir P. de M. G. EGERTON, and T. H.
HUXLEY. 1872.

1855.

DE LA BECHE, SIR H. T., and T. REEKS.—Catalogue of specimens illustrative of
the Composition and Manufacture of British Pottery and Porcelain. 8vo. *London.*
————.—Ed. 2, 1871, and Ed. 3, 1876, by T. REEKS and F. W. RUDLER.

1856.

SMYTH, W. W.—The Iron Ores of Great Britain. Part I. The Iron Ores of the
North and Midland Counties of England. Analyses by J. SPILLER and A. B. DICK.
8vo. *London.*

1857.

HULL, E.— The Geology of the country around Cheltenham. 8vo. *London.*
HUNT, ROBERT.—A Descriptive Guide to the Museum of Practical Geology. 8vo.
London.
————.—Ed. 2, 1859, by R. HUNT;
————.—Ed. 3, 1867, and Ed. 4, 1877, by R. HUNT and F. W. RUDLER;
————.—Ed. 5, 1895, by F. W. RUDLER.

1858.

RAMSAY, A. C., W. T. AVELINE, and E. HULL.—Geology of parts of Wiltshire
and Gloucestershire. 8vo. *London.*
RAMSAY, A. C., H. W. BRISTOW, H. BAUERMAN, and A. GEIKIE.—A Descriptive
Catalogue of the Rock Specimens in the Museum of Practical Geology, with
Explanatory Notices of their Nature and mode of occurrence in place.
————.—2nd Edit. 1859; 3rd Edit. 1862. 8vo. *London.*

1859.

HULL, E.—The Geology of the country around Woodstock, Oxfordshire. 8vo.
London.
HOWELL, H. H.—The Geology of the Warwickshire Coal-field. 8vo. *London.*
[Lias, p. 45.]

1860.

AVELINE, W. T., and R. TRENCH.—The Geology of part of Northamptonshire.
8vo. *London.*
AVELINE, W. T., and H. H. HOWELL.—The Geology of part of Leicestershire.
8vo. *London.*
HUNT, ROBERT.—Mineral Statistics for 1858. Part II. Clay, Building and other
Stones, Fuller's Earth, &c. 8vo. *London.*

1861.

AVELINE, W. T.—The Geology of parts of Northamptonshire and Warwickshire.
8vo. *London.*
HULL, E., and W. WHITAKER.—The Geology of Parts of Oxford and Berkshire.
8vo. *London.*

1864.

GREEN, A. H.—The Geology of the country round Banbury, Woodstock, Bicester,
and Buckingham. 8vo. *London.*
HUXLEY, PROF. T. H.—On the Structure of the *Belemnitidæ*; with a description
of a more complete Specimen of *Belemnites* than any hitherto known, and an
account of a New Genus of *Belemnitidæ, Xiphoteuthis.* Monograph II. 8vo.
and fol. plates. *London.*

1865.

HUXLEY, PROF. T. H., and R. ETHERIDGE.—A Catalogue of Fossils, as arranged in the cases of the Museum of Practical Geology. 8vo. *London.*

1875.

JUDD, J. W.—The Geology of Rutland and the parts of Lincoln, Leicester, Northampton, Huntingdon, and Cambridge included in sheet 64 of the one-inch map of the Geological Survey; with an Introductory Essay on the Classification and Correlation of the Jurassic Rocks of the Midland District of England. 8vo. *London.*

TOPLEY, W.—The Geology of the Weald. (Notes by H. W. BRISTOW; W. T. AVELINE; F. DREW; C. GOULD; and C. LE NEVE FOSTER.) 8vo. *London.*

WHITAKER, W.—Guide to the Geology of London and the neighbourhood, Edit. 5. 1889. 8vo. *London.* [Underground Rocks, pp. 16-24.]

1876.

WOODWARD, H. B.—Geology of East Somerset and the Bristol Coal-fields. With Notes by H. W. BRISTOW, W. A. E. USSHER, and J. H. BLAKE. 8vo. *London.*

1877.

SKERTCHLY, S. B. J.—The Geology of the Fenland. 8vo. *London.*

1880.

STRANGWAYS, C. FOX-.—The Geology of the Oolitic and Cretaceous Rocks south of Scarborough. 8vo. *London.*

1881.

PENNING, W. H., and A. J JUKES-BROWNE.—The Geology of the neighbourhood of Cambridge. 8vo. *London.*

STRANGWAYS, C. FOX-.—The Geology of the Oolitic and Liassic Rocks to the North and West of Malton. 8vo. *London.*

1882.

STRANGWAYS, C. FOX-, and G. BARROW.—The Geology of the country between Whitby and Scarborough. 8vo. *London.*

1884.

STRANGWAYS, C. FOX-.—The Geology of the country north-east of York and south of Malton. 8vo. *London.*

1885.

JUKES-BROWNE, A. J.—The Geology of the South-west part of Lincolnshire, with parts of Leicestershire and Nottinghamshire. 8vo. *London.*

STRANGWAYS, C. FOX-, C. REID, and G. BARROW.—The Geology of Eskdale, Rosedale, &c. 8vo. *London.*

1886.

DAKYNS, J. R. and C. FOX-STRANGWAYS.—The Geology of the country around Driffield. 8vo. *London.*

DAKYNS, J. R., C. FOX-STRANGWAYS, and A. G. CAMERON.—The Geology of the country between York and Hull. 8vo. *London.*

STRANGWAYS, C. FOX-, A. G. CAMERON, and G. BARROW.—The Geology of the country around Northallerton and Thirsk. 8vo. *London.*

1887.

JUKES-BROWNE, A. J.—The Geology of part of East Lincolnshire, including the country near the towns of Louth, Alford, and Spilsby. 8vo. *London.*

1888.

BARROW, G.—The Geology of North Cleveland. 8vo. *London.*

USSHER, W. A. E., A. J. JUKES-BROWNE, and AUBREY STRAHAN.—The Geology of the country around Lincoln. [In part from Notes by W. H. PENNING, W. H. DALTON, and A. C. G. CAMERON.] 8vo. *London.*

1889.

Whitaker, W.—The Geology of London and of part of the Thames Valley. 8vo. *London.* [Underground rocks, vol. i. . pp. 34–39.]

1890.

Ussher, W. A. E.—The Geology of North Lincolnshire and South Yorkshire. (Parts by C. Fox-Strangways, A. C. G. Cameron, C. Reid, and A. J. Jukes-Browne). 8vo. *London.*

1891.

Whitaker, W., H. B. Woodward, F. J. Bennett, S. B. J. Skertchly, and A. J. Jukes-Browne.—The Geology of parts of Cambridgeshire and of Suffolk (Ely, Mildenhall, Thetford). 8vo. *London.*

1893.

Rudler, F. W.—Handbook to the Collection of British Pottery and Porcelain, in the Museum of Practical Geology. 8vo. *London.*

Whitaker, W., S. B. J. Skertchly, and A. J. Jukes-Browne.—The Geology of South-western Norfolk and of Northern Cambridgeshire. 8vo. *London.*

II.—LIST OF WORKS OTHER THAN THOSE OF THE GEOLOGICAL SURVEY.

[For Index of Authors, *see* p. 403.]

1671.

Lister, M.—A Letter . . . on that of M. Steno, concerning Petrify'd Shells *Phil. Trans.*, vol. v. (No. 76), pp. 2282–2284.

1674.

Lister, Martin.—A Description of certain Stones figured like Plants, and by some observing men esteemed to be Plants petrified. [Echinodermata from Bugthorpe.] *Phil. Trans.*, vol. viii., No. 100, p. 6181.

1675.

Lister, M.—Glossoptera tricuspis non-serrata (Malton, &c.). *Phil. Trans.*, vol. ix., No. 110, p. 223.
————.—Of certain Dactili Idæi [*Cidaris* spines], or the true Lapides Judaici, for kind found with us in Britain (Malton, &c.) *Ibid.*, p. 224.
————.—A letter containing his observations of the Astroites, or Star-Stones. [Pentacrinites from Bugthorpe and Leppington.] *Ibid.*, vol x., No. 112, p. 274.

1677.

Plot, Dr. R.—The Natural History of Oxfordshire, being an Essay toward the Natural History of England. [Ed. 2 in 1705.] fol. *Oxford.*

1678.

Lister, M.—Historiæ Animalium Angliæ. 4to. *London.*

1679.

Colwal, D.—An Account of English Alum-Works. *Phil. Trans.*, vol. xii., No. 142, p. 1052.

1684.

Lister, M.—An ingenious Proposal for a new Sort of Maps of Countries; together with Tables of Sands and Clays, such as are chiefly found in the North Parts of England, drawn up about Ten Years since, and delivered to the Royal Society, March 12, 1683. *Phil. Trans.*, vol. xiv., No. 164, pp. 739.

1693.

Lister, M.—An Account of certain transparent Pebbles, mostly of the shape of the Ombriæ or Brontiæ. [Echini from Filey.] *Phil. Trans.*, vol. xvii., No. 201, p. 778.

LHWYD, E.--Epistola in qua agit de lapidibus aliquot perpetua figura donatis, quos nuperis annis in Oxoniensi et Vicinis agris adinvenit. *Phil. Trans.*, vol. xvii., No. 200, pp. 746–754.

1698.

LHWYD, E.—Part of a letter concerning several regularly figured stones lately found by him. *Phil. Trans.*, vol. xx., p. 279.

1699.

LHWYD, E. [LUIDIUS, EDVARDUS.] — Lithophylacii Britannici Ichnographia. [8vo. *London.* Ed. 2. *Oxford*, 1760.]

1700.

PRYME, REV. A. DE LA.—A letter concerning Broughton in Lincolnshire, with observations on the Shell fish observed in the quarries about that place. *Phil. Trans.*, vol. xxii., pp. 677–687.

1712.

MORTON, REV. JOHN.—The Natural History of Northamptonshire. fol. *London.*

1719.

STUKELEY, Dr. W.—An Account of the Impression of the almost Entire Skeleton of a large Animal in a very hard Stone . . . . from [Elston] Nottingham-shire. *Phil. Trans.*, pp. 963-968 (No. 360), pl. i.

1727.

STRACHEY, JOHN.—Observations on the different Strata of Earths, and Minerals, More Particularly of such as are found in the Coal-mines of Great Britain. [Reprinted from the *Phil. Trans.*, 1719 and 1725.] 4to. *London.*

1728.

LEWIS, REV. J.—An Account of the several Strata of Earths and Fossils found in sinking the mineral Wells at Holt [Wiltshire]. *Phil. Trans.*, vol. xxxv., No. 403, pp. 489–491. [*See also Phil. Trans.*, vol. xxxvi., p. 43.]

1728–29.

WOODWARD, DR. JOHN.—An Attempt towards a Natural History of the Fossils of England. 2 vols. 8vo. *London.*

1748.

HILL, J.—A History of Fossils. Folio. *London.*

1754.

BAKER, H.—An Account of some uncommon fossil Bodies. [Oxford, &c.] *Phil. Trans.*, vol. xlviii., pp. 117–123.

OWEN, E.—Observations on the Earths, Rocks, Stones, and Minerals, for some miles about Bristol. . . 8vo. *London.*

1757.

DA COSTA, EMANUEL MENDES—A Natural History of Fossils. 4to. *London.*

1758.

DA COSTA, E. M.--An Account of the Impressions of Plants on the Slates of Coals. [Ferns from Robin Hood's Bay.] *Phil. Trans.*, vol. l., p. 228.

1759.

CHAPMAN, W.—An Account of the fossil Bones of an Alligator found on the Sea-shore, near Whitby, in Yorkshire. *Phil. Trans.*, vol. l., pt. ii., p. 688; also *Gent. Mag.*, vol. xxx. (1760), p. 452.

PLATT, J.—An Account of the fossile Thigh-bone of a large Animal, dug up at Stonesfield, near Woodstock, in Oxfordshire. *Phil. Trans.*, vol. l., pt. 2, p. 524.

WOOLER.—A description of the fossil Skeleton of an Animal found in the Alum Rock near Whitby. *Phil. Trans.*, vol. l., pt. ii. p. 786.

1760.

MICHELL, REV. JOHN—On the Causes and Phenomena of Earthquakes [Alludes to the Lias of Whitby.] *Phil. Trans.*, vol. li., p. 566.

1762.

STEPHENS, J.—An Account of an uncommon Phenomenon in Dorsetshire. [Burning cliff near Charmouth]. *Phil. Trans.*, vol. lii., p. 119.

1765.

PLATT, J.—An Attempt to account for the Origin and the Formation of the Extraneous Fossil commonly called the Belemnite. *Phil. Trans.*, vol. liv., p. 38.

1779.

WALCOTT, JOHN—Description and Figures of Petrifactions found in the Quarries, Gravel Pits, &c., near Bath. 8vo. *London.*

1781.

ENGLEFIELD, SIR H.C.—Account of the Appearance of the Soil at opening a Well at Hanby [Lenton] in Lincolnshire. *Phil. Trans.*, vol. lxxi., pp. 345, 346.

1787.

' LIMBIRD, J.—An Account of the Strata observed in sinking for water at Boston in Lincolnshire. *Phil. Trans.*, vol. lxxvii., pp. 50–54.

1798.

HINDERWELL, T.—The History and antiquities of Scarborough and the Vicinity. Petrefactions &c., by W. TRAVIS. 4to. *York.* [Ed. 2. 8vo. London, 1811.] [Ed. 3, by B. EVANS, parts by F. PALGRAVE, DR. TRAVIS, DR. MURRAY, and W. BEAN. 8vo. Scarborough, 1832.] [Another edition in 1825 not numbered.]

HATCHETT, C.—Observations on Bituminous Substances, with a Description of the Varieties of the Elastic Bitumen. [Kimeridge, p. 136.] *Trans. Linn. Soc.* vol. iv., p. 129.

1804–1811.

PARKINSON, JAMES—Organic Remains of a Former World. An Examination of the Mineralized remains of the Vegetables and Animals of the Antediluvian World ; generally termed Extraneous Fossils. 3 vols. 4to. *London.*

1807.

BOYD C.—Chemical Analysis of Soils. *Letters and Papers Bath and West of England Soc.*, vol. xi., p. 275.

1810.

BOYD, C.—Chemical Analysis of Soils. *Letters and Papers Bath and West of England Soc.*, vol. xii., p. 379.

FARRY, J.—. . . On the Geological Characters and Relations of the Alum Shales on the Northern Coasts of Yorkshire. . . . *Phil. Mag.*, vol. xxxv., pp. 256–261.

WINTER, R.—A mineralogical outline of the district containing the Aluminous Schistus in the County of York, &c. *Nat. Phil. Chem. and Arts*, ser. 2, vol. xxv., pp. 241–257.

1812–1846.

SOWERBY, J., and J. DE CARLE SOWERBY.—The Mineral Conchology of Great Britain. 7 vols. 8vo. *London.*

1813.

TOWNSEND, REV. JOSEPH.—The Character of Moses established for veracity as an Historian, recording events from the Creation to the Deluge. 4to. *Bath and London.*

1814.

HOME, SIR EVERARD.—Some Account of the fossil remains of an Animal more nearly allied to the Fishes than any of the other Classes of animals. *Phil. Trans.*, vol. civ., pp. 571–577.

1815.

FAREY, JOHN.—An alphabetical Arrangement of the Places from whence Fossil Shells have been obtained by Mr. James Sowerby, and drawn and described in vol. i. of his " Mineral Conchology "; with the geographical and stratigraphical Situations of those Places, and a list of their several Fossil Shells, &c. *Phil. Mag.*, vol. xlvi., pp. 211–224.

SMITH, W.—Map of the Strata (in fifteen sheets), scale five miles to an inch, accompanied by a " Memoir " of fifty-one quarto pages. *London.*

1816.

BOGG, EDWARD.—A Sketch of the Geology of the Lincolnshire Wolds. *Trans. Geol. Soc.*, vol. iii., pp. 392–398.

ENGLEFIELD, SIR H. C.—A Description of the Principal Picturesque Beauties, Antiquities, and Geological Phenomena of the Isle of Wight. With additional observations on the Strata of the Island, and their continuation in the adjacent parts of Dorsetshire by THOMAS WEBSTER. Fol. *London.*

HOWE, SIR E.—Some farther account of the fossil remains of an animal, of which a description was given to the Society in 1814. *Phil Trans.*, vol. cvi., pp. 318–321.

HORNER, LEONARD.—Sketch of the Geology of the South-Western part of Somersetshire. *Trans. Geol. Soc.*, vol. iii., pp. 366–379.

1816–19.

SMITH, WILLIAM.—Strata identified by Organized Fossils, containing Prints on Coloured Paper of the most characteristic specimens in each stratum. 4 parts. 4to. *London.*

1817.

BRIGHT, DR. R.—On the Strata in the Neighbourhood of Bristol. *Trans. Geol. Soc.*, vol. iv., p. 193.

SMITH, WILLIAM.—Stratigraphical System of Organised Fossils, with reference to the specimens of the original geological collection in the British Museum. 4to. *London.*

YOUNG, REV. G.—A History of Whitby. [p. 779 *Teleosaurus* found in 1791 near Staithes] 8vo *Whitby.* Also *Phil. Mag.*, vol. li., pp. 206–214, 1818.

1818.

FITTON, DR. W. H.—Notes on the History of English Geology. *Edin. Review*, Feb. 1818.

HOME, SIR E.—Additional facts respecting the fossil remains of an animal. showing that the bones of the sternum resemble those of the *ornithorhynchus paradoxus*. *Phil. Trans.*, vol. cviii., pp. 24–32.

PHILLIPS, WILLIAM.—A Selection of Facts from the best authorities, arranged so as to form an outline of the Geology of England and Wales. 8vo. *London.*

YOUNG, REV. G. and J. BIRD.—Geological and Mineralogical Survey of part of the Yorkshire Coast. *Phil. Mag.*, vol. li., p. 206.

1819.

FAREY, JOHN.—On the importance of knowing and accurately discriminating Fossil Shells, as the means of identifying particular Beds of the Strata in which they are inclosed ; with a list of 279 Species or Varieties of Shells, of which the several Statigraphical and Geographical Localities are mentioned. . . *Phil. Mag.*, vol. liii., pp. 112–132.

————.—A Stratigraphical or Smithian Arrangement of the Fossil Shells which were described (in Latin) by Martin Lister, in 1678, in the 3rd Tract of his " Historiæ Animalium Angliæ." *Phil. Mag.*, vol. liv., pp. 133–138.

HOME, SIR E.—An Account of the fossil skeleton of the *Proteo-Saurus*. *Phil. Trans.*, vol. cix., pp. 209–211.

————.—Reasons for giving the name *Proteo-Saurus* to the fossil skeleton which has been described. *Phil Trans.*, vol. cix., pp. 212–216.

YOUNG, REV. G.—Letter on the discovery of fossil remains [*Ichthyosaurus?*] near Whitby. *Ann. Phil.*, vol. xiii., p. 379.

1820.

HOME, SIR E.—On . . . . the *Proteo-saurus.* *Phil. Trans.*, vol. cx., pp. 159-164.

1821.

CONYBEARE, REV. W. D.—Notice of the discovery of a new Fossil Animal, forming a link between the *Ichthyosaurus* and Crocodile, together with general remarks on the Osteology of the *Ichthyosaurus*; from the Observations of H. T. DE LA BECHE and the Rev. W. D. CONYBEARE. [*Plesiosaurus.*] *Trans. Geol. Soc.*, vol., v., pp. 559-594.

CUMBERLAND, GEORGE—On a new Pentacrinus from Lyme Regis, a new Encrinus, and a Briarean Pentacrinus. *Trans. Geol. Soc.*, vol. v., pp. 379-381.

MILLER, J. S.—A Natural History of the Crinoidea. . . . 4to. *Bristol.*

WINCH, N. J.—Observations on the Eastern part of Yorkshire. *Trans. Geol. Soc.*, vol. v., pp. 545-557.

YOUNG, REV. G.—Account of a Singular Fossil Skeleton [*Ichthyosaurus communis*] discovered at Whitby. *Mem. Wernerian Soc.*, vol. iii., p. 450.

1821-4.

CUVIER, G.—Recherches sur les Ossemens Fossiles. 4to. *Paris.* Ed. 2.

1822.

CONYBEARE, REV. J. J.—On Siliceous Petrifactions imbedded in Calcareous Rock. *Ann. Phil.*, vol. xx. (ser. 2, vol. iv.), pp. 335-337.

CONYBEARE, REV. W. D., and W. PHILLIPS.—Outlines of the Geology of England and Wales. Part i. 8vo. *London.*

CONYBEARE, REV. W. D.—Additional Notice of the Fossil Genera *Ichthyosaurus* and *Plesiosaurus.* *Trans. Geol. Soc.*, vol. i., pp. 103.

DE LA BECHE, [SIR] H. T.—Remarks on the Geology of the South Coast of England, from Bridport Harbour, Dorset, to Babbacombe Bay, Devon. *Trans. Geol. Soc.*, ser. 2, vol. i., p. 40.

SEDGWICK, REV. PROF. A.—On the Geology of the Isle of Wight. [Dorset, p. 342.] *Ann. of Phil.*, ser. 2, vol. iii., p. 329.

WINCH, N. J.—On the Geology of the Eastern Part of Yorkshire. *Ann. Phil.*, ser. 2, vol. iii., p. 374. (For Reply and Counter-reply, see vol. iv., pp. 247, 339.)

WOODS, H.—Account of some Vegetable Remains found in a Quarry near Bath. *Ann. Phil.*, ser. 2, vol. iii., p. 35.

YOUNG, REV. G., and J. BIRD.—A Geological Survey of the Yorkshire Coast. 4to. *Whitby.* [Ed. 2 in 1828.]

1823.

BUCKLAND, REV. PROF. W.—Reliquiæ Diluvianæ. [Lias quarries near Axminster, pp. 241-244; Geol. map of the Vale of Pickering.] 4to. *London.* [Edit. 2. 1824.]

CONYBEARE, REV. W. D., and H. T. DE LA BECHE.—Map of 24 miles round Bath. Geologically coloured. *Bath.*

HARLAN, DR. R. On a new fossil genus, of the order Enaliosauri, Conybeare. *Journ. Acad. Nat. Sci. Philad.*, vol. iii., pp. 331-337.

———.—On a new extinct fossil species of the Genus *Ichthyosaurus.* *Journ. Acad. Nat. Sci. Philad.*, vol. iii., pp. 338, 339.

1824.

BUCKLAND, REV. PROF. W.—Notice on the *Megalosaurus*, or Great Fossil Lizard of Stonesfield. *Trans. Geol. Soc.*, ser. 2, vol. i., pp. 390-396.

BUCKLAND, REV. PROF. W., and REV. W. D. CONYBEARE. Observations on the South-western Coal District of England. *Trans. Geol. Soc.*, ser 2, vol. i., pp. 210-316.

CONYBEARE, REV. W. D.—Additional Notices on the Fossil Genera *Ichthyosaurus* and *Plesiosaurus.* *Trans. Geol. Soc.*, ser. 2, vol. i., pp. 103-123.

———.—On the Discovery of an almost perfect Skeleton of the *Plesiosaurus.* [Lyme Regis.] *Trans. Geol. Soc.*, ser. 2, vol. i., pp. 381-389.

CUMBERLAND, GEORGE.—Remarks on the Strata at Stinchcombe, near Dursley, in Gloucestershire. *Trans. Geol. Soc.*, ser. 2, vol. i., pp. 369, 370.

1824.

FITTON, DR. W. H.—Inquiries respecting the Geological Relations of the Beds between the Chalk and the Purbeck Limestone in the South-east of England. *Ann. Phil.*, vol. xxiv. (ser. 2, vol. viii.), pp. 365–383, 458. [Issued in 4to. in 1836.]

PHILLIPS, R.—Aberthaw Limestone. [Analysis.] *Ann. Phil.*, vol. xxiv. (ser. 2, vol. viii.), p. 72.

SEDGWICK, REV. PROF. A.—On the Phenomena connected with some Trap Dykes in Yorkshire and Durham. *Trans. Cambridge Phil. Soc.*, vol. ii., pt. 1., pp. 21–44; and *Phil. Mag.*, vol. lxvii., pp. 211–219, 249–259 (1826).

WEAVER, T.—Geological Observations on Part of Gloucestershire and Somersetshire. *Trans. Geol. Soc.*, scr. 2, vol. i., pp. 317–368.

YOUNG, REV. G.—A Picture of Whitby and its Environs. [Geological Observations.] 8vo. *Whitby.*

1825.

BRONGNIART, AD.—Note sur les Végétaux fossiles de l'Oolite à Fougères de Mamers. *Ann. des Sciences Nat.*, vol. iv., pp. 422, &c. [Figures some British Liassic plants.]

DESNOYERS, J.—Observations sur quelques systèmes de la formation Oolitique du nord-ouest de la France, &c. [Refers to Stonesfield.] *Ann. des Sciences Nat.*, vol. iv., pp. 353–388.

KÖNIG, CHARLES.—Icones Fossilium Sectiles. Fol. *London.*

KINGDON, J.—Letter on Bones from Chipping Norton. *Ann. Phil.*, ser. 2, vol. x., p. 229.

PRÉVOST, CONSTANT.—Observations sur les Schistes calcaires Oolitiques de Stonesfield en Angleterre, dans lesquels ont été trouvés plusieurs Ossemens fossiles de Mammifères. *Ann. des Sciences Nat.*, vol. iv., pp. 389–417.

YOUNG, REV. G.—Account of a Fossil Crocodile (*Teleosaurus Chapmanni*) recently discovered in the Alum Shale near Whitby. *Edin. Phil. Journ.*, vol. xiii., pp. 76–81; also *Ann. Phil.*, ser. 2, vol. ix., p. 469.

1826.

ANON.—Observations regarding the Position of the Fossil *Megalosaurus* and *Didelphis* or Opossum at Stonesfield. *Edin. Phil. Journ.*, vol. xiv., p. 303.

DE LA BECHE, [SIR] H. T.—On the Lias of the Coast in the Vicinity of Lyme Regis, Dorsetshire. *Trans. Geol. Soc.*, ser. 2, vol. ii., pp. 21–30.

MILLER, J. S.—Observations on Belemnites. *Trans. Geol. Soc.*, ser. 2, vol. ii., pp. 45–62.

SEDGWICK, REV. PROF. A.—On the Classification of the Strata which appear on the Yorkshire Coast. *Ann. Phil.*, vol. xxvii., (ser. 2, vol. xi.), pp. 339–362.

VERNON, REV. W.—An Account of the Strata North of the Humber, near Cave. *Ann. Phil.*, ser. 2., vol. xi., pp. 435–439.

WEBSTER, T.—Observations on the Purbeck and Portland Beds. *Trans. Geol. Soc.*, ser. 2, vol. ii., pp. 37–44.

WILTON, REV. C. P.—Geological Survey of the Shores of the Severn. *Ann. Phil.*, ser. 2, vol. xi., p. 151.

———.—Geology of the Severn. *Quart. Journ. Sci. Lit. & Arts*, vol. xx., p. 413.

1827.

FITTON, DR. W. H.—Remarks on some of the Strata between the Chalk and the Kimmeridge Clay, in the South-east of England. *Proc. Geol. Soc.*, vol. i., pp. 26, 27.

LONSDALE, W.—On the Occurrence of Galena in the Inferior Oolite. *Phil. Mag.*, ser. 2, vol. ii., pp. 234, 235.

1828.

BRODERIP, W. J.—Observations on the Jaw of a Fossil Mammiferous Animal found in the Stonesfield Slate. *Zool. Journ.*, vol. iii., pp. 408–412.

BRONGNIART, A.—Histoire des Végétaux Fossils. 4to. *Paris.*

BUCKLAND, REV. PROF. W.—On the Cycadeoideæ, a Family of Fossil Plants found in the Oolite Quarries of the Isle of Portland. *Trans. Geol. Soc.*, ser. 2, vol. ii., pp. 395–401; *Proc. Geol. Soc.*, vol. i., 1827, pp. 80, 81.

FITTON, DR. W. H.—On the Strata [at Stonesfield in Oxfordshire] whence the Fossil described in the preceding notice was obtained. *Zool. Journ.*, vol. iii., pp. 412–418.

GRAY, DR. J. E.—Description of a new kind of Pear-Encrinite in England [*Millericrinus Prattii*]. *Phil. Mag.*, vol. iv., pp. 219, 220.

MURRAY, DR. P.—Account of a Deposit of Fossil Plants, discovered in the Coal Formation near Scarborough. *Edin. New Phil. Journ.*, vol. v., pp. 311–317.

PHILLIPS, J.—Remarks on the Geology of the North Side of the Vale of Pickering. *Phil. Mag.*, ser. 2, vol. iii., pp. 243–249.

1829.

ANON.—Fossil Asteria. [Horsington.] *Mag. Nat. Hist.*, vol. ii., p. 73.

BUCKLAND, REV. PROF. W.—On the Discovery of Coprolites, or Fossil Fæces, in the Lias at Lyme Regis, and in other Formations : with letter from Dr. W. Prout. *Trans. Geol. Soc.*, ser. 2, vol. iii., pp. 223–238.

BUCKLAND, REV. DR. W.—On the discovery of a new Species of Pterodactyle; and also of the Fæces of the Ichthyosaurus; and of a black substance resembling Sepia, or Indian Ink, in the Lias at Lyme Regis. *Proc. Geol. Soc.*, vol. i., pp. 96–98; *Trans. Geol. Soc.*, ser. 2, vol. iii., pp. 217–222.

CONYBEARE, REV. W. D.—On the Hydrographical Basin of the Thames, with a view more especially to investigate the causes which have operated in the formation of the valleys of that river and its tributary streams. *Proc. Geol. Soc.*, vol. i., pp. 145–149.

CUMBERLAND, G.—Some Account of the Order in which the Fossil Saurians were discovered. *Quart. Journ. Sci. Lit. & Arts*, p. 345.

LONSDALE, WILLIAM.—On the Oolitic District of Bath. *Proc. Geol. Soc.*, vol. i., pp. 98, 99.

PHILLIPS, Prof. J.—Illustrations of the Geology of Yorkshire, or a Description of the Strata and Organic Remains of the Yorkshire Coast. 4to. *York.* [Ed. 2 in 1835. Ed. 3, by R. ETHERIDGE, in 1875.]

STOKES, C.—Letter explanatory of three Drawings of Echini. [One from Stonesfield.] *Trans. Geol. Soc.*, ser. 2, vol. ii., p. 406.

VERNON. REV. W. V.—Analysis of an aluminous Mineral in the Collection of the Yorkshire Philosophical Society [Scarbroite]. *Phil. Mag.*, ser. 2, vol v., pp. 178–189.

1830.

BUCKLAND, Rev. Dr. W., and H. T. DE LA BECHE.—On the Geology of Weymouth, and the adjacent parts of the coast of Dorsetshire. *Proc. Geol. Soc.*, vol. i., pp. 217–221.

DE LA BECHE, [SIR] H. T.—Sections and Views illustrative of Geological Phœnomena. 4to. *London.*

PREVOST, C.—Sur un mémoire de MM. Buckland et Conybeare sur la géologie de la baie de Weymouth. *Bull. Soc. Géol. Fr.*, tome i., pp. 68, 69.

SHARPE, DANIEL,—Description of a New Species of Ichthyosaurus. [*I. grandipes*, Lias, near Stratford-on-Avon.] *Proc. Geol Soc.*, vol. i., pp. 221, 222.

WILTON, REV. C. P.—Memoir on the Geology of the Shore of the Severn, in the Parish of Awre, Gloucestershire. *Quart. Journ. Sc. Lit. & Arts.* N.S. No. 13. pp. 64–73.

1831.

DUNN, J.—On a large species of *Plesiosaurus* in the Scarborough Museum. *Proc. Geol. Soc.*, vol. i., pp. 336, 337.

FLEWKER, J.—A petrified tree [Portland]. *Mag. Nat. Hist.*, vol. iv., p. 73.

MURCHISON, [SIR] R. I.—Notes on the Secondary Formations of Germany as compared with those of England. *Proc. Geol. Soc.*, vol. i., p. 325.

PHILLIPS, PROF. J.—An extemporaneous account of the most remarkable phenomenon in the Geology of Yorkshire. *Rep. Brit. Assoc.* for 1830, p. 56.

SCROPE, G. P.—On the rippled markings of many of the forest marble beds north of Bath, and the *foot tracks* of certain animals occurring in great abundance on their surfaces. *Proc. Geol. Soc.*, vol. i., pp. 317, 318; See also *Journ. Roy. Inst.*, vol. i., p. 538.

WITHAM, H.—Observations on Fossil Vegetables, accompanied by Representations of their Internal Structure, as seen through the Microscope. 4to. *Edinburgh* and *London.*

1831-7.

LINDLEY, J., and W. HUTTON.—The Fossil Flora of Great Britain. 3 vols. 8vo. *London.* [Supplementary volume, 1877, edited by G. A. LEBOUR.]

1832.

GREAVES, J.—A fossilized Fish and *Ichthyosaurus* found in a Stone Quarry near Stratford-upon-Avon. *Mag. Nat. Hist.*, vol. v., p. 549

LONSDALE, WILLIAM.—On the Oolitic District of Bath. *Trans. Geol. Soc.*, ser. 2, vol. iii., pp. 241-276.

MURCHISON, [SIR] R. I.—Structure of the Cotteswold Hills and district around Cheltenham. *Proc. Geol. Soc.*, vol. i., pp. 388-90.

————.—On the Occurrence of stems of fossil plants in vertical positions in the sandstone of the inferior oolite of the Cleveland Hills. *Proc. Geol. Soc.*, vol. i., p. 391.

SMITH, W.—Geological Map of Hackness. [Scale about 6 inches to a mile.] Lithographed by Day, *London.*

1832-33.

FITTON, DR. W. H.—Notes on the Progress of Geology in England. Reprinted from the *Phil. Mag.*, vols. i. and ii. 8vo. *London.*

1833.

CONYBEARE, REV. W. D.—On the alleged Discovery [of Coal at Billesdon Leicestershire. *Phil. Mag.*, ser. 3, vol. iii., p. 112.

FITTON, DR. W. H.—A Geological Sketch of the vicinity of Hastings. Small 8vo. *London.*

HOLDSWORTH, J.—Notice of the Discovery of Coal-Measures and of Fossil Fruits at Billesdon Coplow, in Leicestershire. *Phil. Mag.*, ser. 3, vol. iii., p. 76.

JELLY, REV. H.—The Lansdown Encrinite. *Bath and Bristol Mag.*, No. 5, vol. ii, pp. 36-47.

LONSDALE, WILLIAM.—Report of a Survey of the Oolitic Formations of Gloucestershire. *Proc. Geol. Soc..*, vol. i. pp. 413-415.

MANTELL, DR. G. A.—The Geology of the South-east of England. 8vo. *London.*

PEARCE, J. CHANING.—On the Oolitic Formation and its Contents, as occurring in a Quarry at Bearfield, near Bradford, Wilts. *Proc. Geol. Soc.*, vol. i., pp. 484, 485.

WITHAM, H.—The Internal Structure of Fossil Vegetables found in the Carboniferous and Oolitic Deposits of Great Britain. 4to. *Edinburgh* and *London.*

WOODWARD, S.—An Outline of the Geology of Norfolk. 8vo. *Norwich.*

1833-43.

AGASSIZ, L.—Recherches sur les Poissons fossiles. 5 vols. 4to. and folio (Atlas). *Neufchatel.*

1834.

HASTINGS, DR. CHARLES.—Illustrations of the Natural History of Worcestershire, [Geology, pp. 89-115; Mineral Waters, pp. 115-121.] 8vo. *London.*

HAWKINS, T.—Memoirs of Ichthyosauri and Plesiosauri, extinct Monsters of the Ancient Earth, &c. Fol. *London.*

HOLDSWORTH, J.—A Notice of some important Geological Discoveries at Billesdon Coplow, Leicestershire: with Observations on the Nature of their Relations to the modern System of Geology. *Mag. Nat. Hist.*, vol. vii., p. 38.

MITCHELL, DR. JAMES.—On the Strata of Quainton and Brill in Buckinghamshire. *Proc. Geol. Soc.*, vol. ii., pp. 6, 7.

E E 2

MURCHISON, [SIR] R. I.—Outline of the Geology of the Neighbourhood of Cheltenham. 12mo. Cheltenham. [Ed. 2, by J. BUCKMAN and H. E. STRICKLAND. 8vo. London. 1845.]

NICHOL, W.—Observations on the Structure of Recent and Fossil Coniferæ (Wernerian Soc.) *Edinburgh New Phil. Journ.*, vol. xvi., pp. 137-158.

ROBERTS, G.—The History and Antiquities of the Borough of Lyme-Regis and Charmouth. [Geological notice, with a description of our fossils : revised by DE LA BECHE, with Geological map., pp. 316-332.] 8vo. *London.*

STRICKLAND, H. E.—Letter accompanying a Map of the New Red Marl and Lias in the districts adjacent to Pershore, Evesham, &c. *Proc. Geol. Soc.*, vol. ii., p. 5 ; and *Trans. Geol. Soc.*, ser. 2, vol. v., p. 260. (1837.)

1835.

AGASSIZ, PROF. L.—Lettre sur les Ossemens fossiles de Stonesfield qu'on avait cru pouvoir rapporter à des Didelphes. *Comptes Rendus*, vol. vii., p. 537 ; *Neues Jahrbuch*, vol. iii., p. 185.

BRODERIP, W. J.—Description of some Fossil Crustacea and Radiata. [From the Lias of Lyme Regis.] *Proc. Geol. Soc.*, vol. ii., pp. 201, 202.

BUCKLAND, REV. PROF. W.—Notice of a newly discovered gigantic Reptile. [Buckingham.] *Proc. Geol. Soc.*, vol. ii., p. 190.

BUCKLAND, REV. PROF. W., and [SIR] H. T. DE LA BECHE. On the Geology of the Neighbourhood of Weymouth and the adjacent Parts of the Coast of Dorset. *Trans. Geol. Soc.*, ser. 2, vol. iv., pp. 1-46.

EGERTON, SIR P. DE M. G.—On a peculiarity of Structure in the Neck of Ichthyosauri, not hitherto noticed. *Proc. Geol Soc.*, vol. ii., pp. 192, 193.

FITTON, Dr. W. H.—Notice on the Junction of the Portland and the Purbeck Strata on the Coast of Dorsetshire. *Proc. Geol. Soc.*, vol. ii., pp. 185-187.

HUNTER, W. P.—Some account of the Limestone Quarries and Petrifying Spring at Pounceford, in Sussex ; with Preliminary Remarks on the Wealden Rocks. *Mag. Nat. Hist.*, vol. viii., p. 597. [Reprinted, with additions.]

MURCHISON, [SIR] R. I.—On an outlying basin of Lias on the borders of Salop and Cheshire, with a short account of the lower Lias between Gloucester and Worcester. *Proc. Geol. Soc.*, vol. ii., pp. 114, 115.

NICHOL, W.—On the Anatomical Structure of recent and fossil Woods. *Rep. Brit. Assoc.* for 1834, pp. 660-666.

STRICKLAND, H. E.—Memoir on the Geology of the Vale of Evesham. *Analyst*, vol. ii., p. 1 ; *Memoirs* of H. E. Strickland, pp. 79-89.

1835-36.

ROSE, C. B.—A Sketch of the Geology of West Norfolk. *Phil. Mag.*, ser 3., vol. vii., pp. 171, 274, 370 ; vol. vii., p. 28.

1836.

AGASSIZ, PROF. L.—On Ichthyolites (*Tetragonolepis* from near Stratford-on-Avon). *Analyst*, vol. ii., p. 132.

BEAN, W.—Description and Figures of *Unio distortus* and *Cypris concentrica* from the Upper Sandstone and Shale of Scarborough. *Mag. Nat. Hist.*, vol. ix. pp. 376, 377.

BUCKLAND, REV. PROF. Dr. W.—Geology and Mineralogy considered with reference to Natural Theology. (Bridgewater Treatise.) 2 vols. 8vo. *London.* [Ed. 2. 1858.]

————.—A notice on the Fossil Beaks of four extinct species of Fishes, referrible to the genus *Chimæra*, which occur in the oolitic and cretaceous formations of England. *Proc. Geol. Soc.*, vol. ii., pp. 205, 206 ; *Phil. Mag.*, ser. 3., vol. viii., p. 4.

EGERTON, SIR P. DE M. G.—Further notice on certain peculiarities of Structure in the Cervical Region of the *Ichthyosaurus*. *Proc. Geol. Soc.*, vol. ii., pp. 418, 419.

FITTON, Dr. W. H.—Observations on some of the Strata between the Chalk and the Oxford Oolite, in the South-east of England. *Trans. Geol. Soc.*, ser. 2, vol. iv., pp. 103-388.

HUNTER, W. P.—Rough Notes made during a visit to the Freestone Quarries of the Isle of Portland. . . . *Mag. Nat. Hist.*, vol. ix., p. 97.

WILLIAMSON [PROF]. W. C.—A notice of two hitherto undescribed Species of Radiaria, from the Marlstones of Yorkshire; and Remarks on the Organic Remains in that Stratum. *Mag. Nat. Hist.*, vol. ix., pp. 425, 429, 554.

1837.

ANON.—(Abstract of a paper) on the Fossil *Ichthyosaurus* (from Barrow-on-Soar.) *Analyst*, vol. vii., p. 233.

CHARLESWORTH, E.—Illustrated Zoological Notices. 2. On the recent discovery of a Fossil Crocodile at Whitby. *Mag. Nat. Hist.*, vol. x., p. 531, fig. 65.

EGERTON, SIR P. DE M. G.—A systematic and statigraphical catalogue of the fossil fish in the Cabinets of Lord Cole and Sir Philip Grey Egerton, &c. 4to. *London*.

HUNTON, L.—Remarks on a Section of the Upper Lias and Marlstone of Yorkshire, showing the limited vertical range of the species of Ammonites and other Testacea, with their value as Geological Tests. *Trans. Geol. Soc.*, ser. 2., vol. v., pp. 215-222; and *Proc. Geol. Soc.*, vol. ii., p. 416.

MITCHELL, Dr. J.—On the Strata near Swanwich, in the Isle of Purbeck. *Mag. Nat. Hist.*, vol. x., p. 587.

WILLIAMSON, [PROF.] W. C.—On the Distribution of Fossil Remains on the Yorkshire Coast, from the Lower Lias to the Bath Oolite inclusive. *Trans. Geol. Soc.*, ser. 2., vol. v., pp. 223-242; also, under different title, *Proc. Geol. Soc.*, vol. ii., pp. 82, 429.

1838.

D'ARCHIAC, E. J. A.—Mémoire sur les étages inferieurs de la formation crétacée dans le nord de la France et en Angleterre [Kimeridge of Speeton, p. 262]. *Bull. Soc. Géol. France*, t. ix., pp. 245, 259.

BUCKLAND, REV. PROF. W.—On the discovery of a fossil wing of a Neuropterous Insect in the Stonesfield slate. *Proc. Geol. Soc.*, vol. ii., p. 688.

BURR, FREDERICK—Notes on the Geology of the line of the proposed Birmingham and Gloucester Railway. *Proc. Geol. Soc.*, vol. ii., pp. 539-595.

CLARKE, REV. W. B.—Illustrations of the Geology of the South-east of Dorsetshire. No. 2. On the Strata between Durlstone Head and Old Harry Rocks. *Mag. Nat. Hist.*, ser. 2., vol. ii., pp. 79, 128.

JUKES, J. B.—A Popular Sketch of the Geology of the County of Leicester. *Analyst*, vol. viii., p. 1.

OWEN, [SIR] RICHARD—On the Dislocation of the Tail, at a certain point, observable in the skeletons of many Ichthyosauri. *Proc. Geol. Soc.*, vol. ii., pp. 660-662.

————.—A description of Viscount Cole's specimen of *Plesiosaurus macrocephalus* (Conybeare). *Ibid.* pp. 663-666.

SMEE, ALFRED—On the Strata in which Animal Matter is usually found in Fossils. *Proc. Geol. Soc.*, vol. ii., pp. 672-674.

VALENCIENNES, PROF. A.—Observations sur les mâchoires fossiles des Schistes de Stonesfield, . . *Comptes Rendus*, t. vii. Sept., p. 572. ; *Mag. Nat. Hist.*, ser. 2., vol. iii., p. i. (1839).

1838-39.

BLAINVILLE, H. D. DE.—Doutes sur le prétendu Didelphe fossile de Stonesfield . . *Comptes Rendus*, vii., 402, 727, 749; *Mag. Nat. Hist.*, ser. 2., vol. ii., p. 639, and iii., 49.

1839.

BEAN, W.—A Catalogue of Fossils found in the Cornbrash Limestone of Scarborough; with Figures and Descriptions of some of the undescribed Species. *Mag. Nat. Hist.*, ser. 2., vol. iii., pp. 57-62.

BEDFORD, W.—An Account of the Strata of Lincoln from a recent survey commencing North of the Cathedral and descending to the Bed of the River. *Mag. Nat. Hist.*, ser. 2., vol. iii., pp. 553-556.

BRODERIP, W. J.—Description of some Fossil Crustacea and Radiata, found at Lyme Regis, in Dorsetshire. *Trans. Geol. Soc.*, ser. 2., vol. v., pp. 171–174.

BRODIE, REV. P. B.—A notice on the discovery of the remains of Insects, and a new genus of Isopodous Crustacea belonging to the family Cymothoidæ in the Wealden [Purbeck] Formation in the Vale of Wardour, Wilts. *Proc. Geol. Soc.*, vol. viii., pp. 134, 135.

BUCKLAND, REV. PROF. DR. W., and C. PRÉVOST—(Communications on the " Dirt-bed" of Portland). *Bull. Soc. Geol, France*, tome x , p. 428.

CHARLESWORTH, EDWARD—Illustrated Zoological Notices.—On the Fossil Remains of species of *Hybodus* from Lyme Regis. *Mag. Nat. Hist.*, ser. 2., vol. iii., pp. 242–248. [Note by Miss MARY ANNING, p. 605.]

EGERTON, SIR P. DE M. G.—On certain peculiarities in the Cervical Vertebræ of the *Ichthyosaurus*, hitherto unnoticed. *Trans. Geol. Soc.*, ser 2., vol. v., pp. 187–193.

JELLY, REV. H. On the Fossil Shells of the genus *Modiola* being frequently found in the Bath Oolite, enclosed in the Shells of the genus *Lithodomus*. *Mag. Nat. Hist.*, ser. 2., vol. iii.. pp. 551–553.

MURCHISON, [SIR] R. I.—The Silurian System. [Inferior Oolite and Lias., pp. 13–26, 449, 450, 34–36.] 4to. *London.*

OGILBY, WILLIAM—Observations on the Structure and Relations of the presumed Marsupial Remains from the Stonesfield Oolite. *Proc. Geol. Soc.*, vol. iii., pp. 21–23.

OWEN, PROF. [SIR] R.—On the Jaws of the *Thylacotherium Prevostii* (Valenciennes) from Stonesfield. *Proc. Geol. Soc.*, vol. iii., pp. 5–9.

———.—On the *Phascolotherium. Ibid.*, pp. 17–21.

REPORT (Addressed to the Commissioners of Her Majesty's Woods, Forests, Land Revenues, Works, and Buildings) as the Result of an Inquiry, undertaken under the authority of the Lords Commissioners of Her Majesty's Treasury, by CHARLES BARRY, H. T. DE LA BECHE, WILLIAM SMITH, and CHARLES H. SMITH, with reference to the Selection of Stone for building the new Houses of Parliament. Fol. *London.*

STRICKLAND, H. E.—Notices of the Red Marl and Lias of Worcestershire ; of a Fault by which they are affected ; and of fossil freshwater Shells at Shotover Hill. *Trans. Geol. Soc.*, ser. 2., vol. v., pp. 260, 261.

STUTCHBURY, S.—Description of a new fossil *Avicula* [*longicostata*], from the Lias Shale of Somersetshire. *Mag. Nat. Hist.*, ser. 2., vol. iii., pp. 163, 164.

WOODS, H.—Letter respecting the supposed Frontal Spine of *Hybodus* in the Bath Museum. *Mag. Nat. Hist.*, ser. 2., vol. iii., pp. 282, 283.

1840.

BOWERBANK, DR. J. S.—On the Siliceous Bodies of the Chalk, Greensand and Oolites. *Proc. Geol. Soc.*, vol. iii. pp. 278–281.

CONYBEARE, REV. W. D.—Extraordinary Landslip and Great Convulsion of the Coast of Culverhole Point, near Axmouth. *Edin. New Phil. Journ.*, vol. xxix. p. 160.

CONYBEARE, REV. W. D., and the REV. PROF. W. BUCKLAND.—Ten Plates, comprising a Plan, Sections, and Views representing the changes produced on the Coast of East Devon, between Axmouth and Lyme Regis, by the Subsidence of the Land and Elevation of the Bottom of the Sea, on the 26th December 1839, and 3rd February 1840 ; from Drawings by W. DAWSON, the REV. W. D. CONYBEARE, and Mrs. BUCKLAND ; with a Geological Memoir and Sections descriptive of these and similar Phenomena. Oblong. *London.*

HAWKINS, THOMAS—The Book of the Great Sea-Dragons, Ichthyosauri and Plesiosauri. (30 plates.) Fol. *London.*

MARRAT, W.—On the Discovery of an *Ichthyosaurus* [at Strensham.] *Rep. Brit. Assoc.*, for 1839, sections, p. 70.

OWEN, PROF. R.—Report on British Fossil Reptiles. *Rep. Brit. Assoc.*, for 1839, pp. 43–126.

———.—Note on the Dislocation of the Tail at a certain point observable in the skeleton of many *Ichthyosauri. Trans. Geol. Soc.* ser. 2., vol. v., p. 511.

———.—A Description of a Specimen of the *Plesiosaurus macrocephalus*, Conybeare, in the Collection of Viscount Cole, M.P., &c. *Trans. Geol. Soc.*, ser. 2., vol. v., pp. 515–535.

ROBERTS, GEORGE—An Account of and Guide to the Mighty Land-slip of Dowlands and Bindon, in the Parish of Axmouth, near Lyme Regis, December 25, 1839. (5 Editions in 1840). 8vo. *Lyme Regis.*

STRICKLAND, H. E.—On the occurrence of a Fossil Dragon-fly, from the Lias of Warwickshire. [*Æschna liassina.*] *Mag. Nat. Hist.*, ser. 2, vol. iv., p. 301.

1840–44.

SMITH, C. H.—Lithology; or, observations on Stone used for building. *Royal Inst. Brit. Architects*, 1840, pp. 129–168; 1844, pp. 1–36. 4to.

1840–45.

OWEN, PROF. R.—Odontography: or a Treatise on the Comparative Anatomy of the Teeth. 2 vols. 4to. *London.*

1841.

BOWERBANK, DR. J. S.—On the Siliceous Bodies of the Chalk, Greensands, and Oolites. *Trans. Geol. Soc.*, ser. 2, vol. vi., pp. 181–194.

OWEN, PROF. R.—Observations on the Fossils representing the *Thylacotherium Prevostii*, Valenciennes, with reference to the Doubts of its Mammalian and Marsupial Nature, recently promulgated; and on the *Phascolotherium Bucklandi.* *Trans. Geol. Soc.*, vol. vi., pp. 47–65.

———.—A description of a portion of the skeleton of the *Cetiosaurus*, a gigantic extinct Saurian Reptile occurring in the Oolitic formations of different portions of England. *Proc. Geol. Soc.*, vol. iii., pp. 457–462.

———.—A description of some of the Soft Parts, with the Integument, of the Hind-fin of the *Ichthyosaurus*, indicating the shape of the Fin when recent. *Trans. Geol. Soc.*, ser. 2, vol. vi. pp. 199–201; see also *Proc. Geol. Soc.*, vol. iii., pp. 157, 158, (1840).

PRATT, S. P.—Description of some new Species of Ammonites found in the Oxford Clay, on the line of the Great Western Railway, near Christian Malford. *Ann. and Mag. Nat. Hist.*, vol. viii. pp. 161–165.

THORP, REV. W.—Report on the Agricultural Geology of Part of the Wold District of Yorkshire, and of the Oolite in the neighbourhood of North and South Cave, &c. *Proc. Yorksh. Geol. Soc.*, vol. i., pp. 207–286.

WESTWOOD, PROF. J. O.—(Exhibition of a fossil beetle from Stonesfield.) *Trans. Entom. Soc.*, vol. iv., (Proc.) p. 40.

WILLIAMSON, [PROF.] W. C.—On the Distribution of Organic Remains in the Strata of the Yorkshire Coast, from the Upper Sandstone to the Oxford Clay inclusive. *Trans. Geol. Soc.*, ser. 2, vol. vi., pp. 143–152; also *Proc. Geol. Soc.*, vol. ii., p. 671.

1842.

BRODIE, REV. P. B.—Notice on the Discovery of Insects in the Wealden of the Vale of Aylesbury, Bucks, with some additional observations on the wider distribution of these and other Fossils in the Vale of Wardour, Wiltshire. *Proc. Geol. Soc.*, vol. iii. pp. 780–782.

BUCKMAN, JAMES.—On the Lias Beds near Cheltenham. *Geologist* (Moxon's), pp. 14–20.

———.—Sketch of the Oolite formation of the Cotteswold Range of Hills, near Cheltenham. *Geologist* (Moxon's), pp. 199–208.

JUKES, J. BEETE.—The Geology of the Charnwood Forest: in T. R. Potter's History and Antiquities of Charnwood Forest. 4to. *Nottingham.* [Notes on Lias of Barrow and Hoton, Appendix, pp. 4, 5.] Reprinted by R. Allen in "An Illustrated Hand Book to Charnwood Forest." 8vo. *Nottingham and London*, 1857.

OWEN, PROF. R.—Report on British Fossil Reptiles. Part 2. *Rep. Brit. Assoc.* for 1841, pp. 60–204.

PEARCE, J. CHANING.—On the Mouths of Ammonites, and on Fossils contained in laminated beds of the Oxford Clay, discovered in cutting the Great Western Railway, near Christian Malford in Wiltshire. *Proc. Geol. Soc.*, vol. iii., pp. 592–594.

SANDERS, W.—Notice of Sections of the Railway between Bristol and Bath, a distance of twelve miles, prepared by direction of a Committee of the British Association. *Rep. Brit. Assoc.*, for 1841, p. 67.

STRICKLAND, H. E.—On the genus *Cardinia*, Agassiz, as characteristic of the Lias Formation. *Rep. Brit. Assoc.*, for 1841, pp. 65, 66.

———.—Memoir descriptive of a Series of coloured Sections of the Cuttings on the Birmingham and Gloucester Railway. *Trans. Geol. Soc.*, ser. 2, vol., vi., pp. 545-555; see also *Proc. Geol. Soc.*, vol. iii., pp. 313-317. (1840).

STUTCHBURY, S.—On a new genus of Fossil Bivalve Shells. *Ann. and Mag. Nat. Hist.*, ser. 1, vol. viii. pp. 481-485.

1843.

BEDFORD, W.—The Geology of Lincoln. Papers *Lincolnsh. Topogr. Soc.*, 1841-2, pp. 15-28.

BRODIE, REV. P. B.—Notice on the discovery of the Remains of Insects in the Lias of Gloucestershire, with some remarks on the Lower Members of this Formation. *Proc. Geol. Soc.*, vol. iv., pp. 14-16; *Rep. Brit. Assoc.*, for 1842, Sections, p. 58.

BROWN, CAPT. T.—Description of some New Species of the Genus, *Pachyodon*. *Ann. Nat. Hist.*, vol. xii., pp. 390-396.

———.—The Elements of Fossil Conchology, according to the arrangement of Lamarck, with the newly established genera of other authors. 8vo. *London.*

BUCKMAN, PROF. J.—On an *Ichthyosaurus* from Bredon, Gloucester. *Geologist.* (Moxon's), p. 22.

———.—List of the Fossils found in the neighbourhood of Cheltenham, with remarks. *Geologist* (Moxon's), pp. 104-111.

———.—On the Nature and Origin of the Saline Waters of Cheltenham. *Geologist* (Moxon's), pp. 146-148.

———.—Geological Chart of the Oolitic Strata of the Cotteswold Hills, and the Lias of the Vale of Gloucester. Folio. *Cheltenham.*

———.—On the occurrence of the remains of Insects in the Upper Lias of the county of Gloucester. *Proc. Geol. Soc.*, vol. iv., pp. 211, 212.

EGERTON, SIR P. DE M. G.—On some new species of Fossil Chimæroid Fishes, with remarks on their general affinities. *Proc. Geol. Soc.*, vol. iv., pp. 153-157.

———.—On some new Ganoid Fishes. *Proc. Geol. Soc.*, vol. iv., pp. 183, 184.

GOWEN, R.—Account of the effect of a Bituminous Shale at Christian Malford. *Journ. Roy. Agric. Soc.*, vol. iv., p. 276.

JOHNSTONE, SIR J.—On the application of Geology to Agriculture. *Journ. Roy. Agric. Soc.*, vol. i., pp. 270-274.

MILNE-EDWARDS, H.—Note sur deux Crustacés fossiles de l'Ordre des Isopodes. [*Archæoniscus Brodiei*]. *Ann. des Sciences Naturelles*, ser. 2, vol. xx. p. 326; See also *Ann. Nat. Hist.*, vol. xiii., 1844, p. 110.

MORRIS, PROF. J.—A Catalogue of British Fossils. 8vo. *London.* [Edit. 2, in 1854.]

OWEN, PROF. R.—Report on the British Fossil Mammalia. *Rep. Brit. Assoc.*, for 1842, pp. 54-74.

SIMPSON, MARTIN.—A Monograph of the Ammonites of the Yorkshire Lias. 12mo. *London.*

1844.

AGASSIZ, L.—Synoptical Table of British Fossil Fishes, arranged in the order of the Geological Formations. *Rep. Brit. Assoc.* for 1843, pp. 194-209.

FORBES, PROF. E.—On the Fossil Remains of Starfishes of the order *Ophiuridæ*, found in Britain. *Proc. Geol. Soc.*, vol. iv., pp. 232-234.

OWEN, PROF. R.—A Description of certain Belemnites, preserved, with a great proportion of their soft parts, in the Oxford Clay, at Christian Malford, Wilts. [Note by S. P. PRATT.] *Phil Trans.*, vol. cxxxiv., p. 65-85.

PHILLIPS, PROF. J.—Memoirs of WILLIAM SMITH. 8vo. *London.*

STRICKLAND, H. E.—On *Cardinia*, Agassiz, a fossil genus of Mollusca characteristic of the Lias. *Ann. Nat. Hist.*, vol. xiv., pp. 100-108.

THOMSON, J. G.—Account of the land-slip in Ashley Cutting, on the line of the Great Western Railway. *Proc. Inst. Civ. Eng.*, vol. iii., pp. 129-133.

1845.

BRODIE, REV. P. B.—A History of the Fossil Insects in the Secondary Rocks of England, accompanied by a particular Account of the Strata in which they occur. 8vo. *London.*

BRODIE, REV. P. B., and J. BUCKMAN.—On the Stonesfield Slate of the Cotteswold Hills. *Quart. Journ. Geol. Soc.*, vol. i., pp. 220-225; *Proc. Geol. Soc.*, vol., iv. pp. 437-442.

CARPENTER, DR. W. B.—On the Microscopic Structure of Shells. *Rep. Brit. Assoc.* for 1844, pp. 1-24.

CHARLESWORTH, C.—Notice of the Discovery of a large Specimen of *Plesiosaurus* found at Kettleness on the Yorkshire Coast. *Rep. Brit. Assoc.* for 1844, Sections, pp. 49, 50.

DE LA BECHE, SIR H. T.—Report on the State of Bristol, Bath, Frome, Swansea, Merthyr Tydfil, and Brecon. (Health of Towns Commission.) 8vo. *London.*

EGERTON, SIR P. DE M. G.—On some New Species of Fossil Fish, from the Oxford Clay at Christian Malford. *Quart. Journ. Geol. Soc.*, vol. i., pp. 229-232. *Proc. Geol. Soc.*, vol. iv., pp. 446-449.

LUCAS, W. On the Limestones of Yorkshire [Analyses of Malton limestone, &c.]. *Rep. Brit. Assoc.* for 1844, Sections, pp. 30, 31.

MORRIS, PROF. J.—On the Occurrence of the Genus *Pollicipes* in the Oxford Clay. [Christian Malford]. *Ann. and Mag. Nat. Hist.*, vol. xv., pp. 30, 31.

————.—Description of some new species of the genus *Ancyloceras*. *Ibid.*, p. 31.

STRICKLAND, H. E.—On certain Calcareo-corneous Bodies found in the outer chambers of Ammonites. [*Aptychus.*] *Quart. Journ. Geol. Soc.*, vol. i., pp. 232-235. *Proc. Geol. Soc.*, vol iv., pp. 449-452.

————On an Anomalous Structure in the Paddle of a Species of *Ichthyosaurus*. *Rep. Brit. Assoc.* for 1844, p. 51.

1846.

CHARLESWORTH, E.—On the Fossil Bodies regarded by M. Agassiz as the Teeth of a Fish, and upon which he has founded his supposed genus *Sphenonchus*. *Rep. Brit. Assoc.*, for 1845, Sections, p. 56.

COLEMAN, REV. W. H.—On the Geology of the County of Leicester. (Article in W. White's History, Gazetteer, &c. of Leicestershire and Rutland). With a map. 12mo. *Sheffield.*

MANTELL, DR. G. A.—A few Notes on the Prices of Fossils. *Lond. Geol. Journ.* pp. 13-17.

MORRIS, PROF. JOHN.—On the Subdivision of the genus *Terebratula*. *Quart. Journ. Geol. Soc.*, vol. ii., pp. 382-389.

OWEN, PROF. R.—A History of British Fossil Mammals and Birds. 8vo. *London.*

SEDGWICK, REV. PROF. A.—On the Geology of the Neighbourhood of Cambridge, including the Formations between the Chalk Escarpment and the Great Bedford Level. *Rep. Brit. Assoc.*, for 1845, pp. 40-47.

STRICKLAND, H. E.—On the results of recent Researches into the Fossil Insects of the Secondary Formations of Britain. *Rep. Brit. Assoc.*, for 1845, Sections, p. 58.

————.—On two Species of Microscopic Shells found in the Lias. *Quart. Journ. Geol. Soc.*, vol. ii., pp. 30, 31.

PEARCE, J. C.—Notice of what appears to be the Embryo of an *Ichthyosaurus* in the Pelvic cavity of *Ichthyosaurus* (*communis?*). *Ann. and Mag. Nat. Hist.*, ser. 1, vol. xvii., p. 44.

STUTCHBURY, SAMUEL.—Description of a New Species of *Plesiosaurus* in the Museum of the Bristol Institution. *Quart. Journ. Geol. Soc.*, vol. ii., pp. 411-417.

1847.

CHARLESWORTH, E.—New Species of *Pentacrinus* (*P. gracilis*) in the Lias of Yorkshire. *Lond. Geol. Journ.*, pp. 96, 131. (Figures of *Asterias* and *Ophiura*, plates 17, 20.)

BRODIE, REV. P. B.—Notice on the existence of Purbeck Strata with remains of Insects and other Fossils, at Swindon, Wilts. *Quart. Journ. Geol. Soc.*, vol. iii., pp. 53, 54.

CUNNINGTON, WILLIAM.—On the Fossil Cephalopoda from the Oxford Clay constituting the genus *Belemnoteuthis* (Pearce). *Lond. Geol. Journ.*, pp. 97–99.

DAVIDSON, THOMAS.—Remarks on some Species of Brachiopoda figured in Plate 18 of the "London Geological Journal." *Lond. Geol. Journ.*, pp. 109–117.

DAVIDSON, T. and PROF. J. MORRIS. Description of some Species of Brachiopoda. *Ann. and Mag. Nat. Hist.*, vol. xx., pp. 250–257.

EGERTON, SIR P. DE M. G.—On the Nomenclature of the Fossil Chimæroid Fishes. *Quart. Journ. Geol. Soc.*, vol. iii., pp. 350–353.

IBBETSON, CAPT. L. L. B.—On three Sections of the Oolitic Formations on the Great Western Railway, at the West End of Sapperton Tunnel. *Rep. Brit. Assoc.*, for 1846, Sections, p. 61.

MANTELL, DR. G. A.—Geological Excursions round the Isle of Wight, and along the adjacent coast of Dorsetshire. 8vo. *London*. (Ed. 2 in 1851; Ed. 3 in 1854.)

PEARCE, J. C.—On the fossil Cephalopoda constituting the genus *Belemnoteuthis*, Pearce. *Lond. Geol. Journ.*, pp. 75–78. (See also notes by E. CHARLESWORTH, pp. 79–85, and pp. 127, 128.

SANDERS, W.—On Railway Sections made on the Line of the Great Western Railway, between Bristol and Taunton. *Rep. Brit. Assoc.*, for 1846, pp. 59, 60.

WALTON, JOHN.—On the Laws of Development of Existing Vegetation and the application of these laws to certain Geological Problems. *Quart. Journ. Geol. Soc.*, vol. iii., pp. 64, 65. (Refers to Portland.)

1848.

BOWERBANK, DR. J. S.—Microscopical Observations on the Structure of the Bones of *Pterodactylus giganteus* and other Fossil Animals. *Quart. Journ. Geol. Soc.*, vol. iv., pp. 2–10.

CARPENTER, DR. W. B.—Report on the Microscopic Structure of Shells. Part II. *Rep. Brit. Assoc.*, for 1847, pp. 93–134.

IBBETSON, CAPT. L. L. B. and PROF. J. MORRIS.—Notice of the Geology of the Neighbourhood of Stamford and Peterborough. *Rep. Brit. Assoc. for 1847*, pp. 127–131.

LYCETT, JOHN.—On the Mineral Character and Fossil Conchology of the Great Oolite, as it occurs in the neighbourhood of Minchinhampton. *Quart. Journ. Geol. Soc.*, vol. iv., pp. 181–191.

MANTELL, DR. G. A.—Observations on some Belemnites and other Fossil Remains of Cephalopoda, discovered by Mr. R. N. MANTELL in the Oxford Clay near Trowbridge, Wiltshire. *Phil. Trans.*, vol. cxxxviii., pp. 171–183.

McCOY, F.—On some new Mesozoic Radiata. *Ann. and Mag. Nat. Hist.*, ser. 2, vol. ii., pp. 397–420.

WESTON, C. H.—On the Geology of Ridgway, near Weymouth. *Quart. Journ. Geol. Soc.*, vol. iv., pp. 245-256.

WOODWARD, S. P.—On the Geology of the district explored by the Cotteswold Club, and more particularly the Clay subsoil of the [Royal Agricultural] College Farm. *Proc. Cotteswold Club*, vol. i., pp. 2–8.

1849.

BRODIE, REV. P. B.—Notice on the Discovery of a Dragon-fly and a new Species of *Leptolepis* in the Upper Lias near Cheltenham, with a few remarks on that Formation in Gloucestershire. *Quart. Journ. Geol. Soc.*, vol v., pp. 31–37.

BROWN, CAPT. T.—Illustrations of the Fossil Conchology of Great Britain and Ireland; with Description and Localities of all the Species. 4to. *London*.

BUCKMAN, PROF. J.—On the Discovery of some Remains of the Fossil Sepia in the Lias of Gloucestershire. *Rep. Brit. Assoc.*, for 1848, p. 66.

———.—On the Plants of the "Insect Limestone" of the Lower Lias. *Ibid.*, pp. 66, 67.

———.—On some Experimental Borings in search of Coal [Malmesbury]. *Ibid.*, p. 67.

DAVIDSON, T.—Note sur quelques espèces de *Leptæna* du lias et du marlstone de France et d'Angleterre. *Bull. Soc. Géol. France*, ser. 2, tome vi., pp. 275–277.

HAGEN, H. A.—Ueber die fossile Odonate, *Heterophlebia dislocata*, Westwood. *Stettin, Entom. Zeit.*, vol. x,, pp. 226-231.

LYCETT, JOHN.—A few general Remarks on the Fossil Conchology of the Great Oolite of Minchinhampton in comparison with that of the same Formation in other localities. *Proc. Cotteswold Club*, vol. i., pp. 17-20; *Ann. and Mag. Nat. Hist.*, ser. 2, vol. i., p. 115.

——.—Notes on the distribution of the Fossil Conchology of the Oolitic Formations in the vicinity of Minchinhampton, Gloucestershire. *Proc. Cottesw. Club*, vol. i., pp. 21-28.

McCOY, PROF. F.—On the Classification of some British Fossil Crustacea, with Notices of new Forms in the University Collection at Cambridge. *Ann. Nat. Hist.*, ser. 2, vol. iv., pp. 171-173, 392.

MOBERLY, REV. DR.—Letter describing a large *Plesiosaurus* discovered in Lias at Kettleness, near Whitby. *Rep. Brit. Assoc.* for 1848, Sections, p. 78.

MORRIS, PROF. J.—On *Neritoma*, a fossil genus of Gasteropodous Mollusks allied to *Nerita*. *Quart. Journ. Geol. Soc.*, vol. v., pp. 332-335.

STRICKLAND, H. E.—On the Geology of the Oxford and Rugby Railway. *Proc. Ashmolean Soc.*, vol. ii., p. 192.

WESTON, CHARLES H.—Further Observations on the Geology of Ridgway near Weymouth. *Quart. Journ. Geol. Soc.*, vol. v., pp. 317-319.

WILLIAMSON, PROF. W. C.—On the Microscopic Structure of the Scales and Dermal Teeth of some Ganoid and Placoid Fish. *Phil. Trans.*, vol. cxxxix., p. 435.

——.—On the scaly vegetable heads or collars from Brunswick Bay, supposed to belong to *Zamia gigas*. *Proc. York Phil. Soc.*, vol. i., part 1, pp. 45-70.

YATES, J.—Notice of *Zamia gigas*. *Proc. Yorksh. Phil. Soc.*, vol. i., pp. 37-42.

1850.

BRODIE, REV. P. B.—Sketch of the Geology of the neighbourhood of Grantham, Lincolnshire; and a comparison of the Stonesfield Slate at Collyweston in Northamptonshire with that in the Cotswold Hills. *Proc. Cotteswold Club*, vol. i., pp. 52-61; *Ann. & Mag. Nat. Hist.*, ser. 2, vol. i., p. 256.

——.—On certain Beds in the Inferior Oolite, near Cheltenham; with Notes on a Section of Leckhampton Hill, by H. E. STRICKLAND. *Quart. Journ. Geol. Soc.*, vol. vi., pp. 239-251.

BRONGNIART, AD.—Chronological Exposition of the Periods of Vegetation and the different Floras which have successively occupied the surface of the Earth. *Ann. and Mag. Nat. Hist.*, ser. 2, vol. vi., pp. 73-85, 192-203, 348-370; and *Edin. N. Phil. Journ.*, vol. xlviii., pp. 320-330, xlix., pp. 72-97. [Translated from *Ann. Sci. Nat. Bot.*, sér. 3, tome xi., pp. 285-338, 1849.]

BUCKMAN, PROF. JAMES.—On some Fossil Plants from the Lower Lias. *Quart. Journ. Geol. Soc.*, vol. vi., pp. 413-418.

DAVIDSON, T.—Sur quelques Brachiopodes nouveaux ou peu connus. *Bull. Soc. Géol. France*, ser. 2, tome vii., p. 62.

EGERTON, SIR P. DE M. G.—Palichthyologic Notes. No. 3. On the Ganoidei Heterocerci. *Quart. Journ. Geol. Soc.*, vol. vi., pp. 1-10.

FISHER, REV. O.—(Notes on the Geology of the Ridgeway cutting, Dorset.) In "Guide to Dorchester," published by Wm. Barclay. 8vo. *Dorchester*. [1850 or 1851.]

FORBES, PROF. E.—On the Succession of Strata and Distribution of Organic Remains in the Dorsetshire Purbecks. *Rep. Brit. Assoc.*, for 1850, Sections, pp. 79-81; *Edin. Phil. Journ.*, vol. xlix., pp. 311-313, 391 (1850).

GODWIN-AUSTEN, R. A. C.—On the Age and Position of the Fossiliferous Sands and Gravels of Farringdon. *Quart. Journ. Geol. Soc.*, vol. vi., pp. 454-478. [Swindon, p. 464.]

HAGEN, H. A.—Revue des Odonates ou Libellules d'Europe. *Mem. Soc. Roy. des Sci. Liége*, vol. vi.

LYCETT, J.—On *Trichites*, a fossil genus of Bivalve Mollusks. *Proc. Cotteswold Club*, vol. i., pp. 42-46; *Ann. and Mag. Nat. Hist.*, sec. 2, vol. vi., p. 42.

——.—Tabular view of Fossil Shells from the middle division of the Inferior Oolite in Gloucestershire. *Proc. Cotteswold Club*, vol. i., pp. 62-86; *Ann. and Mag. Nat. Hist.*, ser. 2, vol. vi., p. 401.

MANTELL, DR. G. A.—Supplementary Observations on the Structure of the Belemnite and *Belemnoteuthis*. *Phil. Trans.*, vol. cxl., pp. 393–398.

MANTELL, R. N.—An Account of the Strata and Organic Remains exposed in the Cuttings of the Branch Railway, from the Great Western Line near Chippenham, through Trowbridge, to Westbury in Wiltshire. [With list of Fossils, and Descriptions of new Species, by JOHN MORRIS.] *Quart. Journ. Geol. Soc.*, vol. vi., pp. 310–319.

MORRIS, PROF. J., and J. LYCETT.—On *Pachyrisma*, a Fossil genus of Lamelli-branchiate Conchifera. *Quart. Journ. Geol. Soc.*, vol. vi., pp. 399–402.

ORLEBAR, A. B.—On a *Pygaster* from the Coral Rag, Bullingdon. *Proc. Ash-molean Soc.*, vol. ii., p. 242.

1851.

BRODIE, REV. P. B.—On the Basement Beds of the Inferior Oolite in Gloucester-shire. *Quart. Journ. Geol. Soc.*, vol. vii., pp. 208–212.

————.—Remarks on the Stonesfield State at Collyweston, near Stamford, and the Great Oolite, Inferior Oolite and Lias, in the Neighbourhood of Grantham. *Rep. Brit. Assoc.*, for 1850, Sections, pp. 74–76.

BUNBURY, C. J. F.—On some Fossil Plants from the Jurassic Strata of the Yorkshire Coast. *Quart. Journ. Geol. Soc.*, vol. vii., pp. 179–194.

DELESSE, M.—Recherches sur les Roches Globuleuses. *Mem. Soc. Géol. France*, ser. 2, vol. iv., pp. 336–362.

FRAAS, OSCAR.—On the Comparison of the German Jura Formations with those of France and England. *Quart. Journ. Geol. Soc.*, vol. vii., pp. 42–83.

LYCETT, J.—On the Hinge of the Fossil genus *Platymya Agassiz*; with the description of a new species. *Ann. and Mag. Nat. Hist.*, ser. 2, vol. viii., p. 81.

MILNE-EDWARDS, H., and JULES HAIME.—A Monograph of the British Fossil Corals. [Oolitic]. *Palæontograph. Soc.* 4to. *London.*

ORLEBAR, A. B.—On the Geology of the neighbourhood of Oxford, within the limits of Mr. Stacpoole's map. *Proc. Ashmolean Soc.*, vol. ii., p. 253.

SORBY, H. C.—On the Microscopical Structure of the Calcareous Grit of the Yorkshire Coast. *Quart. Journ. Geol. Soc.*, vol. vii., pp. 1–6; *Proc. Yorksh. Geol. Soc.*, vol. iii., pp. 197–205.

————.—On the Excavation of the Valleys in the Tabular Hills, as shown by the configuration of Yedmandale, near Scarborough. *Proc. Yorksh. Geol. Soc.* vol. iii., pp. 169–172.

WRIGHT, DR. THOMAS.—Contributions to the Palæontology of Gloucestershire:—On the *Strombidæ* of the Oolites. With the description of a new and remarkable *Pteroceras*, by J. LYCETT. *Proc. Cotteswold Club*, vol. i., pp. 115–119; *Ann. and Mag. Nat. Hist.*, ser. 2, vol. vii., p. 306.

1851–86.

DAVIDSON, DR. T.—British Fossil Brachiopoda. [Oolitic and Liassic. With Supplements.] *Palæontograph. Soc.* 4to. *London.*

1851–63.

MORRIS, PROF., and JOHN LYCETT. A Monograph of the Mollusca from the Great Oolite, chiefly from Minchinhampton and the Coast of Yorkshire. (With Supplement, by J. LYCETT.) *Palæontograph. Soc.* 4to. *London.*

1851–52.

WRIGHT, DR. T.—On the *Cidaridæ* of the Oolites, with a description of some new species of that family. *Proc. Cotteswold Club*, vol. i., pp. 134–173; *Ann. Nat. Hist.* ser. 2, vol. viii., p. 241–280.

————.—On the *Cassidulidæ* of the Oolites, with descriptions of some new species of that family. *Ibid.*, pp. 174–227; *Ibid.*, vol. ix., pp. 81, 294.

1852.

AUSTEN, REV. J. H. A Guide to the Geology of the Isle of Purbeck, and the South-west coast of Hampshire. 8vo. *Blandford.*

MANTELL, DR. G. A.—A few Notes on the Structure of the Belemnite. *Ann. and Mag. Nat. Hist.*, ser. 2, vol. x., p. 14.

MAY, C.—Iron-Ore near Middlesbro'-on-Tees. *Proc. Inst. Civ. Eng.* vol. xi. p. 28.

MOORE, C.—On the *Aptychus*. *Proc. Somerset Arch. Soc.*, vol. ii., pp. 111–115.

SORBY, H. C.—On the Direction of Drifting of the Sandstone Beds of the Oolitic Rocks of the Yorkshire Coast. *Proc. York. Phil. Soc.*, vol. i., pp. 111–113.

———.—On the Occurrence of Non-Gymnospermous Exogenous Wood in the Lias near Bristol. [Keynsham]. *Trans. Micros. Soc.*, vol. iii., p. 91.

WESTON, C. H.—On the Sub-escarpments of the Ridgeway Range, and their Contemporaneous Deposits in the Isle of Portland. *Quart. Journ. Geol. Soc.*, vol. viii., pp. 110–120.

1853.

BRODIE, REV. P. B.—Remarks on the Lias at Fretherne, near Newnham, and Purton, near Sharpness; with an Account of some new Foraminifera discovered there, &c. [Note by Prof. T. R. JONES.] *Proc. Cotteswold Club*, vol. i, pp. 241–246; *Ann. and Mag. Nat. Hist.*, ser. 2, vol. xii., p. 272.

———.—Notice of the Occurrence of an Elytron of a Coleopterous Insect in the Kimmeridge Clay at Ringstead Bay, Dorsetshire. *Quart. Journ. Geol. Soc.*, vol. ix., pp. 51, 52.

BUCKMAN, PROF. JAMES.—On the Cornbrash of the neighbourhood of Cirencester. *Proc. Cotteswold Club*, vol. i., pp. 262–267; *Ann. and Mag. Nat. Hist.*, ser. 2, vol. xii., pp. 324–329.

———.—Remarks on *Libellula Brodiei* (Buckman), a Fossil Insect from the Upper Lias of Dumbleton, Gloucestershire. *Proc. Cotteswold Club*, vol. i., pp. 268–270; *Ann. and Mag. Nat. Hist.*, ser 2, vol. xii. pp. 436–438.

COLES, HENRY.—On the Skin of the *Ichthyosaurus*. *Quart. Journ. Geol. Soc.*, vol. ix., pp. 79–81.

GAUDRY, ALBERT.—Note sur Stonesfield près Oxford (Angleterre). *Bull. Soc. Géol. France*, ser 2, tome x., pp. 591–596.

GAVEY, G. E.—On the Railway Cuttings at the Mickleton Tunnel, and at Aston Magna, Gloucestershire. *Quart. Journ. Geol. Soc.*, vol. ix., pp. 29–37.

LYCETT, J.—Note on the *Gryphæa* of the Bed called Gryphite Grit in the Cotteswolds. *Proc. Cotteswold Club*, vol. i., pp. 235, 236; *Ann. and Mag. Nat. Hist.*, ser. 2, vol. xi., p. 200.

———.—Additional Notice of the genus *Tancredia* (Lycett), *Hettangia* (Terquem). *Proc. Cotteswold Club*, vol. i., pp. 237–240.

———.—On some new species of *Trigonia* from the Inferior Oolite of the Cotteswolds, with preliminary Remarks upon that Genus. *Proc. Cotteswold Club*, vol. i., pp. 247–261; *Ann. and Mag. Nat. Hist.*, ser. 2, vol. xii., pp. 225–240.

MERRYWEATHER, DR. GEO.—A Lecture on Gold and Iron Ore with especial reference to the Ironstone of the Vale of Esk, of Staithes and Cleveland. (*Whitby Lit. and Phil. Soc.* March, 1853; also Pp. iv., 86.) 12mo. *London*.

MOORE, C.—On the Palæontology of the Middle and Upper Lias. *Proc. Somerset Arch. Soc.*, vol. iii., pp. 61–76.

MORRIS, PROF. JOHN.—On some Sections in the Oolitic District of Lincolnshire. *Quart. Journ. Geol. Soc.*, vol. ix., pp. 317–344.

PHILLIPS, PROF. J.—The Rivers, Mountains, and Sea-coast of Yorkshire. 8vo. *London.* [Ed. 2 in 1855.]

———.—A Map of the Principal Features of the Geology of Yorkshire (scale 5 miles to an inch). *York.* [Ed. 2 in 1862.]

TRIMMER, JOSHUA.—Notes on the Geology of the Keythorpe Estate, and its relations to the Keythorpe System of Drainage. *Journ. Roy. Agric. Soc.*, vol. xiv., pp, 96–105.

WATTS, J. K.—On the Geology of St. Ives, Huntingdonshire, and its Neighbourhood. *Rep. Brit. Assoc.*, for 1852, Sections, p. 63.

1854.

BRODIE, REV. P. B.—On the Insect Beds of the Purbeck Formation in Wiltshire and Dorsetshire. *Quart. Journ. Geol. Soc.*, vol. x., pp. 475–482.

BUCKMAN, PROF. J.—On the Cornbrash of Gloucestershire and part of Wilts. *Rep. Brit. Assoc.*, for 1853, Sections, pp. 50, 51.

EGERTON, SIR P. DE M. G.—On some new Genera and Species of Fossil Fishes. *Ann. and Mag. Nat. Hist.*, ser. 2, vol. xiii., p. 433.

HAIME, JULES.—Description des Bryozoaires Fossiles de la Formation Jurassique. *Mem. Soc. Géol. France*, ser. 2, tome v., p. 157.

OWEN, PROF. R.—On some Fossil Reptilian and Mammalian Remains from the Purbecks. *Quart. Journ. Geol. Soc.*, vol. x., pp. 420-433.

PHILLIPS, PROF. J.—On a new *Plesiosaurus* in the York Museum. *Rep. Brit. Assoc.* for 1853, Sections, p. 54.

SÆMANN, B.—(Notes on the Oolites of the Cotteswolds, &c.) *Bull. Soc. Géol. France*, ser. 2, tome xi., p. 261.

SORBY, H. C.—On Yedmandale as illustrating the Excavation of some Valleys in the Eastern Part of Yorkshire. *Quart. Journ. Geol. Soc.*, vol. x., pp. 328-333.

WALTER, RICHARD.—Hamdon Hill. *Proc. Somerset Arch. Soc.*, vol. iv., p. 78.

WESTWOOD, PROF. J. O.—Contributions to Fossil Entomology. *Quart. Journ. Geol. Soc.*, Vol. x., pp. 378-396.

WRIGHT, DR. T.—Contributions to the Palæontology of Gloucestershire:—A description, with Figures, of some new Species of Echinodermata from the Lias and Oolites. *Proc. Cotteswold Club*, vol. ii., pp. 17-48; *Ann. Nat. Hist.*, ser. 2, vol. xiii., pp. 312, 376.

1855.

BUCKMAN, PROF. J.—On some Coal Mining Operations at Malmesbury. *Mag. Wilts Arch. and Nat. Hist. Soc.*, vol. ii., pp. 159-161.

KNOX, R.—Descriptions Geological, Topographical, and Antiquarian in Eastern Yorkshire, between the rivers Humber and Tees, with a trigonometrical-surveyed map extending 25 miles from Scarborough, accompanied by a map of all Yorkshire, two of England, and eighteen more descriptive plates of diagrams, &c. (Geology pp. 1-50) 8vo., *London*.

LECKENBY, J.—On the Geological Position of certain Clay Beds in Filey Bay, hitherto referred to the Cretaceous Group, and the probable existence of extensive deposits of Oxford Clay and Lias at a former period. 23rd *Rep. Scarb. Phil. Soc.*, p. 49.

LYCETT, J.—On *Perna quadrata*, Sow. *Proc. Cotteswold Club*, vol. ii., pp. 118-120; *Ann. and Mag. Nat. Hist.*, ser. 2, vol. xv., p. 427.

———.—Note on the Subgenus *Limea*, Brown. *Proc. Cotteswold Club*, vol. ii., p. 131; *Ann. and Mag. Nat. Hist.*, ser. 2, vol. xvi., p. 256.

MOORE, C.—On new Brachiopoda from the Inferior Oolite of Dundry, &c.—*Proc. Somerset Arch. Soc.*, vol. v., pp. 107-128.

MURRAY, P.—The Minerals of Scarborough. 23rd *Rep. of the Scarb. Phil. Soc.*, p. 25.

OWEN, PROF. R.—Notice of a new species of an extinct genus of Dibranchiate Cephalopod (*Coccoteuthis latipinnis*) from the Upper Oolitic Shales at Kimmeridge. *Quart. Journ. Geol. Soc.*, vol. xi., pp. 124, 125.

———.—Notice of some New Reptilian Fossils from the Purbeck Beds near Swanage. *Quart. Journ. Geol. Soc.*, vol. xi., pp. 123, 124.

PHILLIPS, PROF. JOHN.—The neighbourhood of Oxford and its Geology. Oxford Essays for 1855. 8vo. *Oxford*.

———.—Manual of Geology: Practical and Theoretical. 8vo. *London and Glasgow*.

ROSE, C. B.—On the Discovery of Parasitic Borings in Fossil Fish-scales. *Trans. Micros. Soc.*, ser. 2, vol. iii., p. 7.

SIMPSON, M.—The Fossils of the Yorkshire Lias, described from Nature, with a short outline of the Yorkshire Coast. Pp. 149. 12mo. *London*. [Edition 2 in 1884. See also Simpson, 1859.]

TAYLOR, H.—Analyses of Rocks of the Coal Formations. (Cleveland, p. 24.) *Trans. N. Inst. Mining Eng.*, vol. iii., pp. 11-25.

TRIGER.—Sur les terrains jurassiques d'Angleterre dans les environs de Weymouth (ile de Portland). *Bull. Soc. Géol. France*, ser. 2, tome xii., p. 723.

———.—Sur l'Oolite inférieure d'Angleterre et celle du Département de la Sarthe. *Bull. Soc. Géol. France*, ser. 2, tome xii., p. 73.

WILLIAMSON, [PROF.] W. C.—On the Restoration of *Zamites gigas* from the Lower Sandstone and Shale of the Yorkshire Coast. *Rep. Brit. Assoc.* for 1854, Sections, pp. 103, 104.

WRIGHT, DR. T.—On a New Genus of Fossil *Cidaridæ*, with a Synopsis of the Species included therein. *Proc. Cotteswold Club*, vol. ii., pp. 121–127; *Ann. & Mag. Nat. Hist.*, ser. 2, vol. xvi., p. 94.

———.—On some new species of *Hemipedina* from the Oolites. *Proc. Cotteswold Club*, vol. ii., pp. 128–130; *Ann. Nat. Hist.*, ser. 2., vol. xvi., p. 196.

1856.

BAKER, T. B. LL.—Address to Cotteswold Club, 1856. (Section at Crickley Hill, by J. BUCKMAN, p. iv.) *Proc. Cotteswold Club*, vol. ii., (Supplement.)

BLACKWELL, J. K.—The present position of the Iron Industry of Great Britain with reference to that of other countries. *Journ. Soc. Arts*, vol. iv., pp. 65–73 and pp. 113–126.

CROWDER, W.—On the Chemical Composition of the Cleveland Ironstone Beds. *Edin. N. Phil. Journ.*, ser. vol. iii., pp., 286–296.

DICK, A. B.—Analysis of the Cleveland Iron-ore from Eston. *Quart. Journ. Geol. Soc.*, vol. xii., pp. 357, 358.

FISHER, REV. O.—On the Purbeck Strata of Dorsetshire, (read 1854). *Trans. Cambridge Phil, Soc.*, vol. ix., pp. 555–581.

GIEBEL, C. G.—Fauna der Vorwelt. Band 2. Abth. 1. [Fossil Insects.] 8vo. *Leipzig*.

GODWIN-AUSTEN, R. A. C.—On the Possible Extension of the Coal-Measures beneath the South-Eastern Part of England. *Quart. Journ. Geol. Soc.*, vol. xii., pp. 38–73.

MORRIS, J.—General Sketch of the Geology of Hartwell. *Lond. University Mag.*, p. 102.

THORP, REV. W.—On the Ironstone in the Oolite District of Yorkshire. *Proc. Yorksh. Geol. Soc.*, vol. iii., pp. 394–899.

WRIGHT, DR. T.—On the Palæontological and Stratigraphical Relations of the so-called "Sands of the Inferior Oolite." *Quart. Journ. Geol. Soc.*, vol. xii., pp. 292–325.

1856–8.

OPPEL, DR. A.—Die Juraformation Englands, Frankreichs, und des Süd-westlichen Deutschlands. 8vo. *Stuttgart*.

1857.

BRODIE, REV. P. B.—On some Species of Corals in the Lias of Gloucestershire, Worcestershire, Warwickshire, and Scotland. *Edin. New Phil. Journ.*, ser. 2, vol. v., p. 260; *Rep. Brit. Assoc.*, for 1856, Sections, p. 64.

———.—On the Occurrence of some new Species of *Pollicipes* in the Inferior Oolite and Lias of Gloucestershire. *Ann. and Mag. Nat. Hist.*, ser. 2, vol. xix., p. 102; *Rep. Brit. Assoc.*, for 1856, sections, p. 64.

———.—Remarks on the Inferior Oolite and Lias in parts of Northamptonshire, compared with the same Formations in Gloucestershire. *Proc. Cotteswold Club*, vol. ii., pp. 132–134; *Ann. and Mag. Nat. Hist.*, ser. 2, vol. xix. p. 56.

———.—Remarks on the Lias of Barrow in Leicestershire, compared with the lower part of that Formation in Gloucestershire, Worcestershire, and Warwickshire. *Proc. Cotteswold Club*, vol. ii., pp. 139–141; *Ann. and Mag. Nat. Hist.*, ser. 2, vol. xx., pp. 190–192.

BUCKMAN, PROF. J.—On the Basement Beds of the Oolite. *Rep. Brit. Assoc.*, for 1856, Sections, pp. 64, 65.

———.—Address to Cotteswold Club, Jan. 1857. (Notes on Leckhampton Hill p. viii.) *Proc. Cotteswold Club*, vol. ii. (Supplement.)

CROWDER, W.—An attempt to determine the Average Composition of the Rosedale, Whitby and Cleveland Ironstones. *Edin. New Phil. Journ.*, ser. 2, vol. v., pp. 35–53.

———.—The Chemistry of the Iron Manufacture of Cleveland District. *Ibid.*, pp. 264–276, vol. vi., pp. 234–256.

DENNIS, REV. J. B. P.—The existence of Birds [Pterosaurians] during the deposition of the Stonesfield Slate proved by a comparison of the Microscopic Structure of certain Bones of that formation with that of Recent Bones. *Quart. Journ. Micros. Soc.*, vol. v., p. 63.

FALCONER, DR. H.—Description of Two Species of the Fossil Mammalian Genus
Plagiaulax from Purbeck. *Quart. Journ. Geol. Soc.*, vol. xlii., pp. 261–282;
Palæontol. Memoirs, vol. ii., pp. 408–429 (1868).

HULL, PROF. E.—On the South-easterly Attenuation of the Oolitic, Liassic,
Triassic, and Permian Formations. *Rep. Brit. Assoc.* for 1856, Sections, pp. 67, 68.

LYCETT, J.—The Cotteswold Hills. Handbook Introductory to their Geology.
and Palæontology. 8vo. *London & Stroud.*

————-—Notes on the Genus *Quenstedtia.* *Proc. Cotteswold Club*, vol. ii.,
pp. 135–137; *Ann. and Mag. Nat. Hist.*, ser. 2, vol. xix., pp. 53, 54.

————.—On the Sands intermediate the Inferior Oolite and Lias of the
Cotteswold Hills, compared with a similar Deposit upon the Coast of Yorkshire.
Proc. Cotteswold Club, vol. ii., pp. 142–149; *Ann. Nat. Hist.*, ser. 2, vol. xx.,
pp. 170–177.

————.—Note on the Presence of the Fossil genus *Isodonta*, Buv. (*Sowerbya*,
d'Orb.), in the English Jurassic Rocks. *Proc. Cotteswold Club*, vol. ii., pp. 153,
154; *Ann. and Mag. Nat. Hist.*, ser. 2, vol. xix., pp. 367, 368.

1857.

MARLEY, J.—Cleveland Ironstone. Outline of the Main or Thick Stratified Bed,
its Discovery, Application, and Results, in Connexion with the Iron-works in the
North of England. *Trans. N. Inst. Mining Eng.*, vol. v., pp. 165–223. Discussion,
vol. vi., pp. 7, 187 (1858).

MOORE, C.—On the Skin and Food of Ichthyosauri and Teleosauri. *Rep. Brit.
Assoc.*, for 1856, Sections, pp. 69, 70.

————.—On the Middle and Upper Lias of the West of England. *Ibid.*, p. 70.

OWEN, PROF. R.—On a Fossil Mammal (*Stereognathus Ooliticus*) from the
Stonesfield Slate. *Rep. Brit. Assoc.*, for 1856, Sections, p. 73.

————.—On the Affinities of the *Stereognathus ooliticus* (Charlesworth), a
Mammal from the Oolitic Slate of Stonesfield. *Quart. Journ. Geol. Soc.*, vol. xiii.,
pp. 1–11.

PHILLIPS, PROF. J.—On the Geological Structure of Shotover Hill. *Proc.
Ashmolean Soc.*, pp. 142, 143.

SORBY, H. C.—On the Origin of Cleveland Hill Ironstone. *Proc. Yorksh. Geol.
Soc.*, vol. iii., pp. 457–461.

WOODALL, J.—On the Evidence of a Reef of Lower Lias Rock, extending from
Robin Hood's Bay to the neighbourhood of Flamborough Head. *Rep. Brit. Assoc.*
for 1856, Sections, p. 80.

WRIGHT, DR. T.—On the Stratigraphical Distribution of the Oolitic Echinodermata. *Rep. Brit. Assoc.* for 1856, pp. 396–404.

————.—On the Occurrence of Upper Lias Ammonites in the (so-called) Basement Beds of the Inferior Oolite. *Ibid.*, Sections, pp. 80–82.

1857–60.

MARCOU, JULES.—Lettres sur les Roches du Jura et leur Distribution Géogra.
phique dans les Deux Hémisphères. 8vo. *Paris.*

1858.

BEWICK, J.—Remarks on the Ore and Ironstone of Rosedale Abbey. *Trans. N.
Inst. Mining Eng.*, vol. vi., pp. 15, 187.

BRODIE, REV. P. B.—Contributions to the Geology of Gloucestershire, intended
chiefly for the use of Students. *Geologist*, vol. i., pp. 41–48, 80–88, 227–233, 289–
291, 369.

BUCKMAN, PROF. J.—On the Oolite Rocks of Gloucestershire and North Wilts.
Quart. Journ. Geol. Soc., vol. xiv., pp. 98–130.

EGERTON, SIR P. DE M. G.—On *Chondrosteus*, an Extinct Genus of Sturionidæ,
found in the Lias Formation at Lyme Regis. *Phil. Trans.*, vol. cxlviii., p. 871.

HUXLEY, PROF. T. H.—On a New Species of *Plesiosaurus* from Street, near
Glastonbury; with Remarks on the Structure of the Atlas and Axis Vertebræ,
and of the Cranium, in that Genus. *Quart. Journ. Geol. Soc.*, vol. xiv., pp. 281
–294.

JARDINE, SIR WILLIAM.—Memoirs of Hugh Edwin Strickland. 8vo. *London.*

JONES, JOHN.—On *Rhynchonella acuta* and its affinities. *Geologist,* vol. i., pp. 313–318; printed also as Supplement to vol. ii. of *Proc. Cotteswold Club,* (1860).

LYCETT, J.—On some Sections of Upper Lias recently exposed at Nailsworth, Gloucestershire. *Proc. Cotteswold Club,* vol. ii., pp. 155–163; *Ann. and Mag. Nat. Hist.,* ser. 3, vol. ii., p. 255.

MACKIE, S. J.—Gems from Private Collections. *Ammonites communis,* from the Lias of Whitby, Yorkshire. *Geologist,* vol. i., pp. 110–113.

NORWOOD, REV. T. W.—The Comparative Geology of Hotham, near South Cave, Yorkshire. *Geologist,* vol. i., pp. 420–424, 472–479; *Rep. Brit. Assoc.* for 1858, 1859, pp. 96, 97.

PHILLIPS, PROF. J. On some Comparative Sections in the Oolitic and Ironstone Series of Yorkshire. *Quart. Journ. Geol. Soc.,* vol. xiv., pp. 84–98.

PHILLIPS, PROF. JOHN.—On the Estuary Sands in the upper part of Shotover Hill. *Quart. Journ. Geol Soc.,* vol. xiv., pp. 236–241.

SCROPE, G. P.—Geology of Wiltshire. *Mag. Wilts Arch. & Nat. Hist. Soc.,* vol. v., pp. 89–113.

SORBY, H. C.—On the Ancient Physical Geography of the South-east of England. *Edin. New Phil. Journ.,* ser. 2, vol. vii., p. 226.

1858–1860.

WRIGHT, DR. T.—A Monograph on the British Fossil Echinodermata from the Oolitic Formations. *Palæontograph. Soc.* 4to. *London.*

1859.

BINNEY, E. W.—Notice of Lias Deposits at Quarry-Gill and other places near Carlisle. *Quart. Journ. Geol. Soc.,* vol. xv., pp. 549–551.

CUNNINGTON, W.—The Bradford Clay and its Fossils. *Mag. Wilts Arch. & Nat. Hist. Soc.,* vol. vi., p. 1.

GREENWELL, G. C.—On the Ironstone of Wilts and Somerset. *Proc. S. Wales Inst. Eng.,* vol. i., pp. 307–323.

HUXLEY, PROF. T. H.—On *Rhamphorhynchus Bucklandi,* a Pterosaurian from the Stonesfield Slate. *Quart. Journ. Geol Soc.,* vol. xv., pp. 658–670.

LECKENBY, J.—On the Kelloway Rock of the Yorkshire Coast. *Quart. Journ. Geol. Soc.,* vol. xv., pp. 4–15, pls. i.–iii.

————.—Note on the Speeton Clay of Yorkshire. *Geologist,* vol. ii., pp. 9–11.

MARCOU, JULES.—On the Neocomian and the Wealden Rocks in the Jura and in England. *Geologist,* vol. ii., pp. 1–8.

OWEN, PROF. R.—On a New Genus (*Dimorphodon*) of Pterodactyle, with remarks on the Geological Distribution of Flying Reptiles. *Rep. Brit. Assoc.* for 1858, pp. 97, 98.

————.—On the Vertebral Characters of the Order Pterosauria, as exemplified in the Genera *Pterodactylus* and *Dimorphodon.* *Phil. Trans.,* vol. cxlix., p. 161.

SIMPSON, M.—A Guide to the Geology of the Yorkshire Coast. [New edition, called Ed., 3. Another edition, called Ed., 4 in 1868. This is merely a reprint of the stratigraphical details from "the Fossils of the Yorkshire Lias."] 12mo. *London.*

WOOD, N.—On the Deposit of Magnetic Ironstone in Rosedale. *Trans. N. Inst. Mining Eng.,* vol. vii., pp. 85–104.

1860.

BAILY, W. H.—Description of a new Pentacrinite from the Oxford Clay of Weymouth, Dorsetshire. *Ann. and Mag. Nat. Hist.,* ser. 3, vol. vi., pp. 25, 152.

BUCKMAN, PROF. J.—On some Fossil Reptilian Eggs from the Great Oolite of Cirencester. *Quart. Journ. Geol. Soc.,* vol. xvi., pp. 107–110.

DAMON, ROBERT.—Handbook to the Geology of Weymouth and the Island of Portland. 8vo. *London.* [Ed. 2, 1884.] Supplement, with 9 plates., 1864; Ed. 2, 1880; Ed. 3, with 18 plates, 1888.

GUISE, W. V.—Notes on the Inferior Oolite Beds in the neighbourhood of Bath. *Proc. Cotteswold Club,* vol. ii., pp. 170–175.

———.—Annual Address. *Ibid.,* p. 176. (Notes on Cleeve Hill, pp. 179, 180; on Dursley, pp. 182–184; Remarks, by DR. T. WRIGHT, on the " Roadstones " of Cleeve Hill, pp. 184–187; Notes on Swindon, p. 192.)

HORTON, WILLIAM S.—On the Geology of the Stonesfield Slate and its associate formations. *Geologist,* vol. iii., pp. 249–258; *Proc. Liverpool Geol. Soc.,* Session 3, p. 6 (1863).

HULL, PROF. E.—On the Blenheim Iron-ore and the thickness of the formations below the Great Oolite at Stonesfield, Oxfordshire. *Geologist,* vol. iii., pp. 303–305; and *Rep. Brit. Assoc.* for 1860, Sections, pp. 81–83 (1861).

———.—On the South-easterly Attenuation of the Lower Secondary Formations of England; and the probable Depth of the Coal-measures under Oxfordshire and Northamptonshire. *Quart. Journ. Geol. Soc.,* vol. xvi., pp. 63–81.

JONES, J.—Rambles round Chard with a Hammer : or a familiar account of the principal geological facts connected with the district twenty miles round Chard. 8vo. *Chard.*

MOORE, CHARLES.—On the so-called Wealden Beds at Linksfield, and the Reptiliferous Sandstones of Elgin. *Quart. Journ. Geol. Soc.,* vol. xvi., pp. 445–447. [Sections at Pylle Hill, Bristol, and Uphill.]

OWEN, PROF. R.—On some small Fossil Vertebræ from near Frome, Somersetshire. *Quart. Journ. Geol. Soc.,* vol. xvi., pp. 492–497.

———.—On the Orders of Fossil and Recent Reptila, and their Distribution in Time. *Rep. Brit. Assoc.* for 1859, pp. 153–166.

PHILLIPS, PROF. J.—Address to Geological Society. *Quart. Journ. Geol. Soc.,* vol. xvi., pp. 31–55.

———.—Notice of some Sections of the Strata near Oxford. *Quart Journ. Geol. Soc.,* vol. xvi., pp. 115–119, 307–311.

ROBERTS, GEORGE E. The Rocks of Worcestershire : their Mineral Character and Fossil Contents. 8vo. *London.*

WOODWARD, [DR.] S. P.—On an Ammonite with its operculum in situ. [*Amm. subradiatus,* Sow., Dundry.] *Geologist,* vol. iii., p. 328.

WRIGHT, DR. T.—On the Zone of *Avicula contorta* and the Lower Lias of the South of England. *Quart. Journ. Geol. Soc.,* vol. xvi., pp. 374–411.

———.—On the subdivisions of the Inferior Oolite in the South of England, compared with the Equivalent Beds of that Formation on the Yorkshire Coast. [Notes on Dundry Hill, by R. ETHERIDGE.] *Quart. Journ. Geol. Soc.,* vol. xvi., pp. 1–48.

1860–61.

MOORE, CHARLES.—On new Brachiopoda and on the Development of the Loop in *Terebratella. Geologist,* vol. iii., pp. 438–445; vol. iv., pp. 96–99, 190–194; *Proc. Somerset Arch. & Nat. Hist. Soc.,* vol. x., p. 155.

1861.

BEWICK, J.—Geological Treatise on the District of Cleveland, in North Yorkshire, &c. 8vo. *London.*

BRODIE, REV. P. B.—On the Stratigraphical Position of certain Species of Corals in the Lias. *Rep. Brit. Assoc.* for 1860, sections, pp. 73, 74.

———.—On the Geology of South Northamptonshire. *Proc. Warwicksh. Nat. and Arch. Field. Club,* p. 4.

———.—On the Distribution of Corals in the Lias. *Quart. Journ. Geol. Soc.,* vol. xvii., pp. 151, 152.

BROWN, W.—The Iron Ores of Northamptonshire. *Proc. S. Wales. Inst. Engineers.* vol. ii., pp. 193–204.

CLUTTERBUCK, REV. J. C.—On the Course of the Thames from Lechlade to Windsor, as ruled by the Geological Formations over which it passes. *Rep. Brit. Assoc.* for 1860, p. 75.

HORTON, W. S.—On the Oolite Beds of Yorkshire, as compared with their equivalent deposits in Wilts and Gloucestershire. *Geologist*, vol. iv., p. 35 (abstract only). Under different title in *Proc. Liverpool Geol. Soc.* Sessions 1 and 2, p. 8 (Oct. 23, 1860).

HUNT, R.—On the Iron Ore Deposits of Lincolnshire. *Proc. Geol. Soc. W. Riding Yorkshire*, vol. iv., pp. 97–102.

MACALISTER, J. H.—Notes on the Geology of the Country round Newport Pagnall. *Geologist*, vol. iv., pp. 214–216, 263.

MOORE, C.—On the Zones of the Lower Lias and the *Avicula contorta* Zone, *Quart. Journ. Geol. Soc.*, vol. xvii., pp. 483–516.

PORTER, DR. H. — The Geology of Peterborough and its vicinity. 8vo. *Peterborough*.

PRATT, C.—Notes on the Geology of Cleveland. *Geologist*, vol. iv., pp. 81–95, 160.

SEDGWICK, REV. PROF. A.—A Lecture on the Strata near Cambridge and the Fens of the Bedford Level. 8vo. (Privately printed).

SEELEY, PROF. HARRY G.—On the Fen Clay Formation. *Ann. Nat. Hist.* ser. 3. vol. viii., pp. 503–505 ; *Geologist*, vol. iv., pp. 552, 553.

WATSON, J.—On the Geology of the Esk Valley (near Whitby). *Proc. Yorksh. Geol. Soc.*, vol. iv., pp. 91–96.

WHITEAVES, J. F.—On the Invertebrate Fauna of the Lower Oolites of Oxfordshire. *Rep. Brit. Assoc.* for 1860, Sections, pp. 104–108.

———. —On the Palæontology of the Coralline Oolites of the Neighbourhood of Oxford. *Ann. Nat. Hist.* ser. 3., vol. viii., pp. 142–147.

———.— On the Oolitic Echinodermata of the Neighbourhood of Oxford. *Geologist*, vol. iv,, pp. 174, 175.

WILSON, DR. GEORGE, and ARCHIBALD GEIKIE. Memoir of Edward Forbes, F.R.S. 8vo. *London.* [Notes on the Purbeck Strata, pp. 460–469.]

WORTHY, G. S.—On the Geology of Aust Cliff, Gloucestershire. *Proc. Liverpool Geol. Soc.*, p. 10.

WRIGHT, DR. T.—On the *Avicula contorta* Beds and Lower Lias in the South of England. *Rep. Brit. Assoc.* for 1860, Sections, p. 108.

1861–62.

MACALISTER, J. H.—Fossils of North Bucks and the Adjacent Counties. *Geologist*, vol. iv., pp. 481–490 ; *see also* Note on Northampton Sands, vol. v., pp. 66, 67, 1862.

1861–63.

FISHER, REV. O.—Fissures in Portland Strata. *Geologist*, vol. iv., pp. 556, 557 ; vol. vi., pp. 250, 251 (1863.)

1861–81.

OWEN, PROF. R.—Monograph on the Fossil Reptilia of the Oolitic Formations, *Palæontograph. Soc.* 4to. *London.*

1862.

ANON.—[Note of Meeting at Avon Dassett.] *Rep. Warwicksh. Nat. Hist. Soc.* p. 5.

———.—Section at Cleeve Hill. *Ibid.*, p. 10.

BADCOCK, P.—Excursion to Oxford. *Proc. Geol. Assoc.*, vol. i., pp. 155–157.

FALCONER, DR. H.—On the Disputed Affinity of the Mammalian Genus *Plagiaulax*, from the Purbeck Beds. *Quart. Journ. Geol. Soc.*, vol. xviii., pp. 348–369 ; *Palæontol. Memoirs*, vol. ii., pp. 430–451.

GRAY, W.—On the Geology of the Isle of Portland. *Proc. Geol. Assoc.*, vol. i., pp. 128–147.

GUISE, W. V.—Address to Cotteswold Nat. Club, 1861. (Notes on Swift's Hill and Rodborough, by DR. T. WRIGHT, p. 21 ; on Leckhampton Hill, pp. 25, 26). *Proc. Cotteswold Club*, vol. iii., p. 15.

HAGEN, H. A.—A comparison of the fossil insects of England and Bavaria. *Entomol. Ann.*, pp. 1–10 ; *Rep. Brit. Assoc.* for 1861, pp. 113, 114.

HULL, PROF. E.—On Iso-diametric Lines, as means of representing the Distribution of Sedimentary Clay and Sandy Strata, as distinguished from Calcareous Strata, with special reference to the Carboniferous Rocks of Britain. *Quart. Journ. Geol. Soc.*, vol. xviii., pp. 127-146.

JONES, PROF. T. RUPERT.—A Monograph of the Fossil *Estheriæ*. *Palæontograph. Soc.* 4to. *London.*

LUCAS, REV. S.—Section of the Lias Clay in a railway-cutting near Stow-on-the-Wold. *Geologist*, vol. v., pp. 127, 128.

LYCETT, DR. J.—Notes on the Ammonites of the Sands intermediate the Upper Lias and Inferior Oolite. *Proc. Cotteswold Nat. Club*, vol. iii., pp. 1–10.

OWEN, PROF. R.—On a Dinosaurian Reptile (*Scelidosaurus Harrisoni*) from the Lower Lias of Charmouth. *Rep. Brit. Assoc.* for 1861, pp. 121, 122.

ROBERTS, G. E.—Saurian Remains in the Lower Lias [near Droitwich]. *Geologist*, vol. v., p. 150.

SEELEY, PROF. HARRY G.—Notes on Cambridge Geology. I. Preliminary Notice of the Elsworth Rock and associated Strata. *Ann. Nat. Hist.*, ser. 3, vol. x., pp. 97–110; *Rep. Brit. Assoc.* for 1861, Sections, pp. 132–133.

SMITHE, REV. F.—Geology of Churchdown Hill. (Part I.) *Proc. Cotteswold Nat. Club*, vol. iii., pp. 40–49.

WITCHELL, E.—On some Sections of the Lias and Sands exposed in the sewerage works recently executed at Stroud. *Proc. Cotteswold Nat. Club*, vol. iii., pp. 11–14.

WOOD, SEARLES V., Jun.—On the Form and Distribution of the Land-Tracts during the Secondary and Tertiary Periods respectively; and on the Effects upon Animal Life which great changes in geographical configuration have probably produced. *Phil. Mag.*, ser. iv., vol. xxiii., pp. 161-171, 269-282, 382-393.

WRIGHT, DR. T.—Northampton Sands. *Geologist*, vol. v., pp. 38, 39.

1863.

ANON.—*Stricklandinia acuminata* [Buckm.]. *Geologist*, vol. vi., p. 395 and plate.

——.—(Note of Meeting at Dumbleton). *Proc. Warwicksh. Nat. & Archæol. Field Club*, pp. 39, 40.

ABEL, F. A.—Memorandum . . . . of the Results of Experiments into the Comparative Qualities and Fitness for Building Purposes of samples of stone from different Quarries in the Island of Portland. *Papers of the Corps of Roy. Eng.*, ser. 2, vol. xii., p. 6.

BAKER, J. G.—North Yorkshire; Studies of its Botany, Geology, Climate and Physical Geography. [Jurassic Geology, pp. 23–32.] 8vo. *London.* [Ed. 2 in 1888.]

BLAKE, C. CARTER.—On Chelonian Scutes from the Stonesfield Slate. *Geologist.* vol. vi., pp. 183, 184.

CARTE, DR. A., and W. H. BAILY.—On a new Species of *Plesiosaurus* from the Lias near Whitby, Yorkshire. *Rep. Brit. Assoc.* for 1862, Sections, pp. 68, 69; and *Journ. Roy. Dublin Soc.*, vol. iv., p. 160, pl. v.

CLUTTERBUCK, REV. J. C.—The Perennial and Flood Waters of the Upper Thames (with a Geological Map of the Watershed). *Proc. Inst. Civ. Eng.*, vol. xxii., p. 336.

DAY, E. C. H.—On the Middle and Upper Lias of the Dorsetshire Coast. *Quart. Journ. Geol. Soc.*, vol. xix., pp. 278–297.

GRATEAU, E.—Notice sur l'Exploitation du Minerai de Fer dans le Cleveland. *Ann. Génie Civil.*, t. ii., p. 2, pp. 47–54.

GREGSON, C. S.—On a Fossil Elytra from the Stonesfield Slate. *Proc. Liverpool Geol. Soc.*, Session 3, p. 8.

GUISE, W. V.—Address to Cotteswold Nat. Club 1862. (Notes on Geology of Tewkesbury, pp. 53, 54; on Alderton and Dumbleton Hills, pp. 58–60.) *Proc. Cotteswold Club*, vol. iii., p. 51.

HOLL, DR. H. B.—On the Correlation of the several Subdivisions of the Inferior Oolite in the Middle and South of England. *Quart. Jour. Geol. Soc.*, vol. xix., pp. 306–317.

JONES, JOHN.—On Gryphæa Incurva and its Varieties. *Proc. Cotteswold Nat. Club*, vol. iii. pp. 81–95. With 4to Supplement of Plates.

JONES, PROF. T. R.—On Fossil *Estheriæ* and their Distribution. *Quart. Journ. Geol. Soc.*, vol. xix., pp. 140–157.

MACKIE, S. J —Turtles in the Stonesfield Slate. *Geologist*, vol. vi., pp. 41–43.

MOORE, C.—On the Palæontology of Mineral Veins ; and on the Secondary Age of some Mineral Veins in the Carboniferous Limestone. *Rep. Brit. Assoc.* for 1862, Sections, pp. 82, 83.

PHILLIPS, PROF. J.—Notices of Rocks and Fossils, in the University Museum, Oxford. 8vo. *Oxford.*]

PORTER, DR. H.—On the Occurrence of large Quantities of Fossil Wood in the Oxford Clay, near Peterborough. *Quart. Journ. Geol. Soc.*, vol. xix., pp. 317, 318.

WOOD, S. V., JUN.—On the Events which produced and terminated the Purbeck and Wealden Deposits of England and France, and on the Geographical Conditions of the Basin in which they were accumulated. *Phil. Mag.*, ser. 4., vol. xxv., pp. 268–289.

WOODWARD, HENRY.—On a new Macrurous Crustacean (*Scapheus ancylochelis*) from the Lias of Lyme Regis. *Quart. Journ. Geol. Soc.*, vol. xix., pp. 318–321.

1864.

ANON.—Hand Book to the Geological Collection of CHARLES MOORE, Esq., F.G.S., deposited at the Royal Literary and Scientific Institution, Bath. 8vo. *Bath.*

ANON.—Discovery of a New Species of *Plesiosaurus. Geol. Mag.*, vol. i., p. 47.

BRADY, HENRY B.—On *Involutina liassica* (*Nummulites liassicus*, Rupert Jones). *Geol. Mag.* vol. i., pp. 193–196.

CHURCH, PROF. A. H.—On the Colouring Matter of Blue Forest Marble. *Quart. Journ. Chem. Soc.*, ser. 2, vol. ii., pp. 376–386.

DAY, E. C. H.—On *Acrodus Anningiæ, Agass.* ; with Remarks upon the Affinities of the Genera *Acrodus* and *Hybodus. Geol. Mag.*, vol. i., pp. 57–65.

ESKRIGGE, R. A.—On the Lias of Cheshire and Shropshire. *Trans. Manchester Geol. Soc.*, vol. iv., No. 14, pp. 318–331.

GUISE, W. V.—Address to Cotteswold Nat. Club, 1863. (Notes on Glastonbury Tor, by E. C. H. Day, pp. 121, 122 ; on Frocester, pp. 122, 123 ; on Kemble, pp. 123, 124.) *Proc. Cotteswold Club*, vol. iii. p. 113.

GUNN, REV. JOHN.—A Sketch of the Geology of Norfolk. Reprinted from White's History and Directory of the County. 8vo. *Sheffield.* (New Edition 1883.)

HEER, REV. DR. O.— Ueber die fossilen Kakerlaken. *Vierteljahrsschr. nat. gesellsch, Zürich*, vol. ix., pp. 273–302.

HORTON, W. S.—On the Ironstone of the Middle Lias. *Proc. Liverpool Geol. Soc.*, session 5, pp. 8–12.

JONES, JOHN.—On the Natural History, Geology, &c., of Sharpness Point District. *Proc. Cotteswold Nat. Club*, vol. iii., pp. 128–152.

JONES, J., and R. F. TOMES.—[Letters]. On the position of *Gryphæa incurva* in the Lower Lias at Bridgend. *Proc. Cotteswold Nat. Club*, vol. iii. pp. 191–194.

LECKENBY, J.—On the Sandstones and Shales of the Oolites of Scarborough with Descriptions of some New Species of Fossil Plants. *Quart. Journ. Geol. Soc.*, vol. xx., pp. 74–82.

LYCETT, DR. J.—On the Outer Tegument of a Section of the Genus *Trigonia. Geologist*, vol. vii., pp. 217, 218.

MOORE, C.—Geology of the Neighbourhood of Bath. (In the Historic Guide to Bath, by the Rev. G. N. Wright, pp. 387–392.) 8vo. *Bath.*

MORTON, G. H.—On the Lias Formation as developed in Shropshire. *Proc. Liverpool Geol. Soc.*, session 5, pp. 2–6.

PATTINSON, J.—On Zinc, Nickel, and Cobalt, in the Cleveland Ironstone. *Rep. Brit. Assoc.* for 1863, Sections, p. 49.

RAMSAY, [SIR] A. C.—The Breaks in Succession of the British Mesozoic Strata. Anniversary Address to Geological Society. *Quart. Journ. Geol. Soc.*, vol. xx., pp. xl.–lx.

RICHARDSON, T., J. C. STEVENSON, and R. C. CLAPHAM.—On the Chemical Manufactures of the Northern Districts. [Alum Industry, p. 709 ; Mulgrave Cement, p. 713.] *Rep. Brit. Assoc.* for 1863, p. 701.

RILEY, E.—On the occurrence of Vanadium in Pig-iron, smelted from the Wiltshire Oolitic Iron-ore. *Quart. Journ. Chem. Soc.*, ser. 2, vol. ii., p. 21.

ROSS, C. B., and W. DAVIES.—On the Occurrence of Cycloid Fish-scales, &c. in the Oolitic Formation. *Geol. Mag.* vol. i., pp. 92–94.

WRIGHT, DR. T.—A Monograph on the British Fossil Echinodermata from the Cretaceous Formations. vol. i., part i. [Note by J. LECKENBY on the Speeton Clay, p. 9.] 4to. *Palæontograph. Soc.* (vol. for 1862).

————.—Report on Miss Holland's Collection of Lias Fossils. *Proc. Cotteswold Nat. Club*, vol. iii., pp. 153–156.

1864–65.

WRIGHT, DR. T.—On the Ammonites of the Lias Formation. *Proc. Cotteswold Nat. Club*, vol. iii., pp. 162–179, 235–245.

1865.

ANON.—A New Coalfield in Yorkshire [near Thirsk]. *Geol. Mag.*, vol. ii., p. 140.

ANON.—Geology of Wiltshire. *Mag. Wilts Arch. & Nat. Hist. Soc.*, vol. xi. pp. 315–333.

BRADY, DR. H. B.—On the Foraminifera of the Middle and Upper Lias of Somersetshire. *Rep. Brit. Assoc.* for 1864, p. 50.

BRODIE, REV. P. B.—On the Lias Outliers at Knowle and Wootton Wawen in South Warwickshire, and on the Presence of the Lias or Rhætic Bone bed at Copt Heath, its furthest Northern Extension hitherto recognised in that County. *Quart. Journ. Geol. Soc.*, vol. xxi., pp. 159–161; *Rep. Brit. Assoc.* for 1864, p. 52; *Proc. Warwickshire Nat. & Arch. Field Club*, for 1864, p. 26.

————.—Excursion to Fenny Compton, &c. *Proc. Warwick Nat. & Arch. Field Club*, p. 6.

————.—Remarks on three outliers of Lias in North Shropshire and South Cheshire, Staffordshire and Cumberland, and their correlation with the main range. *Proc. Warwickshire Field Club*, p. 6.

BUCKMAN, PROF. J.—The Geology of Gloucestershire in reference to Agriculture and Rural Economy. *Journ. Bath and West of England Agric. Soc.*, ser. 2. vol. xiii., p. 201.

DAY, E. C. H.—On the Lower Lias of Lyme Regis. (Brit. Assoc). *Geol. Mag.* vol. ii., pp. 518, 519; *Geol. and Nat. Hist. Repertory*, vol. i., p. 193.

GUISE, W. V.—Address to Cotteswold Nat. Club, 1864. (Notes on Rolling Bank Quarry, pp. 204–207.)

————.—.————. 1865. (Notes upon the Geology of the Tor Hill, near Glastonbury, by E. C. H. DAY, p. 113; Note on Swift's Hill, &c., by DR. T. WRIGHT, p. 21.) *Proc. Cotteswold Club*, vol. iii., pp. 195, 246,

HEER, DR. OSWALD.—Die Urwelt der Schweiz. [British Jurassic Insects.] 8vo. *Zurich.*

HULL, PROF. E.—On the Iron-bearing Deposits of Oxfordshire.—*Quart. Journ. Science*, vol. ii., p. 360.

MOORE, C.—On the Geology of the South-West of England. *Rep. Brit. Assoc.* for 1864, p. 59.

PHILLIPS, PROF. JOHN.—Note on *Xiphoteuthis elongata. Geol. Mag.*, vol. ii., pp. 57, 58.

SALTER, J. W., and H. WOODWARD—Catalogue and Chart of Fossil Crustacea. 8vo. *London.*

SCOTT, W. L.—On some probable New Sources of Thallium (Whitby). *Rep. Brit. Assoc.*, for 1864, Sections, p. 41.

SEELEY, H. G.—On the significance of the sequence of Rocks and Fossils: Theoretical considerations of the Upper Secondary Rocks, as seen in the section at Ely. *Geol. Mag.*, vol. ii., pp. 45. 262–265.

————.—On the fossil neck bones of a whale (*Palæocetus Sedgwicki*) from the neighbourhood of Ely. *Geol. Mag.*, vol. ii., pp. 54–57.

————.—On *Plesiosaurus macropterus*, a new species from the Lias of Whitby. *Ann. and Mag. Nat. Hist.*, ser. 3., vol. xv., pp. 49–53, 232.

SIMPSON, M.—The Plant Strata of Gristhorpe Bay, near Scarborough. *Geol. and Nat. Hist. Repertory*, vol. i., p. 71.

STODDART, W. W.—Palæontologia Bristoliensis : or the principal fossils of the Bristol District, named and described with a Photograph illustrative of each species, 8vo. *Bristol.*

WAAGEN, DR. W.—Versuch einer allgemeinen Classification der Schichten des Oberen Jura. 8vo. *Munich.* (Notice in *Quart. Journ. Geol. Soc.*, vol. xxi., part 2., p. 14.)

WRIGHT, DR. T.—On the development of Ammonites. *Rep. Brit. Assoc.* for 1864., pp. 73, 74.

1865-70.

PHILLIPS, PROF. JOHN.—A Monograph of British *Belemnitidæ*. *Palæontograph. Soc.* 4to. London.

1866.

ANON—Account of an Excursion to Saltburn. *Geol. Mag.*, vol. iii., p. 517.

ANON—Excursion to Bedminster. *Proc. Bristol Nat. Soc.*, ser. 2., vol. i., pp. 54, 55.

ANSTED, PROF. D. T.—Physical Geography and Geology of the County of Leicester. (Map). 4to. *Westminster.*

BRODIE, REV. P. B.—Notes on a Section of Lower Lias and Rhætic Beds, near Wells, Somerset. *Quart. Journ. Geol. Soc.*, vol. xxii., pp. 93-95.

———.—On a Section of Lower Lias at Harbury, near Leamington. *Rep. Brit. Assoc.*, for 1865, p. 48.

———.—On two New Species of Corals in the Lias of Warwickshire. *Ibid.* p. 49.

———.—On the Geology of Warwick, Leamington, and its neighbourhood. *Rep. Warwicksh. Nat. Hist. & Arch. Soc.*

BUCKMAN, PROF. J.—On the Geology of the County of Dorset in reference to Agriculture and Rural Economy. *Journ. Bath & W. Eng. Agric. Soc.*, ser. 2., vol. xiv., p. 36.

CARRUTHERS, W.—On Araucarian Cones from the Secondary Rocks of Britain. *Geol. Mag.*, vol. iii., pp. 249-252.

———.—On some Fossil Coniferous Fruits. *Ibid.*, pp. 537-546.

HULL, [PROF.] E.—The New Iron fields of England. *Quart. Journ. Science*, vol. iii., pp. 323-332.

HULL, E., & W. BROCKBANK— On the Liassic and Oolitic Iron Ores of Yorkshire and the East Midland Counties. *Proc. Lit. Phil. Soc. Manchester*, vol. v., pp. 119-122.

OWEN, [PROF.] R.—Description of Part of the Lower Jaw and Teeth of a small Oolitic Mammal. (*Stylodon pusillus*, Owen.) *Geol. Mag.*, vol. iii., pp. 199-201.

———.—On a Genus and Species of Sauroid Fish. (*Thlattodus suchoides*, Ow.) from the Kimmeridge Clay of Norfolk. *Geol. Mag.*, vol. iii., pp. 55-57.

———.—On a Genus and Species of Sauroid Fish (*Ditaxiodus impar*, Ow.) from the Kimmeridge Clay of Culham, Oxfordshire. *Geol. Mag.*, vol. iii., pp. 107-109.

PHILLIPS, PROF. JOHN—Oxford Fossils. No. 2. *Geol. Mag.*, vol. iii., pp. 97-99.

SIMPSON, M.—Inferior Oolite, Lias, Belemnites, &c., of the Yorkshire Coast. *Geol. and Nat. Hist. Repertory*, vol. i., p. 215.

TAWNEY, E. B.—On the Western Limit of the Rhætic Beds in South Wales, and on the Position of the '' Sutton Stone.'' With a note on the Corals, by P. MARTIN DUNCAN. *Quart. Journ. Geol. Soc.*, vol. xxii., pp. 69-93.

TAYLOR, W.— Handbook of the Northampton Museum. *Northampton.*

WARRINGTON, R.—On the presence of Manganese in Oolite and Lias. *Quart. Journ. Chem. Soc.*, ser. 2. vol. iii. pp. 206-210.

WOODWARD, DR. HENRY—On a New Crustacean (*Æger Marderi*, H.W.), from the Lias of Lyme Regis, Dorsetshire. *Geol. Mag.*, vol. iii., pp. 10-13.

———.—On the oldest known British Crab (*Palæinachus longipes*) from the Forest Marble, Malmesbury. Wilts. *Quart. Journ. Geol. Soc.*, vol. xxii., pp. 493, 494.

WOODWARD, H.—Notes on the Species of the Genus *Eryon*, Desm., from the Lias and Oolite of England and Bavaria. *Ibid.*, pp. 494–502.

1867.

AUSTIN, FORT-MAJOR T.—On the Occurrence of *Otopteris* in the Lower Lias. [Membury near Axminster, and near Bridgwater.] *Proc. Bristol Nat. Soc.*, ser. 2. vol ii., pp. 43–45.

BRISTOW, H. W.—On the Lower Lias or Lias-Conglomerate of a part of Glamorganshire. *Quart. Journ. Geol. Soc.*, vol. xxiii., pp. 199–207.

BRODIE, REV. P. B.— On the presence of the Purbeck Beds at Brill, in Buckinghamshire; and on the Superior Estuarine Sands there and at certain places in Oxfordshire and Wiltshire. *Quart. Journ. Geol. Soc.*, vol. xxiii., pp. 197–199.

————.— On the correlation of the Lower Lias at Barrow-on-Soar, Leicestershire, with the same strata in Warwick, Worcester, and Gloucester-shires, and on the Occurrences of the Remains of Insects at Barrow. *Rep. Brit. Assoc.*, for 1866, Sections, p. 51; *Ann. Nat. Hist.*, ser. 3., vol. xix., pp. 31–34.

BUCKMAN, PROF. J.—On Stone Roof Tiles of Roman Date. *Proc. Cotteswold Club*, vol. iii., pp. 93–96.

CARRUTHERS, WILLIAM—On an Aroideous Fruit from the Stonesfield Slate. *Geol. Mag.* vol. iv., pp. 146, 147.

————.—On Gymnospermatous Fruits from the Secondary Rocks of Britain *Journ. Botany*, vol. v., pp. 1–21.

————.—On some Cycadean Fruits from the Secondary Rocks of Britain. *Geol. Mag.*, vol. iv., pp. 101–106.

DAVIES, W.—New Species of *Plesiosaurus* [from Charmouth]. *Geol. Mag.*, vol. iv., p. 144.

GUISE, SIR W. V.—Address to the Cotteswold Naturalists' Field Club, 1866. (Notes on Stonehouse and Nailsworth Railway, by Rev. F. SMITH, p. 12; on Bradford Abbas, by Prof. J. BUCKMAN, pp. 18–21.)

————.—.————1867. (Notes on Haresfield, pp. 77–79; on Painswick and other building stones, by W. H. HYETT, pp. 79, 80; on Sutton and Southerndown, by C. MOORE, pp. 83–87.) *Proc. Cottesw. Club*, vol. iv., pp. 3, 75.

KENNEDY, LOCKHART.—On the occurrence of *Ancyloceras annulatus* in Dorsetshire. *Proc. Cotteswold Club*, vol. iv., pp. 54, 55.

MOORE, C.—On the Middle and Upper Lias of the South-west of England. *Proc. Somerset Arch. and Nat. Hist. Soc.*, vol. xiii., pp. 119–244.

————.—On Abnormal Conditions of Secondary Deposits when connected with the Somersetshire and South Wales Coal-Basin. *Quart. Journ. Geol. Soc.*, vol. xxiii., pp. 449–568.

STODDART, W. W.—Geology of Dundry Hill. *Proc. Bristol Nat. Soc.*, ser. 2. vol. ii., pp. 29–33.

TATE, PROF. R.—On the Liassic affinities of the *Avicula contorta* series. *Geol. and Nat. Hist. Repertory*, vol. i., pp. 364–369.

————.—On the oldest known species of *Exogyra*, with a description of the species. *Ibid.*, pp. 378–380.

————.—On the Occurrence of the genus *Crassatella* in the Oolitic Rocks, with a description of a new species. *Geol. and Nat. Hist. Repertory*, vol. i., pp. 396, 397.

————.—On the Fossiliferous Development of the Zone of *Ammonites angulatus*, Schloth., in Great Britain. *Quart. Journ. Geol. Soc.*, vol. xxiii., pp. 305–314.

WITCHELL, E.—On a Section of the Lias and Recent Deposits in the Valley of the River Frome, at Stroud. *Proc. Cotteswold Club*, vol. iv., pp. 56–59.

WRIGHT, DR. T.—Additional Notes on Cleeve Hill Section. *Proc. Cotteswold Club*, vol. iv., pp. 60–74.

————.—On Coral Reefs Present and Past. *Proc. Cotteswold Club*, vol., iv., pp. 97–173.

1867–72.

DUNCAN, DR. P. MARTIN—Supplement to the Monograph of the British Fossil Corals, by Milne Edwards and Haime. [Liassic.] *Palæontograph. Soc.* 4to. London.

1868.

ANON.— Important Discovery of Remains of *Dimorphodon macronyx* in the Lower Lias of Lyme Regis, Dorsetshire. *Geol. Mag.*, vol. v.. p. 536.

ANON.—Excursions to Paulton and Garden Cliff. *Proc. Bristol Nat. Soc.*, ser. 2., vol. iii., pp. 50-52.

ANON.—Excursion to Dundas. *Ibid.*, pp. 67 or 68.

BRAVENDER, JOHN.—The Watershed of the Upper Thames. *Proc. Cottesw. Club*, vol. iv., pp. 240-248.

BRODIE, REV. P. B.—A Sketch of the Lias generally in England, and of the "Insect and Saurian beds," especially in the lower division in the counties of Warwick, Worcester, and Gloucester, with a particular account of the fossils which characterize them. *Proc. Warwickshire Field Club*, p. 1.

BRUCE, A. C.—On the accurate Division of the Local Lias at Rugby into Zones, by their Fossils, more especially by their *Ammonites*. *Rep. Rugby School Nat. Hist. Soc.*, for 1867, pp. 19-21.

BUCKMAN, PROF. J.—On the Geology of the County of Somerset in reference to Agriculture and Rural Economy. *Journ. Bath and West of England Agric. Soc.*, ser. 2., vol. xvi., pp. 108- .

CARRUTHERS, WILLIAM.—British Fossil *Pandaneæ*. *Geol. Mag.*, vol. v., pp. 153-156.

CLEMINSHAW, E.- On the Natural History of the Rugby Lias.—*Rep. Rugby School Nat. Hist. Soc.*, for 1867, pp. 31-37.

[CLEMINSHAW, E.].—A List of Local Lias Fossils. *Rep. Rugby School Nat. Hist. Soc.*, for 1867, pp. 55-57.

EGERTON, SIR P. DE M. G.—On the Characters of some new Fossil Fish from the Lias of Lyme Regis. *Quart. Journ. Geol. Soc.*, vol. xxiv., pp. 499.505.

FALCONER, DR. H.—Palæontological Memoirs and Notes. 2 vols. Edited by C. Murchison. 8vo. *London*.

GROOM-NAPIER, C. O.—On the Lower Lias beds occurring at Cotham, Bedminster, and Keynsham, near Bristol. *Quart. Journ. Geol. Soc.*, vol. xxiv.,. pp. 204-206.

GUISE, SIR W. V.—Address to the Cotteswold Naturalists Field Club. 1868. (Notes on Sutton and Southerndown, pp. 198-201; 208-210; on Stroud and Bussage, pp. 210-212.) *Proc. Cottesw. Club*, vol. iv., p. 196.

HUNT, R.—The Iron Ores of Great Britain. *Quart. Journ. Soc.*, vol. v., pp. 31-39.

HUXLEY, PROF. T. H.—On the Animals which are most nearly intermediate between Birds and Reptiles. *Pop. Sc. Review*, vol. vii., pp. 237-247.

JUDD, J. W.—On the Speeton Clay. *Quart. Journ. Geol. Soc.*, vol. xxiv., pp. 218-250.

MAW, GEORGE.—On the Disposition of Iron in variegated Strata. *Quart. Journ. Geol. Soc.*, vol. xxiv., pp. 351-400.

MORRIS, PROF. J.—Geological Excursion to Bath and its neighbourhood. *Geol. Mag.*, vol. v., pp. 233-236.

ROBINSON, H.—A Lecture on the Cleveland Ironstone District of Yorkshire. 8vo. *London*.

SANDERS, W.—(Geological features of Brent Knoll). *Proc. Bristol Nat. Soc.*, ser. 2., vol. iii., p. 44.

STODDART, W. W.—Notes on the Lower Lias Beds of Bristol. *Quart. Journ. Geol. Soc.*, vol. xxiv., pp. 199-204.

———.--(Excursion to Dundry). *Proc. Bristol Nat. Soc.*, ser. 2., vol. iii. p. 67.

TAUNTON, J. H.—Remarks on the Watershed of the Cotteswolds, in connexion with the Water Supply to the Metropolis. *Proc. Cottesw. Club*, vol. iv., pp. 249-254.

TOPLEY WILLIAM.—On the Lower Cretaceous Beds of the Bas-Boulonnais, with notes on their English Equivalents. *Quart. Journ. Geol. Soc.*, vol. xxiv., pp. 472-483.

WILSON, [REV.] J. M.—On the Objects of the Geological Section of the Natural History Society. *Rep. Rugby School Nat. Hist. Soc.*, for 1867, pp. 11-16.

———.—A List of Local Lias Fossils. *Ibid.* p. 55.

WITCHELL, E.—On the Denudation of the Cotteswolds. *Proc. Cottesw. Club*, vol. iv., pp. 214–230; (Notes on Landslips near Nailsworth, by G. F. PLAYNE, pp. 230–232.)

WOODWARD, Dr. H.—Third Report on the Structure and Classification of the Fossil Crustacea. *Rep. Brit. Assoc.*, for 1867, pp. 44–47.

————.—On a new Brachyurous Crustacean (*Prosopon mammillatum*), from the Great Oolite, Stonesfield. *Geol. Mag.*, vol. v., pp. 3–5.

————.—Contribution to British Fossil Crustacea. *Geol. Mag.*, vol. v., pp. 260, 261; 353–356.

1869.

ALLISON, T.—On the Cleveland Ironstone. *Trans. S. Wales Inst. of Eng.*, vol. vi., p. 236.

————.—The Whin Dyke of Cleveland. *Trans. Cleveland Lit. and Phil. Soc.*, vol. i., 4 pp. 2 plates.

ANON.—Cleveland and its Metropolis (with an account of its Geology and Analyses). 8vo. *Middlesborough*.

ANON.—Additions to the List of Local Lias Fossils. *Rep. Rugby School Nat. Hist. Soc.* for 1868, p. 43.

BRISTOW, H. W.—Table showing the Thickness of the Secondary Strata in the Southern Counties of England. *Rep. Roy. Coal Commission*, fol. *London*.

BRODIE, REV. P. B.—Geological Notes on Northamptonshire, &c. *Geol. Mag.*, vol. vi., pp. 236, 237.

————.—The Oldest British Belemnite. *Ibid.*, p. 239.

BRUCE, A. C.—On the accurate Division of the Local Lias at Rugby into Zones, by their Fossils, more especially by their Ammonites. *Rep. Rugby School Nat. Hist. Soc.*, for 1868, p. 19.

CARRUTHERS, WM.—On some Undescribed Coniferous Fruits from the Secondary Rocks of Britain. *Geol. Mag.*, vol. vi., pp. 1–7.

————.—On *Beania*, a New Genus of Cycadean Fruit, from the Yorkshire Oolites. *Geol. Mag.*, vol. vi., pp. 97–99.

CLEMINSHAW, E.—Geological Notices. [Additions to List of Local Lias Fossils]. *Rep. Rugby School Nat. Hist. Soc.*, for 1868, p. 43.

CUNNINGTON, W.—Geology of Wiltshire, Inferior and Great Oolite. *Mag. Wilts Arch and Nat. Hist. Soc.*, vol. xi. pp. 315–333.

DUNCAN, DR. P. M.—First Report on the British Fossil Corals. *Rep. Brit. Assoc.* for 1868, pp. 99–113.

EGERTON, SIR P. DE M. G.—On two New Species of *Gyrodus*. *Quart. Journ. Geol. Soc.*, vol. xxv., pp. 379–386.

ENNISKILLEN, EARL OF.—Alphabetical catalogue of the type-specimens of fossil fishes in the collection of the Earl of Enniskillen. 8vo. *London*.

GUISE, SIR W. V.—Address Cotteswold Nat. Club, 1869. (Notes on thinning of Oolites, by W. C. LUCY, pp. 7, 8; on Bredon and the Bambury Stone, pp. 16–17.) *Proc. Cottesw. Club*, vol. v., p. 5.

HULKE, J. W.—Note on a large Saurian Humerus from the Kimmeridge Clay of the Dorset Coast. *Quart. Journ. Geol. Soc.*, vol. xxv., pp. 386–389.

————.—Notes on some Fossil Remains of a Gavial-like Saurian from Kimmeridge Bay, collected by J. C. Mansel, Esq. . . . *Ibid.*, pp. 390–400.

HUXLEY, PROF T. H.—On the Upper Jaw of *Megalosaurus*. *Quart. Journ. Geol. Soc*, vol. xxv., pp. 311–314.

JECKS, CHARLES.—On the Ferruginous Sandstone of the Neighbourhood of Northampton. *Rep. Brit. Assoc.*, for 1868, Sections, pp. 69, 70.

MASON, J. WOOD.—On *Dakosaurus* from the Kimmeridge Clay of Shotover Hill. *Quart. Journ. Geol. Soc.*, vol. xxv., pp. 218–220.

MORRIS, PROF. J.—On the Genus *Æchmodus* from the Lias of Lyme Regis. *Geol. Mag.*, vol. vi., pp. 337–341.

————.—Geological Notes on Parts of Northampton and Lincoln-shires. *Geol. Mag.*, vol. vi., pp. 99–105.

PLAYNE, G. F.—On the Physical Geography of the District drained by the River Frome and its Tributaries. *Proc. Cottesw. Club*, vol. v., pp. 21–38.

SEELEY, H. G.—Note on the *Pterodactylus macrurus* (Seeley), a new species from the Purbeck Limestone, indicated by caudal vertebræ five inches long. Note on the thinning away to the Westward in the Isle of Purbeck of the Wealden and Lower Greensand Strata. *Proc. Camb. Phil. Soc.*, Parts vii.-x., p. 130.

————.—Index to the Fossil Remains of Aves, Ornithosauria, and Reptilia, from the Secondary System of Strata arranged in the Woodwardian Museum of the University of Cambridge; with a Prefatory Notice by the Rev. Prof. A. SEDGWICK. 8vo. *Cambridge.*

SHARP, SAMUEL.—Notes on the Northampton Oolites. *Geol. Mag.*, vol. vi., pp. 446-448.

SMITH, T. McD.—Account of the Rugby Well. Report by the Engineer (Metrop. Board of Works) on the Boring Operations at Crossness Pumping Station. 8vo. *London.*

TATE, [PROF.] R.—Additions to the List of Brachiopoda of the British Secondary Rocks. *Geol. Mag.*, vol. vi., pp. 550-556.

————.—Contributions to Jurassic Palæontology. *Ann. and Mag. Nat. Hist.*, ser. 4, vol. iv., p. 417.

THORNE, JAMES.—Excursion to Oxford, May, 1869. Reports of the Excursions to Oxford, &c. *Geol. Assoc.* (separately printed). 8vo. *London.*

WALKER, J. F.—On Secondary Species of Brachiopoda. *Proc. Yorksh. Nat. Club*, p. 214 ; *Geol. Mag.*, vol vii., pp. 560-564 (1870).

WILSON, [REV.] J. M.—On the Victoria Works. [Rugby Blue Lias Lime and Cement Works.] *Rep. Rugby School. Nat. Hist. Soc.*, for 1868, p. 9.

————.—Rugby Water-works. Remarks to Accompany the Section of the Well. *Rep. Rugby School. Nat. Hist. Soc.*, for 1868, pp. 41, 42. (Plate.)

WOODWARD, DR. H.—Fourth Report on the Structure and Classification of the Fossil Crustacea. *Rep. Brit. Assoc.* for 1868, pp. 72-73.

1870.

ANON.—Excursion to Stroud, May Hill, and Swindon, May, 1870. *Proc. Geol. Assoc.*, vol. ii., pp. 33, 34.

ANON.—Excursion to Aylesbury, June, 1870. *Proc. Geol. Assoc.*, vol. ii., p. 36.

BOTT, THOMAS D.—On the Geology of the Neighbourhood of Swanage. *Proc. Geol Assoc.*, vol. ii., pp. 30-32.

BRODIE, REV. P. B.—On the Geology of Warwickshire. *Rep. Warwickshire Nat. Hist. and Arch. Soc.*, 1869-70., pp. 10-34.

CARRUTHERS, WILLIAM. On the Fossil Cycadean Stems from the Secondary Rocks of Britain. *Trans. Linn. Soc.*, vol. xxvi., pp. 675-708.

COCKBURN, W.—The Ironstone of the Cleveland District. *Proc. Cleveland Inst. Eng.*

DUNCAN, DR. P. M.—Second Report on the British Fossil Corals. *Rep. Brit. Assoc.* for 1869, pp. 150-170.

HULKE, J. W.—Note on a Crocodilian Skull from Kimmeridge Bay, Dorset. *Quart. Journ. Geol. Soc.*, vol. xxvi., pp. 167-172.

————.—Note on some Teeth associated with Two Fragments of a Jaw from Kimmeridge Bay. *Ibid.*, pp. 172-174.

————.—Note on some Plesiosaurian Remains obtained by J. C. Mansel, Esq., F.G.S., in Kimmeridge Bay, Dorset. *Quart. Journ. Geol. Soc.*, vol. xxvi., pp. 611-622.

JUDD, J. W.—Additional Observations of the Neocomian Strata of Yorkshire and Lincolnshire; with Notes on their Relations to the Beds of the same age throughout Northern Europe. *Quart. Journ. Geol. Soc.*, vol xxvi., pp. 326-348, pl. xxiii.

LOBLEY, J. LOGAN.—On the Stratigraphical Distribution of the British Fossil Brachiopoda. *Proc. Geol. Assoc.*, vol. ii., pp. 77-140.

MARLEY, J.—On the Magnetic Ironstone of Rosedale Abbey. *Trans. N. Inst. Mining Eng.*, vol. xix., p. 193.

MOORE, C.—On a Specimen of *Teleosaurus* from the Upper Lias. *Rep. Brit. Assoc.* for 1869, Sections, p. 97.

PATTINSON, J.—Analysis of the Cleveland Ironstone in a paper by E. Williams " On the Blast Furnaces at the Cleveland Ironworks." *Trans. S. Wales Inst. Eng.*, vol. vi., p. 284.

PHILLIPS, PROF. J.—The gigantic Oolitic Lizard (*Cetiosaurus*). *Geol. Mag.*, vol. vii., p. 240.

SEELEY, H. G.—On *Zoocapsa dolichorhamphia*, a Sessile Cirripede from the Lias of Lyme Regis. *Ann. and Mag. Nat. Hist.*, ser. 4, vol. v., p. 283.

SHARP, SAMUEL.—The Oolites of Northamptonshire. *Quart. Journ. Geol. Soc.*, vol. xxvi., pp. 354–391.

TATE, PROF. R.—The Fuller's-Earth in the South-west of England. *Quart. Journ. Science*, vol vii., pp. 68–71.

———.—On the Palæontology of the Junction Beds of the Lower and Middle Lias in Gloucestershire. *Quart. Journ. Geol. Soc.*, vol xxvi., pp. 394–408.

WILLIAMSON, PROF. W. C.—Contributions towards the History of *Zamia gigas*, L. and H. *Trans. Linn. Soc.*, vol. xxvi., pt. iv., pp. 663–674, pls. 52, 53.

WOODWARD, DR. HENRY.—British Fossils (*Dimorphodon*.) *Geol. Mag.*, vol. vii, pp. 97–100.

WOODWARD, HORACE B.—The Railway-cutting at Uphill, Weston-super-Mare. *Geol. Mag.*, vol. vii. pp. 239, 240.

WRIGHT, DR. T.—Notes on a New Species of Starfish from the Ironstone Beds of the Inferior Oolite of Northampton. *Quart. Journ. Geol. Soc.*, vol. xxvi., pp. 391–393.

                                    1871.

ANON.—Excursion to Moreton-in-the-Marsh. *Proc. Warwicksh. Nat. and Arch. Field Club*, p. 29.

ANON.—Excursion to Warwickshire. *Proc. Geol. Assoc.*, vol. ii., pp. 284–287.

ANON.— Excursion to Cambridge. *Proc. Geol. Assoc.*, vol. ii., pp. 219–226.

ANON.—Excursion to Oxford. *Proc. Geol. Assoc.*, vol. ii., pp. 243, 244.

ARMITAGE, J.—Localities [of fossils] new to the List. *Rep. Rugby School Nat. Hist. Soc.* for 1870, p. 45.

CARRUTHERS, W.—On some supposed Vegetable Fossils. *Quart. Journ. Geol. Soc.*, vol. xxvii., pp. 443-448 (plate).

DAVIES, WILLIAM.—Alphabetical Catalogue of Type Specimens of Fossil Fishes in the British Museum. *Geol. Mag.*, vol. viii., pp. 208–216.

EGERTON, SIR P. DE M. G.—On a new Chimæroid Fish from the Lias of Lyme Regis (*Ischyodus orthorhinus*). *Quart. Journ. Geol. Soc.*, vol. xxvii., pp. 275–279.

HERMAN, W. DOUGLAS.—On Allophane and an Allied Mineral found at North-ampton. *Quart. Journ. Geol. Soc.*, vol. xxvii, pp. 234–237.

HULKE, J. W.—Note on an Ichthyosaurus (*I. enthekiodon*) from Kimmeridge Bay, Dorset. *Quart. Journ. Geol. Soc.*, vol. xxvii., pp. 440, 441.

———.—Note on a Fragment of a Teleosaurian Snout from Kimmeridge Bay, Dorset. *Quart. Journ. Geol. Soc.*, vol. xxvii., pp. 442, 443.

JONES, PROF. T. R.—On the Range of Foraminifera in Time. *Proc. Geol. Assoc.*, vol. ii., pp. 175–182.

JONES, T. R., and W. K. PARKER.—On Terquem's Researches on the Foraminifera of the Lias and Oolites. *Ann. Mag. Nat. Hist.*, ser. 4., vol. viii., pp. 361–365.

JUDD, [PROF.] J. W. On the Anomalous Mode of Growth of certain Fossil Oysters. *Geol. Mag.*, vol. viii., pp. 355–359.

LOBLEY, J. LOGAN.—Excursion to the Yeovil District. *Proc. Geol. Assoc.*, vol. ii., pp. 247–250.

———.—On the Principal Features of the Stratigraphical Distribution of the British Fossil Lamellibranchiata. *Quart. Journ. Geol. Soc.*, vol. xxvii., pp. 411–418.

MITCHELL, W. STEPHEN.—Some Remarks on the Denudation of the Oolites of the Bath District, with a Theory on the Denudation of Oolites generally. *Quart. Journ. Geol. Soc.*, vol. xxvii., pp. 228–231.

OWEN, PROF. R.—Monograph of the Fossil Mammalia of the Mesozoic Formations. *Palæontograph. Soc.*, 4to *London*. [Section of Durdlestone Bay, by R. ETHERIDGE, p. 22.]

PHILLIPS, PROF. JOHN.—Geology of Oxford and the Valley of the Thames. 8vo. *Oxford*.

RAMSAY, PROF. A. C.—On the Physical Relations of the New Red Marl, Rhætic Beds, and Lower Lias. *Quart. Journ. Geol. Soc.*, vol. xxvii., pp. 189–199.

SEELEY, H. G.—On a new species of *Plesiosaurus* from the Portland Limestone. *Ann. and Mag. Nat. Hist.*, ser. 4, vol. viii., p. 181.

SHARP, SAMUEL.—Note on a Futile Search for Coal near Northampton. *Geol Mag.*, vol. viii., pp. 505, 506.

STODDART, W. W.—Notes on the Geology of Weymouth. *Proc. Bristol Nat. Soc.*, ser. 2, vol. v., p. 66.

TATE, PROF. R.—A Census of the Marine Invertebrate Fauna of the Lias. *Geol. Mag.*, vol. viii., pp. 4–11.

TUTE, REV. J. S.—Fossil Oolitic Plants (Scarborough). *Science Gossip*, No. 79, p. 157.

WILLS, H.—The Lias of Whitby. *Trans. Clifton Coll. Sci. Soc.*, part 1, p. 92.

WOODWARD, H. B.—Notes on Metamorphism of Strata in the Mendip Hills. *Geol. Mag.*, vol. viii., pp. 400–405.

WRIGHT, DR. T.—On the Correlation of the Jurassic Rocks in the Department of the Cote-d'Or, France, with the Oolitic formations in the counties of Gloucester and Wilts, England. *Proc. Cotteswold Club*, vol. v., pp. 143–237.

1872.

BEESLEY, THOMAS.—A sketch of the Geology of the neighbourhood of Banbury. *Proc. Warwicksh. Nat. and Arch. Field Club*, 1872, pp. 11–34; see also *Geol. Mag.* vol. ix., pp. 279–282.

BELL, I. L.—On the Cleveland Ironstone—Its Discovery and its Influence on the Trade of the District. *Coll. Guard.*, vol. xxiii., pp. 607, 608.

BLAKE, REV. J. F.—The Yorkshire Lias and the Distribution of its Ammonites. *Rep. Brit. Assoc.* for 1871, sections, pp. 90–92.

————.—On the Infralias in Yorkshire. *Quart. Journ. Geol. Soc.*, vol. xxviii., pp. 132–147.

BRANNON, PHILIP.—The Visitor's Guide to Swanage and the Isle of Purbeck, with a clear digest of the geology, and a minute description of the coast from Bournemouth Bay to White Nore. 8vo. *Bournemouth.*

CUNNINGTON, W.—On the geology of the neighbourhood of Westbury Station. *Mag. Wilts Arch. and Nat. Hist. Soc.*, vol. xiii., p. 306.

DAVIES, WM.—On the Rostral Prolongations of *Squaloraia polyspondyla*, Ag. *Geol. Mag.*, vol. ix., pp. 145–150.

EGERTON, SIR P. DE M. G.—On *Prognathodus Güntheri*, Egerton, a new Genus of Fossil Fish from the Lias of Lyme Regis. *Quart. Journ. Geol. Soc.*, vol. xxviii. pp. 233–237.

ETHERIDGE, R.—On the Physical Structure of the Watchet area, and the relation of the Secondary Rocks to the Devonian series of West Somerset. *Proc. Cottesw. Club*, vol. vi., pp. 35–48.

GUISE, SIR W. V.—Address to Cotteswold Nat. Club, 1871. (On analysis of certain Oolitic Rocks, by A. H. CHURCH, pp. 239, 240; on Sapperton tunnel, pp. 241–243; on Painswick, pp. 245–247; on Watchet, pp. 249–253.) *Proc. Cotteswold Club*, vol. v., p. 239.

————.—Address to Cotteswold Nat. Club, 1872. (Notes on Swindon, pp. 8, 9; on Tetbury, pp. 14–19.) *Proc. Cotteswold Club*, vol. vi., p. 1.

HULKE, J. W.—Note on some Ichthyosaurian Remains from Kimmeridge Bay, Dorset. *Quart. Journ. Geol. Soc.*, vol. xxviii., pp. 34, 35.

LOWE, W. B.—Geological Report. (Addition to List of Local Lias Fossils). *Rep. Rugby School Nat. Hist. Soc.* for 1871, p. 52.

STODDART, W. W.—A walk to the Cotteswolds. *Proc. Bristol Nat. Soc.*, ser. 2, vol. vii., pp. 7–9.

TATE, R—Note on the Discovery of the oldest known Trigonia (*Trigonia lingonensis*, Dum.) in Britain. *Geol. Mag.*, vol. ix., p. 306.

TAUNTON, J. H.—Sapperton Tunnel on the Thames and Severn Canal. *Proc. Cottesw. Club*, vol. v., pp. 255–270.

WILSON, [REV.] J. M.—Catalogue of Local Fossils in the Museum [Rugby]. *Rep. Rugby School Nat. Hist. Soc.* for 1871, pp. 55–58.

WINWOOD, REV. H. H.—Excursion to Bath, *Proc. Geol. Assoc.*, vol. iii., pp. 89–92.

——.—Excursion to Frocester Hill. *Proc. Bath Nat. Hist. Club*, vol. ii., pp. 359, 360.

WOODWARD, H. B.—List of Minerals found in Somersetshire. *Geol. Mag.*, vol. ix., p. 129.

——.—Note on the Midford Sands. *Geol. Mag.*, vol. ix., pp. 513–515.

WOODWARD, HORACE B., and J. H. BLAKE.—Notes on the Relations of the Rhætic Beds to the Lower Lias and Keuper Formations in Somersetshire. *Geol. Mag.* vol. ix.., pp. 196–202.

1872–1883.

LYCETT, DR. JOHN.—A Monograph of the British Fossil *Trigoniæ*. (With supplements.) *Palæontograph Soc.*, 4to. London.

1873.

ANON.—Excursion to Weymouth. *Proc. Warwickshire Nat. and Arch. Field Club* for 1872, p. 41.

BEESLEY, THOMAS.—Excursion to Banbury. *Proc. Geol. Assoc.*, vol. iii. pp. 197–204.

BLAKE, REV. J. F.—The oldest known British *Trigonia. Geol. Mag.*, vol. x., p. 186.

BROWN, THOS. C.—Geologising on the Cotteswolds. *Geol. Mag.*, vol. x., pp. 134, 135.

BUTLER, ARTHUR G.—On a Fossil Butterfly belonging to the Family *Nymphalidæ*, from the Stonesfield Slate near Oxford. *Geol. Mag.*, vol. x., pp. 2, 3.

EGERTON, SIR P. DE. M. G.—On *Platysiagum sclerocephalum*, Egerton, and *Palæospinax priscus*, Egerton. *Quart. Journ. Geol. Soc.*, vol. xxix., pp. 419–421.

GODWIN-AUSTEN, R. A. C. Address to the Geological Section of the British Association. *Rep. Brit. Assoc.*, for 1872, pp. 90–96.

GUISE, SIR W. V.—Address to Cotteswold Nat. Club, 1873. (Notes on Moreton-in-the-Marsh, pp. 78–80). *Proc. Cotteswold Nat. Club*, vol. vi., p. 68.

LECKENBY, J.—The oldest known British Trigonia. *Geol. Mag.*, vol. x., p. 135.

LOWE, W. B.—Geological Report. (Section of Victoria Lime Works, &c.) *Rep. Rugby Scool Nat. Hist. Soc.*, for 1872, pp. 47–49.

MANSEL-PLEYDELL, J. C.—A Brief Memoir on the Geology of Dorset. *Geol. Mag.*, vol. x., pp. 402–413 ; 438–447 ; Separately printed, 8vo. *Blandford.*

MOORE, C.—On the Presence of Naked Echinodermata (Holothuria) in the Inferior Oolite and Lias. *Rep. Brit. Assoc.* for 1872, sections, pp. 117, 118.

——.–(Geology of Swindon). *Wilts. Arch. Nat. Hist. Mag.*, vol. xiv. no. xli., pp. 137-139.

MORRIS, PROF. J.—Excursion to Aylesbury. *Proc. Geol. Assoc.*, vol. iii., pp. 210, 211.

PEYTON, J. E. H.—The Boring in Sussex. *Nature*, vol. vii., p. 162.

PHILLIPS, PROF. J.—The Sub-Wealden Exploration. *Geol. Mag.*, vol. x., p. 527 ; *Nature*, vol. viii., p. 487.

SHARP, SAMUEL.—Sketch of the Geology of Northamptonshire, *Proc. Geol. Assoc.*, vol. iii., pp. 243–252.

——.—The Oolites of Northamptonshire. Part II. *Quart. Journ. Geol. Soc.*, vol. xxix., pp. 225–302.

TAWNEY, E. B.—On the Occurrence of *Zoophycos scoparius* (Thioll.) in the Inferior Oolite of Dundry. *Proc. Bristol Nat. Soc.*, ser, 2, vol. vii., pp. 40, 41.

TOPLEY, W.—The Sub-Wealden Exploration. *Rep. Brit. Assoc.* for 1872, pp. 122, 123 ; in full in the *Brighton Daily News*, Aug. 16, 1872.

WINKLER, DR. T. C.—Le *Plesiosaurus dolichodeirus*, Conyb., du Musée Teyler. *Archiv. du Musée Teyler*, vol. iii.

1874.

ANON.—Note of Meeting at Burton Dasset and Kineton. *Proc. Warwicksh. Nat. and Arch. Field Club*, p. 17.

ANON.—Note of Excursion to Northampton. *Ibid.*, pp. 18–20.

ANON.—Note of Meeting at Eatington. *Ibid.*, pp. 21, 22.

ARGALL, W.—The Ironstone Works at Easton Neston, Northamptonshire. *Rep. Miners' Assoc. Cornwall & Devon*, pp. 27, 28.

BARR, T. M.—On the Geology of the North-Eastern District of Yorkshire. *Trans. Glasgow Geol. Soc.*, vol. iv., Part 3., p. 291.

BRODIE, REV. P. B.—Notes on a Railway-section of the Lower Lias and Rhætics between Stratford-on-Avon and Fenny Compton, on the Occurrence of the Rhætics near Kineton, and the Insect-beds near Knowle, in Warwickshire, and on the Recent Discovery of the Rhætics near Leicester. *Quart. Journ. Geol. Soc.*, vol. xxx., pp. 746–749.

————.—The Distribution and Correlation of Fossil Insects, and the supposed occurrence of Lepidoptera and Arachnidæ in British and Foreign Strata, chiefly in the Secondary rocks. *Rep. Warwick Nat. Hist. & Arch. Soc.* for 1873, p. 12; and *Proc. Warwick Field Club*, 1874, pp. 16–38.

BUTLER, A. G.—Notes on the Impression of *Palæontina oolitica* in the Jermyn Street Museum. *Geol. Mag.*, Dec. II., vol. i., pp. 446–449.

GUISE, SIR W. V.—Address to Cotteswold Nat. Club, 1874. (Notes on Birdlip, pp. 95–98; on Willsbridge, pp. 99, 100). *Proc. Cottenwold Club*, vol. vi., p. 91.

HUDLESTON, W. H.—Excursion to Northamptonshire. *Proc. Geol. Assoc.*, vol. iv., pp. 123–134.

HUDLESTON, W. H., and JAMES PARKER.—Excursion to Oxford. *Proc. Geol. Assoc.*, vol. iv., pp. 91–97.

HULKE, J. W.—Note on a very Large Saurian Limb-bone adapted for progression upon Land, from the Kimmeridge Clay of Weymouth, Dorset. *Quart. Journ. Geol. Soc.*, vol. xxx., pp. 16, 17.

HYATT, PROF. A.—Evolution of the *Arietidæ*. *Proc. Nat. Hist. Soc. Boston*, vol. xvi., pp. 166–170.

LOBLEY, J. L.—Excursion to the Cheltenham District. *Proc. Geol. Assoc.*, vol. iv., pp. 167–174.

REPORT.—6th Report of the Commissioners appointed in 1868 to inquire into the best means of preventing the pollution of rivers. The Domestic Water Supply of Great Britain. Folio. *London.*

SCUDDER, S. H.—On an English fossil insect described as lepidopterous. *Proc. Boston Soc. Nat. Hist.*, vol. xvi., p. 112.

SEELEY, H. G.—On *Murænosaurus Leedsii*, a Plesiosaurian from the Oxford Clay. Part I. *Quart. Journ. Geol. Soc.*, vol. xxx., pp. 197–208.

————.—Note on some of the Generic Modifications of the Plesiosaurian Pectoral Arch. *Quart. Journ. Geol. Soc.*, vol. xxx., pp. 436–449.

————.—On the Pectoral Arch and Fore Limb of *Ophthalmosaurus*, a new Ichthyosaurian Genus from the Oxford Clay. *Quart. Journ. Geol. Soc.*, vol. xxx., pp. 696–707.

TAWNEY, E. B.—Museum Notes. Dundry Gasteropoda. *Proc. Bristol Nat. Soc.*, ser. 2, vol. i., pp. 2.

TOPLEY, WILLIAM.—On the Correspondence between some Areas of Apparent Upheaval and the Thickening of Subjacent Beds. *Quart. Journ. Geol. Soc.*, vol. xxx., pp. 186–195; *Rep. Brit. Assoc.* for 1873, p. 91.

WINWOOD, REV. H. H.—(Excursions to Broadfield Down, Old Down, and Radstock.) *Proc. Bath Nat. His. Club*, vol. iii., pp. 90–102.

WOODWARD, H.—New Facts bearing on the Inquiry concerning Forms intermediate between Birds and Reptiles. *Quart. Journ. Geol. Soc.*, vol. xxx., pp. 8–15.

WOODWARD, H. B.—Notes on the Geology of the neighbourhood of Wells, Somerset. *Proc. Somerset Arch. & Nat. His. Soc.*, vol. xix., pp. 50–64.

1874–75.

PLANT, J.—Geology of Leicestershire. *Rep. Leicester Lit. Phil. Soc.* 1874, pp. 37–41, and 1875, pp. 42–46.

1874-76.

TOPLEY, W.—Geological Reports on the Sub-Wealden Exploration. *Rep. Brit. Assoc.* for 1873, pp. 491–495; for 1874, pp. 22–27; for 1875, pp. 347–349. *Nature,* vol. xi., 1875, pp. 284, 285.

————.—The Sub-Wealden Exploration. *Trans. N. of Eng. Inst. Eng.,* vol. xxxiii., pp. 185–188.

1874-76-78.

HUDLESTON, W. H.—The Yorkshire Oolites. *Proc. Geol. Assoc.,* vol. iii., No. 7, pp. 283–333; vol. iv., No. 6, pp. 353–410, pl. iv.; vol. v. No. 8., pp. 407–494, pls. iii.-vi.

1875.

ANON.—Bristol and its Environs. Historical, Descriptive, and Scientific. (Brit. Assoc.) 8vo. *London and Bristol.* [Rhætic and Liassic Formations, by R. TATE, pp. 367–377. Inferior Oolite, &c., by E. B. TAWNEY, pp. 377, &c.]

BLAKE, REV. J. F.—On the Kimmeridge Clay of England. *Quart. Journ. Geol. Soc.,* vol. xxxi., pp. 196–233.

BONNEY, REV. PROF. T. G.—Cambridgeshire Geology. 8vo. *Cambridge.*

BRODIE, REV. P. B.—The Lower Lias at Eatington and Kineton, and on the Rhætics in that neighbourhood, and their further extension in Leicestershire, Nottinghamshire, Lincolnshire, Yorkshire, and Cumberland. *39th Rep. Warwick Nat. Hist. Soc.,* pp. 6-17: reprinted. 8vo. *Warwick.*

————.—On the Correlation of Fossil Insects. *Proc. Warwickshire Nat. Field Club,* for 1874, pp. 16–38. (See also *Rep. Warwicksh. Nat. Hist. & Arch. Soc.* 1873, pp. 12–28.)

BUCKMAN, PROF. J.—On the Cephalopoda Bed and the Oolite Sands of Dorset and part of Somerset. *Proc. Somerset Arch. and Nat. Hist. Soc.,* vol. xx., pp. 140–164.

CARRUTHERS, W.—(Sketch of the plant life of the Oolites of Yorkshire.) Address during the Excursion to East Yorkshire. *Proc. Geol. Assoc.,* vol. iv., p. 330.

CROSS, REV. J. E.—The Geology of North-west Lincolnshire. *Quart. Journ. Geol. Soc.,* vol. xxxi., pp. 115–130. (Appendix by R. ETHERIDGE, pp. 126–129.)

CROSSE, H.—Sur les caractères de l'opercule dans le genre *Neritopsis. Journ. Conchyl.,* pp. 57–66.

DAGLISH, T., and R. HOWSE.—Some Remarks on the Beds of Ironstone occurring in Lincolnshire. *Trans. N. Engl. Inst. Eng.,* vol. xxiv., pp. 23–33.

GUISE, SIR W. V.—Address to Cotteswold Nat. Club, 1875. (Notes on Weston, Bath, pp. 133–138; on Radstock, pp. 138–141). *Proc. Cotteswold Club,* vol. vi., p. 127.

HAWKSLEY, T.—Report upon the Boring for Water at Rugby, 1862: with notes on the composition by DR. W. ODLING. *Rep. Rugby School Nat. Hist. Soc.* for 1874, p. 71.

HOPEWELL, E. W.—Two Sections of Wells [Newbold]. *Rep. Rugby School Nat. Hist Soc.,* for 1874, pp. 51, 52.

HYATT, PROF. A.—Genetic Relations of the *Angulatidæ. Proc. Boston Soc. Nat. Hist.,* vol. xvii., pp. 15–23.

————.—On the Biological Relations of the Jurassic Ammonites. *Proc. Boston Nat. Hist. Soc.,* vol. xvii., pp. 236–241.

JONES, PROF. T. R., and W. K. PARKER.—Lists of some English Jurassic Foraminifera. *Geol. Mag.,* Dec. II., vol. ii., pp. 308–311.

MOORE, C.—On the presence of the genera *Plicatocrinus, Cotylederma* and *Solanocrinus* in British Strata. *Geol. Mag.,* Dec. II., vol. ii., pp. 626, 627.

MORRIS, PROF. J.—On the Occurrence of Boring Mollusca in the Oolitic Rocks. *Geol. Mag.,* Dec. II., vol. ii., pp. 267–272.|

OLDHAM, R. D.—Geological Report. (Notes on Rugby.) *Rep. Rugby School Nat. Hist. Soc.,* for 1874, pp. 47–54.

SCUDDER, SAMUEL H.—Fossil Butterflies. *Mem. Amer. Assoc.,* vol. i., pp. xi., 99.

SEELEY, PROF. H. G.—On the Femur of *Cryptosaurus eumerus*, Seeley, a Dinosaur from the Oxford Clay of Great Gransden. *Quart. Journ. Geol. Soc.*, vol. xxxi, pb. 149-151.

——.—Note on *Pelobatochelys Blakei* and other Vertebrate Fossils exhibited by the Rev. J. F. Blake in illustration of his paper on the Kimmeridge Clay. *Quart. Journ. Geol. Soc.*, vol. xxxi., pp. 234-237.

——.—On the Maxillary Bone of a new Dinosaur (*Priodontognathus Phillipsi*), contained in the Woodwardian Museum of the University of Cambridge. *Quart. Journ. Geol. Soc.*, vol. xxxi. pp. 439-443.

——.—On an Ornithosaurian (*Doratorhynchus validus*) from the Purbeck Limestone of Langton, near Swanage. *Quart. Journ. Geol. Soc.*, vol. xxxi., pp. 465-468.

TATE, PROF. RALPH.—(List of Species from the Yellow Lias, Hewlett's Hill, Cheltenham.) *Quart. Journ. Geol. Soc.*, vol. xxxi., p. 508.

——.—(Section, Railway-cutting, Fenny Compton, Warwickshire, eight miles north of Banbury.) *Quart. Journ. Geol. Soc.*, vol. xxxi., p. 509.

——.—On the Lias about Radstock. *Quart. Journ. Geol. Soc.*, vol. xxxi., pp. 493-510.

——.—On Some New Liassic Fossils. *Geol. Mag.*, Dec. II., vol. ii. pp. 203-206.

WALKER, J. F.—Oolitic Brachiopoda. *Geol. Mag.*, Dec. II., vol. ii., p. 572.

WILSON, REV. J. M.—On the propable Existence of a considerable Fault in the Lias near Rugby, and of a new Outlier of the Oolite. *Quart. Journ. Geol. Soc.*, vol. xxxi., pp. 355, 356.

——.—Contributions to the Geology of Hillmorton. *Rep. Rugby School Nat. Hist. Soc.*, for 1874, pp. 8-13.

——.—*Ichthyosaurus* from New Bilton. *Ibid.* p. 48.

——.—Boring at Lodge Farm, [E. of Clifton.] *Ibid.*, p. 52.

WINWOOD, REV. H. H.—Notes on some Railway Sections near Bath. *Proc. Bath Nat. Hist. Club*, vol. iii., pp. 129-135.

WITCHELL, E.—Observations upon a Bed of Fuller's Earth at Whiteshill, near Stroud. *Proc. Cottesw. Club*, vol. vi., pp. 144-145.

WRIGHT, DR. T.—On the Occurrence of the Genus *Cotylederma* in the Middle Lias of Dorsetshire.—*Geol. Mag.*, Dec. II., vol. ii., pp. 505, 506 ; see also vol. iii., pp. 94, 95 (1876.)

1876.

ANON.—Excursion to Chipping Norton, 1876. *Proc. Warwickshire Nat. and Arch. Field Club*, pp. 29-32.

BLAKE, PROF. J. F.—On *Renulina Sorbyana*. *Monthly Micros. Journ.*, vol. xv. pp. 262-264.

BOULGER, PROF. G. S.—Note on a Cetiosauroid Tooth. *Proc. W. Lond. Sci. Assoc.*, vol. i., pp. 99, 100.

BRODIE, W. R.—Notes on the Kimmeridge Clay of the Isle of Purbeck. *Proc. Geol. Assoc.*, vol. iv., pp. 517, 518.

CARRUTHERS, WILLIAM.—Address to Geologists' Association, 1875. *Proc. Geol. Assoc.*, vol. v., pp. 1-16.

DAVIES, WILLIAM.—On the Exhumation and Development of a large Reptile (*Omosaurus armatus*, Owen), from the Kimmeridge Clay, Swindon, Wilts. *Geol. Mag.*, Dec. II., vol. iii., pp. 193-197.

DOVE, GEORGE, JUN.—The Frodingham Iron Field, North Lincolnshire. *Journ. Iron and Steel Inst.*, pp. 318-341.

EGERTON, SIR P. DE M. G.—Notice of (*Harpactes*) *Harpactira velox*, a Predaceous Ganoid Fish of a new Genus, from the Lias of Lyme Regis. *Geol. Mag.*, Dec. II., vol. iii., pp. 441, 442, 576.

HUDLESTON, W. H.—Excursion to Swindon and Faringdon. *Proc. Geol. Assoc.*, vol. iv., pp. 543-554.

HYATT, PROF. A.—Genetic Relations of *Stephanoceras*. *Proc. Boston Soc. Nat. Hist.*, vol. xviii., pp. 360-400.

MEADE, R.—The Iron Industries of Lincolnshire. *Mining Journal*, October 14, 28, and December 30.

PRESTWICH, PROF. J.—Thickness of the Oxford Clay. *Geol. Mag.*, December 2, vol. iii., pp. 237-239.

————.—On the Mineral Water discovered in sinking the Artesian Well, at St. Clement's, Oxford, and on certain geological inferences suggested by the character of the water. *Proc. Ashmolean Soc.*

————.—On the Geological Conditions affecting the Water Supply to Houses and Towns. 8vo. *Oxford.*

READE, T. M.—The Lower Lias of Street, Somerset. *Proc. Liverpool Geol. Soc.*, vol. iii., pp. 97-99.

SEELEY, H. G.—On the Organization of the Ornithosauria. *Journ. Linn. Soc.*, vol. xiii., pp. 84-107.

————.—Similitudes of the Bones in the Enaliosauria. *Journ. Linn. Soc.*, vol. xii., pp. 296-329.

TATE, R., and J. F. BLAKE.—The Yorkshire Lias. 8vo. *London.*

TOPLEY, W.—Table of the Cretaceous and Oolitic Rocks in the South East of England (Scale 1,000 feet. to 1 inch.), distinguishing those discovered in the Sub-Wealden Boring. Section from London to St. Leonard's through Sub-Wealden and Kentish Town Borings (Scale an inch to 8 miles). Appended to H. Willett's Report on the Sub-Wealden exploration.

TRIBOLET, M. DE.—Sur les terrains jurassiques supérieur de la Haute-Marne comparés à ceux du Jura Suisse et Français. *Bull. Soc. Géol.*, sér. 3., tome iv., pp. 259-285. (Comparisons with England.)

WALKER, J. F.—New British Brachiopoda. [Corallian, Dorset.] *Geol. Mag.*, Dec. II., vol. iii., p. 574.

WINKLER, DR. T. C.—Étude sur le genre *Mystriosaurus* et description de deux exemplaires nouveaux de ce genre. *Arch. Mus. Teyler*, tome iv., pp. 49-132.

WINWOOD, REV. H. H.—Notes on a Rhætic and Lower Lias Section on the Bath and Evercreech Line, near Chilcompton. *Proc. Bath Field Club*, vol. iii.; pp. 300-304.

WOODWARD, DR. H.—On some New Macrurous Crustacea from the Kimmeridge Clay of the Sub-Wealden Boring, Sussex, and from Boulogne-Sur-Mer. *Quart. Journ. Geol. Soc.*, vol. xxxii., pp. 47-50.

WRIGHT, DR. T.—Address to the Geological Section. *Rep. Brit. Assoc.*, for 1875, Sections, pp. 47-62.

1876-95.

DE RANCE, C. E.—Reports of the Committee for investigating the Circulation of the Underground Waters. *Rep. Brit. Assoc.*, for 1875-95.

1877.

*AMMON, L. VON.—[Note on the Purbeck Beds of Dorset). *Zeit. Deutsche Geol. Ges.*, band xxix., pp. 198, 199.

BARROW, GEORGE.—On a New Marine Bed in the Lower Oolites of East Yorkshire. *Geol. Mag.*, Dec. II., vol. iv. pp. 552-555.

BEESLEY, T.—The Lias of Fenny Compton, Warwickshire. *Proc. Warwickshire Nat. & Arch. Field Club*, pp. 1-22; and 8vo. *Banbury.* (Reprinted from "*Banbury Guardian.*" Supplement, 1886 ?

————.—On the Geology of the Eastern Portion of the Banbury and Cheltenham Direct Railway. *Proc. Geol. Assoc.*, vol. v., pp. 165-185.

BLAKE, PROF. J. F.—History of the Restoration of Extinct Animals. *Proc. Geol. Assoc.*, vol. v., pp. 91-103.

BLAKE, REV. J. F., and W. H. HUDLESTON.—On the Corallian Rocks of England. *Quart. Journ. Geol. Soc.*, vol. xxxiii., pp. 260-404.

BONNEY, PROF. T. G.—The Coral Rag of Upware. *Geol. Mag.*, Dec. II., vol. iv., pp. 476, 477.

————.—Colouring of Oolitic Rocks. *Ibid.*, p. 576.

BUCKMAN, PROF. J.—On the Fossil Beds of Bradford Abbas and its Vicinity. *Proc. Dorset Nat. Hist. Club*, vol. i., pp. 64-72.

————.—The Cephalopoda-beds of Gloucester, Dorset, and Somerset. *Quart. Journ. Geol. Soc.*, vol. xxxiii., pp. 1-9.

CARPENTER, P. H.—On some Points in the Anatomy of *Pentacrinus* and *Rhizocrinus*. *Journ. Anat. Physiol.*, vol. xii., pp. 35–53.

——.—On the Genus *Actinometra*, Müll, &c. *Journ. Linn. Soc.*, vol. xiii., pp. 440–457.

DAVIDSON, DR. T.—On the Brachiopoda of the Inferior Oolite of Bradford Abbas and its vicinity. With Notes on the Fossil Beds of Bradford Abbas and its vicinity. By PROF. J. BUCKMAN. *Proc. Dorset Nat. Hist. and Antiq. Club*, vol. i., pp. 73–88.

DUNCAN, DR. P. M.—On the *Salenidæ*, Wright, Part II. *Ann. Mag. Nat. Hist.*, ser. 4, vol. xx., pp. 245–257.

ETHERIDGE, R.—Report on the probability of finding Coal under the Estate of the Evington Coal Boring Company, Limited, at Evington, near Leicester. P. 1. Folio. Jan. 12. (Privately printed.)

PLANT, J.—Report on the Boring for Coal on the Estate, near Evington, Leicestershire, of the Evington Coal Boring Company, Limited. Pp. 3. Folio. May 7. (Privately printed.)

GRAHAM, W. B.—On the Geology of Sharnbrook. *Trans. Bedford Nat. Hist. Soc.*, for 1875–76, pp. 45, 46.

GUISE, SIR W. V.—Address to Cotteswold Nat. Club, 1877. (Notes on Malmesbury, pp. 288–290; on Purton, pp. 290–293; on Dundry, pp. 293–296.) *Proc. Cotteswold Club*, vol. vi., p. 279.

HARRISON, W. J.—A Sketch of the Geology of Leicestershire and Rutland. (Reprinted from White's History, &c., of the Counties, pp. 67, 12 photographs.) 8vo. *Sheffield.*

——.—On the Geology of Leicestershire. *Proc. Geol. Assoc.*, vol. v., pp. 126–136.

——.—Excursion to Leicestershire. *Ibid.*, pp. 142–148.

HUDLESTON, W. H., and J. F. WALKER.—On the Distribution of the Brachiopoda in the Oolitic strata of Yorkshire. *Ann. Rep. Yorksh. Phil. Soc.*, for 1876, pp. 7–12.

HULL, PROF. E.—On a Deep Boring for Coal at Scarle, Lincolnshire, &c. *Proc. Inst. Civ. Eng.*, vol. xlix., p. 160; *Rep. Brit. Assoc.*, for 1876, pp. 91, 92.

MACDAKIN, CAPT.—The Northampton Ironstone Beds in Lincolnshire. *Geol. Mag.*, Dec. II., vol. iv., pp. 406–410.

MANSEL-PLEYDELL, J. C.—Notes on a Gavial Skull from the Cornbrash of Closeworth. *Proc. Dorset Nat. Hist. Club*, vol. i., p. 28.

MOORE, C.—The Liassic and other Secondary Deposits of the Southerndown Series. *Trans. Cardiff Nat. Soc.*, vol. viii., pp. 53–60.

SEELEY, H. G.—On the Vertebral Column and Pelvic Bones of Pliosaurus Evansi (Seeley), from the Oxford Clay of St. Neots, in the Woodwardian Museum of the University of Cambridge. *Quart. Journ. Geol. Soc.*, vol. xxxiii., pp. 716–723.

SMITHE, DR. F.—On the Occurrence of *Plicatula lævigata* of d'Orbigny in the Middle Lias of Gloucestershire. *Proc. Cottesw. Club*, vol. vi., pp. 341–347.

——.—On the Middle Lias of North Gloucestershire. The Spinatus Zone. *Ibid.*, pp. 349–405.

STODDART, W. W.—List of the Characteristic Fossils of the Dundry Oolite. *Proc. Cottesw. Club*, vol. vi., pp. 297–300.

TAUNTON, J. H.—Description of the Malmesbury Water Works, with remarks on the Flow of Streams in the Cotteswold District. *Proc. Cottesw. Nat. Club*, vol. vi., pp. 301–306.

WOOD, REV. H. H.—Notes on some Cornbrash Sections in Dorset. *Proc. Dorset Field Club*, vol. i., pp. 22–27.

WOODWARD, DR. H.—A Catalogue of British Fossil Crustacea. 8vo. *London.*

1878.

BARRETT, C.—The Geology of Swyre, Puncknowle, Burton Bradstock, Loders, Shipton Gorge, Litton Cheney, Longbredy, Littlebredy, and Abbotsbury, Dorset. (With Map.) 8vo. *Bridport.*

BLAKE, PROF. J. F., and W. H. HUDLESTON.—The Coral Rag of Upware. *Geol. Mag.*, Dec. II.. vol. v., pp. 90-92.

BUCKMAN, PROF. J.—On some Slabs of *Trigonia clavellata*, [Sow.] from Osmington Mills, Dorset. *Trans. Dorset Nat. Hist. Club.*, vol. ii., pp. 19, 20. [Figures a slab with 32 specimens.]

BUCKMAN, S. S.—On the Species of *Astarte* from the Inferior Oolite of the Sherborne District. *Proc. Dorset Field Club.*, vol. ii., pp. 81-92.

CARTER, H. J.—On Calcareous Hexactinellid Structure . . . . [Refers to Lias Sponges.] *Ann. Nat. Hist.*, ser. 5, vol. i., pp. 417, 418.

DIXON, FREDERICK.—The Geology of Sussex. Edit. 2. [History of the Sub-Wealden Boring, pp. 151-160, by W. TOPLEY.] 4to. *Brighton.*

GOSS, HERBERT.—The Insect Fauna of the Secondary or Mesozoic Period, and the British and Foreign Formations of that Period in which Insect Remains have been Detected. *Proc. Geol. Assoc.*, vol. vi., pp. 116-150; See also *Entomologists' Monthly Mag.*, vols. xv. and xvi.

GUISE, SIR W. V.—Address to Cotteswold Nat. Club, 1878. (Notes on Cheltenham, pp. 4-7.) *Proc. Cotteswold Club*, vol. vii., p. 1.

HUDLESTON, W. H.—Excursion to Chipping Norton. *Proc. Geol. Assoc.*, vol. v., pp. 378-389.

HULKE, J. W.—Note on two Skulls from the Wealden and Purbeck Formations indicating a new Sub-group of Crocodilia. *Quart. Journ. Geol. Soc.*, vol. xxxiv., pp. 377-382.

JONES, PROF. T. R.—Notes on some Fossil Bivalved Entomostraca. *Geol. Mag.*, Dec. II., vol. v., pp. 100-110.
———.— On the Wealden Entomostraca. *Ibid.*, pp. 277, 278.

KEEPING, W.—On *Pelanechinus*, a new Genus of Sea-urchins from the Coral Rag. *Quart. Journ. Geol. Soc.*, vol. xxxiv., pp. 924-930.

MANSEL-PLEYDELL, J. C.—Fossil Cycads. *Proc. Dorset Field Club*, vol. ii., pp. 1-11.

MOORE, C.—Notes on the Palæontology and some of the Physical Conditions of the Meux-Well Deposits. *Quart. Journ. Geol. Soc.*, vol. xxxiv., pp. 914-923.

MORRIS, PROF. J.—Address to the Geologists' Association. *Proc. Geol. Assoc.*, vol. v., pp. 191-230.

NEWTON, E. T.—Notes on a Crocodilian Jaw from the Corallian Rocks of Weymouth. *Quart. Journ. Geol. Soc.*, vol. xxxiv., pp. 398-400.

OWEN, PROF. R.—On the Fossils called "Granicones"; being a Contribution to the Histology of the Exo-skeleton in Reptilia. [Dermal appendages of *Nuthetes destructor*.] *Journ. R. Micros. Soc.*, vol. i., pp. 233-236.

PRESTWICH, PROF. J.—On the Section of Messrs. Meux and Co.'s Artesian Well in the Tottenham Court Road, with Notices of the Well at Crossness, and of another at Shoreham, Kent; and on the probable Range of the Lower Greensand and Palæozoic Rocks under London. *Quart. Journ. Geol. Soc.*, vol. xxxiv., pp. 902-913.

SAUVAGE, DR. H. E.—Sur les *Lepidotus* et *Sphærodus gigas*. *Bull. Soc. Géol. France*, sér. 3, tome v., pp. 626-629.

TAWNEY, E. B.—On the supposed Inferior Oolite at Branch Huish, Radstock. *Proc. Bristol Nat. Soc.*, ser. 2, vol. ii., pp. 175-178.

———.—On an Excavation at the Bristol Water Works Pumping Station, Clifton. *Ibid.*, pp. 179-182.

TOMES, ROBERT F.—A List of the Madreporaria of Crickley Hill, Gloucestershire, with Descriptions of some New Species. *Geol. Mag.*, Dec. II., vol. v., pp. 297-305.

———.—On the Stratigraphical Position of the Corals of the Lias of the Midland and Western Counties of England and of South Wales. *Quart. Journ. Geol. Soc.*, vol. xxxiv., pp. 179-195.

WALFORD, E. A.—On some Middle and Upper Lias Beds in the neighbourhood of Banbury. *Proc. Warwickshire Nat. and Arch. Field Club*, pp. 1-23.

WALKER, J. F.—On the occurrence *Terebratula Morierei* in England. *Geol. Mag.*, Dec. II., vol. v., pp. 552-556; and *Proc. Dorset Nat. Hist. Club.*, vol. iii., pp. 42- . (1879.)

WHITAKER, W.—The Well-section at Holkham Hall, Norfolk. *Proc. Norwich Geol. Soc.*, vol. i., pp. 30, 31.

WILLETT, HENRY.—The Record of the Sub-Wealden Exploration. 8vo. *Brighton.*

WINWOOD, REV. H. H.—Notes on an Oolitic Quarry at Bathford. *Proc. Bath Nat. Hist. Club,* vol. iv., pp. 82–87.

WOODHOUSE, REV. THOMAS.—Some notes on the Geology of Otterhampton. *Proc. Somerset Arch. and Nat. Hist. Soc.,* vol. xxiii., pp. 65–69.

WOODWARD, DR. HENRY.—On *Penæus Sharpii,* a Macrurous Decapod Crustacean, from the Upper Lias, Kingsthorpe, near Northampton. *Geol. Mag.,* Dec. II., vol. v., pp. 164, 165.

————.—On a New and Undescribed Macrouran Decapod Crustacean, from the Lower Lias, Barrow-on-Soar, Leicestershire. *Geol. Mag.,* Dec. II., vol. v., pp. 289–291.

YOUNG, JOHN T.—On the Occurrence of a Freshwater Sponge in the Purbeck Limestone. *Geol. Mag.,* Dec. II, vol. v., pp, 220, 221.

1878–86.

WRIGHT, DR. T.—Monograph on the Lias Ammonites of the British Islands. *Palæontograph. Soc.* 4to. *London.*

1879.

ANON.—Excursions to Williamscote and Chipping Warden, to Blisworth, &c. *Proc. Warwick. Field Club,* pp. 55, 56, &c.

BLAKE, PROF. J. F.—Geological Episodes. *Rep. Brit. Assoc.* for 1879, Sections, pp. 335, 336; *Nature,* September 4, p. 470.

BLAKE, REV. J. F., and W. H. HUDLESTON.—Excursion to Weymouth and Portland. *Proc. Geol. Assoc.,* vol. vi., pp. 172–174.

BUCKMAN, PROF. J.—On the so-called Midford Sands. *Quart. Journ. Geol. Soc.,* vol. xxxv., pp. 736–743.

————.—On a Series of Sinistral Gasteropods from Somerset and Dorset. *Proc. Dorset Nat. Hist. Field Club,* vol. iii., pp. 135–140 (plate).

————.—On the *Belemnoteuthis Montefiorei. Proc. Dorset Nat. Hist. Field Club,* vol. iii., pp. 141–143 (plate).

ETHERIDGE, ROBERT.—The Position of the Silurian, Devonian, and Carboniferous Rocks in the London area. *Pop. Science Rev.,* ser. 2, vol. iii., pp. 279–296.

EVANS, [SIR] JOHN.—On some Fossils from the Northampton Sands. *Rep. Brit. Assoc.* for 1878, pp. 534, 535.

FISCHER, P.—Sub-divisions des Ammonites. *Journ. Conchyl.,* sér. 3, tome xix., pp. 217–260.

GUISE, SIR W. V.—Address to Cotteswold Nat. Club, 1879. (Notes on Stroud, pp. 104–106.) *Proc. Cotteswold Club,* vol. vii., p. 94.

HARRIS, W. H.—Sketch of the Geology of Cardiff and surrounding district. *Sci. Gossip,* No. 173, pp. 99–101.

HUGHES, PROF. T. McK.—On the Relation of the Appearance and Duration of the various Forms of Life upon the Earth to the breaks in the continuity of the Sedimentary Strata. *Proc. Cambridge Phil. Soc.,* vol. iii., pp. 246–258.

HULKE, J. W.—Note on *Poikilopleuron Bucklandi* of Eudes Deslongchamps (père), identifying it with *Megalosauraus Bucklandi. Quart. Journ. Geol. Soc.,* vol. xxxv., pp. 233–238.

KENT, A. U.—The Finding *Terebratula Morierei* [Deslongchamps] at Bradford Abbas. *Proc. Dorset Nat. Hist. Field Club,* vol. iii., pp. 39–41.

MANSEL-PLEYDELL, J. C.—On the Dorset *Trigoniæ. Proc. Dorset Nat. Hist. Field Club,* vol. iii., pp. 111–134 (5 plates).

MOORE, C.—Excursion to Bath. *Proc. Geol. Assoc.,* vol. vi., pp. 196–201.

————.—On Ammonites and Aptychi. *Rep. Brit. Assoc.* for 1879, pp. 341–343.

OLDHAM, T. B.—Geology of the Neighbourhood of Rugby. *Rep. Rugby School Nat. Hist. Soc.* for 1878, pp. 39–46.

————.—Geological Report. (List of additional Species of Rugby Fossils, p. 54. Note on the Middle Lias as exposed in Railway Cuttings near Crick, p. 54.) *Rep. Rugby School Nat. Hist. Soc.* for 1878, pp. 54–58.

Owen, Prof. R.—On the Association of Dwarf Crocodiles (*Nannosuchus* and *Theriosuchus pusillus*, e.g.) with the Diminutive Mammals of the Purbeck Shales. *Quart. Journ. Geol. Soc.*, vol xxxv., pp. 148–155.

Parsons, Dr. H. F.—Geology of the District around Bruton. *Proc. Somerset Arch. & Nat. Hist. Soc.*, ser. 2, vol. iv., pp. 38–42.

Prestwich, Prof. J.—On the Discovery of a Species of *Iguanodon* in the Kimmeridge Clay near Oxford; and a Notice of a very Fossiliferous Band of the Shotover Sands. *Geol Mag.*, Dec. II., vol. vi., pp. 193–195.

Seeley, Prof. H. G.—On the Evidence that certain Species of *Ichthyosaurus* were Viviparous. *Quart. Journ. Geol. Soc.*, vol. xxxv. (Proc.), p. 104.

———.—Note on a Femur and a Humerus of a small Mammal from the Stonesfield Slate. *Quart. Journ. Geol. Soc.*, vol. xxxv., pp. 456–463.

———.—On the Dinosauria. *Proc. Geol. Assoc.*, vol vi., pp. 175–185.

Sorby, Dr. H. C.—Address to the Geological Society. *Quart. Journ. Geol. Soc.*, vol xxxv., pp. 56–95.

Stoddart, W. W.—Geology of the Bristol Coal-field. Part VI. (misnamed Part V.). Jurassic. *Proc. Bristol. Nat. Soc.*, ser. 2, vol. ii., pp. 279–291.

Tomes, R. F.—On the Fossil Corals obtained from the Oolite of the Railway Cuttings near Hook Norton, Oxfordshire. *Proc. Geol. Assoc.*, vol. vi., pp. 152–165.

Wilson, E.—On the South Scarle Section. *Quart. Journ. Geol. Soc.*, vol. xxxv., pp. 812, 813.

Winwood, Rev. H. H.—Excursion to Haresfield Beacon, with Notes by E. Witchell. *Proc. Bath Field Club*, vol. iv., pp. 170–172.

1880.

Anon.—Account of Excursion to Rugby. *Proc. Dudley Geol. Soc.*, vol. iv., pp. 3, 4.

Barrow, G.—On the Cleveland Ironstone. *Proc. Cleveland Inst. Eng.*, session, 1879–80, pp. 108–112, and 180–187.

Blake, Rev. J. F.—On the Portland Rocks of England. *Quart. Journ. Geol. Soc.*, vol. xxxvi., pp. 189–235.

———.—The Portland Building Stone. *Pop. Sc. Rev.*, n. s., vol. iv., pp. 205–212.

Buckman, Prof. J.—On a New Genus of Bivalve Shells, *Curvirostrum striatum*. *Proc. Dorset Nat. Hist. Field Club*, vol. iv., pp. 102, 103.

Buckman, S. S.—The Brachiopoda from the Inferior Oolite of Dorset and a portion of Somerset. *Proc. Dorset Nat. Hist. Field Club*, vol. iv., pp. 1–52.

———.—Some new Species of Ammonites from the Inferior Oolite. *Proc. Dorset Nat. Hist. Field Club*, vol. iv., pp. 137–146. (4 plates.)

Carpenter, P. H.—On some undescribed *Comatulæ* from the British Secondary Rocks. *Quart. Journ. Geol. Soc.*, vol. xxxvi., pp. 36–55.

———.—On the Genus *Solanocrinus*, Goldfuss, and its Relations to recent Comatulæ. *Journ. Linn. Soc.*, (Zool.), vol. xv., pp. 187–216.

Casley, George.—Geology of Lyme Regis. 8vo. *Lyme Regis*. [Geological Map of the Environs of Lyme Regis, and Section of the Coast from Bridport Harbour to Salcombe Hill, by H. T. De la Beche.]

Cobbold, Edgar S.—Notes on the Strata exposed in laying out the Oxford Sewage-farm, at Sandford-on-Thames. *Quart. Journ. Geol. Soc.*, vol. xxxvi., pp. 314–320.

Guise, Sir W. V.—Address to Cotteswold Nat. Club, 1880. (Notes on Alderton, pp. 148–150; on Dudbridge, pp. 150–153; F. D. Longe, on the Polyzoa of the Cotteswold Oolites, pp. 153–155.) *Proc Cotteswold Club*, vol. vii. p. 138.

Hudleston, W. H.—Excursion to Oxford. *Proc. Geol. Assoc.*, vol. vi., pp. 338–344.

———.—Excursion to Aylesbury. *Proc. Geol. Assoc.*, vol. vi., pp. 344–352.

Hulke, J. W.—*Iguanodon Prestwichii*, a new Species from the Kimmeridge Clay, . . . founded on numerous fossil remains lately discovered at Cumnor, near Oxford. *Quart. Journ. Geol. Soc.*, vol. xxxvi., pp. 433–456.

Moore, C.—The Hedgemead Landslip [Bath]. *Proc. Bath Nat. Hist. Club*, vol. iv., pp. 249–258.

MORIÈRE, J.—Considérations générales sur la flore fossile et spécialement sur celle du Lias. *Bull. Soc. Linn. Norm.*, sér. 3, tome iv., pp. 361–374.

NATHORST, A. G.—Berättelse, ... om en med understöd af allmänna medel utförd vetenskaplig resa till England. *Ofversigt K. Vetens. Akad. Förhandlingar*, 1880, No. 5, pp. 23–84.

———.—Nagra anmärkningar om *Williamsonia*, Carruthers. *Ibid.*, No. 9, pp. 33–51.

OLDHAM, T. B.—On the Annelids found in the Lias near Rugby. *Rep. Rugby School Nat. Hist. Soc.* for 1879, pp. 10, 11.

———. The Middle Lias as exposed near Crick. *Rep. Rugby School Nat. Hist. Soc.*, for 1879, pp. 23, 24.

———.—Geological Report. List of additional species of Rugby Fossils, p. 54. *Rep. Rugby School Nat. Hist. Soc.*, for 1879, pp. 54–57.

PRESTWICH, PROF. J.—Notes on the Occurrence of a new Species of *Iguanodon* in a Brick-pit of the Kimmeridge Clay at Cumnor Hurst, three miles W.S.W. of Oxford. *Quart. Journ. Geol. Soc.*, vol. xxxvi., p. 430.

SAUVAGE, H. E.—Sur le genre *Machimosauré*. *Bull. Soc. Geol. France*, sér. 3. tome vii., pp. 623–637.

SAUVAGE and LIENARD.—Mémoire sur le genre *Machimosaurus*. 4to. *Paris.*

SEELEY, PROF. H. G.—On *Rhamphocephalus Prestwichi*, Seeley, an Ornithosaurian from the Stonesfield Slate of Kineton [Gloucestershire]. *Quart. Journ. Geol. Soc.*, vol. xxxvi., pp. 27–35.

———.—On the skull of an *Ichthyosaurus* from the Lias of Whitby, apparently indicating a new species (*I. Zetlandicus*, Seeley), preserved in the Woodwardian Museum of the University of Cambridge. *Quart. Journ. Geol. Soc.*, vol. xxxvi.. p. 635, pl. xxv.

———.—On the Cranial Characters of a large Teleosaur from the Whitby Lias preserved in the Woodwardian Museum of the University of Cambridge indicating a new species, *Teleosaurus eucephalus*. *Quart. Journ. Geol. Soc.*, vol xxxvi., p. 627, pl. xxiv.

———.—The Dinosauria. *Pop. Sc. Rev.*, ser. 2. vol. iv., pp. 44–60.

SLADEN, W. P.—On traces of Ancestral Relations in the Structure of the *Asteroidea*. *Proc. Yorksh. Geol. Soc.*, vol. vii., pp. 275–284.

SOLLAS, PROF. W. J.—On the Geology of the Bristol District. *Proc. Geol. Assoc.*, vol. vi., pp. 375–391.

———.—Excursion to Bristol. *Ibid.*, pp. 396–402.

SORBY, DR. H. C.—On the Structure and Origin of Non-calcareous Stratified Rocks. Address to the Geological Society. *Quart. Journ. Geol. Soc.*, vol. xxxvi., pp. 46–92.

VINE, GEORGE ROBERT.—A Review of the Family *Diastoporidæ* for the purpose of Classification. *Quart. Journ. Geol. Soc.*, vol. xxxvi., pp. 356–361.

WITCHELL, E.—On a Section of Stroud Hill, and the Upper Ragstone Beds of the Cotteswolds. *Proc. Cottesw. Club*, vol. vii., pp. 117–135.

WRIGHT, DR. THOMAS. Modern Classification of the *Ammonitidæ*. *Proc. Cotteswold Nat. Club*, vol. vii., pp. 169–219.

———.—On a new species of *Ophiurella*. *Proc. Dorset Nat. Hist. Field Club*, vol. iv., pp. 56–58 (plate).

1880–87.

THOMPSON, BEEBY. Local Geology. [Lower and Middle Lias.] *Journ. Northamptonsh. Nat. Hist. Soc.*, vol. i., pp. 11, 48, 84, 142, 222, 280, 232 ; vol. ii., pp. 57, 147, 202, 239.

1881.

ANDREWS, REV. W. R.—Outline of the Geology of the Vale of Wardour. *Proc. Dorset Nat. Hist. Field Club*, vol. v., pp. 57–68 (map and sections).

ANON.—Section of Strata near Shapwick. *Proc. Somerset Arch. Soc.*, vol. xxvi., p. 126.

BAXTER, R. C.—A short Description of the Sub-Wealden Gypsum Company's Works. *Proc. Holmesdale Nat. Hist. Club*, for 1879 and 1880, pp. 19–22.

BIRD, C.—A Short Sketch of the Geology of Yorkshire. Pp. 196. 8vo. *Bradford.*

BLAKE, REV. J. F.—On the Correlation of the Upper Jurassic Rocks of England with those of the Continent. Part I. The Paris Basin. *Quart. Journ. Geol. Soc.,* vol. xxxvii., pp. 497–587.

BUCKMAN, PROF. J.—On the *Trigonia Bella* [Lycett], from Eype, near Bridport, Dorset. *Proc. Dorset Nat. Hist. Field Club,* vol. v., pp. 154–156 (plate).

————.—On the Terminations of some Ammonites from the Inferior Oolite of Dorset and Somerset. *Quart. Journ. Geol. Soc.,* vol. xxxvii., pp. 57–66.

BUCKMAN, S. S.—A Descriptive Catalogue of some of the Species of Ammonites from the Inferior Oolite of Dorset. *Quart. Journ. Geol. Soc.,* vol. xxxvii., pp. 588–608.

ETHERIDGE, R.—On a New Species of *Trigonia* from the Purbeck Beds of the Vale of Wardour. *Quart. Journ. Geol. Soc.,* vol. xxxvii., pp. 246–251. (With a Note on the Strata. By the Rev. W. R. ANDREWS, pp. 251–253).

FOSTER, C. LE NEVE, and R. J. FRECHEVILLE.—Report of the Inspectors of Mines, to Her Majesty's Secretary of State, for the year 1880. Folio. *London.*

GUISE, SIR W. V.—Address to the Cotteswold Nat. Club, 1881. (Notes on Berkeley, pp. 4–7.) *Proc. Cotteswold Club,* vol. viii., p. 1.

HOLMES, T. V.—The Permian, Triassic, and Liassic Rocks of the Carlisle Basin. *Quart. Journ. Geol. Soc.,* vol. xxxvii., pp. 286–297.

HUDLESTON, W. H.—Note on some Gasteropoda from the Portland Rocks of the Vale of Wardour and of Bucks. *Geol. Mag.,* Dec. II., vol. viii., pp. 385–395.

————.—Excursion to Salisbury, Stonehenge, and Vale of Wardour. *Proc. Geol. Assoc.,* vol. vii., pp. 134–142.

————.—On the Geology of the Vale of Wardour. *Ibid.,* 161–185.

LEE, JOHN EDWARD.—Note-Book of an Amateur Geologist. 8vo. *London.*

LONGE, FRANCIS D.—On the Relation of the Escharoid Forms of Oolitic Polyzoa to the Cheilostomata and Cyclostomata. *Geol. Mag.,* Dec. II., vol. viii., pp. 23, 34.

LYCETT, DR. JOHN.—Note on the Generic Distinctness of *Purpuroidea* and *Purpura,* with remarks upon the Purpuroid Shells figured in the Geol. Mag. Plate VIII. Decade II., vol. vii., 1880. *Geol. Mag.,* Dec. II., vol. viii., pp. 483–502.

MANSEL-PLEYDELL, J. C.—Note on the Cone from the Inferior Oolite Beds of Sherborne. *Proc. Dorset. Nat. Hist. Field Club,* vol. v., pp. 141–143 (plate).

MOORE, CHARLES.—On Abnormal Geological Deposits in the Bristol District. *Quart. Journ. Geol. Soc.,* vol. xxxvii., p. 67–82.

NEWTON, E. T.—Notes on the Mandible of an *Ischyodus Townsendii,* found at Upway, Dorsetshire, in the Portland Oolite. *Proc. Geol. Assoc.,* vol. vii., pp. 116–119.

PHILLIPS, J. ARTHUR.—On the Constitution and History of Grits and Sandstones. *Quart. Journ. Geol. Soc.,* vol. xxxvii., pp. 6–27.

QUILTER, H. E.—Exposure of the Middle Series of the Bucklandi Beds in Leicestershire. *Midland Nat.,* vol. iv., p. 265.

SAUVAGE, H. E.—Sur les *Dinosauriens* jurassiques. *Bull. Soc. Géol. France,* sér. 3, tome viii., pp. 522–524.

SHARP, S.—Huge Fossil at Twywell. *Journ. Northamptonsh. Nat. Hist. Soc.,* vol. i., p. 230.

————.—Local Wells and Borings. *Ibid.,* p. 291.

————.—*Stellaster Sharpii. Ibid.* p. 322.

SOLLAS, PROF. W. J.—On a Species of *Plesiosaurus* (*P. Conybeari*) from the Lower Lias of Charmouth; with Observations on *P. megacephalus,* Stutchbury, and *P. brachycephalus,* Owen. Accompanied by a Supplement on the Geographical Distribution of the Genus Plesiosaurus. By G. F. WHIDBORNE. *Quart. Journ. Geol. Soc.,* vol. xxxvii., pp. 440–480.

STRUCKMANN, C.—On the Parallelism of the Hanoverian and English Upper Jurassic Formations. [Translated by W. S. Dallas.] *Geol. Mag.,* Dec. II., vol. viii., pp. 546–556.

VINE, G. R.—Further Notes on the Family *Diastoporidæ,* Busk. Species from the Lias and Oolites. *Quart. Journ. Geol. Soc.,* vol. xxxvii., pp. 381–390.

WILLETT, E. W.—Notes on a Mammalian Jaw from the Purbeck Beds at Swanage, Dorset. With an Introduction by H. WILLETT. *Quart. Journ. Geol. Soc.*, vol. xxxvii., pp. 376–379.

WOODWARD, DR. H.—Contributions to the Study of Fossil Crustacea. *Geol. Mag.*, Dec. II., vol. viii., pp. 530–534.

WORTHINGTON, J. K.—Geological Report. Lists of Rugby Fossils. *Rep. Rugby School Nat. Hist. Soc.*, for 1880, pp. 48–57.

WRIGHT, DR. T.— The Physiography and Geology of the country around Cheltenham. *Midland Nat.*, vol. iv., pp. 145–159.

1881–82.

WOODWARD, C. J.—The Minerals of the Midlands. *Midland Nat.*, vol. iv., pp. 87, 90, 112; vol. v., pp. 11–13, 203.

1882.

CARPENTER, P. H.—On some new or little-known Jurassic Crinoids. *Quart. Journ. Geol. Soc.*, vol. xxxviii., pp. 29–43.

ETHERIDGE, ROBERT.—Address to the Geological Society—On the Analysis and Distribution of the British Jurassic Fossils. *Quart. Journ. Geol. Soc.*, vol. xxxviii., pp. 59–236.

EUNSON, JOHN.—The Boring at the Kettering Road (Northampton). *Journ. Northamptonsh. Nat. Hist. Soc.*, vol. ii., p. 69.

GUISE, SIR W. V.—Address to Cotteswold Nat. Club, 1882. (Notes on Andoversford, pp. 70–72; on Birdlip, pp. 72–76.) *Proc. Cotteswold Club*, vol. viii., p. 60.

HARRISON, W. J.—Geology of the Counties of England and of North and South Wales. 8vo. *London.*

HUDLESTON, W. H.—Excursion to the Isle of Purbeck. *Proc. Geol. Assoc.*, vol. vii., pp. 377–390.

HULL, PROF. E.—Palæo-Geological and Geographical Maps of the British Isles and adjoining parts of the Continent of Europe. *Trans. Roy. Dublin Soc.*, n.s., vol. i., part xix., pp. 257–296.

————.—.Contributions to the Physical History of the British Isles. 8vo. *London.*

JONES, PROF. T. R.—Catalogue of the Fossil Foraminifera in the Collection of the British Museum. 8vo. *London.*

KEEPING, H.—On some Sections of Lincolnshire Neocomian. *Quart. Journ. Geol. Soc.*, vol. xxxviii., pp. 239–244.

LUCY, W C.—On the Minerals of Gloucestershire, with part of the adjacent Counties of Somerset and Worcestershire. *Proc. Cotteswold Nat. Club*, vol. viii., pp. 30–34.

MEADE, RICHARD.—The Coal and Iron Industries of the United Kingdom. 8vo. *London.*

MOORE, CHARLES.—Notes on Wiltshire Geology and Palæontology. *Mag. Wilts Arch. and Nat. Hist. Soc.*, vol. xx., pp. 45–54.

PARKIN, C.—On Jet Mining. *Trans. N. Eng. Inst. Mining Eng.*, vol. xxxi., pp. 51–58.

ROBINSON, C.E.—A Royal Warren ; or Picturesque Rambles in the Isle of Purbeck. [Notes on Purbeck and Portland Beds, &c.] 4to. Etching Co. *London.*

SLATTER, T. J.—On the Foramifera from the Lias. *Proc. Warwickshire Field Club* for 1881.

SOLLAS, W. J.—On a Rare Plesiosaur from the Lias at Bridport. *Proc. Bristol Nat. Soc.*, ser. 2, vol. iii., pp. 322, 323.

TOMES, R. F.—Description of a New Species of Coral from the Middle Lias of Oxfordshire. *Quart. Journ. Geol. Soc.*, vol. xxxviii., pp. 95, 96.

————.—On the Madreporaria of the Inferior Oolite of the Neighbourhood of Cheltenham and Gloucester. *Quart. Journ. Geol. Soc.*, vol. xxxviii., pp. 409–450.

TOPLEY, W.—Excursion to Battle and Hastings. *Proc. Geol. Assoc.*, vol. vii., pp. 356–359 ; *Record of Excursions*, 1891, pp. 127–130.

WALFORD, E. A.—On *Natica cincta*, its surface markings, and variations in growth. *Banbury Nat . Hist. Soc.*, July 31.

WITCHELL, E.—The Geology of Stroud and the area drained by the Frome. 8vo. *Stroud.*

———.—On the Pisolite and the Basement Beds of the Inferior Oolite of the Cotteswolds. *Proc. Cotteswold Nat. Club*, vol. viii. pp. 35–49.

WORTHINGTON, J. K.—A few Notes on the Geology of Rugby. *Rep. Rugby School Nat. Hist. Soc.*, for 1881, pp. 17–23.

———.— Geological Report. List of additional species of Rugby Fossils. *Rep. Rugby School Nat. Hist. Soc.* for 1881, pp. 49–53.

WRIGHT, Dr. T.—On a New Species of Star Fish from the Forest Marble, Wilts. *Proc. Cotteswold Nat. Club*, vol. viii. pp. 50–52.

———.—On a New Species of Brittle Star, from the Coral Rag of Weymouth. *Proc. Cotteswold Nat. Club*, vol. viii., pp. 53–55.

———.—On a new Astacamorphous Crustacean, from the Middle Coral Reef of Leckhampton Hill. *Proc. Cotteswold Nat. Club*, vol. viii., pp. 56–59.

———.— On a New Species of *Ophiurella* [Corallian, Weymouth.] *Proc. Dorset Field Club*, vol. iv. pp. 56, 57.

1883.

ADDY, JOHN.—The Water-Supply of Peterborough. *Proc. Inst. Civil Eng.*, vol. lxxiv. pp. 146–189.

BEESLEY, THOMAS.—A new local Fossil. [*Discina Gunnii*, Inferior Oolite, Hook Norton.] *Banbury Guardian*, Aug. 9.

BUTLIN, W. H.—The Northampton Iron Industry. *Iron*, vol xxi., p. 450; *Engineering*, vol. xxxv., p. 579.

———.—On the Northampton Iron Ore District. *Journ. Iron and Steel Inst.*, pp. 188–212; *London Iron Trades Exchange*, vol. xxxii., p. 641.

CARR, W. D.—The Lincoln Lias. *Geol. Mag.*, Dec. II,, vol. x., pp. 164–169.

CRICK, W. D.—Notes on the Geology of Wymington Tunnel; with list of Fossils by T. J. GEORGE. *Journ. Northamptonsh. Nat. Hist. Soc.*, vol. ii., p. 272.

EUNSON, HENRY J.—On a Deep Boring at Northampton. *Proc. Inst. Civ. Eng.*, vol. lxxiv., p. 270.

GUISE, SIR W. V.—Address to Cotteswold Nat. Club, 1883. (Notes on Stroud, pp. 93–97.) *Proc. Cotteswold Club*, vol. viii., p. 89.

HINDE, DR. G. J.—Catalogue of the Fossil Sponges in the Geological Department of the British Museum. 4to. *London.*

HULKE, J. W.—Address to the Geological Society. *Quart. Journ. Geol. Soc.*, vol. xxxix., pp. 41–65.

LUCY, W. C.—Hock Crib, Fretherne. *Proc. Cotteswold Club*, vol. viii., pp. 131–133.

McMURTRIE, J.—Notes on Autumn Excursions on the Mendips. *Bath. Nat. Hist. & Antiq. Field Club*, pp. 98–110.

OWEN, PROF. SIR R.—On the Skull of *Megalosaurus*. *Quart. Journ. Geol. Soc.*, vol. xxxix., pp. 334–347.

———.—On Generic Characters in the Order *Sauropterygia*. *Quart. Journ. Geol. Soc.*, vol. xxxix., pp. 133–138.

PAUL, J. D.—On a Section exposed on the Great Northern Railway near Thurnby and Scraptoft. *Trans. Leicester Lit. & Phil. Soc.*, for 1882–83, pp. 50, 51.

QUILTER, H. E.—On an Exposure of Lower Lias on the Crown Hill in a Pit of the Evington Lime Company. *Trans. Leicester Lit. and Phil. Soc.* for 1882–83, pp. 51, 52.

SAUVAGE, H. E.—Note sur le genre *Pachycormus*. *Bull. Soc. Linn. Norm.*, sér. 3, tome vii., pp. 144–149.

SMITHE, DR. F.—On the Occurrence of the Mineral Vivianite in the Cotteswolds; with Remarks. *Proc. Cotteswold Club*, vol. viii., pp. 112–117.

SOLLAS, PROF. W. J.—Descriptions of Fossil Sponges from the Inferior Oolite, with a Notice of some from the Great Oolite. *Quart. Journ. Geol. Soc.*, vol. xxxix., pp. 541–554.

Tomes, R. F.—On some new or imperfectly known Madreporaria from the Coral Rag and Portland Oolite of the Counties of Wilts, Oxford, Cambridge, and York. *Quart. Journ. Geol. Soc.*, vol. xxxix., pp. 555–565.

——.—On the Fossil Madreporaria of the Great Oolite of the Counties of Gloucester and Oxford. *Quart. Journ. Geol. Soc.*, vol. xxxix., pp. 168–196.

Vine, G. R.—Third Report of the Committee on Fossil Polyzoa. *Brit. Assoc.*, Rep. for 1882, p. 249.

Walford, E. A.—Fossils from the Transition Bed between the Middle and Upper Lias. *Journ. Northamptonsh. Nat. Hist. Soc.*, vol. ii., p. 297.

——.—On some Crinoidal and other Beds in the Great Oolite of Gloucestershire, and their probable equivalents in North Oxon. *Proc. Warwickshire Nat. Field Club* for 1882, p. 20.

——.—On the Relation of the so-called "Northampton Sand" of North Oxon to the Clypeus-Grit. *Quart. Journ. Geol. Soc.*, vol. xxxix., pp. 224–245.

Whidborne, Rev. G. F.—Notes on some Fossils, chiefly Mollusca, from the Inferior Oolite. *Quart. Journ. Geol. Soc.*, vol. xxxix., pp. 487–540.

Williamson, Prof. W. C.—On some Anomalous Oolitic and Palæozoic Forms of Vegetation. *Proc. R. Inst.*, vol. x., pp. 220–232.

1884.

Bate, C. Spence.—*Archæastacus Willemæsii*, a New Genus of Eryonidæ. [Lower Lias, Lyme Regis.] *Rep. Brit. Assoc.*, for 1883, p. 511.

Carr, W. D.—Excursion to Lincoln. *Proc. Geol Assoc.*, vol. viii., pp. 383–385.

Cope, E. D.—The Tertiary Marsupialia. *Amer. Nat.*, vol. xviii., p. 691. (Refers to Purbeck Mammals.)

Davis, J. W.—On a New Species of *Heterolepidotus* from the Lias. *Proc. Yorksh. Geol. Soc.*, vol. viii., pp. 403–407.

——.—Description of a New Genus of Fossil Fishes from the Lias. *Ann. and Mag. Nat. Hist.*, ser. 5, vol. xiii., pp. 448–453.

Duncan, Dr. P. M.—A Revision of the Families and Genera of the Sclerodermic Zoantharia, Ed. and H., or Madreporaria. *Journ. Linn. Soc.*, vol. xviii., (Zool.) pp. 1–204.

Eunson, Henry John.—The range of the Palæozoic Rocks beneath Northampton. *Quart. Journ. Geol. Soc.*, vol. xl., pp. 482–496.

——.—On a Probable Fault in the Lias under Northampton. *Journ. Northamptonshire Nat. Hist. Soc.*, vol. iii., p. 169.

Guise, Sir W. V.—Address to the Cotteswold Naturalists' Field Club, 1884. (Notes on Section of Kellaways Rock at South Cerney, pp. 137–139, 151–153; Section of Cephalopoda-bed at Uley Bury, p. 143; sections near Notgrove Station, pp. 144, 145.) *Proc. Cotteswold Club*, vol. viii., p. 135.

Harker, Prof. Allen.—On a Remarkable Exposure of the Kellaway's Rock in a recent Cutting near Cirencester. *Proc. Cotteswold Club*, vol. viii., pp. 176–187.

Hinde, Dr. G. J.—On some Fossil Calcisponges from the well-boring at Richmond, Surrey. *Quart. Journ. Geol. Soc.*, vol. xl., pp. 778–788.

Hughes, Prof. T. McK.—Diagram Section from Elsworth, by Cambridge, to Balsham. *Proc. Geol. Assoc.*, vol. viii., p. 401.

——.—On some Tracks of Terrestrial and Freshwater Animals. *Quart. Journ. Geol. Soc.*, vol. xl., pp. 178–186.

Jones, Prof. T. R.—Notes on the Foraminifera and Ostracoda from the Deep Boring at Richmond. *Quart. Journ. Geol. Soc.*, vol. xl., pp. 765–777.

Judd, Prof. J. W.—On the Nature and Relations of Jurassic Deposits which underlie London. With an Introductory Note on a Deep Boring at Richmond, Surrey, by Collett Homersham. *Quart. Journ. Geol. Soc.*, vol. xl., pp. 724–764.

——.—Jurassic Rocks under London. *Nature*, vol. xxix., p. 329.

Jukes-Browne, A. J.—The Classification of the Jurassic System. *Geol. Mag.*, Dec. III., pp. 525, 526.

Lucy, W. C.—Section of Birdlip. Some Remarks on a Boring for Water near Birdlip for the City of Gloucester. *Proc. Cotteswold Club*, vol. viii., pp. 161–166.

Marriott, J.—The Beds exposed in the Railway-cutting at Slawston Hill. *Trans. Leicester Lit. and Phil. Soc.*, for 1883–84, pp. 80, 81.

MOORE, C. A.—The Railway Cutting north of Market Harborough Station
Trans. Leicester Lit. and Phil. Soc., pp. 82, 83.

OWEN, SIR R.—History of British Fossil Reptiles. 4 vols. 4to. *London.*

————.—On the Cranial and Vertebral Characters of the Crocodilian Genus
Plesiosuchus, Owen. *Quart. Journ. Geol. Soc.*, vol. xl., pp. 153–159.

PARKER, JAMES.—Map and Sections of Strata south of Oxford (prepared for the
Meeting of the Excursion of the Geol. Assoc. 1880; revised for meeting of Warwickshire Field Club, 1884). (Privately printed.) 8vo. *Oxford.*

PAUL, J. D.—The Cores from a Boring made near Evington. *Trans. Leicester
Lit. and Phil. Soc.*, for 1883–84, pp. 83–86.

PLANT, JAMES.—On the Map of South-east Leicestershire, just issued by the
Government Geological Survey. *Trans. Leicester Lit. and Phil. Soc.*, pt. ix., p. 327.

PILBROW, J.—Some Particulars of an Artesian Well bored through the Oolitic
Rocks at Bourn, Lincolnshire, in 1856. *Proc. Inst. Civ. Eng.*, vol. lxxv., p. 245.

QUILTER, H. E.—The Cutting near Market Harborough. *Trans. Leicester Lit.
and Phil. Soc.*, for 1883–84, pp. 86, 87.

RICHARDS, J. THEODORE.—Synopsis of British Fossil Cycadaceous Leaves.
(Reprint?) 8vo. *Edinburgh.*

THOMPSON, B.—On Swallow Holes and Dumb Wells. *Journ. Northamptonshire
Nat. Hist. Soc.*, vol. iii., p. 159.

TOMES, R. F.—A Comparative and Critical Revision of the Madreporaria of the
White Lias of the Midland and Western Counties of England, and of those of the
Conglomerate at the base of the South Wales Lias. *Quart. Journ. Geol. Soc.*,
vol. xl., pp. 353–374.

VINE, G. R.—Polyzoa (Bryozoa) found in the Boring at Richmond, Surrey,
referred to by Prof. Judd, F.R.S. *Quart. Journ. Geol. Soc.*, vol. xl., pp. 784–794.

WARD, L. F.—On Mesozoic Dicotyledons. *Ann. and Mag. Nat. Hist.*, ser. 5,
vol. xiii., p. 383.

1884–90.

THOMPSON, B.—The Upper Lias of Northamptonshire. *Journ. Northamptonshire
Nat. Hist. Soc.*, vol. iii., pp. 3–14, 183–200, 299–314, vol. iv., pp. 16–28, p. 215,
vol. v., p. 54, 96, vol. vi., pp. 54–85.

1885.

BLAKE, PROF. J. F.—Report on Lias and Rhætic. *Internat. Geol. Congress*
(English Committee), pp. 95–105. [Ed. 2 in 1888].

BLANFORD, W. T.—The Classification of the Jurassic System. *Geol. Mag.*,
Dec. III., vol. ii., pp. 239, 240.

DUNCAN, PROF. P. MARTIN.—On the Structure of the Ambulacra of some Fossil
Genera and Species of Regular Echinoidea. *Quart. Journ. Geol. Soc.*, vol. xli.,
pp. 419–453.

ETHERIDGE, R.—Manual of Geology, Theoretical and Practical: By John Phillips.
New Ed. Part II. Stratigraphical Geology and Palæontology. 8vo. *London.*

GILL, W. H.—Speeton Cliffs to Gristhorpe Bay. *Trans. Leeds Geol. Assoc.*,
Pt. i., p. 21.

GUISE, SIR W. V.—Address to the Cotteswold Naturalists' Field Club, 1885.
(Notes on Section at Lassington, p. 225; Chipping Norton, pp. 225–228; Dursley,
pp. 228–230.) *Proc. Cotteswold Club*, vol. viii., p. 223.

HARKER, ALFRED.—The Oolites of the Cave district. *Naturalist*, pp. 229–232.

HARKER, PROF. ALLEN.—A Weathered Concretion of Sandstone [Kellaways
Rock]. *Agric. Students' Gazette*, ser. 2, vol. ii., pp. 65–67.

HUDLESTON, W. H.—Report on the Oolites. *Internat. Geol. Congress* (English
Committee) pp. 63–93. [Ed. 2 in 1888.]

HUDLESTON, W. H., and H. B. WOODWARD.—Excursion to Sherborne and
Bridport. *Proc. Geol. Assoc.*, vol. ix., pp. 187–209.

JONES, PROF. T. R.—On the Ostracoda of the Purbeck Formation. *Quart.
Journ. Geol. Soc.*, vol. xli., pp. 311–353.

JUDD, PROF. J. W.—A Problem for Cheshire Geologists. (Lias at Wem.) *Proc.
Chester Soc. Nat. Sci.*, vol. iii., pp. 45–49.

JUDD, PROF. J. W., and COLLETT HOMERSHAM.—Supplementary Notes on the
Deep Boring at Richmond, Surrey. *Quart. Journ. Geol. Soc.*, vol. xli, pp. 523–528.

LUCY, W. C.—Section of a Well-Sinking at the Island, Gloucester, by Messrs. Robertson & Co., and some Remarks upon the Thickness of the Lower Lias at Gloucester and the Neighbourhood. *Proc. Cotteswold Club*, vol. viii., pp. 218–221.

MANSEL-PLEYDELL, J. C.— On a Fossil Chelonian Reptile from the Middle Purbecks. *Proc. Dorset Nat. Hist. Club*, vol. vi, pp. 66–69.

MOORE, C. A.—Notes on the Geology of the neighbourhood of Rugby. *Trans. Leicester Lit. und Phil. Soc.* for 1884–85., pp. 117–119.

SOLLAS, PROF. W. J.—On an Hexactinellid Sponge from the Gault, and a Lithistid from the Lias of England. *Scient. Proc. R. Dublin Soc.*, vol. iv., pp. 443–446.

TOMES, R. F.—On some new or imperfectly known Madreporaria from the Great Oolite of the Counties of Oxford, Gloucester, and Somerset. *Quart. Journ. Geol. Soc.*, vol. xli., pp. 170–190.

WALFORD, E. A.—On the Stratigraphical Positions of the *Trigoniæ* of the Lower and Middle Jurassic Beds of North Oxfordshire and adjacent Districts. *Quart. Journ. Geol. Soc.*, vol. xli., pp. 35–47.

WILSON, E.—The Lias Marlstone of Leicestershire as a source of Iron. *Midland Nat.*, vol. viii. pp. 61–66, 94–97, 123–127, 152–158.

WOODWARD, ARTHUR SMITH—On the Literature and Nomenclature of British Fossil Crocodilia. *Geol. Mag.*, Dec. III., pp. 496–510.

WOODWARD, DR. H.—On Recent and Fossil *Pleurotomariæ*. *Geol. Mag.*, Dec. III., vol. ii., pp. 433–439.

1885–87.

LYDEKKER, R.—Catalogue of the Fossil Mammalia in the British Museum. 5 parts. 8vo. *London.*

1886.

BATES, E. F., and L. HODGES.—Notes on a recent Exposure of the Lower Lias and Rhætics in the Spinney Hills, Leicester. *Trans. Lit. and Phil. Soc. Leicester*, ser. 2, part i., pp. 22, 23.

BATHER, F. A.—Note on some recent Openings in the Liassic and Oolitic Rocks of Fawler in Oxfordshire, and on the Arrangement of those Rocks near Charlbury. *Quart. Journ. Geol. Soc.*, vol. xlii., pp. 143–146.

BUCKMAN, S. S.—Notes on Jurassic Brachiopoda. *Geol. Mag.*, Dec. III., vol. iii. pp. 217–219.

———.—On the Lobe-line of certain species of Lias Ammonites. Described in the Monograph by the late Dr. Wright. *Geol. Mag.*, Dec. III., vol. iii., pp. 442, 443.

CARTER, JAMES.—On the Decapod Crustaceans of the Oxford Clay. *Quart. Journ. Geol. Soc.*, vol. xlii., pp. 542–559.

CHADWICK, S.—[Notice of *Asteracanthus ornatissimus.*] *Malton Nat. Soc.* p. 6.

COLE, REV. E. M.—Notes on the Geology of the Hull, Barnsley, and West Riding Junction Railway and Dock. Pp. 60, Frontispiece, Five Plates, and Map. 8vo. *Hull.*

———.—On some Sections at Cave and Drewton. *Proc. Yorksh. Geol. Soc.* n.s., vol. ix., p. 49.

———.—On the Physical Geography and Geology of the East Riding of Yorkshire. *Ibid.*, p. 113.

DUNCAN, PROF. P. MARTIN—On the *Astrocæniæ* of the Sutton Stone and other Deposits of the Infra-Lias of South Wales. *Quart. Journ. Geol. Soc.*, vol. xlii., pp. 101–112.

———.—On a New Species of *Axosmilia* (*A. elongata*) from the Pea Grit of the Inferior Oolite of England. *Geol. Mag.*, Dec. III., vol. iii., p. 340–342.

———.—On the Structure and Classificatory Position of some Madreporaria from the Secondary Strata of England and South Wales. *Quart. Journ. Geol. Soc.*, vol. xlii., pp. 113–142.

EUNSON, H. J.—Notes on a Deep Boring at Orton, Northamptonshire. *Journ. Northamptsonsh. Nat. Hist. Soc.*, vol. iv., p. 57.

GARDNER, J. STARKIE.—On Mesozoic Angiosperms. *Geol. Mag.*, Dec. III., vol. iii., pp. 193–204; 342–345.

GREEN, BURTON.—Kimmeridge Shale : its Origin, History, and Uses.   8vo. London.

HARRIS, G. F.—Our Building Stones.   *Builder.*   March 13 to Dec. 25.

HUDLESTON, W. H.—Geology of Malton.   *Ann. Rep. Malton Nat. Soc.* for 1884-5, pp. 1-30.

JONES, PROF. T. R., and C. D. SHERBORN—On the Microzoa found in some Jurassic Rocks of England.   *Geol. Mag.,* Dec. III. vol. iii., pp. 271-274.

KENDALL, J. D.—The Iron Ores of the English Secondary Rocks.   *Trans. N. of England Inst. of Mining Engineers,* vol. xxxv. pp. 105-157.

LUCY, W. C.—Southerndown, Dunraven and Bridgend Beds.   *Proc. Cotteswold Club,* vol. viii., pp. 254-264.

PLATNAUER, H. M.—On the occurrence of *Strophodus Rigauri* in the Yorkshire Cornbrash.   *Rep. York. Phil. Soc.,* p. 36.

QUILTER, H. E.—The Lower Lias of Leicestershire.   *Geol. Mag.,* Dec. III., vol. iii., pp. 59-65.

SCUDDER, S. H.—A review of mesozoic cockroaches.   *Mem. Bost. Soc. Nat. Hist.,* vol. iii., pp. 439-485.

TOMES, R. F.—On the Occurrence of Two Species of Madreporaria in the Upper Lias of Gloucestershire.   *Geol. Mag.,* Dec. III., vol. iii., pp. 107-111.

————.—On some New or Imperfectly Known Madreporaria from the Inferior Oolite of Oxfordshire, Gloucestershire, and Dorsetshire.   *Geol. Mag.,* Dec. III., vol. iii., pp. 385-398; 443-452.

VEITCH, W. Y.—Three new Species in the Yorkshire Lias.   *Proc. Yorksh. Geol. Soc.,* n. s., vol. ix., p. 54.

WETHERED, EDWARD—The Pea-grit of Leckhampton Hill.   *Geol. Mag.,* Dec. III., vol. iii., p. 525.

WHITAKER, W.—On Deep Borings at Chatham : a Contribution to the Deep-seated Geology of the London Basin.   *Rep. Brit. Assoc.* for 1885, p. 1041.

————.—On some Borings in Kent.   *Quart. Journ. Geol. Soc.,* vol. xlii., pp. 26-48; *see also* vol. xliii., p. 199.

WILSON, E.—The Lias Marlstone of Leicestershire as a source of Iron.   *33rd Ann. Rep. Nottingham Nat. Soc.,* 1885, p. 43.

WITCHELL, E.—On the Basement-beds of the Inferior Oolite of Gloucestershire.   *Quart. Journ. Geol. Soc.,* vol. xlii., pp. 264-271.

————.—On the Forest Marble and Upper Beds of the Great Oolite, between Nailsworth and Wotton-under-Edge.   *Proc. Cotteswold Club,* vol. viii., pp. 265-280.

WOODWARD, A. SMITH.—The History of Fossil Crocodiles.   *Proc. Geol. Assoc.,* vol. ix., pp. 288-344.

————.—On the Palæontology of the Selachian Genus *Notidanus,* Cuvier.   *Geol. Mag.,* Dec. III., vol. iii., pp. 205-217 ; 253-259.

WOODWARD, H. B.—Account of a Well-sinking made by the Great Western Railway Company at Swindon.   With Lists of Fossils by E. T. NEWTON.   *Quart. Journ. Geol. Soc.,* vol. xlii., pp. 287-308.

————.—Report on Erosion of Sea-coasts.   Axmouth to Eype, Bridport Harbour, &c.   *Rep. Brit. Assoc.,* for 1885, pp. 423-426.

1887.

BLAKE, J. F.—On the new specimen of *Solaster Murchisoni* from the Yorkshire Lias.   *Geol. Mag.,* Dec. III., p. 529. pl. xv.

BUCKMAN, S. S.—Some New Species of Brachiopoda, from the Inferior Oolite of the Cotteswolds.   *Proc. Cotteswold Club,* vol. ix., pp. 38-43.

————.—On *Ammonites serpentinus,* Reinecke, *Am. falcifer,* Sowb., *Am. elegans,* Sowb., *Am. elegans,* Young, etc.   *Geol. Mag.,* Dec. III., vol. iv., pp. 396-400.

CARPENTER, DR. P. HERBERT—On Crinoids and Blastoids.   *Proc. Geol. Assoc.,* vol. x., pp. 19-28.

COPE, PROF. E. D.—Lydekker, Boulenger, and Dollo on Fossil Tortoises.   *Geol. Mag.,* Dec. III., vol. iv., pp. 572, 573.

CRICK, W. D.—Note on some Foraminifera from the Oxford Clay at Keyston, near Thrapston.   *Journ. Northampton Nat. Hist. Soc.,* vol. iv., p. 232.

DALTON, W. H.—The Collingham or Scarle Boring. *Geol. Mag.*, Dec. III., vol., iv., p. 48.

DAVIES, WILLIAM.—On New Species of *Pholidophorus* from the Purbeck Beds of Dorsetshire. *Geol. Mag.*, Dec. III., vol. iv. pp. 337–339.

———.—Notes on Chelonia. *Geol. Mag.*, Dec. III., vol. iv., p. 380.

DAVIS, JAMES W.—On *Chondrosteus acipenseroides*, Agassiz. *Quart. Journ. Geol. Soc.*, vol. xliii , pp. 605–616.

GARDNER, J. S., and others.—Second Report on the Fossil Plants of the Tertiary, and Secondary Beds of the United Kingdom. (From Jurassic of Yorkshire, and Specimens in York Museum). *Rep. Brit. Assoc.* for 1886., p. 241.

GROOM, T. T.—On some New Features in *Pelanechinus corallinus*. *Quart. Journ. Geol. Soc.*, vol. xliii., pp. 703–714.

GROVES, T. B.—The Abbotsbury Iron Deposits. *Proc. Dorset Field Club*, vol. viii., pp. 64–66.

GUISE, SIR W. V.—Address to the Cotteswold Naturalists' Field Club, 1886. (The occurrence of a new Fossil Annelide [*Pachynereis corrugatus*] from the Stonesfield Slate of Eyford, by R. ETHERIDGE, pp. 2, 3.) *Proc. Cotteswold Club*, vol. ix. p. 1.

HAEUSLER, DR. RUDOLF.—Bemerkungen über einige liasische Milioliden. *N. Jahrbuch f. Mineralogie*, bd. i. pp. 190–194.

HULKE, J. W.—Note on some Dinosaurian Remains in the Collection of A. Leeds, Esq., of Eyebury, Northamptonshire. *Quart. Journ. Geol. Soc.*, vol. xliii. pp. 695–702.

LYDEKKER, R., and G. A. BOULENGER.—Notes on Chelonia from the Purbeck, Wealden and London-clay. *Geol. Mag.*, Dec. III., vol. iv., pp. 270–275.

MORGAN, PROF. C. LLOYD.—Bristol Building Stones. *Proc. Bristol Nat. Soc.*, ser. 2., vol. v., pp. 95–115.

ROBERTS, THOMAS.—On the Correlation of the Upper Jurassic Rocks of the Swiss Jura with those of England. *Quart. Journ. Geol. Soc.*, vol. xliii., pp. 229–269.

TAUNTON, J. H.—Some Notes on the Hydrology of the Cotteswolds and the District around Swindon. *Proc. Cotteswold Club.*, vol. ix., pp. 52–69.

———.—Visit to the Boxwell Springs South Cerney, on 20th July, 1886. *Ibid.*, p. 70.

THOMPSON, B.—The Middle Lias of Northamptonshire. *Journ. Northampton Nat. Hist. Soc.*, vol. iv., p. 167.

TRAQUAIR, DR. R. H.—Notes on *Chondrosteus acipenseroides*, Agassiz. *Geol. Mag.*, Dec. III., vol. iv., pp. 248–257.

VINE, G. R.—Jurassic Polyzoa in the Neighbourhood of Northampton. *Journ. Northampton Nat. Hist. Soc.*, vol. iv., p. 202.

———.—Notes on the Polyzoa and other Organisms from the Gayton Boring, Northamptonshire. *Journ. Northampton Nat. Hist. Soc.*, vol. iv., p. 255.

WALFORD, EDWIN A.—Notes on some Polyzoa from the Lias. *Quart. Journ. Geol. Soc.*, vol. xliii., pp. 632–636.

WILSON, E.—British Liassic Gasteropoda. *Geol. Mag*, Dec. III., vol. iv., pp. 193–202 ; 258–262.

WITCHELL, E.—On the Genus *Nerinæa* and its Stratigraphical distribution in the Cotteswolds. *Pro. Cotteswold Nat. Club*, vol. ix. pp. 21–37.

———.—The Pea Grit of Leckhampton Hill. *Geol. Mag.*, Dec. III., vol. iv., pp. 46, 47.

WOODWARD, A. SMITH.—Notes on Some Post-Liassic Species of *Acrodus*. *Geol. Mag.*, Dec. III., vol. iv., pp. 101–105.

———.—On the Anatomy and Systematic Position of the Liassic Selachian, *Squaloraja polyspondyla*, Agassiz. *Proc. Zool. Soc.*, for 1886, pp. 527–538.

WOODWARD, H. B.—Notes on the Geology of Rousdon [near Lyme Regis]. *Meteorol. Observations, Rousdon Observatory*, vol. iii., for 1886, pp. 18, 19.

———.—Notes on the Geology of Brent Knoll, in Somersetshire. *Proc. Bath Nat. Hist. Club*, vol. vi., pp. 125–130.

———.—Notes on the Ham Hill Stone. *Ibid.*, pp. 182–184.

———.—Note on some Pits near Chipping Norton, Oxfordshire. *Essex Naturalist*, vol. i. pp. 265, 266.

WRIGHT, DR. T.—On a New *Ophiurella* from the Calciferous Grit, near Sandsfoot Castle, Weymouth, Dorset. *Geol. Mag.*, Dec. III., vol. iv., pp. 97, 98.

1887–1895.

BUCKMAN, S. S.—A Monograph of the Inferior Oolite Ammonites of the British Islands. *Palæontograph. Soc.* 4to. *London.*

HUDLESTON, W. H.—A Monograph of the British Jurassic Gasteropoda. Part I. Gasteropoda of the Inferior Oolite. Nos. 1-7. General Introduction, pp. 1–15. *Palæontograph. Soc.*, 4to. *London.*

1888.

BLAKE, J. F.—On a Starfish from the Yorkshire Lias. *Rep. Brit. Assoc.* for 1887., p. 716.

BRODIE, REV. P. B.—On the Range, Extent, and Fossils of the Rhætic Formation in Warwickshire. *Proc. Warwickshire Nat. Field Club* for 1887, p. 19.

BUCKMAN, S. S.—The Inferior Oolite between Andoversford and Bourton-on-the-Water. *Proc. Cotteswold Club*, vol. ix., pp. 108-135.

BUCKMAN, S. S., and J. F. WALKER.—On Oolitic Brachiopoda new to Yorkshire. *Rep. Yorks. Phil. Soc.*, p. 41.

CAMERON, A. C. G.—The Clays of Bedfordshire. *Proc. Geol. Assoc.*, vol. x., pp. 446-454.

UISS, SIR W. V.—Address to Cotteswold Naturalists' Club, 1887. *Proc. Cotteswold Club*, vol. ix., pp. 81-95.

HARKER, PROF. ALLEN.—Excursion to Cirencester and Minchinhampton. *Proc. Geol. Assoc.*, vol. x., pp. 157-163. Diagram-Section from Stroud to Cirencester, by H. B. WOODWARD, p. 162.

HINDE, DR. G. J.—On the History and Characters of the Genus *Septastræa*, D'Orbigny (1849), and the Identity of its Type Species with that of *Glyphastræa*, Duncan (1887). *Quart. Journ. Geol. Soc.*, vol. xliv., pp. 200-227.

HUDLESTON, W. H.—Excursion to Aylesbury. *Proc. Geol. Assoc.*, vol. x., pp. 166-172.

HYATT, PROF. A.—Evolution of the Faunas of the Lower Lias. *Proc. Boston Soc. Nat. Hist.*, vol. xxiv., pp. 17-30.

JONES, PROF. T. R.—Ostracoda from the Weald Clay of the Isle of Wight. *Geol. Mag.*, Dec. III., vol. v., pp. 534-539.

JONES, PROF. T. R., and C. D. SHERBORN.—On some Ostracoda from the Fuller's-earth Oolite and Bradford Clay. *Proc. Bath Nat. Hist. Club*, vol. vi., p. 249.

LUCY, W. C.—The Origin of the Cotteswold Club, and an epitome of the Proceedings from its Formation to May, 1887. 8vo. *Gloucester.* (Privately printed.)

LYDEKKER, R.—Note on a New Wealden Iguanodont and other Dinosaurs. *Quart. Journ. Geol. Soc.*, vol. xliv., pp. 46-60.

————.—On the Skeleton of a Sauropterygian from the Oxford Clay, near Bedford. *Quart. Journ. Geol. Soc.*, vol. xliv. (Proc,) pp. 89, 90.

————.—Note on the Classification of the *Ichthyopterygia* (with a Notice of Two New Species). *Geol. Mag.*, Dec. III., vol. v., pp. 309-314.

————.—Notes on the Sauropterygia of the Oxford and Kimeridge Clays, mainly based on the Collection of Mr. Leeds at Eyebury. *Geol. Mag.*, Dec. III., vol. v., pp. 350-356.

MANSEL-PLEYDELL, J. C.—Fossil Reptiles of Dorset. *Proc. Dorset Nat. Hist. Club*, vol. ix., p. 1.

NEWTON, E. T.—On the Skull, Brain, and Auditory Organ of a new Species of Pterosaurian (*Scaphognathus Purdoni*) from the Upper Lias, near Whitby, Yorkshire. *Proc. Roy. Soc.*, xliii., pp. 436–440 ; *Phil. Trans.*, vol. 179 B., pp. 503-537, pls. 77, 78.

————.—Notes on Pterodactyls. *Proc. Geol. Assoc.*, vol. x., pp. 406-424.

OSBORN, H. F.—On the Structure and Classification of the Mesozoic Mammalia. *Journ. Acad. Nat. Sci. Philad.*, vol. ix., p. 282.

PLATNAUER, H. M.—Note on *Hybodus obtusus*. *Rep. York Phil. Soc.*, p. 85.

PRIOR, DR. C. E.—Report on Bedfordshire Well-waters to the Rural Sanitary Authority of the Bedford District. *Bedford.*

SEELEY, PROF. H. G.—On *Cumnoria,* an Iguanodont Genus founded upon the *Iguanodon Prestwichi,* Hulke. *Rep. Brit. Assoc.* for 1887, p. 698.

———.—On the Classification of the Fossil Animals commonly named Dinosauria. *Proc. Roy. Soc ,* vol. xliii., p. 165.

———.—Researches on the Structure, Organisation, and Classification of Fossil Reptilia, Part 3. *Ibid,* p. 172.

SHERBORN, C. D.—Notes on *Webbina irregularis* (d'Orb) from the Oxford Clay at Weymouth. *Proc. Bath Nat. Hist. Club,* vol. vi., p. 332.

STRANGWAYS, C. FOX-, and G. W. LAMPLUGH.—La Géologie de l'Est de Yorkshire. Congrès Géologique International (Explications des Excursions). 8vo. *London.* pp. 131-175.

TOMES, R. F.—On *Heterastræa,* a new Genus of Madreporaria from the Lower Lias. *Geol. Mag.,* Dec. III., vol. v., pp. 207-218.

WALKER, J. F.—On the occurrence of *Terebratula Gesneri* in Yorkshire. *Rep. York. Phil. Soc.,* p. 33.

WETHERED, E.—On the Geology of Bath and the Neighbourhood. *Scientific News,* vol. ii., pp. 219-222.

WINWOOD, REV. H. H.—Geology of the Bath District. (With Geological Map by H. B. WOODWARD.) 8vo. *Bath.*

———.—Handbook to the Geological Museum, Bath. 8vo. *Bath.*

WITCHELL, E.—On a Section of Selsley Hill. *Proc. Cotteswold Club,* vol. ix.,. pp. 96-107.

WOODWARD, A. SMITH.—Palæontological Contributions to Selachian Morphology. *Proc. Zool. Soc.,* Part I., p. 126.

———.—On some Remains of the Extinct Selachian *Asteracanthus* from the Oxford Clay of Peterborough, preserved in the Collection of Alfred N. Leeds, Esq., of Eyebury. *Ann. Nat. Hist.,* ser. vi., vol. ii., pp. 336-342.

———.—Note on the Early Mesozoic Ganoid, *Belonorhynchus,* and on the supposed Liassic Genus *Amblyurus. Ann. Nat. Hist.,* ser. vi., vol. i., pp. 354-356.

——— DR. H.—On a New Species of *Æger* from the Lower Lias, of Wilmcote, Warwickshire. *Geol. Mag.,* Dec. III., vol. v., pp. 385-387.

———.—On *Eryon antiquus,* Broderip, Sp., from the Lower Lias, Lyme-Regis, Dorset. *Geol. Mag.,* Dec. III., vol. v., pp. 433-441.

———- H. B.—The Relations of the Great Oolite to the Forest Marble and Fuller's Earth in the South-west of England. *Geol. Mag.,* Dec: III., vol. v., pp. 467, 468; and *Rep. Brit. Assoc.* for 1888, pp. 651, 652.

— ———.—Note on the Portland Sands of Swindon and elsewhere. *Ibid.,* pp. 469, 470; *Ibid.,* p. 652.

———-.—Further Note on the Midford Sands. *Ibid.,* p. 470; *Ibid.,* pp. 650, 651.

———.--Notes sur la Géologie de la vallée de Wardour. *Internat. Geol. Commission.* 8vo. *London.*

1889-90.

LYDEKKER, RICHARD.—Catalogue of the Fossil Reptilia and Amphibia in the British Museum (Natural History). 4 Parts. 8vo. *London.*

1889.

BRODIE, REV. P. B.—On the Predominance and Importance of the Blattidæ in the Old World. *Warwickshire Nat. Club.*

BROWETT, ALFRED.—The Bath Oolite and Method of quarrying it. *Midland Nat.,* vol. xii., pp. 187-190.

BROWNE, MONTAGU.—The Vertebrate Animals of Leicestershire and Rutland. 8vo. *Birmingham and Leicester.*

———.—On a Fossil Fish (*Chondrosteus*) from Barrow-on-Soar, hitherto recorded only from Lyme Regis. *Trans. Leicester Lit. and Phil. Soc.,* vol. ii. pp. 17-35.

BUCKMAN, S. S.—On the Cotteswold, Midford, and Yeovil Sands, and on the Division between Lias and Oolite. *Quart. Journ. Geol. Soc.*, vol. xlv., pp. 440, 473.

————.—The Descent of *Sonninia* and *Hammatoceras*. *Quart. Journ. Geol. Soc.*, vol. xlv., pp. 651–663.

————.—On Jurassic Ammonites. *Geol. Mag.*, Dec. III., vol. vi., pp. 200–203.

————.—The relations of Dundry with the Dorset-Somerset and Cotteswold areas during part of the jurassic period. *Proc. Cotteswold Club*, vol. ix., pp. 374–387.

CAMERON, A. C. G.—Excursion to Bedford and Clapham. *Proc. Geol. Assoc.*, vol. x., pp. 504–510.

ETHERIDGE, R., and H. WILLETT.—On the Dentition of *Lepidotus maximus*, Wagner, as illustrated by Specimens from the Kimeridge Clay of Shotover Hill, near Oxford. *Quart. Journ. Geol. Soc.*, vol. xlv., pp. 356–358.

HINDE, DR. G. J.—On a true Leuconid Calcisponge from the Middle Lias of Northamptonshire, . . *Ann. Nat. Hist.*, ser. vi., vol. iv., pp. 852–357.

HOLMES, T. V.—The Geology of North-west Cumberland. *Proc. Geol. Assoc.*, vol. xi., pp. 231–257. (Permian, Triassic and Liassic Rocks, pp. 237–247.)

HUDLESTON, W. H.—Excursion to Weymouth. *Proc. Geol. Assoc.*, vol. xi, pp. xlix–lvii.

————.—On the Geological History of Iron-ores. *Proc. Geol. Assoc.*, vol. xi., pp. 104–144.

HULKE, J. W.—Contribution to the Skeletal Anatomy of the Mesosuchia based on Fossil Remains from the Clays near Peterborough in the Collection of A. Leeds, Esq. *Proc. Zool. Soc.*, for 1888, pp. 417–442.

JUKES-BROWNE, A. J.—On the Occurrence of Granite in a Boring at Bletchley. *Geol. Mag.*, Dec. III., vol. vi., pp. 356–361.

LAMPLUGH, G. W.—On the Sub-divisions of the Speeton Clay. *Quart. Journ. Geol. Soc.*, vol. xlv., pp. 575–618.

LUCY, W. C.—Addresses to Cotteswold Nat. Club, 1888 and 1889. (Meetings at Kemble, &c., pp. 313–316.) *Proc. Cotteswold Club*, vol. ix., pp. 171–191, &c.

————.—Notes on the jurassic Rocks at Crickley Hill. *Proc. Cotteswold Club*, vol. ix., p. 289–299.

————.—Remarks on the Dapple Bed of the Inferior Oolite at the Horsepools, and on some Pebbles from the Great Oolite at Minchinhampton. *Proc. Cotteswold Club*, vol. ix., pp. 388–395.

LYDEKKER, R.—On the Remains and Affinities of five Genera of Mesozoic Reptiles. *Quart. Journ. Geol. Soc.*, vol. xlv., pp. 41–58.

————.—On certain Chelonian Remains from the Wealden and Purbeck. *Quart. Journ. Geol. Soc.*, vol. xlv., pp. 511–518.

————.—Note on some Points in the Nomenclature of Fossil Reptiles and Amphibians, with Preliminary Notices of Two New Species. *Geol. Mag.*, Dec. III. vol. vi., pp. 325, 326.

————.—On an Ichthyosaurian Paddle Showing the Contour of the Integuments. *Geol. Mag.*, Dec. III., vol. vi., pp. 388–390.

MANSEL-PLEYDELL, J. C.—On a New Specimen of *Histionotus angularis*, Egerton. *Geol. Mag.*, Dec. III., vol. vi., pp. 241, 242 ; and *Proc. Dorset Nat. Hist. Club*, vol. xi., pp. 91–96. (1890).

————.—*Cimoliosaurus richardsoni*. *Proc. Dorset Nat. Hist. Field Club*, vol. x., p. 171.

NIKITIN, S.—Quelques excursions en ¦Europe occidentale. *Bull. Soc. Géol. Belge*, vol. iii., pp. 29–58. And 8vo. *St. Petersberg*.

PAVLOW, A.—Études sur les Couches Jurassiques et Crétacées de la Russie. I. Jurassique Supérieur et crétacé inférieur de la Russie et de l'Angleterre. *Bull. Soc. Imp. des Nat. Moscow*, sér. 2, tome iii., pp. 61, 176. (Reprinted, 8vo. *Moscow*.)

QUILTER, H. E.—The Life-Zones of the British Lias. *Research*, vol. ii., pp. 97–99.

ROBERTS, THOMAS. The Upper Jurassic Clays of Lincolnshire. *Quart. Journ. Geol. Soc.*, vol. xlv., pp. 545–559.

SEELEY, PROF. H. G.—On the the Origin of Oolitic Texture in Limestone Rocks. *Rep. Brit. Assoc.* for 1888, pp., 674, 675.

———.—Researches on the Structure, Organization, and Classification of the Fossil Reptilia. Parts II. and III., and V. *Phil. Trans.*, Part 13, vol. clxxix., pp. 59, 141, 487.

———.—Note on the Pelvis of *Ornithopsis*. *Quart. Journ. Geol. Soc.*, vol. xlv. pp. 391–396.

STUART, M. G.—The Ridgway Fault. *Proc. Dorset Field Club*, vol. x., pp. 55–70.

THOMPSON, BEEBY.—The Middle Lias of Northamptonshire. [Reprinted from the *Midland Naturalist*, 1885–89.] 8vo. *London.*

TOMES, R. F.—Notes on an Amended List of Madreporaria of Crickley Hill. *Proc. Cotteswold Club*, vol. ix., pp. 300–307.

WALFORD, E. A.—On some Bryozoa from the Inferior Oolite of Shipton Gorge, Dorset. *Quart. Journ. Geol. Soc.*, vol. xlv., pp. 561–574.

WETHERED, E.—On the Microscopic Structure of the Jurassic Pisolite. *Geol. Mag*, Dec. III., vol. vi., pp. 196–200.

WHITAKER, W.—On the Extension of the Bath Oolite under London, as shown by a Deep Boring at Streatham. *Rep. Brit. Assoc.*, for 1888, pp. 656, 657.

WILSON, E., and W. D. CRICK.—The Lias Marlstone of Tilton, Leicestershire. *Geol. Mag.*, Dec. III., vol. vi., pp. 296–305, 337–342.

WOODWARD, A. SMITH.—Palæichthyological Notes. (On a Symmetrical Hybodont Tooth from the Oxford Clay of Peterborough.) *Ann. Nat. Hist.*, ser. vi., vol. iii., pp. 297–302.

———.—On a Head of *Hybodus De la Bechei*, associated with Dorsal Fin-spines, from the Lower Lias of Lyme Regis, Dorsetshire. *Ann. Report Yorkshire Phil. Soc.* for 1888, pp. 58–61.

———.—On the Palæontology of Sturgeons. *Proc. Geol. Assoc.*, vol. xi., pp. 24–44.

———.—Palæontology in the Malton Museum. *Geol. Mag.*, Dec. III., vol. v., p. 361.

———.—On the *Myriacanthidæ*—an Extinct Family of Chimæroid Fishes. *Ann. Nat. Hist.*, ser. vi., vol. iv., pp. 275–280.

———.—Preliminary Notes on some New and little-known British Jurassic Fishes. *Geol. Mag.*, Dec. III., vol. vi., pp. 448–455; *Rep. Brit. Assoc.* for 1889, pp. 585, 586 (1890).

WOODWARD, DR. H.—Visit to the British Museum (Natural History), Department of Geology. Demonstration on Edentata, Marsupialia, &c. *Proc. Geol., Assoc.*, vol. x., pp. 457–468.

WOODWARD, H. B.—Notes on the Rhætic Beds and Lias of Glamorganshire. *Proc. Geol. Assoc.*, vol. x., pp. 529–538.

———.—Excursion to Lyme Regis. *Proc. Geol. Assoc.*, vol. xi., pp. xxvi–xlix.

———.—Notes on the Coast-line from Penarth to Porth Cawl, in Glamorganshire. *Rep. Brit. Assoc.* for 1888, pp. 900–903.

1889–91.

WOODWARD, A. SMITH.—Catalogue of Fossil Fishes in the British Museum. Two Parts. 8vo. *London.*

1890.

BRODIE, REV. P. B.—On the Character, Variety, and Distribution of the Fossil Insects in the Palæozoic (Primary), Mesozoic (Secondary), and Cainozoic (Tertiary) Periods. *Warwicksh. Nat. and Arch. Field Club.*

BROWNE, M.—A Contribution to the History of the Lias and Rhætics in Leicestershire. *Rep. Leicester Lit. and Phil. Soc.*, ser. 2, vol. ii., p. 147 (Abstract only).

———.—Revision of a genus of Fossil Fishes, *Dapedius*. *Ibid.*, p. 196.

BUCKMAN, S. S.—On the so-called "Upper-Lias Clay" of Down Cliffs. *Quart. Journ. Geol. Soc.*, vol. xlvi., pp. 518–521.

———.—The Sections exposed between Andoversford and Chedworth: a comparison with similar strata upon the Banbury line. *Proc. Cotteswold Club*, vol. x., pp. 94–100.

H H 2

CAMERON, A. C. G.—Note on the recent Exposures of Kellaways Rock at Bedford. *Rep. Brit. Assoc.* for 1889, pp. 577, 578.

COLE, REV. E. M.—Notes on the Driffield and Market Weighton Railway. *Proc. Yorksh. Geol. Soc.*, vol. xi., p. 170.

DAWKINS, PROF. W. BOYD.—The Discovery of Coal near Dover. *Nature*, vol. xli., March 6th, pp. 418, 419., vol. xlii., July 31st, pp. 319–322 ; *Contemp. Rev.*, vol. lvii., April, pp. 470–478.

FOORD, ARTHUR H., and G. C. CRICK.—On the Muscular Impressions of some Species of Carboniferous and Jurassic Nautiloids compared with those of the recent Nautilus. *Ann. Nat. Hist.*, ser. vi., vol. v., pp. 220–224.

————.—Descriptions of new and imperfectly-defined Species of Jurassic Nautili contained in the British Museum (Natural History). *Ann. Nat. Hist.*, ser vi., vol. v., pp. 265–291, 388–398.

HARKER, ALFRED.—Notes of North of England Rocks, III. *Naturalist*, pp. 300 –304.

HARKER, PROF. ALLEN.—On the Sections in the Forest Marble, and Great Oolite formations, exposed by the new railway from Cirencester to Chedworth. *Proc. Cotteswold Club*, vol. x., pp. 82–93.

HINDE, DR. G. J.—On a new Genus of Siliceous Sponges from the Lower Calcareous Grit of Yorkshire. *Quart. Journ. Geol. Soc.*, vol. xlvi., pp. 54–61.

JONES, PROF. T. R.—On some Fossil *Estheriæ*. *Geol. Mag.*, Dec. III., vol. vii., pp. 385–390.

KIRBY, W. F.—A Synonymic Catalogue of Neuroptera Odonata, or Dragon Flies, with an Appendix on Fossil Species. 8vo. *London*.

LAMPLUGH, G. W.—The Neocomian Clay at Knapton [considered Jurassic by Judd]. *Naturalist*, p. 336.

LYDEKKER, R.—Contributions to our knowledge of the Dinosaurs of the Wealden and the Sauropterygians of the Purbeck and Oxford Clay. *Quart. Journ. Geol. Soc.*, vol. xlvi., pp. 37–53.

————.—On a Crocodilian Jaw from the Oxford Clay of Peterborough. *Quart. Journ. Geol. Soc.*, vol. xlvi., pp. 284–288.

————.—On Ornithosaurian Remains from the Oxford Clay of Huntingdonshire. *Quart. Journ. Geol. Soc.*, vol. xlvi., pp. 429–431.

MANSEL-PLEYDELL, J. C.—Memoir upon a New Ichthyopterygian from the Kimeridge Clay of Gillingham, Dorset, *Ophthalmosaurus Pleydelli*. *Proc. Dorset Nat. Hist. Club*, vol. xi ; pp. 7–15.

————.—*Histionotus angularis*. *Proc. Dorset Nat. Hist. Club*, vol. xi., p. 91.

QUILTER, H. E.—[Ammonites collected from the Lower Lias of Barrow-on-Soar by Mr. Montagu Browne.] *Rep. Leicester Lit. and Phil. Soc.*, ser. 2, vol. ii., p. 147.

SAUNDERS, JAMES.—Notes on the Geology of South Bedfordshire. *Geol. Mag.*, Dec. III., vol. vii., pp. 117–127.

SOLLY, H. S.—The Geology of Bridport. *Proc. Dorset. Nat. Hist. and Field Club*, vol. xi., pp. 109–117.

SOLLY, H. S., and J. F. WALKER.—Note on the Fault in the Cliff West of Bridport Harbour. *Proc. Dorset Nat. Hist. and Field Club*, vol. xi., pp. 118–121.

THOMPSON, B.—The Bletchley Boring. *Journ. Northamptonshire Nat. Hist. Soc.* vol. v., pp. 20–25.

WETHERED, E.—On the occurrence of the Genus *Girvanella* in Oolitic Rocks, and Remarks on Oolitic Structure. *Quart. Journ. Geol. Soc.*, vol. xlvi., pp. 270–283.

————.—On the occurrence of fossil forms of the Genus *Chara* in the Middle Purbeck Strata of Lulworth, Dorset. *Proc. Cotteswold Club*, vol. x., pp. 101 –103.

WHITAKER, W.—Coal in the South-east of England. *Journ. Soc. Arts*, vol. xxxviii., pp. 543–557.

WILSON, E.—Fossil Types in the Bristol Museum. *Geol. Mag.*, Dec. III., vol. vii., pp. 363–372, 411–416.

WOODWARD, A. SMITH.—On a New Species of Pycnodont Fish (*Mesodon Damoni*) from the Portland Oolite. *Geol. Mag.*, Dec. III., pp. 158, 159.

————.—The Fossil Sturgeon of the Whitby Lias. *Naturalist*, No. 177, p. 101.

WOODWARD, A. SMITH.—A Synopsis of the Fossil Fishes of the English Lower Oolites. *Proc. Geol. Assoc.*, vol. xi., pp. 285–306.

————.—Notes on some Ganoid Fishes from the English Lower Lias. *Ann. Nat. Hist. Soc.*, ser. vi., vol. v., pp. 430–436.

————.—On some British Jurassic Fish-Remains referable to the Genera *Eurycormus* and *Hypsocormus*. *Quart. Journ. Geol. Soc.*, vol. xlvi. (Proc.), p. 8 ; and *Geol. Mag.*, Dec. III., vol. vii., p. 93.

————.—On a Head of *Eurycormus* from the Kimmeridge Clay of Ely. *Geol. Mag.*, Dec. III., vol vii., pp. 289–292.

————.—On the Gill rakers of *Leedsia problematica*—a Gigantic Fish from the Oxford Clay. *Ibid.*, p. 292.

————.—On some new Fishes from the English Wealden and Purbeck Beds, referable to the Genera *Oligopleurus, Strobilodus* and *Mesodon*. *Proc. Zool. Soc.*, pp. 346–353.

WOODWARD, A. SMITH, and C. D. SHERBORN.—A Catalogue of British Fossil Vertebrata. 8vo. *London*. Supplement for 1890. *Geol. Mag.*, Dec. III., vol. viii., pp. 25–34.

WOODWARD, DR. H.—On a New British Isopod (*Cyclosphæroma trilobatum*) from the Great Oolite of Northampton. *Geol. Mag.*, Dec. III., vol. vii., pp. 529–533.

1891.

BLAKE, PROF. J. F.—The Geology of the Country between Redcar and Bridlington. *Proc. Geol. Assoc.*, vol. xii., p. 115.

BLAKE, J. F., G. W. LAMPLUGH, and C. M. COLE.—Excursion to the East Coas of Yorkshire. *Proc. Geol. Assoc.*, vol. xii., p. 207.

BOULENGER, G. A.—On British Remains of *Homæosaurus*. with Remarks on the Classification of the Rhynchocephalia. *Proc. Zool. Soc.*, pp. 167–172.

BROWNE, M.—(Note on some fossils from Loseby.) *Rep. Leicester Lit. and Phil. Soc.*, ser. 2, vol. ii., p. 349.

————.—Notes upon *Colobodus*, a Genus of Mesozoic Fossil Fishes. *Geol. Mag.*, Dec. III., vol. viii., pp. 501, 502 ; *Rep. Brit. Assoc.* for 1891 (1894), p. 644.

BUCKMAN, S. S.—The Ammonite Zones of Dorset and Somerset. *Geol. Mag.*, Dec. III.. vol. viii., pp. 502–504 ; *Rep. Brit. Assoc.* for 1891 (1892), pp. 655, 656.

CAMERON, A. C. [G.]—On the Continuity of the Kellaways Beds over extended areas near Bedford, and on the Extension of the Fuller's Earth Works at Woburn. *Geol. Mag.*, Dec. III., vol. viii., p. 504 ; *Rep. Brit. Assoc.* for 1891, 1892, p. 636.

CLARKE, J. F. M.—The Geology of the Bridgwater Railway. *Proc. Bath. Nat. Hist. Field Club*, vol. vii., p. 127.

CRICK, W. D., and C. D. SHERBORN.—On some Liassic Foraminifera from Northamptonshire. *Journ. Northamptonshire Nat. Hist. Soc.*, vol. vi., pp. 208–214.

DE RANCE, C. E. — On the Underground Waters of Lincolnshire. *Proc. Yorkshire and Polytech. Soc.*, vol. xii.. pp. 22–51.

FOORD, A. H.—Catalogue of the Fossil Cephalopoda in the British Museum. Part II. *Nautiloidea*. 8vo. *London*.

HARKER, PROF. ALLEN.—On the Geology of Cirencester Town ; and a recent discovery of the Oxford Clay in a deep well-boring at the Water Works. *Proc. Cotteswold Club*, vol. x., pp. 178–191.

HAWKESWORTH, E.—From Kettleness to Saltburn. *Trans. Leeds Geol. Assoc.*, pt. vi., p. 55.

HOWES, G. B.—Observations on the Pectoral Fin-Skeleton of the Living Batoid Fishes, and of the Extinct Genus *Squaloraja*, with especial reference to the Affinities of the same. *Proc. Zool. Soc.* for 1890, p. 675.

JAEKEL, OTTO.—*Acanthoteuthis* aus dem unteren Lias von Lyme Regis in England. *Sitz. Ber. Geo. Naturforschender Freunde zu Berlin*, No. 5, p. 88.

JONES, PROF. T. R.—On some more Fossil Estheriæ. *Geol. Mag.*, Dec. III., vol. viii., pp. 1–57.

JUKES-BROWNE, A. J.—On a Boring at Shillingford, near Wallingford (on Thames). *Midland Naturalist*, vol. xiv., pp. 201–208.

JUKES-BROWNE, A. J., and the Rev. W. R. ANDREWS.—The Lower Cretaceous Series of the Vale of Wardour. *Geol. Mag.*, Dec. III., vol. viii., pp. 292–294.

LAMPLUGH, G. W.—On the Speeton Clays and their Equivalents in Lincolnshire. *Rep. Brit. Assoc.* for 1890, pp. 808, 809.

LEIGHTON, T.—Excursion to Oxford. *Proc. Geol. Assoc.*, vol. xi., pp. cxlvi.–cl.

LUCAS, JOSEPH.—Report to Lieut.-Colonel W. G. Dawkins : On the prospect of finding Coal under the Estate of Over Norton, Oxfordshire (in the parish of Chipping Norton). 8vo. *London.*

LYDEKKER, R.—On certain Ornithosaurian and Dinosaurian Remains. *Quart. Journ. Geol. Soc.*, vol. xlvii., pp. 41–44.

——.—Note on a nearly perfect Skeleton of *Ichthyosaurus tenuirostris* from the Lower Lias of Street, Somerset. *Geol. Mag.*, Dec. III., vol. viii., pp. 289, 290.

MILLS, H. M.—A few Notes on the Ironstone Deposits of Leicestershire. *Proc. Chesterfield and Mid. C. Inst. Engineers*, vol. xviii., p. 64 ; *Trans. Fed. Inst. Min. Eng.*

NEWTON, E. T.—Note on the Occurrence of *Ammonites jurensis* in the Ironstone of the Northampton Sands in the Neighbourhood of Northampton. *Geol. Mag.* Dec. III., vol. viii., pp. 493, 494.

PLATNAUER, H. M.—List of Figured Specimens in the York Museum. *Ann. Rep. Yorks. Phil. Soc.* for 1890, p. 56.

ROBERTS, R. W. B.—The Cliff Sections of the Yorkshire Coast. [Flamborough to Whitby.] *Journ. Liverpool Geol. Assoc.*, vol. xi., p. 59.

SCUDDER, S. H.—Index to the known Fossil Insects of the World, including Myriapods and Arachnids. *Bull. U.S. Geol. Survey*, No. 71.

SEELEY, PROF. H. G.—On the Neural Arch of the Vertebræ in the Ichthyosauria. *Rep. Brit. Assoc.* for 1890, p. 809.

——.—The Ornithosaurian Pelvis. *Ann. Nat. Hist.*, ser. 6, vol. vii., pp. 237- .

SMITHE, DR. F. and W. C. LUCY.—Some Remarks on the Geology of Alderton, Gretton, and Ashton-under-Hill. *Proc. Cotteswold Club*, vol. x., pp. 202–211.

THIESSING, J. B.—Notizen über den Lias von Lyme Regis. *Naturforsch. Gesells. Bern.* for 1890, p. 1.

THOMPSON, B.—The Oolitic Rocks at Stowe-nine-churches. *Journ. Northamptonshire Nat. Hist. Soc.*, vol. vi., pp. 294–319.

THOMPSON, B., and W. D. CRICK.—Excursion to Northamptonshire. *Proc. Geol. Assoc.*, vol. xii., pp. 172–190.

WALLIS, A. M.—The Portland Stone Quarries. *Proc. Dorset. Nat. Hist. Club*, vol. xii., pp. 187–194.

WETHERED, EDWARD.—The Inferior Oolite of the Cotteswold Hills, with Special Reference to its Microscopical Structure. *Quart. Journ. Geol. Soc.*, vol. xlvii., pp. 550–569.

WILSON, EDWARD.—On a Specimen of *Waldheimia perforata* (Piette), showing Original Colour-markings. *Geol. Mag.*, Dec. III., vol. viii., pp. 458, 459.

WINWOOD, REV. H. H., and H. B. WOODWARD.—Excursion to the Mendip Hills. *Proc. Geol. Assoc.*, vol. xi., pp. 171–216.

WOODS, HENRY.—Catalogue of the Type Fossils in the Woodwardian Museum, Cambridge. 8vo. *Cambridge.*

WOODWARD, A. S.—*Pholidophorus germanicus* : an addition to the Fish Fauna of the Upper Lias of Whitby. *Geol. Mag.*, Dec. III., vol. viii., p. 545.

WOODWARD, H. B.—Brief Notes on the Geology of the Mendip Hills. *Proc. Geol. Assoc.*, vol. xi., pp. 481–494.

1892.

BLAKE, PROF. J. F.—The Evolution and Classification of the Cephalopoda, an account of Recent Advances. *Proc. Geol. Assoc.*, vol. xii., pp. 275–295.

BLAKE, J. F., H. B. WOODWARD, and others.— Excursion to Devizes, Seend, Swindon, and Faringdon. *Proc. Geol. Assoc.*, vol. xii., pp. 323–333.

BUCKMAN, S. S.—The Reported Occurrence of *Ammonites jurensis* in the Northampton Sands. *Geol. Mag.*, Dec. III., vol. ix., pp. 258-260.

——.—The Morphology of " *Stephanoceras* " zigzag. *Quart. Journ. Geol. Soc.*, vol xlviii., pp. 447–452.

CRICK, W. D., and C. D. SHERBORN.—On some Liassic Foraminifera from North-amptonshire. *Journ. Northamptonshire Nat. Hist. Soc.*, vol. vii., pp. 67–73.

DAWKINS, PROF. W. BOYD.—The Further Discovery of Coal at Dover, and its bearing on the Coal Question. *Trans. Manchester Geol. Soc.*, vol. xxi., pp. 456–474.

HUDLESTON, W. H., and EDWARD WILSON.—A Catalogue of British Jurassic Gasteropoda. 8vo. London.

HULKE, J. W.—On the Shoulder Girdle in Ichthyosauria and Sauropterygia. *Proc. Roy. Soc.*, vol. lii., p. 233.

LORIEUX, E.— Le Sondage de Douvres. *Annales des Mines*, ser. ix., tome ii., pp. 227–232.

MANSEL-PLEYDELL, J. C.—Kimmeridge Coal-Money and other Manufactured Articles from the Kimmeridge Shale. *Proc. Dorset Nat. Hist. Club*, vol. xiii., pp. 178–190.

MURRAY, GEORGE.—On a Fossil Alga belonging to the Genus *Caulerpa* from the Oolite. *Phycolog. Memoirs*, vol. i., pp. 11–15.

PAVLOW, A., and G. W. LAMPLUGH.—Argiles de Speeton et leurs équivalents. *Bull. Soc. Imp. des Nat. Moscow*, for 1891. 8vo. *Moscow.*

ROBERTS, THOMAS.—The Jurassic Rocks of the Neighbourhood of Cambridge. 8vo. *Cambridge.*

THOMPSON, B.—Report of the Committee to work the very Fossiliferous Tran-sition Bed between the Middle and Upper Lias in Northamptonshire, in order to obtain a more clear idea of its Fauna, and to fix the position of certain Species of Fossil Fish, and more fully investigate the Horizon on which they occur. *Rep. Brit. Assoc.* for 1891, pp. 334–331.

WALKER, J. F.—On Liassic Sections near Bridport, Dorsetshire. *Geol. Mag.*, Dec. III., vol ix., pp. 437–443 ; and *Rep. Brit. Assoc.* for 1890, pp. 799, 800 (1891).

———.—The Discovery of *Terebratulina substriata*, Schlotheim, in Yorkshire. *Geol. Mag.*, Dec. III., vol. ix., p. 364.

———.—On Yorkshire Thecidea. *Geol. Mag.*, Dec. III., vol. ix., p. 548.

WINWOOD, REV. H. H.—Charles Moore, F.G.S., and his Work ; with a List of the Fossil Types and described Specimens in the Bath Museum, by E. WILSON. *Proc. Bath Nat. Hist. Club*, vol. vii., pp. 232–292.

WOODWARD, A. S.—On some Teeth of new Chimæroid Fishes from the Oxford and Kimmeridge Clays of England. *Ann. Nat. Hist.*, ser. 6, vol. x., pp. 13–16.

———.—Supplementary Observations on some Fossil Fishes of the English Lower Oolites. *Proc. Geol. Assoc.*, vol. xii., pp. 238–241.

———.—On the Skeleton of a Chimæroid Fish (*Ischyodus*) from the Oxford Clay of Christian Malford, Wiltshire. *Ann. Nat. Hist.*, ser. vi., vol. ix., pp. 94–96.

WOODWARD, DR. H.—On a Neuropterous Insect from the Lower Lias, Barrow-on-Soar, Leicestershire. *Geol. Mag.*, Dec. III., vol. ix., pp. 193–198.

WOODWARD, H. B.—On Geological Zones. *Proc. Geol. Assoc.*, vol. xii., pp. 295–315.

———.—Remarks on the Formation of Landscape Marble. *Geol. Mag.*, pp. 110–114.

———.—Notes on the Geology of Crewkerne. *Proc. Somerset Arch. & Nat. Hist. Soc.*, vol. xxxvii., pp. 60–69.

1893.

BLAKE, PROF. J. F.—On the Bases of the Classification of Ammonites. *Proc. Geol. Assoc.*, vol. xiii., pp. 24–39.

———.—Excursion to Brill. *Proc. Geol. Assoc.*, vol. xiii., pp. 71–74.

BRODIE, REV. P. B.—Reminiscences of a Search for Fossil Insects: Where and How to find them. *Warwickshire Nat. Club*, 1892. 8vo. *Warwick*, p. 20.

BROWNE, M.—A Contribution to the History of the Geology of the Borough of Leicester. *Trans. Leicester Lit. and Phil. Soc.*, vol. iii., pp. 123–240.

BUCKMAN, S. S.—The Bajocian of the Sherborne District : its Relation to Subjacent and Superjacent Strata. *Quart. Journ. Geol. Soc.*, vol. xlix., pp. 479–521.

———.—"The Top of the Inferior Oolite" and a correlation of "Inferior Oolite" deposits. *Proc. Dorset Nat. Hist. Club*, vol. xiv., pp. 37–43.

BURROWS, H. W.—Examination of Building Stones. *Journ. R. Inst. Brit. Archit.*, ser. 2, vol. ix., pp. 284–300.

CUMMING, L.—The Geology of Rugby. *Rep. Rugby School Nat. Hist. Soc.*, for 1892, pp. 4–8.

ETHERIDGE, R.—On the Rivers of the Cotteswold Hills within the Watershed of the Thames, and their importance as supply to the main river and the metropolis. *Proc. Cotteswold Club*, vol. xi., pp. 49–101.

HINDE, DR. G. J.—A Monograph of the British Fossil Sponges. Part III. Sponges of Jurassic Strata. *Palæontograph. Soc.* 4to. *London.*

JUKES-BROWNE, A. J.—On some Recent Borings through the Lower Cretaceous Strata in East Lincolnshire. *Quart. Journ. Geol. Soc.*, vol. xlix., pp. 467–477.

KENDALL, J. D.--The Iron Ores of Great Britain and Ireland. 8vo. *London.*

LYDEKKER, R.—On Two Dinosaurian Teeth from Aylesbury. *Quart. Journ. Geol. Soc.*, vol. xlix., pp. 566–568.

———.—On the Jaw of a New Carnivorous Dinosaur from the Oxford Clay of Peterborough. *Quart. Journ. Geol. Soc.*, vol. xlix., pp. 284–287.

REPORT.—Report of the Royal Commission appointed to inquire into the Water Supply of the Metropolis. Minutes of Evidence Taken before the Royal Commission ; with Appendices, and General Index. Fol. *London.*

RUTLEY, F.—On the Dwindling and Disappearance of Limestones. *Quart. Journ. Geol. Soc.*, vol xlix., pp. 372–382.

SEELEY, PROF. H. G.—On *Omosaurus Phillipsi* (Seeley). *Ann. Rep. Yorkshire Phil. Soc.*, for 1892, pp. 52–57.

———.—Further Observations on the Shoulder-Girdle and Clavicular Arch in the *Ichthyosauria* and *Sauropterygia*. *Proc. Roy. Soc.*, vol. liv., pp. 149–168.

———.—On a Reptilian Tooth with two Roots. [*Nuthetes.*] *Ann. Nat. Hist.*, ser. 6., vol. xii., pp. 227–230.

———.—Supplemental Note on a Double-rooted Tooth from the Purbeck Beds, in the British Museum. *Ibid.*, pp. 274–276.

TOMES, R. F.—Observations on the Affinities of the Genus *Astrocænia*. *Quart. Journ. Geol. Soc.*, vol. xlix., pp. 569–573

———.—Description of a New Genus of Madreporaria from the Sutton Stone of South Wales. *Quart. Journ. Geol. Soc.*, vol. xlix., pp. 574–578.

WALKER, J. F.—On Brachiopoda recently discovered in the Yorkshire Oolites. *Ann. Rep. Yorkshire Phil. Soc.* for 1892, pp. 47–51.

WHITAKER, W., and H. B. WOODWARD.—Notes of some Somerset Wells. *Proc. Bath Nat. Hist. Club*, vol. vii., pp. 340–345.

WINWOOD, REV. H. H.—On some deep-Well Borings in Somerset and elsewhere. *Proc. Bath Nat. Hist. Club*, vol. vii., pp. 335–340.

WOODWARD, A. S.—Some Cretaceous [and Jurassic] Pycnodont Fishes. *Geol. Mag.*, Dec. III., vol. x., pp. 433–436.

———.—On some British Upper-Jurassic Fish-remains, of the Genera *Caturus*, *Gyrodus*, and *Notidanus*. *Ann. Nat. Hist.*, ser. 6, vol. xii., pp. 398–402.

——— .—On the Cranial Osteology of the Mesozoic Ganoid Fishes, *Lepidotus* and *Dapedius*. *Proc. Zool. Soc.*, pp. 559–565.

WOODWARD, DR. H.—On Fossils applied as Charms or Ornaments. *Geol. Mag.*, Dec. III., vol. x., pp. 246–249.

WOODWARD, H. B , REV. H. H. WINWOOD, W. H. WICKES, and E. WILSON.—Excursion to Bath, Midford, and Dundry Hill, in Somerset, and to Bradford-on-Avon and Westbury, in Wiltshire. *Proc. Geol. Assoc.*, vol. xiii., pp. 125–140.

                                 1894.

ANDREWS, REV. W. R., and A. J. JUKES-BROWNE.—The Purbeck Beds of the Vale of Wardour. *Quart. Journ. Geol. Soc.*, vol. i., pp. 44–59.

BARROIS, C.—Sur les couches traversées par le sondage profond de Douvres, d'après M. Boyd Dawkins. *Ann Soc. Géol. du Nord*, tome xxii., p. 82.

BRODIE, REV. P. B.—Further Remarks on . . . . Insects from . . . . the Lias . . . . *Geol. Mag.*, Dec. IV., vol. i., pp. 167–169.

———.—Notice of a section in the Middle Lias at Napton. *Warwickshire Nat. Club*, 1893.

BRODIE, REV. P. B..—Notice of a section in the Lower Lias at the Cement Works, near Rugby. *Warwickshire Nat. Club*, 1893.

BROWN, A.—On the Structure and Affinities of the Genus *Solenopora*, together with Descriptions of New Species. *Geol. Mag.*, Dec. IV., vol. i., pp. 145–151, 195–203.

BUCKMAN, S. S.—Jurassic Ammonites: Notes on a Pamphlet by Dr. Emile Haug. *Geol. Mag.*, Dec. IV., vol. i., pp. 170–172.

———.—Jurassicc Ammonites: Notes on a Pamphlet by Professor Guido Bonarelli. *Geol. Mag.*, Dec. IV., vol. i., pp. 298–300.

———.—Jurassic Ammonites : On the Genus *Cymbites* (Neumayr). *Ibid.*, pp. 357–363.

DAWKINS, PROF. W. BOYD.—The Probable Range of the Coal-measures under the Newer Rocks of Oxfordshire and the Adjoining Counties. *Geol. Mag.*, Dec. IV., vol. i., pp. 459–462 ; *Rep. Brit. Assoc.* for 1894, pp. 646–648 ; *Trans. Federated Inst. Mining Eng.*, vol. vii.

———.—On the South-eastern Coalfield at Dover. *Trans. Manch. Geol. Soc.*, vol. xxii., p. 488.

———.— On the Deposit of Iron Ore in the Boring at Shakespeare Cliff, Dover, *Geol. Mag*, Dec. IV., vol. i., pp. 512, 513 ; *Rep. Brit. Assoc.* for 1894, pp. 648, 649.

GOODRICH, E. S.—On the Fossil Mammalia from the Stonesfield Slate. *Quart. Journ. Microsc. Sci.*, vol. xxxv. pp. 407–432.

GREEN, PROF. A. H.—Some Points of Special Interest in the Geology of the Neighbourhood of Oxford. *Rep. Brit. Assoc.* for 1894, pp. 644, 645.

GREENWELL, G. C.—On Ironstone Deposits near Dover. *Trans. Manch. Geol. Soc.*, vol. xxii., p. 553.

GREGORY, DR. J. W.—Catalogue of the Jurassic Bryozoa in the York Museum. *Ann. Rep. Yorks. Phil. Soc.* for 1893, pp. 58–61.

JONES, PROF. T. R.—On the Rhætic and some Liassic Ostracoda of Britain. *Quart. Journ. Geol. Soc.*, vol. l., pp. 156–168.

MANSEL-PLEYDELL, J. C.—Kimmeridge Shale. *Proc. Dorset Nat. Hist.*, vol. xv., pp. 172–183.

MARR, J. E., and T. LEIGHTON.—Excursion to Cambridge and Ely. *Proc. Geol. Assoc.*, vol. xiii., pp. 292–295.

MONCKTON, H. W.—On a Specimen of *Eryma elegans* from the Inferior Oolite of Dundry Hill. *Proc. Geol. Assoc.*, vol. xiii., p. 210.

———.—On a Variety of *Ammonites (Stephanoceras) subarmatus*, Young, from the Upper Lias of Whitby. *Quart. Journ. Geol. Soc.*, vol. l. (Proc.) p. 4.

PLATNAUER, H. M.—Appendix to the List of Figured Specimens in the Museum of the Yorkshire Philosophical Society. *Ann. Rep. Yorks. Phil. Soc.* for 1893. pp. 46–56.

SCHWARZ, E. H. L.—The Aptychus. *Geol. Mag.*, Dec. IV., vol. i., pp. 454–459.

SEWARD, A. C.—Catalogue of the Mesozoic Plants in the Department of Geology: British Museum. Part I. 8vo. *London.*

STRANGWAYS, C. Fox-.—Dr. Alex. Brown on *Solenopora*. *Geol. Mag.* Dec. IV., vol. i. p. 236.

———.—The Valleys of North-East Yorkshire and their mode of formation. *Trans. Leicester Lit. and Phil. Soc.*, vol. iii., pp. 333–344.

THOMPSON, B.—Landscape Marble. *Quart. Journ. Geol. Soc.*, vol. l., pp. 393–410.

THOMPSON, B., and W. D. CRICK.—Excursion to Wellingborough. *Proc. Geol. Assoc.*, vol. xiii., pp. 283–291.

WALFORD, E. A.—On some Bryozoa from the Inferior Oolite of Shipton Gorge. Dorset. *Quart. Journ. Geol. Soc.*, vol. l. pp. 72–78.

———.—On Cheilostomatous Bryozoa from the Middle Lias. *Quart. Journ. Geol. Soc.*, vol. l., pp. 79–84.

———.— On the Terraced Hill Slopes of North Oxfordshire. *Geol. Mag.*, Dec. IV., vol. i., pp. 465, 466 ; *Rep. Brit. Assoc.* for 1894, pp. 645, 646.

———.—Report of Committee appointed to open further sections in the neighbourhood of Stonesfield in order to show the relationship of the Stonesfield Slate to the underlying and overlying strata. *Geol. Mag.*, Dec. IV., vol. i., pp. 462–465 ; *Rep. Brit. Assoc.* for 1894, pp. 304–306.

WOODWARD, A. S.—On a Second British Species of the Jurassic Fish *Eurycormus*. *Geol. Mag.*, Dec. IV., vol. i., pp. 214–216.

1895.

ANDREWS, C. W.—On the Development of the Shoulder-girdle of a Plesiosaur (*Cryptoclidus oxoniensis*, Phil.) from the Oxford Clay. *Ann. Nat. Hist.*, ser. 6, vol. xv., pp. 333–346.

———.—Note on a Skeleton of a Young Plesiosaur from the Oxford Clay of Peterborough. *Geol. Mag.*, Dec. IV., vol. ii., pp. 241–243.

BUCKMAN, S. S.—The Bajocian of the Mid-Cotteswolds. *Quart. Journ. Geol. Soc.*, vol. li., pp. 388–462.

FOORD, A. H.—A Short Account of the Ammonites and their Allies, as exhibited in the Cephalopod Gallery at the British Museum (Natural History). *Geol. Mag.*, Dec. IV., vol. ii., pp. 391–400.

HARRIS, G. F.—On the Analysis of Oolitic Structure. *Proc. Geol. Assoc.*, vol. xiv., pp. 59–79.

STRAHAN, A.—On Overthrusts of Tertiary Date in Dorset. *Quart. Journ. Geol. Soc.*, vol. li., pp. 549–562 (plates).

THOMPSON, B., and W. D. CRICK. Excursion to Brigstock and Geddington. *Proc. Geol. Assoc.*, vol. xiv., pp. 144–118.

WALFORD, E. A.—The making of the Dassett and Edge Hills of South Warwickshire. (Privately printed.) 3 pp. 8vo. *Banbury*.

WETHERED, E. B.—The Formation of Oolite. *Quart. Journ. Geol. Soc.*, vol. li., pp. 196–209.

WINWOOD, REV. H. II.—Well Boring at Bitton. *Proc. Bath Nat. Hist. Club*, vol. viii., pp. 141–146.

———.—Excavations at the Bath Waterworks, Monkswood. *Ibid.*, pp. 146–156.

WOODWARD, A. SMITH.—Note on Megalosaurian Teeth discovered by Mr. J. Alstone in the Portlandian of Aylesbury. *Proc. Geol. Assoc.*, vol. xiv., pp. 31, 32.

———.— A Contribution to Knowledge of the Fossil Fish Fauna of the English Purbeck Beds. *Geol. Mag.*, Dec. IV., vol. ii., pp. 145–152.

———.—On the Liassic Fish *Osteorachis macrocephalus*. *Geol. Mag.*, Dec. IV., vol. ii., pp. 204–206.

———.—A Description of *Ceramurus macrocephalus*, a small Fossil Fish from the Purbeck Beds of Wiltshire. *Ibid.*, pp. 401, 402.

INDEX.

N.B.—Names of Persons (authors, observers, and informants) are in small capitals.
Names of Fossils of which there are figures, or to which special reference is made,
are in *italics*.

R.

S.

Lightning Source UK Ltd.
Milton Keynes UK
UKHW041241070119
334726UK00009BA/539/P